The Human Hippocampus

Mehmet Turgut • Niyazi Acer
Ahmet T. Turgut • Concezio Di Rocco
Editors

The Human Hippocampus

Development, Neuroanatomy, Neurophysiology, Neuropathology and Surgery

Springer

Editors
Mehmet Turgut
Department of Neurosurgery
Aydın Adnan Menderes University
Faculty of Medicine
Aydın, Türkiye

Ahmet T. Turgut
Department of Radiology
Ankara Medipol University Faculty
of Medicine
Ankara, Türkiye

Niyazi Acer
Department of Anatomy
Sanko University Faculty of Medicine
Gaziantep, Türkiye

Concezio Di Rocco
Department of Neurosurgery
INI - International Neuroscience
Institute
Hannover, Germany

ISBN 978-3-031-88906-6 ISBN 978-3-031-88907-3 (eBook)
https://doi.org/10.1007/978-3-031-88907-3

This Springer imprint is published by the registered company Springer Nature Switzerland AG
The registered company address is: Gewerbestrasse 11, 6330 Cham, Switzerland

Contents

Part I

Fundamental Information About the Human Hippocampus

1 History of Human Hippocampus

Nikolaos Syrmos and Mehmet Turgut

1.1 Introduction

The development of neuroanatomical knowledge, including terminology as a distinct scientific field, is closely related to the development of medicine as an autonomous science, since the primitive–ancient periods of humanity on Earth and of neurosurgery as a medical discipline [1–4]. The hippocampus is one of the most studied parts of the human brain, both anatomically-morphologically and physiologically-functionally [1–5]. It also remains a very important brain compartment and a challenging study subject in anatomy, physiology, neuroanatomy, neurophysiology, and neurosurgery [1–6]. Its normal function is affected in numerous morbidities of the central nervous system (CNS) in humans [1–6].

N. Syrmos (✉)
Aristotle University of Thessaloniki, Thessaloniki, Macedonia, Greece

Human Performance and Health, Master Science Program, Aristotle University of Thessaloniki, Thessaloniki, Macedonia, Greece
e-mail: milanako76@yahoo.gr

M. Turgut
Department of Neurosurgery, Aydın Adnan Menderes University Faculty of Medicine, Aydın, Türkiye

Department of Histology and Embryology, Aydın Adnan Menderes University Health Sciences Institute, Aydın, Türkiye

1.2 Ancient Period

In medical history, the first documented physician and the first neurological surgeon was the Hellenic Hippocrates of Kos (460–370 BC), who performed anatomical studies and pioneered surgical and neurosurgical operations [1–6] (Fig. 1.1). In his marvelous books, he described many structures of the brain [1–5]. He had many students and managed, through them, to deliver the spirit of the true study of the human body as a scientific entity [1–6].

Fig. 1.1 Hippocrates of Kos (460–370 BC)

M. Turgut et al. (eds.), *The Human Hippocampus*, https://doi.org/10.1007/978-3-031-88907-3_1

1.3 Hellenic-Roman-Islamic Era

Alcmeone of Crotone (sixth century BC) was one of the first documented physicians to attempt to describe CNS structures, using the knowledge of his period [1–3]. Galen of Pergamon (131–201 BC) was a skilled surgeon and anatomist of his era. He and his pupils used neuroanatomical terms in the colloquial Hellenic language of his time [1–4] (Fig. 1.2). The Arab physicians Avicenna (980–1037) and Albucasis (936–1013) codified the medical and neuroanatomical terminology of their period [1–5]. Paul of Aegina (625–690) was the scientist who documented, through his studies, the works and scientific efforts of Hippocrates, Galen, Avicenna, and Albucasis in order to describe well the morphology and function of the human body [6–13].

Fig. 1.2 Galen of Pergamon (131–201 BC)

1.4 First Description of the Hippocampus

The first description and documentation belong to Gulio Cesare Arantio (Latin version of the name Julios Caesar Arantius) [7–10]. Arantio (1530–1589) was an Italian physician, polymath, anatomist, and pioneer surgeon [11–16]. He studied anatomy and human morphology under his teacher, Andreas Vesalius (Andries van Wezel) (31 December 1514–15 October 1564), the so-called "Founder of Modern Human Anatomy" [1, 9, 16, 17] (Fig. 1.3). Vesalius was born in Brussels and worked as a professor at the University of Padua in Padova, Italy (1537–1542), and he was an imperial doctor for Emperor Charles V [1, 9, 16, 17]. Vesalius, through his students, made very important contributions to achieving true scientific knowledge of the human body [1, 9, 16, 17]. In 1587, Arantio published, in a single volume, three very important anatomical works-studies [1, 9, 16, 17]: (a) The first issue of the *Book of Anatomical Observations* (Latin version: *Anatomicarium Observationum Liber*); (b) The first version of the *Book on Tumors According to the Affected Site* (Latin version: *De Tumoribus Secundum Locus Affectus Liber*); and (c) The third revised edition of the *Book on the Human Fetus* (Latin version: *De Humano Foetu Liber*). In the *Book of Anatomical Observations*, there are five chapters: (1) The first two chapters describe the ventricles and the choroid plexuses. (2) The third chapter provides an accurate description of the intraventricular location of the hippocampus (inferior to the temporal bone). (3) The fourth chapter mentions the procedures required to access the target structures of the human brain. (4) The fifth chapter comments on the ventricles and the hippocampus functions.

Figs. 1.3 Andreas Vesalius (1514–1564)

1.5 First Illustration of the Hippocampus

The first documented design of the hippocampus was attributed to Johann Georg Duvernoy (in German, Johannes Georgius Duvernoi) [1, 9, 18, 19]. Duvernoy (1691–1759) was a German physician, anatomist, and botanist [1, 9–11, 18]. In 1729, he published a short essay entitled *De Sinibus Cerebri* in the *Commentaries of the Imperial Academy of Saint Petersburg*, where he presented an accurate illustration with explanations, a sort of anatomical table, and also a short description of the morphology of the hippocampus [1, 9, 11, 19].

1.6 Important Notes About Studies on the Hippocampus

The term "hippocampus" was infrequently used during the seventeenth century, and further morphological characteristics of the hippocampus remain unclear [1, 10, 11]. In the eighteenth century, however, the hippocampus was referred to in many animals [1, 9, 18, 19]. These facts provoked huge discussions in the Royal French Academy of Sciences (*Académie Royale des Sciences*), culminating in the seminal definition of the hippocampus in the 1780s by Félix Vicq d'Azyr (1748–1794) [1, 7–9]. Félix Vicq d'Azyr (23 April 1748–20 June 1794) was a French physician, polymath, and anatomist [1, 8] (Fig. 1.4). He is considered "the originator of comparative anatomy" and the discoverer of the theory of "homology in biology" [1, 8]. Nevertheless, no hypothesis regarding the function of the hippocampus was suggested, and the relationship between the hippocampus and memory was not defined until the twentieth century [1, 8].

Fig. 1.4 Felix Vicq d'Azyr (1748–1794)

1.7 Important Commentaries About the Studies Regarding Hippocampus

Many important physicians and outstanding anatomists, through the ages, such as Galen, Vesalius, and Willis, ignored this anatomical structure as a result of its deep location inside the human brain [1, 9–11]. Arantius, when he described the structure, denominated it *hippocampus* from the Hellenic name "ΙΠΠΟΚΑΜΠΟΣ" (*ΙΠΠΟΣ*—sea horse and *ΚΑΜΠΟΣ*—sea monster) [1, 9, 16–18]. He provided important and very useful details about the neuroanatomical-surgical procedure to access the brain, as well as technical details and information about the approach used, without providing any illustration of the structure [1, 9, 16, 17, 19]. In 1729, the first documented illustration and information about the surrounding structures were provided by Duvernoy [1, 9, 16, 17]. In 1543, however, Vesalius' publication *On the Fabric of the Human Body* (Latin name *De Humani Corporis Fabrica*) presented drawings representing the hippocampus [1, 9, 10, 16, 17]. Also, in 1573, Costanzo Varolio (Costantius Varolius), in his work *On the Optic Nerves* (Latin name *De Nervis Opticis*), included plates with similar anatomical formations without mentioning the name [1, 9, 11, 16, 17]. Costanzo Varolio (1543–1575) was an Italian anatomist [1, 9, 11, 16, 17]. He was also a papal physician to Gregory XIII [1, 9, 11, 16, 17]. He was born in Bologna [1, 9, 16, 17] (Fig. 1.5). He was a pupil of the anatomist Giulio Cesare Aranzio, who was himself a pupil of Vesalius [1, 9, 11, 16, 17]. Thomas Willis (27 January 1621–11 November 1675), an English doctor, had an important role in the history of anatomy and neuroscience [1, 9, 10, 16–19] (Fig. 1.6). In his scientific works, *Brain Anatomy* (*Cerebri Anatome*) and *Anima Brutorum* (*The Soul of Beasts*), he presented tables with designs very close to the hippocampus [1, 9–11, 16, 17]. He labeled the hippocampus as *"D.D. Corporis*

Fig. 1.5 Costanzo Varolio (1543–1575)

Fig. 1.6 Thomas Willis (1621–1675)

Fig. 1.7 Bartolomeo Eustachi (1510–1574)

callosi margo qui caudicem medullarem propr cerebellum amplexabatur" ("D.D. Margin of the corpus callosum embracing the brainstem near the cerebellum") [1, 9, 12–14, 16, 17]. Bartolomeo Eustachi (1510–27 August 1574), called *Eustachio* (Latin name *Bartholomaeus Eustachius*), was an Italian anatomist [1, 9, 15–17]. He was one of the pioneers of human anatomy [1, 9, 15–17] (Fig. 1.7). He published *Tabulae Anatomicae* (*Anatomical Plates*), which included an illustration very close to the hippocampus [1, 9, 16–19]. His work was recognized one and a half centuries after his death by Giovanni Maria Lancisi (1654–1720) [1, 9, 10, 16, 17]. Giovanni Maria Lancisi (26 October 1654–20 January 1720), an Italian anatomist, suggested a relationship between mosquitoes and the prevalence of malaria [1, 9–11, 16, 17]. He was also known for his anatomical studies on the nervous system, especially his detailed investigation of the corpus callosum of the brain [1, 9–11, 16, 17]. He also studied human cardiovascular diseases [1, 9–11, 16, 17].

1.8 Conclusion

Through the ages, many scientific efforts have been made to understand the anatomical, physiological, morphological, and structural aspects of hippocampal structures [20–22]. All of these contributions demonstrate the necessity of reciprocal interaction to obtain useful results in neuroanatomical, morphological, and neurosurgical procedures and approaches within the CNS of humans [20–22]. The modern era provides the tools (diagnostic, radiological, imaging, physiological, etc.) to achieve better results [20–22].

References

1. Engelhardt E. Hippocampus discovery first steps. Dement Neuropsychol. 2016;10(1):58–62.
2. Syrmos NC, Giannouli V, Kottas S, Turgut M. History of cranial subdural hematoma subdural hematoma: past to present to future management. 2021. pp. 33–43.

3. Syrmos NC, Giannouli V, Turgut M. The sutures of the skull: a historical perspective the sutures of the skull: anatomy, embryology, imaging, and surgery. 2021. pp. 7–18.
4. Goodrich JT. How to get in and out of the skull: from tumi to "hammer and chisel" to the Gigli saw and the osteoplastic flap. Neurosurg Focus. 2014;36(4):E6.
5. Syrmos N, Ampatzidis G, Fachantidou A, Mouratidis I, Syrmos C. Historical back training in most important points of neurosurgery. Ann Gen Psychiatry. 2010;9(Suppl 1):S89.
6. Tatu L, Bogousslavsky J. Beasts and gods: hippocampal quarrels before memory. Rev Neurol (Paris). 2022;178(10):991–5.
7. Casillo SM, Luy DD, Goldschmidt E. A history of the lobes of the brain. World Neurosurg. 2020;134:353–60.
8. Parent A. Felix Vicq d'Azyr: anatomy, medicine and revolution. Can J Neurol Sci. 2007;34(1):30–7.
9. Bir SC, Ambekar S, Kukreja S, Nanda A. Julius Caesar Arantius (Giulio Cesare Aranzi) (1530-1589) and the hippocampus of the human brain: history behind the discovery. J Neurosurg. 2015;122:971–5.
10. Duvernoy H, Cattin F, Risold P-Y. The human hippocampus. Berlin: Springer; 2013.
11. Bartsch T. The clinical neurobiology of the hippocampus. An integrative view. Oxford: Oxford University Press; 2012.
12. Andersen P, Morris R, Amaral D, Bliss T, O'Keefe J. The hippocampal formation. In: Andersen P, Morris R, Amaral D, Bliss T, O'Keefe J, editors. The hippocampus book. Oxford: Oxford University Press; 2007. p. 3–6.
13. Angevine J, Jr Isaacson RL, Pribram KH. The development of the hippocampal region. In: The hippocampus, Structure and development, vol. I. New York: Plenum Press; 1975. p. 61–94.
14. Judaš M, Pletikos M. A note on the sea-horse in the human brain. Transl Neurosci. 2010;1(4):335–7.
15. Swanson LW. Neuroanatomical terminology. Oxford: Oxford University Press; 2015. p. 296–9.
16. Brambilla GA. Storia delle scoperte fisico-medico-anatomico-chirurgiche fatte dagli uomini illustri italiani. Tom 2. Giulio Cesare Aranzio. Milano: Nel Imperial Monistero di s. Ambrogio Maggiore; 1781. pp. 188–198.
17. Arantius JC. De Humano Foetu Liber tertio editus, ac recognitus. Anatomicarum Observationum Liber, ac de Tumoribus Secundum Locos Affectos Liber. Venetiis: Apud Iacobum Brechtanum; 1578. pp. 41–46.
18. Tilney F. The hippocampus and its relations to the corpus callosum. J Nerv Ment Dis. 1939;89:433–513.
19. Walther O. Hippocampal terminology: concepts, misconceptions, origins. Endeavour. 2002;26(2):41–4.
20. Kobayashi S, Matsushima T, Sakai T, Matsushima K, Bertalanffy H, Rutka JT. Evolution of microneurosurgical anatomy with special reference to the history of anatomy, surgical anatomy, and microsurgery: historical overview. Neurosurg Rev. 2022;45(1):253–61.
21. Sakai T. Historical evolution of anatomical terminology from ancient to modern. Anat Sci Int. 2007;82(2):65–81.
22. Syrmos NC. The seven miracles of neurosurgery. J Neurosurg Sci. 2020;64(4):407. https://doi.org/10.23736/S0390-5616.19.04855-0.

2 Development of the Hippocampus and Hippocampal Neurogenesis

Züleyha Doğanyiğit and Aslı Okan Oflamaz

2.1 Introduction

At the intersection of neuroscience and wonder lies the phenomenon of hippocampal neurogenesis, a process pivotal for learning, memory, and emotional regulation. This elaborate process not only underscores the brain's remarkable ability to generate new neurons throughout an individual's lifespan but also plays a key part in understanding the complexities of mental health and neurological disorders. The significance of hippocampal neurogenesis in the context of human health and disease underlines the importance of delving deep into its mechanisms, challenges, and potential therapeutic implications.

This section of the book embarks on a comprehensive review of hippocampal neurogenesis, starting with a comprehensive investigation of the embryonic development and structure of the hippocampus. It further navigates through the various stages and phases of neurogenesis. Through an examination of current research advances and future directions, the discussion aims to shed light on the technological innovations in studying neurogenesis and its potential for medical and therapeutic applications.

Z. Doğanyiğit · A. O. Oflamaz (✉)
Department of Histology and Embryology, Yozgat Bozok University Faculty of Medicine, Yozgat, Türkiye
e-mail: asli.okan@yobu.edu.tr

2.1.1 The Development and Structure of the Hippocampus

The hippocampus is a critical component of the brain located in the medial temporal cortex, responsible for differentially regulating emotions, cognition, memory, and learning. Hippocampal development in human infants begins early in pregnancy, and most of its developmental growth occurs during late pregnancy and the neonatal period [1].

2.1.1.1 Embryonic Development

The neural tube and neural crest are the first two stages of the neurulation process that form the entire nervous system. The ectoderm, mesoderm, and endoderm are three germ layers that appear during the process of gastrulation in the third week of embryogenesis. The notochord thickens and stimulates the upper ectoderm, and the neural plate emerges. The nervous system in the center, which includes the brain, cerebellum, and spinal column, is formed by the neural plate folding on itself to form a neural tube. The cerebral cortex and spinal cord are divided into plates by the sulcus limitans, which is found in the brain's fourth ventricle. It separates the lower surface, which produces neurons that regulate movement, from the alar plate, which produces sensory neurons. The neural tube produces the prosen-

M. Turgut et al. (eds.), *The Human Hippocampus*, https://doi.org/10.1007/978-3-031-88907-3_2

cephalon, which is the forebrain, the mesencephalon, which is the midbrain, and the rhombencephalon, which is the hindbrain. The telencephalon and diencephalon are two divisions of the prosencephalon. The dorso-medial portion of the telencephalon houses the hippocampus. The developing medial wall forms two significant structures: the choroid plexus and the cortical hem. Through released substances, the hem designs the medial telencephalon and provides instructions for the creation of the hippocampus [2]. The cortical hem area is characterized by the neuroepithelium, which is typical of the surrounding telencephalic structures. It encodes the proneural genes neurogenin 1 and 2 [3]. The cortical skirt, located at the medial margin of the telencephalon, together with the abundant expression of regulating secreted morphogens such as the signaling molecules Wingless-associated (Wnt) and bone morphogenetic protein (Bmp), highlights the essential role of the cortical skirt as a signaling center that helps establish the hippocampus in its medial position and initiates hippocampal development [4].

2.1.1.2 Fetal Hippocampus Formation

By the ninth week of pregnancy, the hippocampus becomes distinct from different areas of the brain in the dorsal part of the lamina terminalis. The dentate gyrus (DG) and cornu ammonis (CA) structures, which grow and thicken by the tenth week of pregnancy, perform hippocampal inversion by folding around the hippocampal sulcus in the temporal lobe to form two interlocking C-shaped structures in the 13th and 16th weeks [1, 5]. Hippocampal growth and development accelerate until the 20th week [6]. It is stated that towards the middle of pregnancy in humans, the hippocampus resembles the adult form, and as a result of histological analyses performed on human fetal and infant hippocampal samples, pyramidal neurons appear in the first half of pregnancy [5, 6]. It is stated that the branched dendritic structure of hippocampal pyramidal neurons gradually increases and develops by the 33rd week of pregnancy [7].

2.1.1.3 Structure of the Hippocampus

The hippocampal formation consists of four cornu ammonis (CA1–CA4) subfields, DG, subiculum, and entorhinal cortex. It is rich in neurons and has five layers. These are: stratum alveus (ALV), a thin layer of axon-rich white matter, the pyramidal layer with basal dendrites reaching to the stratum oriens (SO), and the apex of dendrites extending to the stratum radiatum (SR) and stratum lacunosum molecular layer (SL-M). It consists of stratum pyramidale (SP). Cells in the SO and SL-M layers are interneurons that support the pyramidal cells in the hippocampus [8].

2.1.2 Neurogenesis

Adult neurogenesis begins with precursor cells in the subventricular zone (SVZ) that line the lateral ventricles and the subgranular zone (SGZ) in the hippocampus's DG. In both these locations, it generates highly proliferative transit progenitors from slowly dividing, glia-like stem cells, which ultimately provide development to fate-dependent neural as well as glial descendants [9]. These precursors, known as neuroblasts or glioblasts, travel to their eventual destination and differentiate into postmitotic neurons, astroglia, and oligodendrocytes. In the context of neurogenesis, newborn neurons undergo further maturation to establish a suitable dendritic shape, integrate through synapses within pre-existing circuits, and undergo an activity-dependent sorting phase [10].

2.1.2.1 Neurogenesis in Subventricular Zone

In the SVZ, astrocytic stem cells are located between the striatal parenchyma and the ependymal cell layer lining the lateral ventricles [11]. These stem cells generate fast-dividing transit-amplifying progenitors, which produce migrating neuroblasts that travel down the rostral migratory stream (RMS) and tangentially to the olfactory bulb (OB) (Fig. 2.1) [12]. During migration, these proliferative cells move in chains and are surrounded by specialized astrocytes [13]. Neuroblasts escape the RMS, change direction,

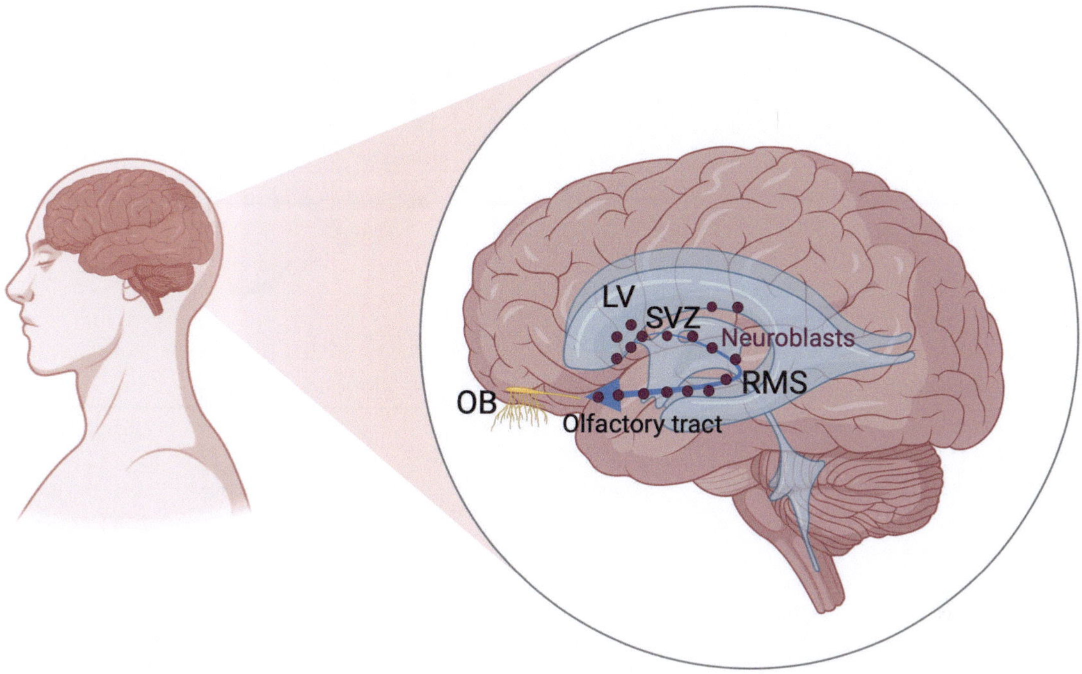

Fig. 2.1 Diagram of RMS and SVZ neuroblasts. Neuroblasts that proliferate from neural stem cells are unique to the ventral SVZ; these neuroblasts move via the RMS to the olfactory bulb, and then develop to adult neurons [17]. (Created with BioRender.com)

and travel radially outward to develop into GABAergic, dopaminergic, and glutamatergic periglomerular and granule interneurons [14]. Granule cells create dendrodendritic reciprocal synapses with mitral and tufted neurons in the bulb and suppress their activity to fine-tune their output, eventually playing a role in odor identification and learning [15]. Periglomerular cells form one-way and reciprocal synapses with the apical dendrites of mitral/tufted cells and the terminals of olfactory nerves that converge into the glomeruli [16].

2.1.2.2 Neurogenesis in Subgranular Zone

Astrocytic stem cells live in the SGZ's inner third of the granule cell layer (GCL), which connects the DG to the hilus. They receive input from the entorhinal cortex and send their axonal projections to area CA3 via the mossy fiber pathway. They offer stimulating stimuli to the pyramidal cells of CA3 (Fig. 2.2a). These shoot relatively seldom, so the way they function is regulated by several interneurons in the DG and hilus [18, 19]. Adult neurogenesis is separated into four stages: precursor cell, early survival, postmitotic maturation, and late survival (Fig. 2.2b).

The Precursor Stage

The precursor stage of neurogenesis in hippocampal regions is defined by the expansion of a cell pool that may differentiate into neurons. This stage arises from a group of primitive cells with glial features, some of which display characteristics of radial glia [22, 23]. These radial glia-like type 1 cells subsequently give rise to intermediate progenitor cells, known as type 2 cells, which have a strong proliferation capability and are crucial for the next phase of neurogenesis. While these cells, termed type 2a, still express glial markers, they no longer display the characteristic morphology of radial cells. The so-called type 2b cells of these contain the transcription factors NeuroD1 and Prox1. Of these, Prox1 was ini-

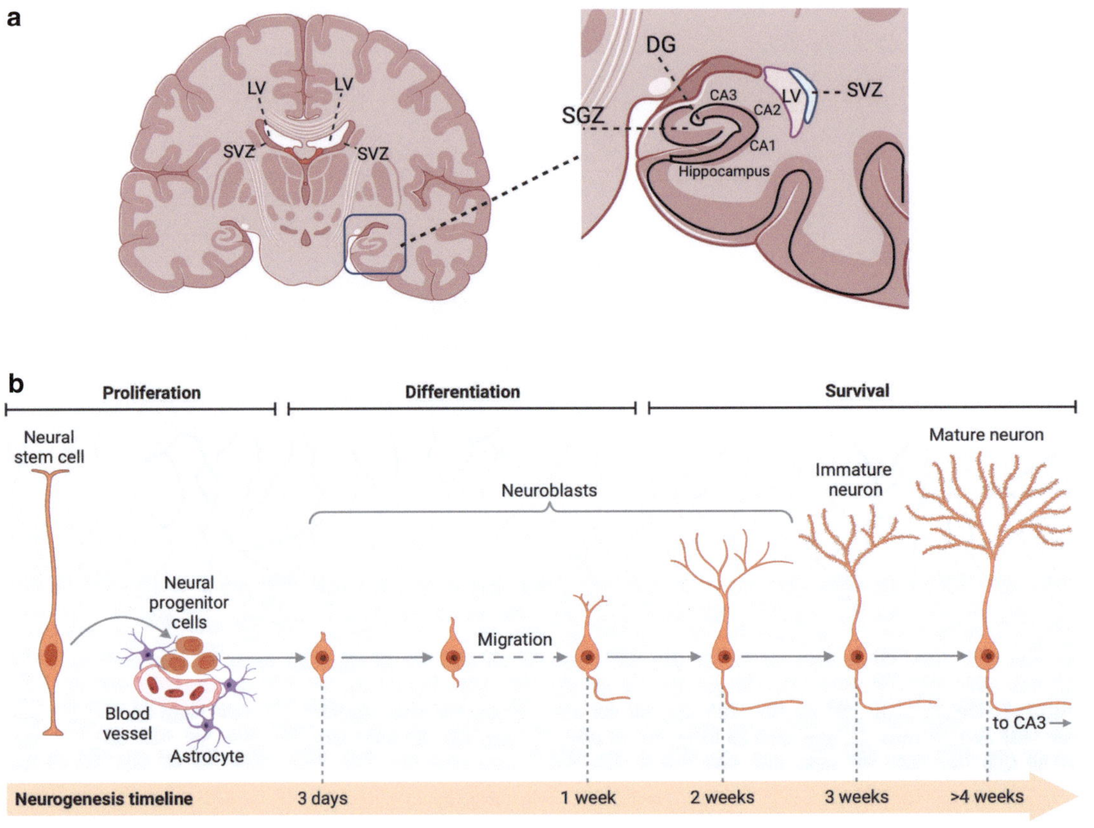

Fig. 2.2 (**a**) A schematic representation of the two areas: SVZ and the SGZ, where adult neurogenesis is found. (**b**) Neural progenitor cells are distinguished by their uneven shape, premature spikes, and synaptic quiet in the first week following birth. In the second week, they have moved into the layer of granule cells, grown spineless dendrites, and developed sluggish GABAergic synaptic inputs. By the third week, they begin to generate efferent interactions to the CA3 and afferent connections from the enterorhinal cortex's perforant pathway. Moreover, a switch occurs in synaptic inputs from glutamatergic to GABAergic. The growing neurons have high membrane resistance, high resting potential, and high excitability at this stage. Finally, compared to mature dentate granule cells, these immature neurons show more synaptic plasticity between weeks 4 and 6. They exhibit larger LTP amplitudes and a lower threshold for LTP induction [20, 21]

tially proven to play a role in granule cell differentiation [24]. Doublecortin (DCX), one of the neuronal lineage markers that emerge in the type 2b stage, is also expressed by type 3 cells, whereas nestin is downregulated. Although type 3 cells typically exhibit low generative activity, cell division accelerates in pathological situations such as epileptic seizures [25].

Expansion and Early Survival Phases

The precursor cell phase serves to expand the pool of cells that can develop into neurons. Neural stem cells (NSCs) go through multiple consecutive precursor stages characterized by increased proliferation during the expansion stage [26]. This is succeeded by the early survival stage, where newly formed neural progenitor cells (NPCs) exit the cell cycle. Because most newborn neurons die within a few days of development, apoptotic mechanisms cause a large decline in the number of neurons during this phase [27]. Within a few days after the cell cycle ends, new cells send their nerve endings to the area of interest CA3 and make appropriate synapses [28]. At this stage, the primary synaptic input to developing cells is GABAergic, and GABA remains excitatory [29]. GABA regulates neuronal development and synaptic integration in these cells [30]. Quantitatively, most control happens at that phase of neuronal growth rather than the extension stage, as is frequently supposed [31]. The reason for this is that some of the excess new neurons formed by precursor cell proliferation, especially those that survive the first 2 weeks, join the DG network for an extended period, while a small part does not survive for a long time [32, 33].

Postmitotic Maturation Stages

Postmitotic differentiation is related to the development of effective interactions, developing axons and dendrites, including synaptogenesis. At this point in time, depolarizing GABA is essential for facilitating the development of glutamatergic synapses, similar to the scenario in the growing cortex [34]. The first spines emerge approximately a week following the marking of growing cells, even if axonal contact with CA3 occurs 10 days later. Axons of developing neurons compete with those of established synapses in the mossy fiber bundles in CA3 to attach to target cells [35].

Late Survival Stages

The adaptive changes that take place in adult neurogenesis late in brain development are little understood. While new neurons display postmitotic markers, such as the transitory marker calretinin, for 3–4 weeks after cell cycle exit, they change from calretinin to calbindin as their calcium-binding protein once they have fully integrated into the network [36]. Once glutamatergic synaptic connections are established, young neurons undergo a period of enhanced plasticity of synapses. Compared to mature granule cells, the threshold for long-term potentiation (LTP) in embryonic neurons is lower [37]. In actuality, under normal (i.e., inhibited) settings, the only LTP measurable through the DG originates from newly formed neurons that have not yet been blocked by nearby interneurons [38]. These specific features direct input towards new neurons [39]. This critical period lasts 1–1.5 months after cells are produced [40].

2.2 Discussion

2.2.1 Technological Innovations in Studying Neurogenesis

Recent advancements in the field of neurogenesis research have been significantly propelled by technological innovations. High-resolution microscopy and live-cell imaging have revolutionized the ability to observe neural stem cell activities, including proliferation, migration, and differentiation in real-time [41, 42]. Furthermore, the advent of novel genetic tools such as transgenic animal models and cell-specific labeling techniques has given greater insights into the behavior and fate of neural stem cells and their offspring [43]. These technological strides have not only enhanced our understanding of neurogenesis but also opened new avenues for detailed study at the cellular level.

2.2.2 Potential for Medical and Therapeutic Applications

The exploration of neurogenesis holds substantial promise for medical and therapeutic applications [44, 45]. Understanding the regulatory mechanisms of neural stem cell functions and identifying factors that influence neurogenesis are crucial for developing treatments for neurodegenerative diseases [46], brain injuries [47], and psychiatric disorders [48]. Current research is exploring the potential of stem cell-based therapies to restore damaged or missing neurons and the use of pharmacological interventions to stimulate endogenous neurogenesis. These therapeutic strategies aim to restore neurological functions and improve the quality of life for individuals suffering from various neurological conditions.

2.3 Conclusion

Throughout this exploration of hippocampal neurogenesis, we have journeyed from the embryonic beginnings of the hippocampus—highlighting its critical role in cognitive and emotional functions—to understanding how its growth and development continue throughout life. This book chapter has underscored the remarkable ability of the hippocampus to create novel neurons and adapt, offering a blueprint for future research in neuroscience and the ongoing quest to unlock the mysteries of the human brain.

References

1. Bajic D, Ewald U, Raininko R. Hippocampal development at gestation weeks 23 to 36. An ultrasound study on preterm neonates. Neuroradiology. 2010;52(6):489–94. https://doi.org/10.1007/s00234-010-0673-x.
2. Subramanian L, Tole S. Mechanisms underlying the specification, positional regulation, and function of the cortical hem. Cereb Cortex. 2009;19(suppl 1):i90–5. https://doi.org/10.1093/cercor/bhp031.
3. Imayoshi I, Shimogori T, Ohtsuka T, Kageyama R. Hes genes and neurogenin regulate non-neural versus neural fate specification in the dorsal telencephalic midline. Development. 2008;135(15):2531–41. https://doi.org/10.1242/dev.021535.
4. Lie DC, Colamarino SA, Song HJ, Désiré L, Mira H, Consiglio A, et al. Wnt signalling regulates adult hippocampal neurogenesis. Nature. 2005;437(7063):1370–5. https://doi.org/10.1038/nature04108.
5. Kier EL, Kim JH, Fulbright RK, Bronen RA. Embryology of the human fetal hippocampus: MR imaging, anatomy, and histology. AJNR Am J Neuroradiol. 1997;18(3):525–32. PMID: 9090416.
6. Ge X, Shi Y, Li J, Zhang Z, Lin X, Zhan J, et al. Development of the human fetal hippocampal formation during early second trimester. Neuroimage. 2015;119:33–43. https://doi.org/10.1016/j.neuroimage.2015.06.055.
7. Nakahira E, Yuasa S. Neuronal generation, migration, and differentiation in the mouse hippocampal primordium as revealed by enhanced green fluorescent protein gene transfer by means of in utero electroporation. J Comp Neurol. 2005;483(3):329–40. https://doi.org/10.1002/cne.20441.
8. Kullmann DM. Interneuron networks in the hippocampus. Curr Opin Neurobiol. 2011;21(5):709–16. https://doi.org/10.1016/j.conb.2011.05.006.
9. Alvarez-Buylla A, Lim DA. For the long run: maintaining germinal niches in the adult brain. Neuron. 2004;41(5):683–6. https://doi.org/10.1016/s0896-6273(04)00111-4.
10. Carleton A, Petreanu LT, Lansford R, Alvarez-Buylla A, Lledo PM. Becoming a new neuron in the adult olfactory bulb. Nat Neurosci. 2003;6(5):507–18. https://doi.org/10.1038/nn1048.
11. Doetsch F, García-Verdugo JM, Alvarez-Buylla A. Cellular composition and three-dimensional organization of the subventricular germinal zone in the adult mammalian brain. J Neurosci. 1997;17(13):5046–61. https://doi.org/10.1523/JNEUROSCI.17-13-05046.1997.
12. Altman J. Autoradiographic and histological studies of postnatal neurogenesis. IV. Cell proliferation and migration in the anterior forebrain, with special reference to persisting neurogenesis in the olfactory bulb. J Comp Neurol. 1969;137(4):433–57. https://doi.org/10.1002/cne.901370404.
13. Lois C, García-Verdugo JM, Alvarez-Buylla A. Chain migration of neuronal precursors. Science. 1996;271(5251):978–81. https://doi.org/10.1126/science.271.5251.978.
14. Brill MS, Ninkovic J, Winpenny E, Hodge RD, Ozen I, Yang R, et al. Adult generation of glutamatergic olfactory bulb interneurons. Nat Neurosci. 2009;12(12):1524–33. https://doi.org/10.1038/nn.2416.
15. Lledo PM, Alonso M, Grubb MS. Adult neurogenesis and functional plasticity in neuronal circuits. Nat Rev Neurosci. 2006;7(3):179–93. https://doi.org/10.1038/nrn1867.
16. Kosaka K, Kosaka T. Synaptic organization of the glomerulus in the main olfactory bulb: compartments of the glomerulus and heterogeneity of the periglomerular cells. Anat Sci Int. 2005;80(2):80–90. https://doi.org/10.1111/j.1447-073x.2005.00092.x.
17. Wang C, Liu F, Liu YY, Zhao CH, You Y, Wang L, et al. Identification and characterization of neuroblasts in the subventricular zone and rostral migratory stream of the adult human brain. Cell Res. 2011;21(11):1534–50. https://doi.org/10.1038/cr.2011.83.
18. Altman J, Das GD. Autoradiographic and histological evidence of postnatal hippocampal neurogenesis in rats. J Comp Neurol. 1965;124(3):319–35. https://doi.org/10.1002/cne.901240303.
19. Rietze R, Poulin P, Weiss S. Mitotically active cells that generate neurons and astrocytes are present in multiple regions of the adult mouse hippocampus. J Comp Neurol. 2000;424(3):397–408. PMID: 10906710.
20. Gault N, Szele FG. Immunohistochemical evidence for adult human neurogenesis in health and disease. WIREs Mech Dis. 2021;13(6):e1526. https://doi.org/10.1002/wsbm.1526.
21. Lazarov O, Hollands C. Hippocampal neurogenesis: learning to remember. Prog Neurobiol. 2016;38–140:1–18. https://doi.org/10.1016/j.pneurobio.2015.12.006.
22. Steiner B, Klempin F, Wang L, Kott M, Kettenmann H, Kempermann G. Type-2 cells as link between glial and neuronal lineage in adult hippocampal neurogenesis. Glia. 2006;54(8):805–14. https://doi.org/10.1002/glia.20407.
23. Fukuda S, Kato F, Tozuka Y, Yamaguchi M, Miyamoto Y, Hisatsune T. Two distinct subpopulations of nestin-

positive cells in adult mouse dentate gyrus. J Neurosci. 2003;23(28):9357–66. https://doi.org/10.1523/JNEUROSCI.23-28-09357.2003.
24. Karalay O, Doberauer K, Vadodaria KC, Knobloch M, Berti L, Miquelajauregui A, et al. Prospero-related homeobox 1 gene (Prox1) is regulated by canonical Wnt signaling and has a stage-specific role in adult hippocampal neurogenesis. Proc Natl Acad Sci U S A. 2011;108(14):5807–12. https://doi.org/10.1073/pnas.1013456108.
25. Jessberger S, Römer B, Babu H, Kempermann G. Seizures induce proliferation and dispersion of doublecortin-positive hippocampal progenitor cells. Exp Neurol. 2005;196(2):342–51. https://doi.org/10.1016/j.expneurol.2005.08.010.
26. Kempermann G, Song H, Gage FH. Neurogenesis in the adult hippocampus. Cold Spring Harb Perspect Biol. 2015;7(9):a018812. https://doi.org/10.1101/cshperspect.a018812.
27. Kuhn HG. Control of cell survival in adult mammalian neurogenesis. Cold Spring Harb Perspect Biol. 2015;7:a018895. https://doi.org/10.1101/cshperspect.a018895.
28. Sun GJ, Sailor KA, Mahmood QA, Chavali N, Christian KM, Song H, Ming G, li. Seamless reconstruction of intact adult-born neurons by serial end-block imaging reveals complex axonal guidance and development in the adult hippocampus. J Neurosci. 2013;33(28):11400–11. https://doi.org/10.1523/JNEUROSCI.1374-13.2013.
29. Tozuka Y, Fukuda S, Namba T, Seki T, Hisatsune T. GABAergic excitation promotes neuronal differentiation in adult hippocampal progenitor cells. Neuron. 2005;47(6):803–15. https://doi.org/10.1016/j.neuron.2005.08.023.
30. Ge S, Goh ELK, Sailor KA, Kitabatake Y, Ming G, li., Song, H. GABA regulates synaptic integration of newly generated neurons in the adult brain. Nature. 2006;39(7076):589–93. https://doi.org/10.1016/j.neuron.2005.08.023.
31. Kempermann G, Chesler EJ, Lu L, Williams RW, Gage FH. Natural variation and genetic covariance in adult hippocampal neurogenesis. Proc Natl Acad Sci U S A. 2006;103(3):780–5. https://doi.org/10.1073/pnas.0510291103.
32. Kempermann G, Gast D, Kronenberg G, Yamaguchi M, Gage FH. Early determination and long-term persistence of adult-generated new neurons in the hippocampus of mice. Development. 2003;130(2):391–9. https://doi.org/10.1242/dev.00203.
33. Crespo D, Stanfield BB, Cowan WM. Evidence that late-generated granule cells do not simply replace earlier formed neurons in the rat dentate gyrus. Exp Brain Res. 1986;62(3):541–8. https://doi.org/10.1007/BF00236032.
34. Chancey JH, Adlaf EW, Sapp MC, Pugh PC, Wadiche JI, Overstreet-Wadiche LS. GABA depolarization is required for experience-dependent synapse unsilencing in adult-born neurons. J Neurosci. 2013;33(15):6614–22. https://doi.org/10.1523/JNEUROSCI.0781-13.2013.
35. Toni N, Laplagne DA, Zhao C, Lombardi G, Ribak CE, Gage FH, Schinder AF. Neurons born in the adult dentate gyrus form functional synapses with target cells. Nat Neurosci. 2008;11(8):901–7. https://doi.org/10.1038/nn.2156.
36. Brandt MD, Jessberger S, Steiner B, Kronenberg G, Reuter K, Bick-Sander A, et al. Transient calretinin expression defines early postmitotic step of neuronal differentiation in adult hippocampal neurogenesis of mice. Mol Cell Neurosci. 2003;24(3):603–13. https://doi.org/10.1016/s1044-7431(03)00207-0.
37. Schmidt-Hieber C, Jonas P, Bischofberger J. Enhanced synaptic plasticity in newly generated granule cells of the adult hippocampus. Nature. 2004;429(6988):184–7. https://doi.org/10.1038/nature02553.
38. Saxe MD, Battaglia F, Wang JW, Malleret G, David DJ, Monckton JE, et al. Ablation of hippocampal neurogenesis impairs contextual fear conditioning and synaptic plasticity in the dentate gyrus. Proc Natl Acad Sci U S A. 2006;103(46):17501–6. https://doi.org/10.1073/pnas.0607207103.
39. Marín-Burgin A, Mongiat LA, Pardi MB, Schinder AF. Unique processing during a period of high excitation/inhibition balance in adult-born neurons. Science. 2012;335(6073):1238–42. https://doi.org/10.1126/science.1214956.
40. Ge S, Yang CH, Hsu KS, Ming GL, Song H. A critical period for enhanced synaptic plasticity in newly generated neurons of the adult brain. Neuron. 2007;54(4):559–66. https://doi.org/10.1016/j.neuron.2007.05.002.
41. Sun B, Wang M, Hoerder-Suabedissen A, Xu C, Packer AM, Szele FG. Intravital imaging of the murine subventricular zone with three photon microscopy. Cereb Cortex. 2022;32(14):3057–67. https://doi.org/10.1093/cercor/bhab400.
42. McGinley LM, Willsey MS, Kashlan ON, Chen KS, Hayes JM, Bergin IL, et al. Magnetic resonance imaging of human neural stem cells in rodent and primate brain. Stem Cells Transl Med. 2021;10(1):83–97. https://doi.org/10.1002/sctm.20-0126.
43. Gilbert EAB, Lakshman N, Lau KSK, Morshead CM. Regulating endogenous neural stem cell activation to promote spinal cord injury repair. Cells. 2022;11(5):846. https://doi.org/10.3390/cells11050846.
44. Zhu T, Wang L, Wang LP, Wan Q. Therapeutic targets of neuroprotection and neurorestoration in ischemic stroke: applications for natural compounds from medicinal herbs. Biomed Pharmacother. 2022;148:112719. https://doi.org/10.1016/j.biopha.2022.112719.
45. Tian Z, Zhao Q, Biswas S, Deng W. Methods of reactivation and reprogramming of neural stem cells for neural repair. Methods. 2018;133:3–20. https://doi.org/10.1016/j.ymeth.2017.08.014.

46. Alipour M, Nabavi SM, Arab L, Vosough M, Pakdaman H, Ehsani E, et al. Stem cell therapy in Alzheimer's disease: possible benefits and limiting drawbacks. Mol Biol Rep. 2019;46(1):1425–46. https://doi.org/10.1007/s11033-018-4499-7.
47. Adugna DG, Aragie H, Kibret AA, Belay DG. Therapeutic application of stem cells in the repair of traumatic brain injury. Stem Cells Cloning. 2022;15:53–61. https://doi.org/10.2147/SCCAA.S369577.
48. Zhang Y, Zhao Y, Song X, Luo H, Sun J, Han C, et al. Modulation of stem cells as therapeutics for severe mental disorders and cognitive impairments. Front Psychiatry. 2020;11:80. https://doi.org/10.3389/fpsyt.2020.00080.

Neuroanatomy and Patterns of Cortical Connectivity of the Human Hippocampus

3

Seher Yılmaz and Şükrü Ateş

3.1 Introduction

The hippocampus, found in the medial temporal lobe and close to the temporal horn of the lateral ventricle, is crucial for memory, learning, spatial orientation, and emotions [1]. The hippocampus transforms temporary memories into lasting ones, aids in remembering locations, and retrieves experiences associated with particular places. Known as the archicortex, it is an ancient brain region crucial for emotions and behavior. Neural plasticity in the hippocampus continues throughout life. The hippocampus, cingulate gyrus, hypothalamus, and amygdala together form the "limbic system" [1, 2].

3.1.1 Neuroanatomy of the Human Hippocampus

This formation, located in the medial part of both temporal lobes (Fig. 3.1), is named after the combination of *hippos* (horse) and *kampos* (sea) due to its resemblance to a seahorse [3]. Because its outer surface resembles a ram's horn, the name *Cornu Ammonis*, meaning the horns (*Cornu*) of Ammon, a ram-headed Egyptian god, was also used. The letters "CA," which are the initials of *Cornu Ammonis*, are used in hippocampus segmentation [4, 5].

The hippocampus extends along the inner boundary of the temporal horn inside the lateral ventricle (Fig. 3.2). In its superior neighborhood, the cingulate gyrus is present; in the inferior neighborhood, the paraterminal gyrus is in the anterior, and the parahippocampal gyrus is in the posterior.

The hippocampal formation consists of the hippocampus, DG, CA, and folds of the subiculum (Fig. 3.3) [5–7]. The subiculum, located in the upper medial part of the parahippocampal gyrus, connects the CA region with the entorhinal cortex. During development, the CA and DG fold into the lateral ventricle, bringing their outer layers closer together. The two hippocampi are linked by commissural pathways [4].

Anatomically, the side of the hippocampus facing the ventricle is convex, while the side facing the lower hemisphere is concave. Its front side is wide and flat and is called *pes hippocampi*. There are two or three finger-like superficial protrusions in the front part. These protrusions are called *digitationes hippocampi* [8, 9]. The entire ventricular surface of the hippocampus is covered with the "alveus" layer formed by axons coming from its own cells. Alveus fibers approach each other medially in the form of a flat band, forming the "fimbria hippocampi". The anterior end of the fimbria hippocampi, uncus, ends in the white matter of the gyri hippocampi. Its posterior

S. Yılmaz (✉) · Ş. Ateş
Department of Anatomy, Yozgat Bozok University Faculty of Medicine, Yozgat, Türkiye
e-mail: sehery38@hotmail.com

M. Turgut et al. (eds.), *The Human Hippocampus*, https://doi.org/10.1007/978-3-031-88907-3_3

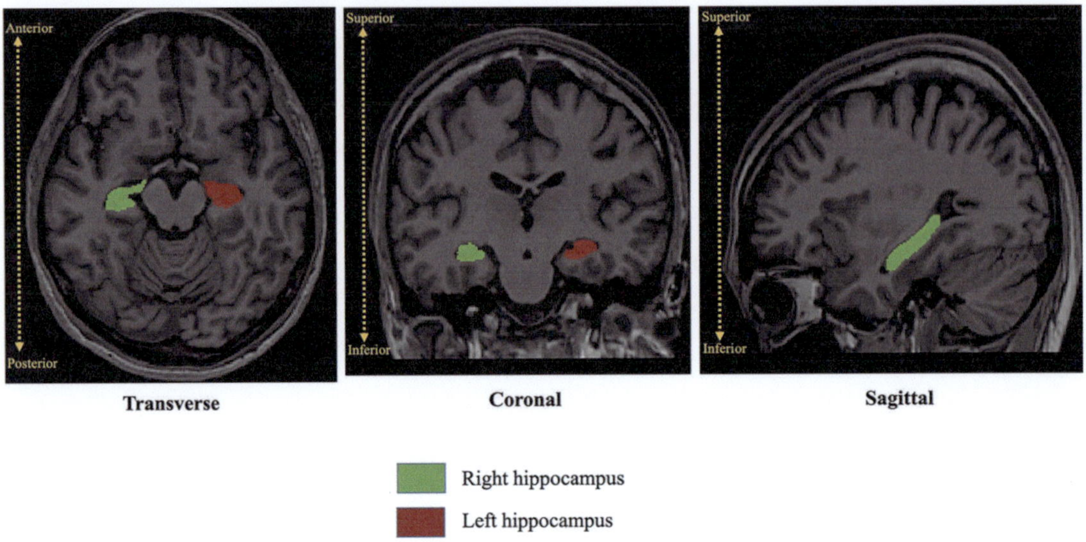

Fig. 3.1 Appearance of the hippocampus in transverse, coronal, and sagittal sections in brain magnetic resonance imaging

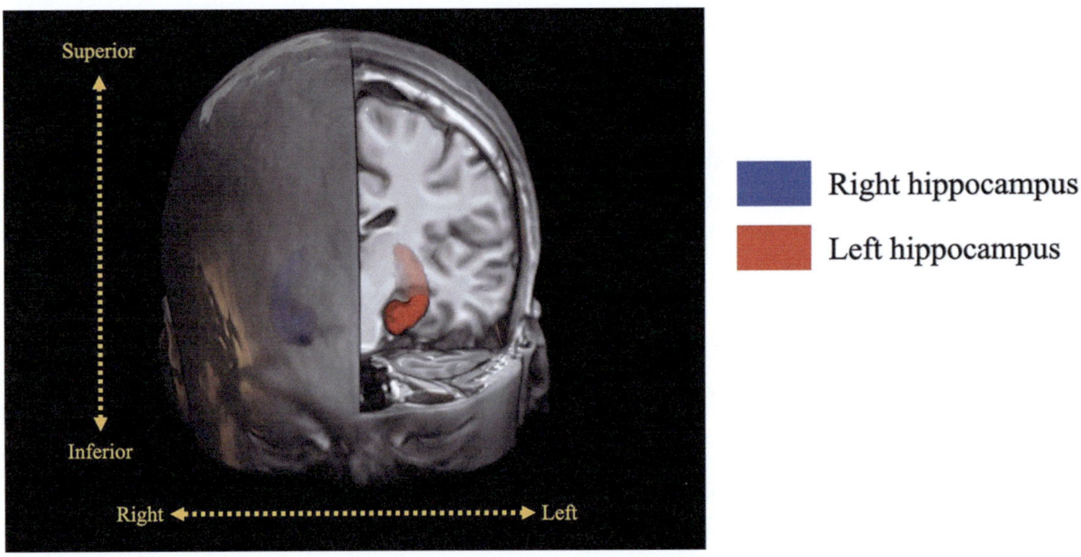

Fig. 3.2 Front view of the whole brain, including the skull (magnetic resonance imaging)

end forms the crus fornicis together with the alveus. The fibers that originate in the alveus and travel into the fimbria mark the initial formation of the fornix. The fornix consists of myelinated fibers. Sensory fibers leave the hippocampus from the fornix [10, 11].

From superficial to deep, cornu ammonis is examined in five layers: stratum lacunosum-moleculare, stratum radiatum, stratum pyramidale, stratum oriens, and alveus. The hippocampus, which can also be referred to as "CA" representing the initials of Cornu Ammonis, is divided into different areas such as CA1, CA2, CA3, and CA4 due to changes in the cell structure (Fig. 3.3) [1, 7, 8]. Among these areas, CA1 is situated nearest to the subiculum, while CA4 is located closest to the dentate gyrus. CA1 is the largest of these regions, bordered on the side by the presubiculum and

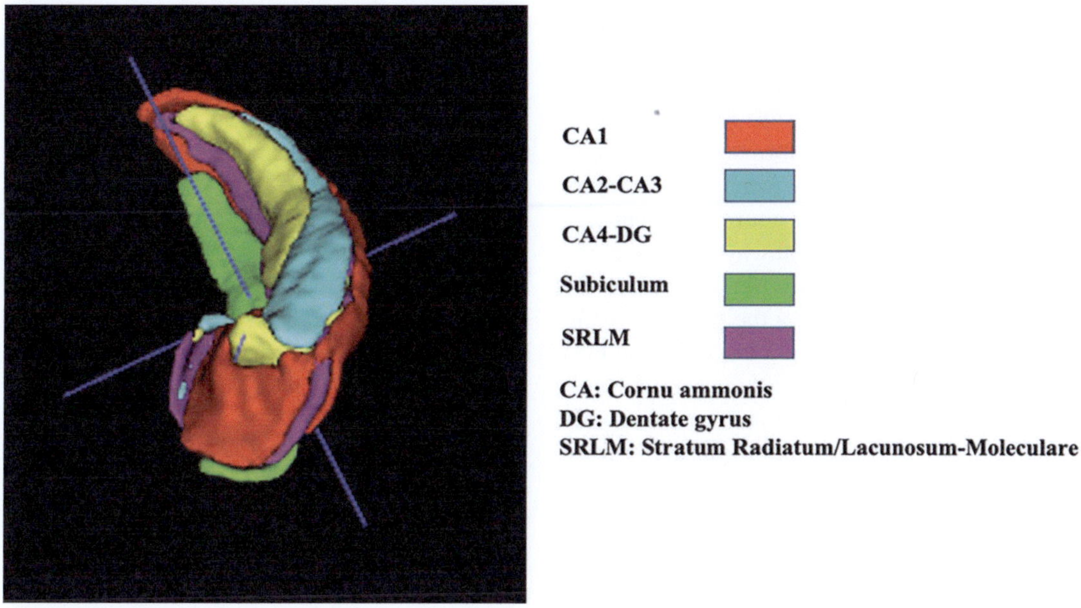

Fig. 3.3 Formations of the hippocampus

medially by CA2 [12, 13]. In CA1, about 90% of neurons are pyramidal cells essential for spatial memory, with the rest being interneurons. CA1 neurons process information from the entorhinal cortex or CA3 region. A strong connection between CA3 and CA1 is crucial for reference memory. The CA2 layer is located adjacent to the dentate gyrus, with CA1 forming its lateral boundary and CA3 defining its medial edge. The CA2 area of the hippocampus receives neural signals from the supramammillary area of the hypothalamus, but it does not get input from the mossy fibers originating in the DG. The CA3 layer, which faces the hilus of the DG, is situated next to CA2, which forms its medial boundary. Additionally, some researchers categorize the outermost neurons in the CA3 region as CA4 [4, 14]. In the CA3 region, mossy fibers from granule cells in the DG send signals to the upper branches of neurons. The axons of CA3 pyramidal neurons extend to the alveus, fimbria, and fornix areas. The CA3 area is linked to CA1 through Schaffer collateral fibers. Outputs from CA1 extend to the subiculum, entorhinal cortex, and prefrontal cortex. CA1 primarily receives input from CA3, with a minor contribution from the entorhinal cortex. The CA1 area acts as a checker for mistakes, spotting differences between the signals it gets from the cortex and the data it receives from CA3 and the entorhinal cortex. While cell numbers in CA1 and CA3 are lower before adolescence, they gradually increase after this period. This shows that spatial learning and memory develop in adolescence [14–16].

3.1.2 Patterns of Cortical Connectivity of the Human Hippocampus

The hippocampus is crucial for various types of memory, including episodic, short-term, and working memory, and for consolidating memories. It primarily receives input from the association cortices and the entorhinal cortex, suggesting that memories are stored in the association and primary cortices rather than in the medial temporal lobe [17]. Stimuli reach the hippocampus through various pathways, including perforant pathways, mossy fibers, Schaffer collaterals, and alvear fibers, all carrying sensory signals from the entorhinal cortex [4, 18].

Cornu ammonis is divided into four regions, called 1–4 (Fig. 3.3). Axons extend from the DG to the CA3 area of the hippocampus. Originating there, axons in the hippocampus can branch out to the CA1 region through the fornix or via Schaffer collateral fibers. Axons originating from CA1 may either enter the fornix or project into the subiculum [19]. Axons from the subiculum either travel to the fornix or return to the entorhinal cortex. The alvear pathway carries nerve fibers straight from the entorhinal cortex to the CA1 and CA3 regions. Like the perforant pathway, these fibers start in CA1 and CA3 and then travel to the subiculum [20, 21].

Efferent axons from the hippocampus mainly come from the subiculum and form the fornix, a white matter structure that curves over the ventricles. In addition, stimulation enters the hippocampus from the thalamus, hypothalamus, and brainstem via the fornix. The hippocampus also receives signals from the parahippocampal gyrus cortex and transmits these signals to the corpus mamillare, area septalis, and hypothalamic nuclei via the fornix [4, 18].

The primary efferent pathway of the hippocampus is the fornix, which also includes commissural fibers linking the hippocampi on both sides. Axons from the hippocampus and subiculum travel through the alveus to the fimbria hippocampi, connecting in the crus fornicis beneath the splenium of the corpus callosum. This pathway continues as the corpus fornicis at the back of the thalamus [14, 22, 23]. After the corpus fornicis, the axons that continue as the columna fornicis curve in the front part of the foramen interventriculare and give fibers to the nucleus anteriores thalami and nucleus dorsalis lateralis thalami. These fibers are called postcommissural fibers. Most fibers from the fornix project to the mammillary bodies and the ventromedial nucleus of the hypothalamus. Some, known as precommissural fibers, travel from the column of the fornix to the anterior commissure, reaching the hypothalamic rostral area, septal area, and substantia innominata. The hippocampus also sends efferent signals to the amygdala, cingulate gyrus, prefrontal cortex, entorhinal cortex, and brainstem [14, 24].

In the 1930s, American neurologist James Papez identified a circuit, known as the Papez circuit, which links these structures in the hippocampus and cortical regions. This route is believed to be involved in experiencing and showing emotions [25]. This phase begins with the fibers coming from the subiculum passing via the fornix and reaching the mammillary bodies. From there, axons continue via the mammillothalamic tract to the anterior nucleus of the thalamus and then pass through the internal capsule to the cingulate gyrus. The cingulate cortex then transmits signals to the parahippocampal gyrus, entorhinal cortex, and hippocampus. While Papez's circuit describes the pathways of the limbic system, some parts play lesser roles in emotions and have different functions [25–27].

Afferent Pathways [1, 4, 28–30]:

- Perforant pathway: Axons originating in the entorhinal cortex proceed to the DG via the subiculum, and these axons are distributed to the entire hippocampus except for the CA4 area.
- Mossy fibers: They go from the DG to the CA3 area.
- Schaffer's collateral fibers: They are pyramidal cell extensions extending from CA3 and CA2 to the CA1 area.
- Alveal fibers: These fibers come from subcortical areas and pass through the alveus to the hippocampus and are distributed to the CA1 part of the hippocampus and the inner layer of the subiculum.
- Subiculum: The subiculum is part of the hippocampus and sends back projections to the cortex. These back projections aid in the processing and integration of hippocampal information by the cortex.
- Cingulum: Carries signals from the gyrus cinguli to the hippocampus.

Efferent Pathways [1, 4, 23, 28–30]:

- Through the fibers of the dorsal fornix: fasciolar gyrus, indusium griseum, cingulate gyrus, and septal nuclei.
- Through the pre-commissural fibers (fornix): paraterminal gyrus, preoptic, and anterior hypothalamic nuclei.

- Through the post-commissural fibers (fornix): anterior nucleus of the thalamus, hypothalamic nuclei, mammillary body.
- Through stria medullaris thalami: habenular nucleus.

3.1.3 Blood Vessels of the Hippocampus

The hippocampus gets its blood flow from the branches of the posterior cerebral artery (PCA) and the anterior choroidal artery (ACA) [31]. In its second part, the PCA divides into the anterior inferior temporal artery and the anterior hippocampal-parahippocampal artery, which deliver blood to the entorhinal region. Additionally, the parieto-occipital trunk of the PCA also supplies blood to the parahippocampal gyrus and the hippocampus [1, 32]. The perforating branches of the anterior choroidal artery supply blood to the amygdala and hippocampus [33, 34].

The anterior hippocampal arteries (AHA) deliver blood to the uncus and the hippocampal head, while the posterior hippocampal arteries (PHA) deliver blood to the body and tail of the hippocampus [1, 4, 35]. The AHA supplies blood to the hippocampal head by descending into the uncal sulcus. The PHA follows the hippocampal sulcus, branching into larger and smaller vessels. The larger branches enter the hippocampus, while the smaller ones supply the margo denticulatus and fimbriodentate sulcus, creating a dense network along the sulcus [1, 31, 35].

Intrahippocampal arteries (IHAs), or deep branches, are divided into large and small ventral and dorsal types. Large ventral IHAs supply the stratum lacunosum, stratum pyramidale, molecular layer of the dentate gyrus, and CA1 and CA2 regions. Large dorsal IHAs target the granular layer of the DG and CA3 and CA4 regions. Small ventral IHAs focus on the proximal DG, while small dorsal IHAs extend into the fibrodentate sulcus, supplying nearby areas [1, 31].

Deep hippocampal veins, also known as intrahippocampal veins (IHVs), are classified into two types: sulcal and subependymal IHVs. Sulcal IHVs emerge originating in the CA1 and CA2 regions and flow into the superficial hippocampal sulcus, where they receive branches originating in the stratum moleculare. In contrast, subependymal IHVs are found on the ventricular surface of the hippocampus [1, 4, 31]. The CA2 and deep intrahippocampal veins from the subiculum drain into the subependymal veins. Superficial veins form two arcades over the fimbriodentate and superficial hippocampal sulci. One arcade collects from the subependymal and deep intrahippocampal areas, while the other gathers from deeper regions. These arcades merge at both ends, draining into the inferior ventricular vein at the front and the medial atrial vein at the back, which both eventually flow into the basal vein [1, 35].

References

1. Pradip Chauhan M, Kinjal Jethwa M, Ashish Rathawa M, Girish Chauhan BDS, Simmi Mehra M. The anatomy of the hippocampus. Exon Publications; 2021. pp. 17–30. https://doi.org/10.36255/exonpublications.cerebralischemia.2021.hippocampus.
2. Bombardi C, Grandis A, Pivac N, Sagud M, Lucas G, Chagraoui A, Lemaire-Mayo V, De Deurwaerdère P, Di Giovanni G. Chapter 3—Serotonin modulation of hippocampal functions: from anatomy to neurotherapeutics. In: Di Giovanni G, De Deurwaerdere P, editors. Progress in brain research. Elsevier; 2021. pp. 83–158. https://doi.org/10.1016/bs.pbr.2021.01.031.
3. Afifi AK, Bergman RA. Functional neuroanatomy: text and atlas. New York: McGraw-Hill; 1998. http://dni.dali.dartmouth.edu/bjt1n15yiwke/18-miss-elseorn-1/read-0071408126-functional-neuroanatomy-text-and-atlas-paperback.pdf. Accessed 23 Apr 2024.
4. İzci Y, Erbaş YC. Hipokampus: yapısı ve fonksiyonları. Türk Nöroşir Derg. 2015;25:287–95.
5. Knierim JJ. The hippocampus. Curr Biol. 2015;25:R1116–21.
6. Hayman LA, Fuller GN, Cavazos JE, Pfleger MJ, Meyers CA, Jackson EF. The hippocampus: normal anatomy and pathology. Am J Roentgenol. 1998;171:1139–46. https://doi.org/10.2214/ajr.171.4.9763010.
7. Özdemir M, Soysal H, Eraslan Ö, Dilli A. Normative hippocampal volumetric measurements using magnetic resonance imaging. Turk J Med Sci. 2019;49:1464–70.
8. Dekeyzer S, De Kock I, Nikoubashman O, Vanden Bossche S, Van Eetvelde R, De Groote J, Acou M, Wiesmann M, Deblaere K, Achten E. "Unforgettable"—a pictorial essay on anat-

omy and pathology of the hippocampus. Insights Imaging. 2017;8:199–212. https://doi.org/10.1007/s13244-016-0541-2.
9. Haładaj R. Anatomical variations of the dentate gyrus in normal adult brain. Surg Radiol Anat. 2020;42:193–9. https://doi.org/10.1007/s00276-019-02298-5.
10. Au CKF, Abrigo J, Liu C, Liu W, Lee J, Au LWC, Chan Q, Chen S, Leung EYL, Ho CL, Ko H, Mok VCT, Chen W. Quantitative susceptibility mapping of the hippocampal fimbria in Alzheimer's disease. J Magn Reson Imaging. 2021;53:1823–32. https://doi.org/10.1002/jmri.27464.
11. Fogwe LA, Reddy V, Mesfin FB. Neuroanatomy, hippocampus. In: StatPearls. Treasure Island: StatPearls Publishing; 2024. http://www.ncbi.nlm.nih.gov/books/NBK482171/. Accessed 23 Apr 2024.
12. Scorza CA, Araujo BHS, Leite LA, Torres LB, Otalora LFP, Oliveira MS, Garrido-Sanabria ER, Cavalheiro EA. Morphological and electrophysiological properties of pyramidal-like neurons in the stratum oriens of Cornu ammonis 1 and Cornu ammonis 2 area of proechimys. Neuroscience. 2011;177:252–68. https://doi.org/10.1016/j.neuroscience.2010.12.054.
13. Piskorowski RA, Chevaleyre V. Hippocampal area CA2: interneuron disfunction during pathological states. Front Neural Circuits. 2023;17:1181032. https://doi.org/10.3389/fncir.2023.1181032.
14. Songur A, Özen OA, Sarsilmaz M. Hipokampus. Turkiye Klinikleri J Med Sci. 2001;21:427–31.
15. Suzuki M, Hagino H, Nohara S, Zhou S-Y, Kawasaki Y, Takahashi T, Matsui M, Seto H, Ono T, Kurachi M. Male-specific volume expansion of the human hippocampus during adolescence. Cereb Cortex. 2005;15:187–93.
16. Tien RD, Felsberg GJ, Crain B. Normal anatomy of the hippocampus and adjacent temporal lobe: high-resolution fast spin-echo MR images in volunteers correlated with cadaveric histologic sections. Am J Roentgenol. 1992;159:1309–13. https://doi.org/10.2214/ajr.159.6.1442407.
17. Lazarov O, Hollands C. Hippocampal neurogenesis: learning to remember. Prog Neurobiol. 2016;138–140:1–18. https://doi.org/10.1016/j.pneurobio.2015.12.006.
18. Anand KS, Dhikav V. Hippocampus in health and disease: an overview. Ann Indian Acad Neurol. 2012;15:239–46.
19. Kantarci K. Fractional anisotropy of the fornix and hippocampal atrophy in Alzheimer's disease. Front Aging Neurosci. 2014;6:316. https://doi.org/10.3389/fnagi.2014.00316.
20. Mercer A, Thomson AM. Cornu ammonis regions–antecedents of cortical layers? Front Neuroanat. 2017;11:83. https://doi.org/10.3389/fnana.2017.00083.
21. Hainmueller T, Bartos M. Dentate gyrus circuits for encoding, retrieval and discrimination of episodic memories. Nat Rev Neurosci. 2020;21:153–68. https://doi.org/10.1038/s41583-019-0260-z.
22. Duvernoy HM, Cattin F, Risold P-Y. The human hippocampus: functional anatomy, vascularization and serial sections with MRI. Berlin: Springer Berlin Heidelberg; 2013. https://doi.org/10.1007/978-3-642-33603-4.
23. Gazes Y, Li P, Sun E, Razlighi Q, Tsapanou A. Age specificity in fornix-to-hippocampus association. Brain Imaging Behav. 2019;13:1444–52. https://doi.org/10.1007/s11682-018-9958-1.
24. Steward O, Scoville SA. Cells of origin of entorhinal cortical afferents to the hippocampus and fascia dentata of the rat. J Comp Neurol. 1976;169:347–70. https://doi.org/10.1002/cne.901690306.
25. Kamali A, Milosavljevic S, Gandhi A, Lano KR, Shobeiri P, Sherbaf FG, Sair HI, Riascos RF, Hasan KM. The cortico-limbo-thalamo-cortical circuits: an update to the original Papez circuit of the human limbic system. Brain Topogr. 2023;36:371–89. https://doi.org/10.1007/s10548-023-00955-y.
26. Aggleton JP, Nelson AJD, O'Mara SM. Time to retire the serial Papez circuit: implications for space, memory, and attention. Neurosci Biobehav Rev. 2022;140:104813. https://doi.org/10.1016/j.neubiorev.2022.104813.
27. Aggleton JP, Pralus A, Nelson AJD, Hornberger M. Thalamic pathology and memory loss in early Alzheimer's disease: moving the focus from the medial temporal lobe to Papez circuit. Brain. 2016;139:1877–90. https://doi.org/10.1093/brain/aww083.
28. Di Ieva A, Fathalla H, Cusimano MD, Tschabitscher M. The indusium griseum and the longitudinal striae of the corpus callosum. Cortex. 2015;62:34–40.
29. Cascella M, Al Khalili Y. Short term memory impairment. 2019. https://europepmc.org/article/nbk/nbk545136. Accessed 23 Apr 2024.
30. Martens M, Filippini N, Masaki C, Godlewska BR. Functional connectivity between task-positive networks and the left precuneus as a biomarker of response to lamotrigine in bipolar depression: a pilot study. Pharmaceuticals (Basel). 2021;14:534.
31. Tatu L, Vuillier F. Structure and vascularization of the human hippocampus. Front Neurol Neurosci. 2014;34:18–25. https://doi.org/10.1159/000356440.
32. Kuybu O, Tadi P, Dossani RH. Posterior cerebral artery stroke. In: StatPearls. Treasure Island: StatPearls Publishing; 2024. http://www.ncbi.nlm.nih.gov/books/NBK532296/. Accessed 23 Apr 2024.
33. Hou K, Yu J. Role of the anterior choroidal artery in the endovascular treatment of brain arteriovenous malformations. Acta Neurol Belg. 2023;123:57–67. https://doi.org/10.1007/s13760-022-01878-4.
34. Isolan GR, Stefani MA, Schneider FL, Claudino HA, Yu YH, Choi GG, Telles JPM, Rabelo NN, Figueiredo EG. Hippocampal vascularization: proposal for a new classification. Surg Neurol Int. 2020;11:378. https://doi.org/10.25259/SNI_708_2020.
35. Xu Y, Mohyeldin A, Nunez MA, Doniz-Gonzalez A, Vigo V, Cohen-Gadol AA, Fernandez-Miranda JC. Microvascular anatomy of the medial temporal region. J Neurosurg. 2021;137:747–59. https://doi.org/10.3171/2021.9.JNS21390.

4 Structural and Functional Properties of Hippocampal Neurons

Feyzullah Beyaz

4.1 Introduction

The two layers that make up the hippocampus—the CA and the DG—are coiled within each other, creating a bilaminar structure. Perhaps as a result of a developmental misalignment, the DG takes on a concave form and slides beneath the medial end of the CA [1]. In their ultimate form, the CA and DG resemble two interlocking U-shaped layers, with the hippocampal sulcus serving as a dividing line. The CA and DG are parts of the most primitive form of the cortex, the allocortex or archaeocortex. Contrast this with the more complex isocortex, sometimes known as the neocortex [2]. Different subfields of the hippocampus have different morphologies, cell shapes and sizes, connections, electrical properties, and susceptibility to damage [3]. The "CA" acronym stands for "central area," while the hippocampus is really divided into many areas called "CA1, CA2, CA3, and CA4" [1–3]. Hippocampal development also involves the entorhinal cortex, parasubiculum, subiculum, and presubiculum. Massive, pyramid-shaped projection neurons and smaller interneurons are neuronal structures present in some parts of the hippocampus formation that are similar to those in other regions of the cortex [4, 5]. Intrahippocampal circuits mostly transmit information in one direction, while intrinsic associative connections are dispersed over three dimensions, making the neuroanatomy of hippocampus development distinct [1–3]. After a Nissl slice of the hippocampus is prepared and stained, it is clear that the so-called hippocampus is actually a complex network of structures with various cellular configurations [1–3] (Fig. 4.1).

Anatomically distinct surrounding regions that make up what is known as the "hippocampal formation" include the entorhinal cortex, subiculum, dentate gyrus, presubiculum, and parasubiculum. Their connectivity through discrete and mostly one-way neuronal connections is the fundamental reason these six locations are categorized as part of the hippocampus formation [6]. The first three connections in the hippocampus's circuitry, sometimes called the "trisynaptic circuit" [2, 7], were the primary focus of early investigations into its genesis. There are three synapses in this circuit: one connecting the entorhinal cortex (EC) to the dentate gyrus (DG), another connecting the DG to the CA3, and a third connecting the CA3 to CA1 [1–3].

F. Beyaz (✉)
Department of Histology and Embryology,
Erciyes University Faculty of Veterinary Medicine,
Kayseri, Türkiye
e-mail: fbeyaz@erciyes.edu.tr

M. Turgut et al. (eds.), *The Human Hippocampus*, https://doi.org/10.1007/978-3-031-88907-3_4

4.2 Cornu Ammonis (CA)

The CA can be seen as having six distinct layers, beginning in the ventricular cavity and ending in the primitive hippocampal sulcus. The alveus, oriens, pyramidale, radiatum, lacunosum, and moleculare strata are the layers that make up the rock formation [1–3] (Fig. 4.1). The inner surface of the ventricles is covered by the alveus, a layer that also contains the axons of the neurons in the subiculum and hippocampus. The principal pathways for information transmission from these areas are these axons. Afterwards, these fibers go into the fibrils. Also, the alveus is supplied with afferent fibers, the majority of which come from the septum. In humans in particular, the stratum oriens merges with the stratum pyramidale, making its boundaries difficult to discern [3, 8]. Axons from pyramidal neurons go via this structure, which is made up of scattered nerve cells called basket cells, on their way to the alveus. In the stratum pyramidale, you'll find the pyramidal neurons that make up the CA (Fig. 4.2). The pyramidal soma is triangular in form, with the alveus at its base and the primitive hippocampus sulcus at its apex. From its point of origin, the axon travels through the oriens layer and finishes at the alveus. As they travel, these axons link with other pyramidal neurons via Schaffer collaterals [9], which retrace their path into the stratum radiatum [3]. Pyramidal neurons have very long apical dendrites that traverse the whole thickness of the CA and terminate at the stratum moleculare in the vestigal hippocampal sulcus. Basal dendrites are produced by the soma's basal angles in addition to the apical dendrites. A few of these formations form branching patterns as they protrude into the stratum oriens [3]. Because they have two distinct dendritic trees, hippocampal pyramidal neurons are often referred to as "double pyramidal" [10]. Basket cells, with their cell bodies in the stratum oriens, give rise to a dense network of branches that encircle the soma. Also, just like the rest of the CA, the stratum pyramidale is uniformly distributed with stellate neurons and basket-type interneurons [11, 12] (Fig. 4.2). The pyramidal cell layer is the main cellular layer. While CA2 and CA3 have lesser densities of the pyramidal cell layer, CA1 has a very high density. Underneath the sparsely cellular pyramidal cell layer lies the thin stratum oriens. Pyramidal cell basal dendrites and other interneurons make up this layer. For example, the CA3–CA3 associational connections and the CA3–CA1 Schaffer collateral connections are located in the stratum oriens, which is the infrapyramidal region. Located under the stratum oriens lies the thin and fibrous alveus. Only in the

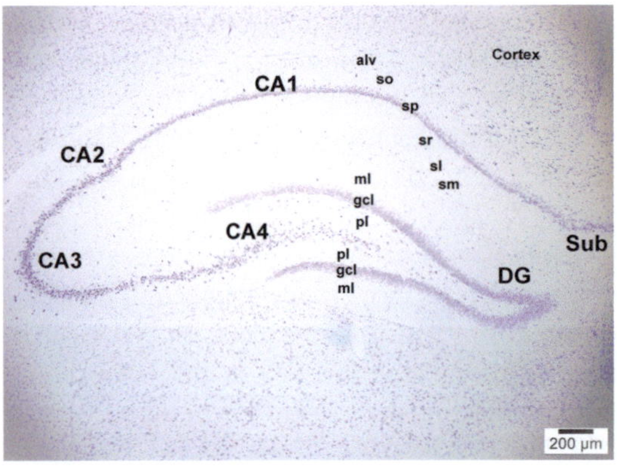

Fig. 4.1 Nissl-stained coronal section of the rat hippocampus. *CA1-CA4* CA1-CA4 fields of cornu ammonis; *DG* dentate gyrus; *Sub* subiculum; *alv* alveus; *so* stratum oriens of CA1; *sp* stratum pyramidale of CA1; *sr* stratum radiatum of CA1; *sl* stratum lacunosum of CA1; *sm* stratum moleculare of CA1; *ml* molecular layer of the dentate gyrus; *gcl* granule cell layer of the dentate gyrus; *pl* polymorphic layer of the dentate gyrus. Bar: 200 μm

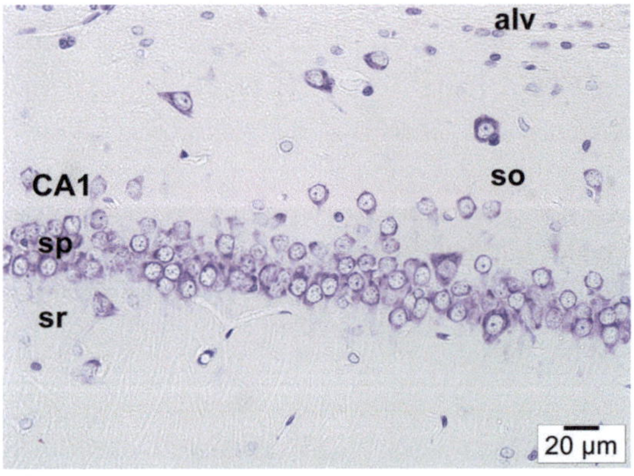

Fig. 4.2 Nissl-stained coronal section of the CA1 of rat hippocampus. *CA1* CA1 field of cornu ammonis; *alv* alveus; *so* stratum oriens of CA1; *sp* stratum pyramidale of CA1; *sr* stratum radiatum of CA1. Bar: 20 μm

CA3 field, and not in CA2 or CA1, is the stratum lucidum, a tiny acellular zone. It is mostly populated by mossy fibers and is located right above the pyramidal cell layer. The stratum lucidum shows a little thickening toward its distal end. Mossy fibers travel longitudinally and bend in a temporal orientation; this only happens at the septal levels of the hippocampus. In contrast to rats, guinea pigs and other small mammals have a more pronounced end bulb, which serves to demarcate the CA3/CA2 border. The stratum radiatum is situated immediately above the pyramidal cell layer in CA2 and CA1, and it is placed superficially relative to the stratum lucidum in CA3. The striated appearance of the stratum radiatum is caused by the parallel organization of the apical dendrites from pyramidal neurons. Schaffer collaterals, which are fibers arising from septal nuclei, and commissural fibers create synapses with apical dendrites in the stratum radiatum [13]. The surface of the CA is home to many parallel axonal fasciculi that make up the stratum lacunosum. The majority of the fasciculi are Schaffer collaterals and perforant fibers. The stratum moleculare is located next to the sulcus of the vestibular hippocampus. The fast erasure of the brain's grooves throughout development makes it difficult for the CA region's outer layer of cells to separate from the DG region's outer layer. A few number of neurons called interneurons make up the stratum moleculare. Pyramidal neurons' first dendritic branching patterns are part of it [1]. Therefore, pyramidal neurons are able to access all layers of the CA due to their elongation. The connections between CA3 and CA3 neurons, as well as the Schaffer collaterals, which are connections between CA3 and CA1 neurons, are located in the stratum radiatum, which is located above the pyramidal layer. The stratum lacunosum-moleculare is the name of the hippocampus's outermost layer. Afferents from many areas, including the midline thalamus's nucleus reuniens, and fibers from the entorhinal cortex terminate in this layer. A wide variety of interneurons are found in both the stratum radiatum and the stratum lacunosum-moleculare.

Because pyramidal neurons in the CA have various properties, the organization of the brain in coronal slices can vary [2, 3, 14]. Lorente de No (1934) [6] classified the CA into four distinct domains, CA1–CA4. The subiculum is the source of CA1. Pyramidal neurons are characterized by small, widely distributed somata that are triangular in form. In contrast to the structure seen in rats, the stratum pyramidale in the CA1 area of humans is notable for its robust, thin, and compact composition. A deep layer, the stratum profundum, and a shallow layer, the stratum superficiale, have been identified inside the stratum pyramidale of human CA1. A sparse population of pyramidal neurons characterizes the first, which is located next to the stratum oriens; in contrast, a dense population of pyramidal neurons distinguishes the second. CA2 is made up of densely packed, thin stratum pyramidale, which is formed of huge, oval-shaped somata. Compared to CA1, this is completely opposite. Although CA3 is definitely present in the hippocampus of humans and simians, its presence is debatable in many other species [15]. When it reaches the concave portion of the gyrus dentatus, the third area of the CA curves, creating a genu. Although they are not as densely packed, the pyramidal somata in this area are similar to those in CA [2].

The dentate gyrus (DG) is the origin of the mossy fibers, which are fragile, unmyelinated fibers that define CA3. Crammed between the layers of radiatum and pyramidale, these fibers encircle the pyramidal cell bodies. This adds a characteristic layer to CA3 known as the stratum lucidum [1]. CA4 is different from CA3 since it is located in the DG's concave section. The somata in this area are characteristic of CA4 and have an oval shape, are large, and are scattered among huge, intertwined, and mossy myelinated fibers. One part of the CA, CA1, is known as the Sommer sector [16], while another, CA3, is known as the Spielmeyer sector [17], because its susceptibility to hypoxia varies across the CA. On the other hand, CA4 is part of the Bratz sector—an intermediate vulnerability sector [18].

As its name suggests, the pyramidal cell layer is the primary structural component of the CA region. Pyramidal cells' cell bodies are found inside this layer [3]. In order to divide the CA region into four separate subfields, Lorente de No

(1934) used the Golgi impregnation technique. In [6], CA1, CA2, CA3, and CA4 are cited. Close to the subiculum is where you'll find the CA1 region. The choroid plexus and fimbria/fornix are close to the CA3 region. While CA4 is situated within the DG's hilus, CA2 acts as a little separating line between CA1 and CA3 [1, 6]. With the exception of the CA2–CA3 subfield boundary, the various hippocampal regions may be distinguished using standard histology stains such as cresyl violet. The hippocampal fissure separates the dentate gyrus from the CA1 region; it is an intrinsic division that does not include any cellular components. Pyramidal neurons are characterized by their cell bodies that are triangular in shape. Three different cortical areas include pyramidal cells that are larger than the one in CA1. Specifically, the pyramidal cells in question measure 40–60 μm at their bases, while the ones found in CA1 measure 20–40 μm. The pyramidal cells' basal dendrites make up the stratum oriens [1]. Pyramidal cell apical dendrites are situated in the stratum moleculare. In contrast to CA1 pyramidal cells, CA3 pyramidal neurons' dendrites are thicker and shorter. Long, thread-like extensions of granule cells called mossy fibers (MFs) interact with CA3 pyramidal cells in the stratum lucidum and extend to the CA3 region. The existence of "thorny excrescences" is a defining feature of the second kind. These are the postsynaptic components of the synapses that link the MF terminals to the apical dendrites of CA3 pyramidal cells [19]. Lorente de No's 1934 study pinpoints the CA2 region as the subdivision of CA that does not have the same stratum lucidum pattern as CA3. The presence of structures called thorny excrescences characterizes this pattern. Mossy fibers (MF) so do not contribute to this region in any way [6]. In the CA2 sector, there has been a great deal of talk and debate. Lorente de Nó first used the word to describe a little region of cells between CA3 and CA1. Similar in size to CA3's somas are CA2's pyramidal cells. Nevertheless, the mossy fibers that originate from the dentate gyrus do not innervate CA2, in contrast to CA1 [1, 6]. Despite occasional skepticism, the vast majority of recent evidence points to the existence of a unique CA2 area inside the hippocampus. Several criteria, including neurochemical markers, can be used to distinguish this part of the hippocampus from others [1]. As a transitional area between the CA1 and CA3 zones, the CA2 has been written off by many authors as being of little significance. It is a place where CA1 and CA3 pyramidal neurons combine, but it doesn't seem to have a distinct identity [1, 3]. These researchers have determined that CA1 and CA3 make up the bulk of the CA region. Quantitative rat studies have shown that there are around 330,000 pyramidal cells in the CA3 area and about 420,000 in the CA1 region. Furthermore, a CA4 area was described by Lorente de Nó. Figures 4.1 and 4.3 show that the region called CA4, which was first characterized by Theodor Blackstad (1956) and subsequently by David Amaral (1978), correlates to the deep layer of the dentate gyrus, also called the polymorphic layer.

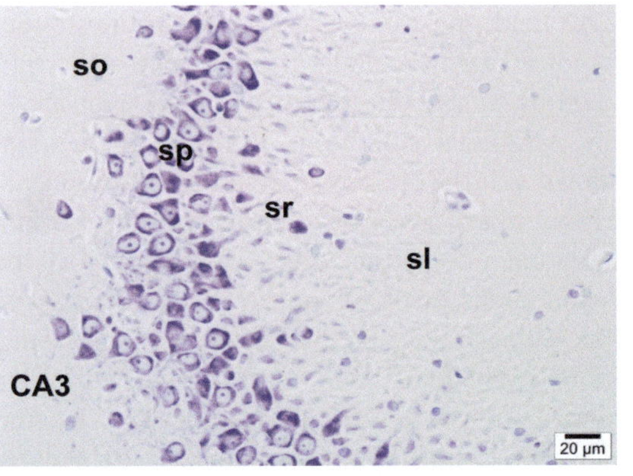

Fig. 4.3 Nissl-stained coronal section of rat the dentate gyrus. *CA4* CA4 field of cornu ammonis; *DG* dentate gyrus; *ml* molecular layer of the dentate gyrus; *gcl* granule cell layer of the dentate gyrus; *pl* polymorphic layer of the dentate gyrus. Bar: 100 μm

4.2.1 Pyramidal Cells of CA3 and CA2

With the bulk of its neurons located in the pyramidal cell layer, pyramidal cells are the most common kind of neuron in the hippocampus. The dendrites of pyramidal cells branch out in two directions: one from the base, into the stratum oriens, and another from the top, into the hippocampal fissure. You may see this data in Figs. 4.4 and 4.5. Dendritic length and structure of CA3 pyramidal cells show great heterogeneity, as reported by Ishizuka et al. (1995). The branches of the dentate gyrus contain the tiny cells, which are around 20 μm in width and 300μm² in soma size. The overall dendritic length of these cells is 8–10 mm. The biggest cells, located near the field's periphery, have a diameter of 30 m and a soma size of around 700 m². The total length of these cells' dendrites is 16–18 mm. In CA3 pyramidal cells, the arrangement of dendritic branches

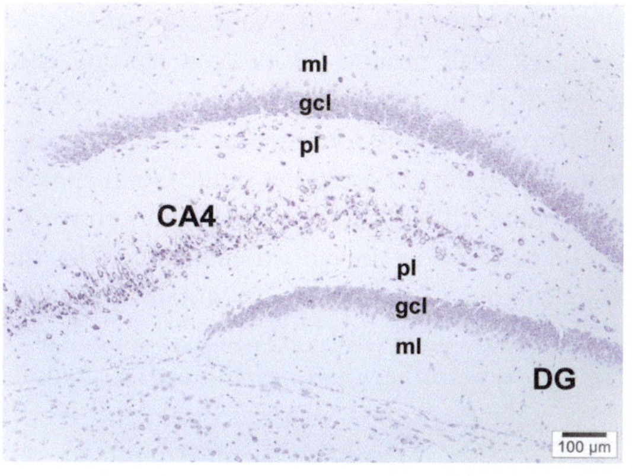

Fig. 4.4 NeuN immuno-stained coronal section of the rat hippocampus. NeuN immunopositive neurons. *CA2-CA4* CA2-CA4 fields of cornu ammonis; *DG* dentate gyrus; *Sub* subiculum; *CC* corpus callosum. Bar: 200 μm

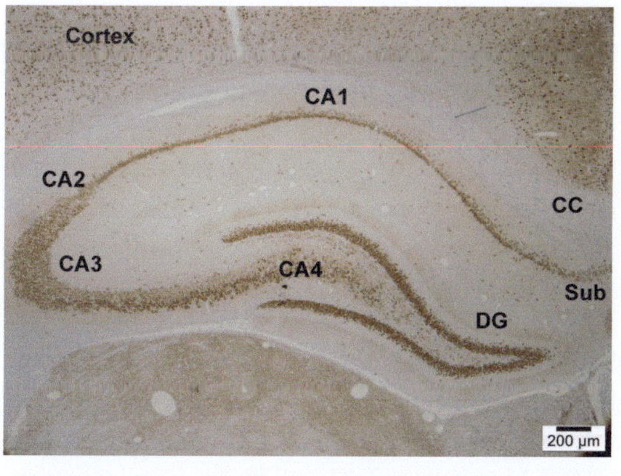

Fig. 4.5 NeuN immuno-stained sagittal section of the rat hippocampus. NeuN immunopositive neurons. *CA2-CA4* CA1-CA4 fields of cornu ammonis; *DG* dentate gyrus; *Sub* subiculum; *CC* corpus callosum. Bar: 100 μm

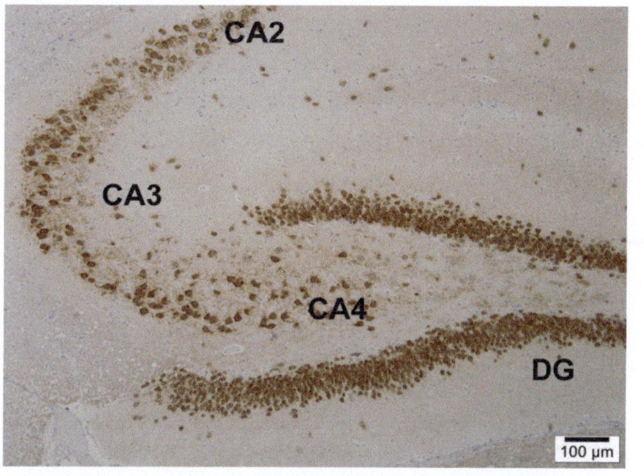

Fig. 4.6 Nissl-stained coronal section of the CA3 of rat hippocampus. *CA3* CA3 field of cornu ammonis; *so* stratum oriens of CA3; *sp* stratum pyramidale of CA3; *sr* stratum radiatum of CA3; *sr* stratum lucidum of CA3. Bar: 20 μm

differs according to the position of the cell soma. Only a small fraction of dentate gyrus branch cells really project into the stratum lacunosum-moleculare. This means the entorhinal cortex isn't sending them any signals at all. A higher density of mossy fiber terminals on the apical and basal dendritic trees is seen in cells that are closer to the granule cells. Since the CA3 cells are further distant and can only get input from mossy fibers on their apical dendrites, they do not have as much of an impact as the granule cells (Fig. 4.6) [2, 20].

4.2.2 Pyramidal Cells of CA1

Pyapali et al. (1998) showed that the dendritic trees of CA1 pyramidal cells are quite consistent, in contrast to CA3 pyramidal cells, which display a great deal of variability in dendrite arrangement [2]. In addition to being more homogeneous, these cells are somewhat smaller than the cells in CA3. The typical dendrite length is around 13.5 mm, while the typical CA1 cell somata are roughly 193 square meters in size, which is about 15 m in circumference. Pyramidal cells always have the same dendritic shape and the same total dendritic length, no matter where they are in CA1. The number of apical dendrites is variable among pyramidal cells; some have just one, while others have two. Dendrites on cells with two apical ends are noticeably longer than those on cells without them. The basal dendritic arbors of neurons that only have one apical dendrite tend to be slightly bigger, however. To sum up, the dendritic tree length is almost same across all CA1 neurons. There are variations in the entorhinal cortex inputs received at different positions along the transverse axis of CA1 (see Figs. 4.2 and 4.4), hence the observed morphological consistency does not necessarily represent functional consistency [1, 2].

4.2.3 Interneurons

The vast majority of the neurons in the hippocampus are pyramidal. Similar to the dentate gyrus (Fig. 4.3), all levels [4, 5] have a large population of interneurons. Every subfield of the hippocampus has the great majority of interneuron types. The pyramidal basket cell is situated inside or near the pyramidal cell layer, and its dendrites extend into the stratum oriens, stratum radiatum, and stratum lacunosum-moleculare. The beaded-appearing, spineless dendrites can develop symmetrical or asymmetrical synapses. Most excitatory inputs are known to go to the pyramidal cells in the hippocampus. The pyramidal basket cell receives about 2000 excitatory impulses that reach its dendritic tree. Since each pyramidal cell only gives a single synapse to a specific basket cell, there is a great deal of convergence between the two types of cells. Neurons with axons that are comparable to basket cells share a number of morphological features. As for the morphology of the basket cells, the stratum oriens contains fusiform ones, while the stratum radiatum is home to stellate ones. The axons of every one of these cell types eventually fuse with the soma and dendrites of the pyramidal cells. The lateral range of the basket cell axonal plexus is 900–1300 μm, as determined by a 400 μm in vitro slice preparation. There could be as many as one node in this complex network. There are around 10,000 synaptic varicosities, and each pyramidal cell can create two to ten synapses with a single basket cell [21]. A normal basket cell may link the nervous system to over a thousand pyramidal cells. So, it's possible for a single basket cell to significantly inhibit several pyramidal cells. Hippocampal interneuron types also include the chandelier cell and the axo-axonic cell. Homologous to dentate gyrus axo-axonic cells are found in the hippocampus. Similar to basket cells, these cells have somas inside or near the pyramidal cell layer, and their dendrites extend over the whole hippocampus. The axons of the chandelier cells are around 1 mm long and spread out horizontally. Just under the pyramidal cell layer are the brain's outer layers, through which axons travel. Their collateral branches connect to the pyramidal cell layer and end on the proximal axons of pyramidal neurons at regular intervals. Each axo-axonic cell forms synapses with around 1200 axon starting segments of pyramidal cells. Four to ten axo-axonic cells are responsible for targeting each initial segment [2]. In the initial studies of the Golgi apparatus, conducted by Ramon y Cajal and Lorente de Nó [2, 6], it was undeniably demonstrated that the hippocampus's stratum oriens, stratum radiatum, and stratum lacunosum-moleculare included several cell types other than pyramidal. The bulk of these neurons had GABAergic markers, which was first discovered in 1978 by Ribak and colleagues [22], indicating that they are likely interneurons. Freund and Buzsaki (1996) provided a detailed description of the location, morphology, and dispersion of these cells' axons [4] using a categorization system based on the particular area they connect to. An axon network unique to the stratum lacunosum-moleculare (also known as cells that terminate in conjunction with entorhinal afferents) defines the O-LM cell subgroup of neurons [23]. In this type of interneuron, the precise region of the hippocampus determines where the soma is located. The prevailing notion seems to be that areas with recurring pyramidal cell collaterals are where dendritic trees and cell bodies are located. The CA3 area contains all strata except the stratum lacunosum-moleculare. Conversely, it is exclusive to the CA1 stratum oriens. The axons of O-LM cells leave the cell body-containing stratum oriens and go directly to the stratum lacunosum-moleculare. From that point on, they branch out to create an intricate web. Symmetrical synapses are formed by the lengthy apical dendrites of pyramidal neurons with the axons. The idea that O-LM cells get most of their stimulating input from the recurrent branches of pyramidal cells raises the possibility that this interneuron type employs feedback processes to suppress the distant branches of pyramidal cells in two ways [2]. A distinct family of interneurons in the hippocampus is represented by the bistratified cell, which is located close to the pyramidal cell layer. These neurons have multipolar dendritic trees, but they never make it to the stratum lacunosum-moleculare. When an axon from a bistratified cell splits into the stratum oriens and the deep portion of the stratum radiatum, a complex network of nerve endings is created. Axons are complex networks of neurons that can have up to 16,000 synaptic varicosities and lengths exceeding 80 mm. Axons protrude from the dendritic spines and dendritic shafts of pyramidal cells in bistratified cells. Little is known about the inputs they receive, despite the fact that their dendrites are located near the sites of associational connections in CA3 and the Schaffer collateral fibers in CA1. As a result, feedforward and feedback mechanisms are likely what motivate them [2]. The stellate or multipolar dendritic plexus encasing interneurons also resides exclusively in stratum radiatum. Following a specific pattern of branching in the stratum radiatum, the axons of these cells frequently finish on the dendrites of adjacent pyramidal cells. A great number of interneurons call the stratum lacunosum-moleculare home; it is located at the boundary between the two layers. The characteristic of LM neurons, also known as stratum lacunosum-moleculare interneurons, is the lateral extension of dendrites from inside the layer. Their branches may still extend to the pyramidal cell layer, though. Axons often branch horizontally and are located in stratum lacunosum-moleculare or the surface part of stratum radiatum. Although symmetrical synapses do develop on distant branches of pyramidal cells by axons from this cluster of neurons, the function of these connections remains unknown [2, 24]. The hippocampus is home to another type of interneuron called an IS neuron, which is in charge of interneuron selection. The cell bodies of interneurons (IS neurons) are uniformly distributed across the hippocampus's layers. The use of antibodies directed against the calcium-binding protein calretinin allows for the identification of these cell bodies [2]. Among the many things that set IS neurons apart is the tendency for their dendrites to cluster together. The fact that their axons end on interneurons is a unifying feature. The input/output properties of these neurons have received minimal attention in the scientific literature. Nevertheless, the dendrites of these cells are structured in a manner that allows them to accept input from every major and inner hippocampus incoming connection. In addition, the axons can send signals to any of the interneurons mentioned before. All of the interneurons that have been characterized thus far exhibit GABA markers, according to immunopositivity [4, 5]. Many neuroactive chemicals have a common location with the interneurons of the hippocampus and other parts of the cerebral cortex. Thus, a large number of these neurons may be studied using antibodies that target peptides such as somatostatin, VIP, cholecystokinin, and neuropeptide Y (NPY) , in addition to calcium-binding proteins such as parvalbumin (Figs. 4.7, 4.8, and 4.9), calbindin, and

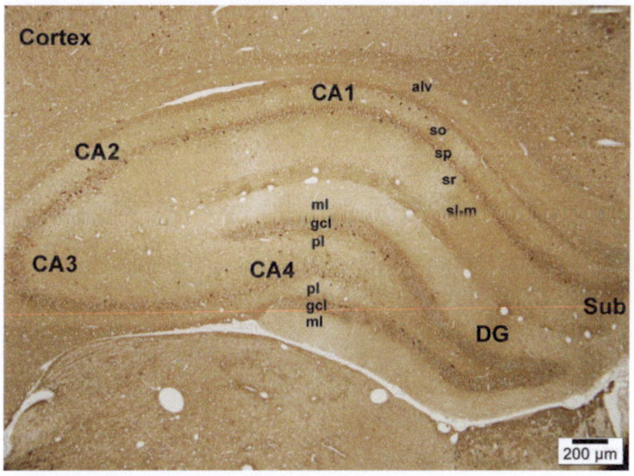

Fig. 4.7 Parvalbumin immuno-stained coronal section of the rat hippocampus. Parvalbumin immunopositive interneurons. *CA1-CA4* CA1-CA4 fields of cornu ammonis; *DG* dentate gyrus; *Sub* subiculum; *alv* alveus; *so* stratum oriens of CA1; *sp* stratum pyramidale of CA1; *sr* stratum radiatum of CA1; *sl* stratum lacunosum of CA1; *sm* stratum moleculare of CA1; *ml* molecular layer of the dentate gyrus; *gcl* granule cell layer of the dentate gyrus; *pl* polymorphic layer of the dentate gyrus. Bar: 200 µm

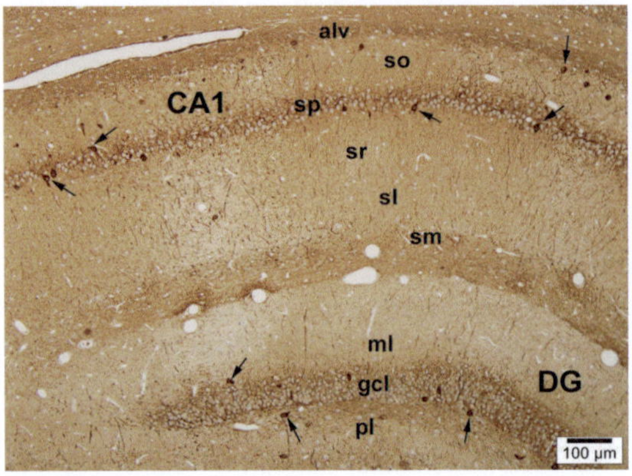

Fig. 4.8 Parvalbumin immuno-stained coronal section of the rat hippocampus. Parvalbumin immunopositive interneurons (arrows). *CA1* CA1 fields of cornu ammonis; *DG* dentate gyrus; *Sub* subiculum; *alv* alveus; *so* stratum oriens of CA1; *sp* stratum pyramidale of CA1; *sr* stratum radiatum of CA1; *sl* stratum lacunosum of CA1; *sm* stratum moleculare of CA1; *ml* molecular layer of the dentate gyrus; *gcl* granule cell layer of the dentate gyrus; *pl* polymorphic layer of the dentate gyrus. Bar: 200 μm

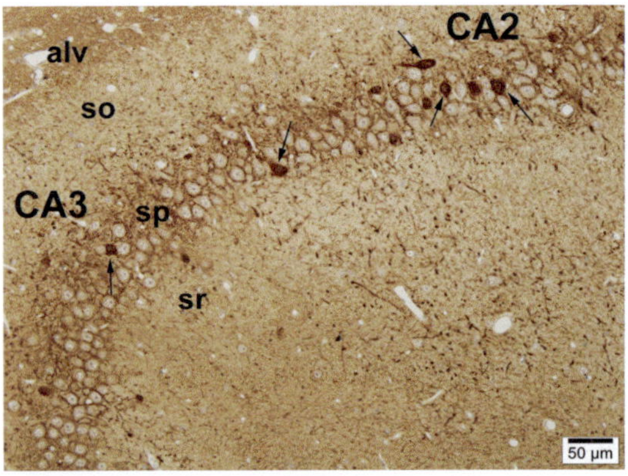

Fig. 4.9 Parvalbumin immuno-stained coronal section of the rat hippocampus. Parvalbumin immunopositive interneurons (arrows). *CA2-CA3* CA2-CA3 fields of cornu ammonis; *alv* alveus; *so* stratum oriens of CA1; *sp* stratum pyramidale of CA1; *sr* stratum radiatum of CA1. Bar: 50 μm

calretinin (Fig. 4.10). What exactly these neuroactive chemicals do to GABAergic neurons is still a mystery. Nonetheless, they are useful indicators for classifying the varied family of GABAergic interneurons [4, 5, 25–27].

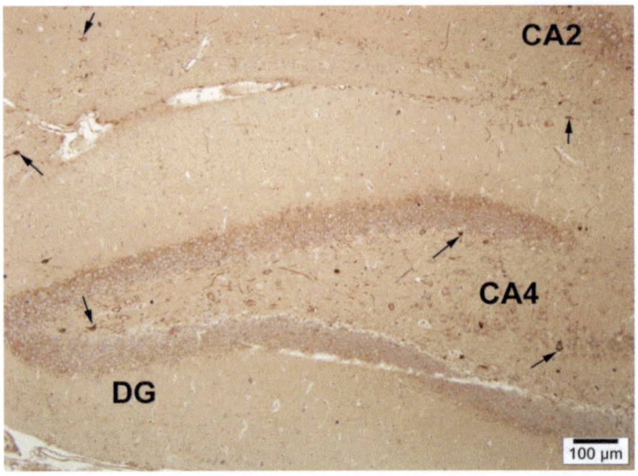

Fig. 4.10 Calretinin immuno-stained coronal section of the rat hippocampus. Calretinin immunopositive interneurons (arrows). *CA2-CA3* CA2-CA3 fields of cornu ammonis; *alv* alveus; *so* stratum oriens of CA1; *sp* stratum pyramidale of CA1; *sr* stratum radiatum of CA1. Bar: 50 μm

4.3 Dentate Gyrus (DG)

A triangle section of the cerebral cortex, the dentate gyrus (DG) usually takes on a V or U form. Dentate gyrus (DG) form and lack of subregional differentiation are maintained throughout hippocampus development. The “superapogeal blade” refers to the region of the granule cell layer situated between the CA3 and CA1 fields, just below the hippocampal fissure. The infrapyramidal blade describes the flip side. The point where the two blades converge at the very top of a V or U shape is known as the crest [1]. A thin, dorsally concave lamina known as the DG may be seen in coronal slices of the hippocampal body. The concavity of this lamina encircles the CA4 portion of the CA, as seen in Fig. 4.11. In terms of anatomy, the vestigial hippocampal sulcus is the sole physical boundary that separates the CA from the DG. It is difficult to differentiate between the cornu ammonis (CA) and dentate gyrus (DG) layers in the deepest recess of this groove because they are so closely packed together [2, 3]. The structural complexity of the DG is lower than that of the CA. The allocortex is composed of three easily distinct layers: the molecular, the granulo-

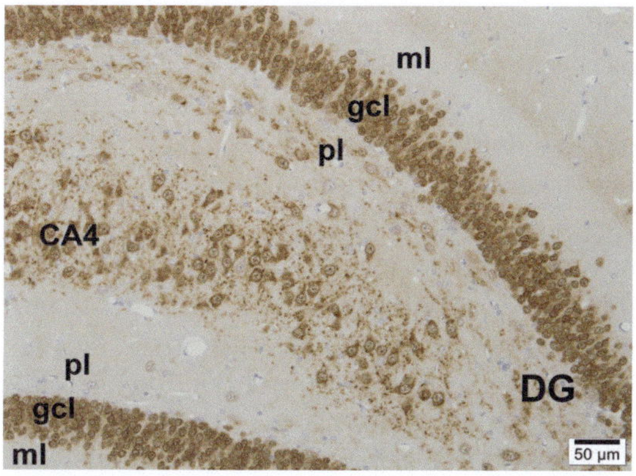

Fig. 4.11 NeuN immuno-stained coronal section of the rat hippocampus. NeuN immunopositive neurons, *CA4* CA4 field of cornu ammonis; *DG* dentate gyrus; *ml* molecular layer of the dentate gyrus; *gcl* granule cell layer of the dentate gyrus; *pl* polymorphic layer of the dentate gyrus. Bar: 50 μm

sum, and the polymorphic. You may find the images in Figs. 4.1, 4.3, 4.8, and 4.11. The cell bodies of granular neurons are typically found in the stratum granulosum. This layer is easily discernible because of the dense packing of tiny, spherical neurons therein. Through the polymorphism layer, visually diverse and texturally variable axons go to CA3 and CA4. From each granular soma's basal pole extends one dendrite into the stratum moleculare. This is seen in Figs. 4.3 and 4.11. Another characteristic that distinguishes this layer from the CA's stratum moleculare is the vestigal hippocampal sulcus, which separates the thick stratum moleculare from it. Fibers surrounding the hippocampal sulcus are received by the outer two-thirds of the polymorphic layer via a perforant tract. On the flip side, the central area that touches the stratum granulosum is populated with commissural and septal fibers [3, 28]. Connecting the granular layer to CA4, axons go via the polymorphic layer—also known as the plexiform layer. The molecular and polymorphism layers contain a negligible number of interneurons. The area dentata, which comprises the DG and CA4, is considered to be an independent entity according to Blackstad (1956). One name for CA4 is the fascia dentata's endfolium, end-blade, or hilus, as seen in Fig. 4.11 [29]. To form the dentate gyrus, three separate layers are used. Due to its small cell density and proximity to the hippocampal fissure, the molecular layer is so called. The typical thickness of the layer in question in rats is around 250 μm. The primary cell layer, also known as the granule cell layer, lies underneath the molecular layer. In a narrow layer, you can find four or eight granule cells. The fascia dentata consists of two layers, one made up of molecules and the other of granule cells, and it is shaped like a V or U. The third layer of the dentate gyrus, a cellular region known as the polymorphic cell layer, is enclosed by this substance. Cell bodies of dentate gyrus (DG) neurons are situated in the stratum granulosum, a granule cell layer. The granule cells' soma has an approximate diameter of 7 μm. The granule cells' dendrites are closest to the surface of the molecular layer. The axons of the granule cells, known as mossy fibers (MFs), go to the pyramidal cells in the CA area, also known as the CA3 region. The mossy fibers' proximal ends go through the polymorphism layer before they reach the CA3 area. One million granule cells make up a rat's DG, according to studies [30].

4.3.1 Neuron Types

4.3.1.1 Dentate Granule Cell

Granule cells are the most common kind of dentate gyrus cell (Figs. 4.3 and 4.11). An oval-shaped cell body, about 10 μm wide and 18 μm tall, characterizes these cells [31]. Granule cells are often quite near to one another, and the absence of a glial sheath between them is common. Spiny dendrites form the granule cell's characteristic cone shape, and all of the branches point toward the molecular layer's outermost portion. Dendrites terminate at the hippocampal fissure or the ventricular surface for the bulk of their length. Superpyramidal leaflets often have dendritic arbors of granule cells that are 3500 m in length, whereas infrapyramidal leaflets typically have them 2800 m in length. The amount of dendritic spines on granule cell dendrites was calculated by Desmond and Levy (1985), as mentioned in reference [32]. Scientists found that the density of cell spines in the suprapyramidal leaf is

1.6/m, whereas in the infrapyramidal leaf it is 1.3/m. It is reasonable to assume that an infrapyramidal granule cell has around 3640 spines and an average suprapyramidal granule cell has about 5600 spines based on the above statistics and the usual dendritic lengths indicated earlier. On the order of 121,066 granule cells, the dentate gyrus of rats contains [33, 34]. Even while the dentate gyrus continues to undergo cell development and neurogenesis as an adult, new stereological studies have demonstrated that the total number of granule cells remains constant in these animals [34]. The dentate gyrus and the amount of granule cells in adult mice are larger in mice exposed to socially complicated environments or to rolling wheels, as opposed to animals kept in standard laboratory cages. Research has shown that similar procedures performed on adult animals can affect the survival and proliferation of freshly generated neurons, although they have little to no impact on dentate gyrus size or total granule cell count [2]. The septotemporal axis of the dentate gyrus shows variation in the packing density and thickness of the granule cell layers. Granule cell packing density is higher in the septal region than in the temporal region. There is a differential in the packing density of CA3 pyramidal cells. Consequently, the ratio of granule cells to CA3 pyramidal cells is around 12:1 at the septal levels of hippocampal formation, and then drops to 2:3 in the temporal pole. The probability of contact is substantially lower in the septal area than in the temporal region, taking into account that the CA3 pyramidal cells receive the majority of their input from granule cells and that the number of synapses from mossy fibers is rather constant along the septotemporal axis. In the dentate gyrus, you'll only find the granule cell, which is considered a "principal" cell [35]. Axons emanating from it join the dentate gyrus to a different part of the hippocampus known as CA3. Mossy cells are neurons that branch out from the dentate gyrus in one side of the brain and join with the dentate gyrus on the other side [2].

New granule neurons are dynamically and intricately controlled throughout the course of several developmental stages when they are generated from adult neural stem cells. During development, brain cells undergo a series of events that culminate in the formation of neural precursors, cells differentiate, proliferate, migrate, and eventually integrate into preexisting hippocampus networks. Hippocampal neural stem cells (also called radial Type I cells) may differentiate into different types of neurons and astrocytes, and they can also self-renew. The cells possess a distinctive radial process that penetrates the molecular layer; they are located in the dentate gyrus' (DG) subgranular zone. In doing so, they prove that the radial glia marker GFAP is present. In the absence of negative selection by microglia in the hippocampus, active Type I cells differentiate into Type II cells, which are transient amplifying neural precursor cells (NPC) . Axons extend toward CA3, which aids in their functional integration, and dendritic trees form in the dentate gyrus (DG) molecular cell layer as newborn granule neurons mature. Changes in the morphological and functional characteristics of freshly generated cells, as well as the sequential expression of stage-specific markers, allow for their identification throughout development [36]. The physically distinct bipolar horizontal processes of progenitors in the subgranular zone allow for their identification. They begin to differentiate into axons that go to the CA3 region and primary apical dendrites that face the molecular cell layer after the cell cycle concludes. In addition, when dendritic trees develop into adults, they get more and more intricate. As morphological changes occur, certain marker proteins undergo dynamic expression. Glial markers, such the Sox2 transcription factor, are present in transient amplifying Type II cells throughout the early stages of the neurogenic lineage. Transcriptional factors like NeuroD are involved in neuronal development and are produced by differentiating progenitor cells in later stages. Conversely, a protein known as doublecortin (DCX) is involved in microtubule production and is expressed by immature neurons [37, 38]. The majority of neurons are inhibitory interneurons; however, there are a few other types [2].

4.3.1.2 Interneurons

Among interneuron types, pyramidal basket cells have been the subjects of the most study. The granule cell layer's deep surface is the usual location for these cells. The cells are situated at a moderate angle in the granule cell layer and feature pyramidal cell bodies that range in diameter from 25 to 35 μm. The axons of these cells form pericellular plexuses that ring and form synapses with the cell bodies of granule cells; this is where the term "basket cell" comes from. It is believed that the majority of these cells are inhibitory since they have molecular markers for the neurotransmitter GABA [22, 39]. Throughout the dentate gyrus' whole transverse or septotemporal range, the number of basket cells fluctuates [40]. At the septum level of the infrapyramidal blade, the ratio of basket cells to granule cells is 1:180, whereas in the suprapyramidal blade, it is 1:100. The infrapyramidal blade has a ratio of 1:300 at temporal levels, while the suprapyramidal blade has a ratio of 1:150 [2]. The number of interneurons seen in the developing rat hippocampus has increased significantly. There is a wide variety of specialized hippocampal interneurons that perform different functions [39]. It is common practice to distinguish between cell types by looking at how their axonal plexus is structured. Although some interneurons terminate their axons at the neuronal bodies of neighboring neurons, others terminate them at the beginning segments of other axons. In addition, interneurons have been classified according on the inputs they receive. Molecular and polymorphism layers are penetrated by several multipolar cells that have numerous spiky dendrites. Conversely, a small subset of cells have a dendritic structure that is comparable to that of a spindle. A number of these cells have similar complex morphological traits, including helical nuclei, rods inside the nucleus, and an abundance of cytoplasm around the cell body with large Nissl structures. To top it all off, it seems like every single one of these granule cell layer cells produces axons that go into the basket plexus. As can be shown in Figs. 4.8 and 4.10, a considerable portion of these neurons have GABA immunoreactivity [4, 5, 39]. As inhibitory interneurons, they form symmetrical synaptic connections with granule cell bodies, proximal dendrites, and occasionally axon starting segments. But there isn't a homogeneity in the neurochemical makeup of these cells; evidence suggests that some subsets cohabit with various classes of neuroactive substances [2].

4.3.1.3 Neurons of the Molecular Layer

Figures 4.3 and 4.11 show that the molecular layer is mostly filled with dendrites of granule, basket, and polymorphic cells, along with axons and terminal axonal arbors that come from many places, including the entorhinal cortex. In addition, there are at least two distinct kinds of neurons in the molecular layer. Deep inside the molecular layer lies the first cell. Either a multipolar or triangular shape characterizes its cell body. A substantial terminal plexus is produced by the axon of the cell and is mostly located in the outer two-thirds of the molecular layer. The molecular layer perforant pathway is associated with the MOPP cell, which is called after its spiny dendrites that are contained inside the molecular layer. In order to provide a systematic naming system for interneurons during hippocampal development, Han et al. (1993) presented this nomenclature [41]. The nomenclature uses a two-pronged technique of labeling, where the cell body is identified and the area of axon dispersion is connected. Another type of molecular layer neuron was found by Frotscher and colleagues. This kind of neuron is comparable to the neocortex-original chandelier cell, sometimes called an axo-axonic cell [42]. Typically, you may find these cells close to or even within the granule cell layer's outermost portion. The axo-axonic cell gets its name from the fact that its axon goes all the way from the molecular layer to the granule cell layer, branches out all over the place, and terminates with symmetrical synaptic connections on the granule cell axon starting segments. Because of this, they resemble chandeliers in appearance. Experts estimate that up to 1000 granule cells can be connected to each axo-axonic cell. It is probable that these cells function as an extra mechanism for inhibitory modulation of granule cell output (see Figs. 4.8 and 4.10) since they form symmetrical connections and show positivity for markers characteristic to GABAergic neurons. Dendrites of axo-axonic cells are usually found in the molecular layer, although their inputs are yet unknown. According to popular belief, these cells get input from the entorhinal cortex through the perforant pathway. There are interneurons in the molecular layer that are interneuron-specific (IS), meaning they have cell bodies and are made to terminate on other interneurons. Immunohistochemistry, by focusing on vasoactive intestinal peptide (VIP), reveals the existence of these IS neurons. In addition, the dendrites of the O-LM and HIPP cells overlap spatially with their axons [2].

4.3.1.4 Neurons of the Polymorphic Cell Layer

A large diversity of neuron types are found in the polymorphic layer. One of the most interesting and ubiquitous cell types is the mossy cell. This is probably the kind of cell that Ramon y Cajal meant when he talked about the "stellate or triangular" cells in his fusiform cell subzone. It is very much the same type of cell that Lorente de Nó called "modified pyramids." The cell bodies of mossy cells are big, measuring 25–35 μm on average, and they frequently take on a triangular or multipolar shape (Figs. 4.2 and 4.11). From the cell body, several stout dendrites protrude and traverse extensive distances inside the polymorphic layer. Typically, a limited number of extra branches are produced by each primary dendrite, which splits into two or more branches once or twice. A dendrite may occasionally extend into the molecular layer from the granule cell layer, while the majority of the progeny dendritic branches are found in the polymorphic layer. Interruptions into the adjacent CA3 field by mossy cell dendrites are rare. Thorny excrescences, which are large, complex spines that run the length of the proximal dendrites of mossy cells, are one of their defining features. The dentate granule cell axons, which are responsible for mossy fibers, end at the precise spots shown by the spines. "Thorny excrescences" are comparable structures that can be observed on the proximal dendrites of pyramidal cells in CA3, although they are not as common as those on mossy cells. Mossy cell distal dendrites, in contrast to pyramidal cell distal dendrites in the hippocampus, show pedunculate spines that are less densely distributed. In both the ipsilateral and contralateral dentate gyrus, axons originating from mossy cells reach to the inner third of the molecular layer, and evidence suggests that these cells have glutamate positivity. The result is that the dendrites of granule cells end up unevenly. Thus, it is reasonable to assume that the dentate gyrus receives its stimulating associational/commissural projection mostly from the mossy cells. Therefore, the mossy cell is not an interneuron in the conventional sense. In addition, there are several fusiform cells in the polymorphic layer; the presence or absence of spines is the primary identifying feature. You have entered the text "[2]." Recently, a distinct cell type known as the HIPP cell has been discovered; it was originally called the long-spined multipolar cell by Amaral (1978). Two or three main dendrites protrude from the cell's ends and, for the most part, line up with the granule cell layer in this type of cell. These projections can extend nearly the whole breadth of a single blade of granule cell stratum. The cell's most striking characteristic is an abundance of long, frequently branching spines distributed over the cell body and dendrites [43]. The axons of these cells reach the perforant route zone, which is located in the outer two-thirds of the molecular layer, as revealed by intracellular staining techniques. On the dendrites of granule cells, these axons terminate with symmetrical synapses, which are presumably inhibitory. The axonal plexus is a unique feature of these neurons that can extend up to 3.5 mm along the septotemporal axis of the dentate gyrus. Note that the length of the dentate gyrus in rats is around 10 mm. An astounding 100,000 synaptic terminals can be generated by the extensive axonal projection. Unlike other inhibitory interneurons, this long-spined multipolar HIPP cell contains a local axonal plexus and spiny dendrites, making it a peculiar interneuron. One possible explanation is that some of these HIPP cells are linked to the somatostatin/GABA cells that innervate the outer molecular layer with somatostatin. Molecular and hilus-extending thin aspiny dendrites are another feature of the polymorphic layer's multipolar and triangular cells. In the inner third of the molecular layer, the axons of these HICAP cells, which are connected to the hilar commissural-associational route, branch out into the granule cell layer. There may be a variety of neurons in the dentate gyrus' polymorphic layer, but nothing is known about their axonal plexus organization. Also, contrary to popular opinion, interneurons do more than only lower neuronal activity [2] (for examples, see Figs. 4.8 and 4.10).

4.4 Subiculum

Research on the structural aspects of the rat subiculum has been limited thus far. At the border between the CA1 and subiculum areas, the layer of pyramidal cells rapidly expands and the staining intensity is elevated when the Timm dye is used. This border also marks the end of the stratum radiatum of CA1, the region that receives projections from CA3, and the beginning of the molecular layer of the subiculum. One half of the layer may be deep and link to the stratum radiatum of CA1, while the other part might be shallow and connect to the stratum lacunosum-moleculare of CA1 and the presubiculum molecular layer [44]. While the entorhinal cortex supplies fibers to the stratum superficially, CA1 supplies fibers to the stratum deepest. Within the subiculum, you won't find the stratum oriens of the CA1 area. Huge pyramidal neurons call the major cell layer of the subiculum home. This layer begins at the very tip of CA1 and continues under the presubiculum's layer II. There is regularity and symmetry in the size and form of the cells. Their dendrites, which can extend into the molecular layer at the tip and into the lower sections of the pyramidal cell layer at the base, are rather remarkable. There is a suggestion to divide cells into two categories based on their firing characteristics: normally spiking cells and intrinsically bursting cells [45]. There may not be any outward signs of differentiation between these two cell types, but their distribution inside the pyramidal cell layer is clearly defined. Typical spiking cells are more prevalent in the upper layers of the pyramidal cell layer, whereas bursting cells are more abundant in the deeper sections. Additionally, bursting cells (somatostatin) and normal spiking neurons (NADPH-diaphorase/nitric oxide synthetase) may be distinguished utilizing selective labeling methods. Although both kinds of neurons are projections, only bursting cells appear to have a different form of connection—namely, to the entorhinal cortex. A large number of smaller neurons, thought to represent the subiculum's interneurons, coexist with the pyramidal cells. The similarities between the interneurons in the subiculum and the hippocampus are not well understood at the moment. Certain subsets of these neurons share characteristics with CA1 neurons, for example, GABAergic cells are identified by the presence of parvalbumin, a calcium-binding protein [2] (Figs. 4.1, 4.4, and 4.7).

4.5 Conclusion

There are several different subfields within the hippocampus, each with its own unique architecture, cell size and shape, connections, electrical properties, and susceptibility to injury. Differentiation according to form and neural network is made possible by the specific arrangement of cells and their connections in various parts of the hippocampus. In the initial cellular layers, the majority of the neuronal cells are organized in a unidirectional pattern with three synapses. The three-synapse network is organized in a lamellar fashion, with fibers that are parallel to one another and almost perpendicular to the hippocampus's long axis. Within the hippocampus, functional units are formed by this arrangement. Here are some distinguishing features of pyramidal cells in the hippocampus: (a) Parallel alignment of the dendrites causes an increase in excitability and current flow beyond the cells, a feature commonly observed in epilepsy. (b) Cortical layers connect at different points along the dendritic tree due to the pyramidal dendrites' orientation at a right angle to the surface of the cortex. (c) Dendritic spines on pyramidal dendrites improve electrical currents, which helps neurons generate action potentials at distant synapses. A pattern of sustained bursting firing would be observed in pyramidal cells in the absence of basket cells. An axon, a pyramidal cell, a basket cell, and an inhibitory link to the pyramidal cell are simplified in the graphic. The images also show the side connections between the two cells. Commonly referred to as recurrent inhibition, this circuit is detailed here. The formation of a secondary link with a specific cell type known as the basket cell is a typical feature of the nervous system. When a pyramidal cell activates a basket cell, the latter responds by inhibiting or providing negative feedback to the former. To put it simply, the pyramidal cell uses the basket cell as a method of controlling its own activity. Basket cells are a kind of inhibitory neuron that employs the neurotransmitter gamma-aminobutyric acid or GABA. Pyramidal cells fire all at once, which is the hallmark of epilepsy. Consciousness impairment, involuntary muscle contractions, and, in extreme cases, seizures, and coma are all results of neurons firing at the same time. Among neurological disorders, epilepsy is second only to stroke in frequency. Some estimates put the lifetime risk of developing epilepsy at around 1% of the population. A localized region of the brain is the only target of focal or partial seizures. Still, epilepsy seizures can start off as localized attacks that spread to other parts of the brain's cortex before becoming more widespread. Convulsions, or epileptic seizures, need the coordinated firing of many neurons all at once. A breakdown in postsynaptic inhibition is likely responsible for this event's occurrence. To better understand inhibition and the basket cell's function, we can disable the connection between the two cells. From a neuroscientific standpoint, there are several reasons why adult hippocampal neurogenesis is an intriguing topic. Primarily, the maintenance of the hippocampus's (the brain region in charge of learning, memory, and mood regulation) optimal functioning depends on the generation of new neurons inside the hippocampus. In addition, certain neuropsychiatric disorders, such as schizophrenia, depression, and age-related memory loss, are associated with hippocampal neurogenesis, the process of creating new neurons in the hippocampus. Autism spectrum disease is only one of numerous neurodevelopmental disorders that may be investigated. It is also possible to see the occurrence of neurogenesis in a mainly constrictive environment. The adult brain provides a real-life example of how neural regeneration works at the molecular and cellular levels. We may be able to utilize this information to develop new therapies for neurological diseases.

References

1. Taupin P. The hippocampus: neurotransmission and plasticity in the nervous system. In: Nova biomedical books. 1st ed. New York; 2007. p. 3–9.
2. Andersen P, Morris R, Amaral D, Bliss T, O'Keefe J. The hippocampus book. 1st ed. New York: Oxford University Press; 2007. p. 37–109.
3. Duvernoy HM. The human hippocampus: functional anatomy, vascularization and serial sections with MRI. 3rd ed. New York: Springer; 2005. p. 5–9.
4. Freund TF, Buzsaki G. Interneurons of hippocampus. Hippocampus. 1996;6:347–470.
5. Pelkey KA, Chittajallu R, Craig MT, Triccoire L, Wester JC, McBain CJ. Hippocampal GABAergic inhibitory neurons. Physiol Rev. 2017;97:1619–747. https://doi.org/10.1152/physrev.00007.2017.
6. Lorente de No R. Studies on the structure of the cerebral cortex. II. Continuation of the study of the ammonic system. J Psychol Neurol. 1934;46:113–77.
7. Anderson P, Bliss TVP, Skrede KK. Lamellar organization of hippocampal excitatory pathways. Exp Brain Res. 1971;13:222–38.
8. Stephan H, Manolescu J. Comparative investigations on hippocampus in insectivores and primates. Z Mikrosk Anat Forsch. 1980;94(6):1025–50.
9. Schaffer K. Beitrag zur Histologie der Ammonshornformation. Arch Mikrosk Anat. 1892;39:611–32.
10. Isaacson RL. The limbic system. New York: Plenum; 1974. p. 292.
11. Olbrich HG, Braak H. Ratio of pyramidal cells versus non-pyramidal cells in sector CA1 of the human Ammon's horn. Anat Embryol. 1985;173(1):105–10.
12. Braak H. On the structure of human archicortex. I. The cornu ammonis. A Golgi and pigment architectonic study. Cell Tissue Res. 1974;152:349–83.
13. Mouritzen DA. The density of neuron in the human hippocampus. Neuropathol Appl Neorobiol. 1979;5:249–64.
14. Stephan H. Evolutionary trends in limbic structures. Neurosci Biobehav Rev. 1983;7:367–74.
15. Amaral DG, Insausti R, Cowan WM. The commissural connections of the monkey hippocampal formation. J Comp Neurol. 1984;224:307–36.
16. Sommer W. Erkrankung des Ammonshorns als aetiologisches moment der epilepsie. Arch Psychiatr. 1880;10:631–75.

17. Spielmeyer W. Die pathogenese des epileptischen krampfes. Z Dtsch Ges Neurol Psychiatry. 1927;109:501–20.
18. Bratz E. Ammonshornbefunde der epileptischen. Arch Psychiatr Nervenkr. 1899;31:820–36.
19. Claiborne BJ, Amaral DG, Cowan WM. A light and electron microscopic analysis of the mossy of rat dentate gyrus. J Comp Neurol. 1986;246:435–58.
20. Ishizuka N, Cowan WM, Amaral DG. A quantitative analysis of the dentritic organization of pyramidal cells in the rat hippocampus. J Comp Neurol. 1995;362:17–45.
21. Pyapali GK, Sik A, Penttonen M, Buzsaki G, Turner DA. Dentritic properties of hippocampal CA1 pyramidal neurons in rat: intracellular staining in vivo and in vitro. J Comp Neurol. 1998;391:335–52.
22. Ribak CE, Vaughn JE, Saito K. Immunocytochemical localization of glutamic acid decarboxylase in neuronal somata following colchicine inhibition of axonal transport. Brain Res. 1978;140:315–32.
23. Lacaille JC, Mueller AL, Kunkel DD, Schwartzkroin PA. Local circuit interactions between oriens/alveus interneurons and CA1 pyramidal cells in hippocampal slices: electrophysiology and morphology. J Neurosci. 1987;7:1979–93.
24. Lacaille JC, Schwartzkroin PA. Stratum lacunosum-moleculare interneurons of hippocampal CA1 region. I. Intracellular response characteristics, synaptic responses, and morphology. J Neurosci. 1988;8:1400–10.
25. Corvino V, Di Maria V, Marchese E, Lattanzi W, Biamonte F, Michetti F, Geleso MC. Estrogen administration modulates hippocampal GABAergic subpopulations in the hippocampus of trimethyltin-treated rats. Front Cell Neurosci. 2015;9:433.
26. Lucas EK, Clem RG. GABAergic interneurons: the orchestra or the conductor in fear learning and memory? Brain Res Bull. 2018;141:13–9.
27. Tzilivaki A, Tukker JJ, Maier N, Poirazi P, Sammons RP. Hippocampal GABAergic interneurons and memory. Neuron. 2023;111:3154–75.
28. Cerbone A, Patacchioli FR, Sadile AG. A neurogenetic and morphogenetic approach to hippocampal functions based on individual differences and neurobehavioral covariations. Behav Brain Res. 1993;55:1–16.
29. Blackstad TW. Commissural connections of the hippocampal region in the rat, with special reference to their mode of termination. J Comp Neurol. 1956;105:417–538.
30. Boss BD, Peterson GM, Cowan WM. On the number of neurons in the dentate gyrus of rat. Brain Res. 1985;338:144–50.
31. Claiborne BJ, Amaral DG, Cowan WM. A quantitative three-dimensional analysis of granule cells dendrites in rat dentate gyrus. J Comp Neurol. 1990;302:206–19.
32. Desmond NL, Levy WB. Granule cell dentritic spine density in the rat hippocampus varies with spine shape and location. Neurosci Lett. 1985;54:219–24.
33. West MJ, Slomianka L, Gundersen HJG. Unbiased stereological estimation of the total number of neurons in the subdivision of the rat hippocampus using the optical fractionator. Anat Rec. 1991;231:482–97.
34. Rapp PR, Gallagher M. Preserved neuron number in the hippocampus of aged rats with spatial learning deficits. Proc Natl Acad Sci USA. 1996;93:9926–30.
35. Kempermann G, Kuhn G, Gage FH. More hippocampal neurons in adult mice living in an enriched environment. Nature. 1997;386:493–5.
36. Kempermann G, Kuhn HG, Gage FH. Experienced-induced neurogenesis in the senescent dentate gyrus. J Neurosci. 1998;18:3206–12.
37. Gaarskjaer FB. Organization of the mossy fiber system of the rat studied in extended hippocampi. I. Terminal area related to number of granule and pyramidal cells. J Comp Neurol. 1978a;178:49–72.
38. Kantak KM, Wettstein JG. Cognitive enhancement. In: Costa V, Lugert S, Jagasia R, editors. Role of adult hippocampal neurogenesis in cognition in physiology and disease: pharmacological targets and biomarkers. Springer; 2015. p. 99–101.
39. Ribak CF, Seress L. Five types of basket cell in the hippocampal dentate gyrus: a combined Golgi and electron microscopic study. J Neurocytol. 1983;12:577–97.
40. Seress L, Pokorny J. Structure of the granular layer of the rat dentate gyrus: a light microscopic and Golgi study. J Anat. 1981;133:181–95.
41. Han ZS, Buhl EH, Lorinczi Z, Somogyi P. A high degree of spatial selectivity in the axonal and dentritic domains of physiologically identified local-circuit neurons in the dentate gyrus of the rat hippocampus. Eur J Neurosci. 1993;5:395–410.
42. Soriano E, Frotscher M. A GABAergic axo-axonic cell in the fascia dentata controls the main excitatory hippocampal pathway. Brain Res. 1989;503:170–4.
43. Amaral DG. A Golgi study of the cell types in the hilar region of the hippocampus in the rat. J Comp Neurol. 1978;182:851–914.
44. O'Mara SM, Commins S, Anderson M, Gigg J. The subiculum: a review of form, physiology and functions. Prog Neurobiol. 2001;64:129–55.
45. Greene JRT, Totterdell S. Morphology and distribution of electrophysiology defined classes of pyramidal and nonpyramidal neurons in rat ventral subiculum in vitro. J Comp Neurol. 1997;380:395–408.

Neurophysiology of the Human Hippocampus

5

Enes Akyüz

5.1 Introduction

The hippocampus is a complex brain structure located below the temporal lobe in the cerebral cortex. It is named after the sea creature of the genus *Hippocampus* because of its resemblance to a seahorse. Although it used to be called only *Cornu Ammonis* (CA), the hippocampus consists of cornu ammonis and dentate gyrus [1]. Externally, it appears as a structure consisting of neurons packed in an S-shaped extension of the temporal lobe. The hippocampus is physiologically and anatomically different from other brain regions. Although the hippocampus is located under the cortex, it is not categorized as a subcortical structure [2]. The hippocampus, which stands out as the most important region for learning and memory, enables the conversion of short-term memory into long-term memory [3, 4]. Different cell densities, such as pyramidal, granular, and basket, are observed in the layers of the hippocampus. The addition of its complexity at the cellular level to its important role in memory has made it the most extensively studied brain region in terms of neurophysiology.

The neurophysiology of the hippocampus depends on the interaction between neurons, excitatory and inhibitory neurotransmitter mechanisms, membrane properties, and function of ion channels. The excitatory neurotransmitter glutamate and the suppressive neurotransmitter gamma-aminobutyric acid (GABA) are responsible for the control of excitability. The GABA neurotransmitter regulates the excitatory effect of glutamate through two postsynaptic receptors, $GABA_A$ and $GABA_B$. Similarly, glutamatergic excitatory synapses bind to two different postsynaptic receptors. These are α-amino-3-hydroxy-5-methyl-4-isoxazolepropionic acid (AMPA) and N-methyl-D-aspartate (NMDA) receptors involved in long-term potentiation (LTP). In addition to these neurotransmitters, calcium (Ca^{2+}) and potassium (K^+) ion channels in hippocampal neurons contribute to the regulation of neural activity. In order for the hippocampus to function, it needs to receive inputs from other surrounding brain regions. Information from cortical, subcortical, and limbic areas, especially the entorhinal cortex, is evaluated in the hippocampus. Afferent neurons reaching the hippocampus are evaluated by pyramidal neurons in the CA1 and CA3 regions and transmitted to the relevant region by efferent fibers. The relationship between the hippocampus and its environment makes it neurophysiologically important. The high density of NMDA receptors involved in LTP in the hippocampus explains its prominence in learning and memory. The continuous addition of new neurons to the dentate gyrus region and the

E. Akyüz (✉)
Department of Biophysics, University of Health Sciences International Faculty of Medicine, İstanbul, Türkiye
e-mail: enesakyuz25@gmail.com

M. Turgut et al. (eds.), *The Human Hippocampus*, https://doi.org/10.1007/978-3-031-88907-3_5

occurrence of neurogenesis indicate that the hippocampus undergoes changes throughout life [5, 6]. Considering all these features, it seems that the hippocampus needs to be explained in detail from a neurophysiological point of view. In this book chapter, we first examined the functional and neuronal connections in the hippocampus and the role of neurotransmitters. Then, the LTP mechanism in the hippocampus was explained, and the neurophysiological role of the hippocampus in memory was presented.

5.2 Functional and Neuronal Connectivity

Functional connections allow different regions in the brain to communicate. The hippocampus needs functional connections to fulfill cognitive functions such as learning and memory. The hippocampus, which contributes to memory encoding and retrieval—critical subprocesses of episodic memory—must be connected to the neocortex. These connections include the entorhinal cortex, the prefrontal cortex (PFC), the limbic system (especially the amygdala), the hypothalamus, and the medial temporal cortex, which includes the hippocampus. Elucidating the functional connectivity between these structures represents an essential step toward an integrative explanation of brain function [7, 8]. The entorhinal cortex, the main input and output structure of the hippocampal formation, provides communication between the hippocampus and cortical structures. The entorhinal cortex processes information from environmental stimuli and transmits it to the hippocampus, contributing to the processing of this information into long-term memory [9]. The connection of the hippocampus with the PFC contributes to information retrieval and long-term memory formation. The phenomenon called memory can be defined as recording daily experiences and recalling these experiences in any situation. This process consists of two steps: learning and recall by stimulation. The hippocampus, together with the PFC, enables the rapid encoding of information in memories and the assembly and organization of memory networks [10]. Similarly, the connections of the hippocampus with the limbic system regulate emotional learning and memory processes. In particular, hippocampal activity and function interact with the amygdala in emotional information processing. As a part of the limbic system, the amygdala is responsible for controlling attention, cognitive function, mood disorders, and, most importantly, emotional memory. The connection of the hippocampus with the amygdala plays a role in the processing and storage of emotional information [11]. The medial temporal lobe, which includes the hippocampus, amygdala, and parahippocampal regions, helps in memory by performing functions such as encoding, consolidation, retrieval, and recall [12].

In addition to the functional connection of the hippocampus with other brain regions in memory, its relationship with the hypothalamus has been found to be effective in reducing stress responses. This effect is associated with increased activity of GABAergic neurons in the hippocampus in response to increased stress [13]. The hippocampus interprets the information it receives from other regions through environmental interaction within its own neuronal connections. Neuronal connections in the hippocampus manage information processing and transmission processes. In particular, connections between pyramidal and granular cells in the hippocampus play a critical role in information processing and storage. The density and types of cells in the CA1, CA2, CA3, dentate gyrus, and subiculum regions of the hippocampus vary. This situation provides each region with unique characteristics [14]. These cells can be simply grouped as pyramidal, granular, chandelier, and basket cells. Pyramidal cells are the most common cell type in the hippocampus involved in information processing and transmission. Granular, basket, and chandelier cells are interneurons that play an inhibitory role (Fig. 5.1a). In addition to all these excitatory and inhibitory neurons, supporting cells such as astroglia and oligodendrocytes form a complex environment for the maintenance of hippocampal functions and the execution of cognitive functions [15].

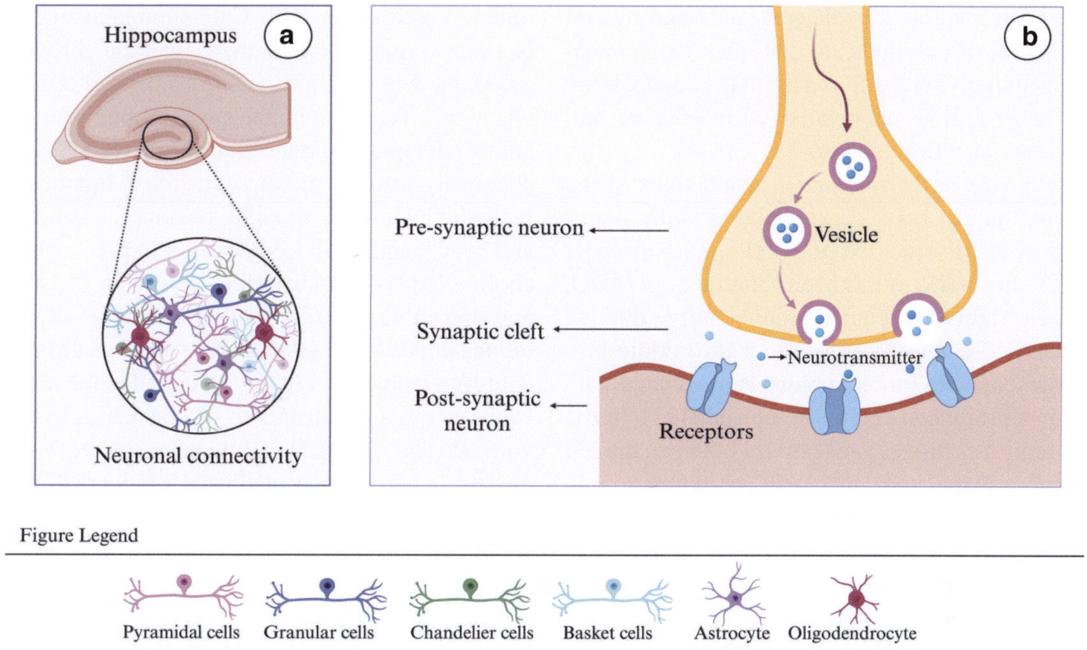

Fig. 5.1 Neuronal connectivity and neurotransmitter release and activity in the hippocampus. In the hippocampus, neuronal connections are formed by excitatory pyramidal and inhibitory granular, chandelier, and basket cells (**a**). This neuronal connection communicates through electrical and chemical stimuli. Chemical stimuli are transmitted from the presynaptic region by releasing neurotransmitters into the synaptic cleft and taking them in through receptors in the postsynaptic region (**b**). The hippocampus can only perform its memory function if neuronal activity is in homeostasis

The hippocampus can be divided into anatomically defined hippocampal subfields (aHPSFs). Hippocampal functional networks are analyzed based on the assumption that function within a subfield is generally similar but differs between different subfields [16]. Various studies have shown that the anterior and posterior hippocampus function through peripheral connections. The posterior hippocampus is more connected with the occipital, inferior parietal, medial, and superior frontal lobes, while the anterior hippocampus has more functional connections with the inferior frontal and lateral temporal lobes [17]. This suggests that different subregions of the hippocampus have specialized connections with different brain regions and that these connections play an important role in memory formation. The subregions of the hippocampus and their connections to functional brain networks vary depending on gender. In particular, marked differences in functional connections in the anterior hippocampus indicate a difference in the function of the hippocampus between men and women [18]. In this case, it can be said that the hippocampus should be investigated as a region with multiple different functions and that the subregion of interest should be examined. In studies examining the functional and neuronal connectivity of the hippocampus, resting-state functional magnetic resonance imaging (rsfMRI) is used to elucidate the connections with different regions and within itself. The development of hippocampal connectivity and working memory performance were investigated in newborns (1–3 years, n = 202) using rsfMRI [19]. This study demonstrates strong local functional connectivity of the hippocampus with subcortical and limbic regions, emphasizing its critical role in working memory. In addition to its role in memory, the hippocampus may contribute to the faster acquisition of mathematical skills in some children. This is consistent with the functional connectivity of the hippocampus with the basal ganglia, dorsolateral, and ventrolateral PFC [20]. It may indicate that quantitative measures of individual differences in the connectivity of brain regions associated with learning and memory may be a marker.

Damage to the functional connections of the hippocampus may be associated with major depressive disorder (MDD) and bipolar disorder (BD). In studies, it has been found that in MDD, lower right amygdalar connectivity due to decreased connectivity of the right middle hippocampus with limbic regions is associated with a more prolonged depressive episode [21]. In BD patients, functional connectivity between the left hippocampus and the left precuneus, and between the right hippocampus and the left posterior cingulate, was found to be reduced [7]. New insights into the dysfunction of the hippocampus at the neural network level could potentially improve the early detection of psychiatric disorders in the future.

In conclusion, the functional and neuronal connectivity of the hippocampus with other brain regions contributes to many mechanisms, such as mood, comprehension, stress, and procedural intelligence, especially memory. This suggests that the hippocampus may not function in isolation but in concert with other brain regions and that information from other regions passes through the hippocampus. Studying aHPSFs to elucidate the neurophysiological role of the hippocampus may provide faster and more reliable results.

5.3 Neurotransmitters in the Hippocampus

The hippocampus, which is responsible for learning and memory formation, involves the interaction of neurotransmitters that regulate neural functions. Neurotransmitters carry out chemical information transfer in the central and peripheral nervous systems. These molecules provide communication between neurons by amplifying and transmitting signals within cells. Neurotransmitters are synthesized in the presynaptic neuron and secreted from synaptic terminals via vesicles with Ca^{+2} stimulation [22]. Neurotransmitters fulfill their function through receptors (Fig. 5.1b). Signal transmission occurs when the receptors in the postsynaptic neuron receive the released neurotransmitter [23]. These chemical stimuli affect numerous functions, including emotion, thought, memory, learning, and movement [24]. Glutamate, GABA, acetylcholine (ACh), noradrenaline, serotonin (5-HT), and dopamine contribute to the functions of the hippocampus [25]. Glutamate is the most abundant free amino acid in the brain and is the main excitatory neurotransmitter of the central nervous system [26, 27]. Glutamate receptors on neuronal and non-neuronal cells are divided into two main classes: ionotropic and metabotropic. These include ionotropic NMDA, AMPA, kainate, and metabotropic mGluR receptors [28]. Among these receptors, NMDA receptors are particularly prominent in the hippocampus. The involvement of NMDA receptors in the LTP mechanism may explain the role of learning and memory in the hippocampus [29]. Glutamate is an essential excitatory neurotransmitter and contributes to synaptic communication, while GABA is an inhibitory neurotransmitter that regulates the activity of neurons. GABA is involved in information processing and memory formation through $GABA_A$ and $GABA_B$ receptors in the hippocampus [30]. The GABAergic system also regulates the activity and flexibility of neural networks, ensuring the correct functioning of neural transmission. Disruption of the balance between glutamate and GABA leads to synaptic loss [31]. Although these two neurotransmitters are prominent in the hippocampus, ACh also has a crucial function in learning and memory formation. Noradrenaline, serotonin, and dopamine are involved in mood, motivation, neurogenesis, and reward systems [32, 33]. The interaction of these neurotransmitters in the hippocampus is of great importance for the regulation of cognitive functions and emotional stabilization. Disruption of the balance between these molecules may cause memory and learning problems and mood disorders.

Preclinical and human data suggest that hippocampal dysfunction plays a critical role in the onset of psychosis. This is thought to be caused by neural hyperactivity in the hippocampus and increased subcortical dopamine function via glutamatergic projections to the striatum. Higher hippocampal glutamate levels are observed in individuals at risk for psychosis [34]. These findings may indicate that hippocampal glutamate levels can be used as a potential marker for psychosis risk and may shed light on the development of future pharmacotherapeutic approaches. Women with lower GABA concentrations exhibited worse memory performance when the relationship between decreased hippocampal GABA concentration and memory was examined [35]. Decreased hippocampal GABA concentration may cause a decline in memory performance as well as an increase in unwanted thoughts. Fronto-hippocampal inhibition, which underlies hippocampal recall activity and thought suppression, is controlled via GABAergic interneurons [36]. Increased fronto-hippocampal activity with decreased GABA may lead to increased unwanted thoughts. Similar to the functions of GABA, ACh regulates arousal, learning, and attention in the neocortex and hippocampus. This effect is mediated by cholinergic neurotransmission. Studies have shown that ACh release coordinated between the PFC and hippocampus is linked to reward and arousal and provides bidirectional mechanisms to support the acquisition of learned behavior during cognitive tasks [37]. In another study, ACh was found to prioritize the flow of information between the hippocampus and the entorhinal cortex. Suppression of prior information in the CA3 region and enhancement of the transfer of information from the entorhinal cortex to the CA1 region are mediated by ACh interacting with MAChR-4 receptors in CA3 while binding to muscarinic ACh receptor-3 (MAChR-3) in the entorhinal cortex [38]. This explains the fact that ACh decreases and increases the transmission of information in two different regions. Interestingly, another study has shown that hippocampal cholinergic terminals interact with GABAergic synapses and co-transduce GABA along with ACh. This co-transduction has an effect on cross-regulated compound postsynaptic potentials [39]. These findings may allow the development of new approaches for alternative pharmacotherapies in neurodegenerative disorders. In addition to glutamate, GABA, and ACh, which are essential in the hippocampus, other neurotransmitters also show activity. These chemical stimuli may be involved in the functional connection between the dorsal hippocampus and the amygdala, contributing to social recognition memory. In the dorsal hippocampus (CA1) and basolateral amygdala, dopaminergic, noradrenergic, and histaminergic systems have been found to be involved in social recognition memory [40]. Alterations in β-noradrenergic, D1/D5 dopaminergic, and H2-histaminergic receptors for these neurotransmitters may be the cause of social deficits in psychiatric disorders such as autism and schizophrenia. In addition, 5-HT also exhibits its mood- and cognition-regulating properties in the hippocampus. Serotonin receptor 4 (5-HT4R) is required to maintain proper excitability of granule cells in the dentate gyrus. Loss of 5-HT4R has been found to lead to antidepressant-like behavioral responses and an increase in baseline anxiety [41]. Hippocampal 5-HT4Rs are needed to modulate mood and anxiety.

Overall, neurotransmitters in the hippocampus are key in neuronal communication and functional connectivity. Glutamate and GABA function as the two main neurotransmitters of the hippocampus, while the effect of ACh on memory and learning emphasizes the importance of hippocampal cholinergic transmission. However, other neurotransmitters, such as 5-HT, dopamine, and noradrenaline, also have effects on the hippocampus. Disruptions in the concentration or receptors of these neurotransmitters may lead to the emergence of psychiatric disorders, including schizophrenia, autism, and psychosis, especially memory problems.

5.4 Long-Term Potentiation in the Hippocampus

One of the major challenges in neuroscience is to identify the molecular and cellular processes underlying learning and memory formation. LTP is the cellular counterpart of learning and memory formation in the hippocampus. The strengthening of connections between neurons is mediated by LTP in the hippocampus, and this mechanism is the leading and best-documented neuronal substrate for memory formation [42, 43]. In the hippocampus, an excitatory signal is sent to the synaptic gap by glutamate release from the presynaptic neuron. Glutamate binding to NMDA receptors in the postsynaptic neuron also suppresses Mg^{2+} ions that block these receptors. Activation of NMDA receptors activates intracellular signaling pathways by allowing the entry of Ca^{2+} ions. Intracellular protein synthesis is increased by stimulation of cellular signal transduction molecules such as protein kinases and calcium-dependent enzymes. In LTP, NMDA receptors, as well as AMPA receptors, play a role in strengthening or weakening the synapse. As a result of activated signaling pathways, the synaptic connection of the postsynaptic neuron is strengthened (Fig. 5.2).

The best-characterized form of NMDA receptor-dependent LTP occurs between CA3 and CA1 pyramidal neurons of the hippocampus, explaining the role of the hippocampus in memory and learning [29, 44, 45].

The study of how LTP occurs in the hippocampus can be realized with high-frequency stimulation (HFS). HFS induces synaptic changes and significantly determines the presence of LTP in different regions of the hippocampus. In particular, it reveals that LTP can be observed even at synapses distant from the afferent pathway stimulated by HFS [46]. These findings may suggest that LTP is a physiological process in the hippocampus. Due to the different distribution of receptors in subregions of the hippocampus, LTP properties may vary. Although higher basal synaptic transmission is detected in the dorsal dentate gyrus, LTP magnitude increases significantly in the ventral direction. However, this situation varies with age. The difference between LTP magnitude and synaptic transmission decreases with age [47]. This may have important implications for the hippocampus and age-related neurological disorders.

Neurogenesis in the hippocampus ensures the continuous development of the region. In addi-

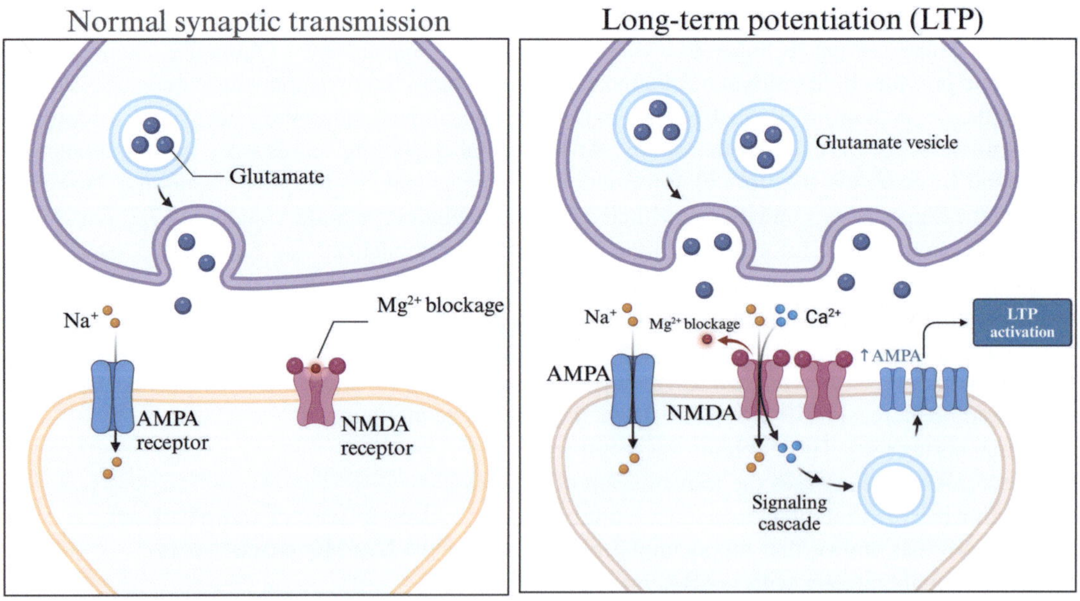

Fig. 5.2 Long-term potentiation in the hippocampus. The main steps of LTP include stimulation, neurotransmitter release, receptor activation, unblocking of Mg2+ from the NMDA receptor, Ca2+ influx, and activation of the signaling pathway. This process strengthens communication between nerve cells and causes long-term changes in synaptic connections, which play a critical role in learning and memory formation

tion, individual and environmental factors also have an impact on the function of the hippocampus. In a study, the effect of exercise on synaptic plasticity and LTP was examined, and exercise was found to be an effective tool in increasing sensitivity and LTP in the dentate gyrus region [48]. In addition, factors such as malnutrition and seizures may negatively affect LTP, which decreases, especially in children, due to decreased expression of NMDA and AMPA receptors [49].

In summary, LTP, an important reflection of synaptic plasticity, is an activity-based long-term increase in the efficiency of excitatory synaptic transmission following the transmission of a short, high-frequency electrical stimulus sequence. It is through LTP that the memory and learning properties of the hippocampus are explained at the cellular level. The function of LTP varies depending on the different densities of NMDA and AMPA receptors in subregions of the hippocampus. In addition, individual habits and environmental factors may affect the memory role of the hippocampus by influencing LTP.

5.5 Conclusion

The hippocampus is a small region of the brain consisting of the CA1–3 and dentate gyrus regions. It has central roles in basic cognitive processes such as memory, learning, emotional behavior, and episodic memory. The hippocampus accomplishes all these functions through functional connections, neuronal activity, and neurotransmitter systems. Neuronal interaction, which forms the basis of information processing, occurs through neurotransmitters in various synaptic connections. In particular, neurotransmitters such as glutamate and GABA, which play key roles in synaptic communication, enable the formation of complex information-processing networks in learning, memory, and emotion processing. GABA stabilizes neural activity, while glutamate contributes to stimulus transmission and learning processes. It accomplishes this function through NMDA and AMPA receptors and ions with which glutamate interacts, called LTP, which ensures synaptic strengthening between neurons. As a result, the hippocampus, known as the memory center of the brain, enables the transfer of information from short-term memory to long-term memory through cellular, chemical, neuronal, and physiological processes.

References

1. Knierim JJ. The hippocampus. Curr Biol. 2015;25(23):R1116–21. https://doi.org/10.1016/j.cub.2015.10.049.
2. Ackerman S. Discovering the brain, vol 2. Major structures and functions of the brain. Washington, DC: National Academies Press (US); 1992. https://www.ncbi.nlm.nih.gov/books/NBK234157/1992.
3. Rao YL, Ganaraja B, Murlimanju BV, Joy T, Krishnamurthy A, Agrawal A. Hippocampus and its involvement in Alzheimer's disease: a review. 3 Biotech. 2022;12(2):55. https://doi.org/10.1007/s13205-022-03123-4.
4. Anand KS, Dhikav V. Hippocampus in health and disease: an overview. Ann Indian Acad Neurol. 2012;15(4):239–46. https://doi.org/10.4103/0972-2327.104323.
5. Knowles WD. Normal anatomy and neurophysiology of the hippocampal formation. J Clin Neurophysiol. 1992;9(2):252–63.
6. Epp JR, Chow C, Galea LA. Hippocampus-dependent learning influences hippocampal neurogenesis. Front Neurosci. 2013;7:57. https://doi.org/10.3389/fnins.2013.00057.
7. Zou W, Song P, Lu W, Shao R, Zhang R, Yau SY, Yuan TF, Wang Y, Lin K. Global hippocampus functional connectivity as a predictive neural marker for conversion to future mood disorder in unaffected offspring of bipolar disorder parents. Asian J Psychiatr. 2022;78:103307. https://doi.org/10.1016/j.ajp.2022.103307.
8. Raud L, Sneve MH, Vidal-Piñeiro D, Sørensen Ø, Folvik L, Ness HT, Mowinckel AM, Grydeland H, Walhovd KB, Fjell AM. Hippocampal-cortical functional connectivity during memory encoding and retrieval. Neuroimage. 2023;279:120309. https://doi.org/10.1016/j.neuroimage.2023.120309.
9. Witter MP, Doan TP, Jacobsen B, Nilssen ES, Ohara S. Architecture of the entorhinal cortex a review of entorhinal anatomy in rodents with some comparative notes. Front Syst Neurosci. 2017;11:46. https://doi.org/10.3389/fnsys.2017.00046.
10. Preston AR, Eichenbaum H. Interplay of hippocampus and prefrontal cortex in memory. Curr Biol. 2013;23(17):R764–73. https://doi.org/10.1016/j.cub.2013.05.041.
11. Song J. Amygdala activity and amygdala-hippocampus connectivity: metabolic diseases, dementia, and neuropsychiatric issues. Biomed

Pharmacother. 2023;162:114647. https://doi.org/10.1016/j.biopha.2023.114647.
12. Cutsuridis V, Yoshida M. Editorial: memory processes in medial temporal lobe: experimental, theoretical and computational approaches. Front Syst Neurosci. 2017;11:19. https://doi.org/10.3389/fnsys.2017.00019.
13. Bang JY, Zhao J, Rahman M, St-Cyr S, McGowan PO, Kim JC. Hippocampus-anterior hypothalamic circuit modulates stress-induced endocrine and behavioral response. Front Neural Circuits. 2022;16:894722. https://doi.org/10.3389/fncir.2022.894722.
14. Cembrowski MS, Spruston N. Heterogeneity within classical cell types is the rule: lessons from hippocampal pyramidal neurons. Nat Rev Neurosci. 2019;20(4):193–204. https://doi.org/10.1038/s41583-019-0125-5.
15. Booker SA, Vida I. Morphological diversity and connectivity of hippocampal interneurons. Cell Tissue Res. 2018;373(3):619–41. https://doi.org/10.1007/s00441-018-2882-2.
16. Chang WT, Langella SK, Tang Y, Ahmad S, Zhang H, Yap PT, Giovanello KS, Lin W. Brainwide functional networks associated with anatomically- and functionally-defined hippocampal subfields using ultrahigh-resolution fMRI. Sci Rep. 2021;11(1):10835. https://doi.org/10.1038/s41598-021-90364-7.
17. Tang L, Pruitt PJ, Yu Q, Homayouni R, Daugherty AM, Damoiseaux JS, Ofen N. Differential functional connectivity in anterior and posterior hippocampus supporting the development of memory formation. Front Hum Neurosci. 2020;14:204. https://doi.org/10.3389/fnhum.2020.00204.
18. Alahmadi AAS, Alotaibi NO, Hakami NY, Almutairi RS, Darwesh AMF, Abdeen R, Alghamdi J, Abdulaal OM, Alsharif W, Sultan SR, Kanbayti IH. Gender and cytoarchitecture differences: functional connectivity of the hippocampal sub-regions. Heliyon. 2023;9(10):e20389. https://doi.org/10.1016/j.heliyon.2023.e20389.
19. Liu J, Chen Y, Stephens R, Cornea E, Goldman B, Gilmore JH, Gao W. Hippocampal functional connectivity development during the first two years indexes 4-year working memory performance. Cortex. 2021;138:165–77. https://doi.org/10.1016/j.cortex.2021.02.005.
20. Supekar K, Swigart AG, Tenison C, Jolles DD, Rosenberg-Lee M, Fuchs L, Menon V. Neural predictors of individual differences in response to math tutoring in primary-grade school children. Proc Natl Acad Sci U S A. 2013;110(20):8230–5. https://doi.org/10.1073/pnas.1222154110.
21. Ge R, Torres I, Brown JJ, Gregory E, McLellan E, Downar JH, Blumberger DM, Daskalakis ZJ, Lam RW, Vila-Rodriguez F. Functional disconnectivity of the hippocampal network and neural correlates of memory impairment in treatment-resistant depression. J Affect Disord. 2019;253:248–56. https://doi.org/10.1016/j.jad.2019.04.096.
22. Kavalali ET. The mechanisms and functions of spontaneous neurotransmitter release. Nat Rev Neurosci. 2015;16(1):5–16. https://doi.org/10.1038/nrn3875.
23. Hansen JY, Shafiei G, Markello RD, Smart K, Cox SML, Nørgaard M, Beliveau V, Wu Y, Gallezot JD, Aumont É, Servaes S, Scala SG, DuBois JM, Wainstein G, Bezgin G, Funck T, Schmitz TW, Spreng RN, Galovic M, Koepp MJ, et al. Mapping neurotransmitter systems to the structural and functional organization of the human neocortex. Nat Neurosci. 2022;25(11):1569–81. https://doi.org/10.1038/s41593-022-01186-3.
24. Teleanu RI, Niculescu AG, Roza E, Vladâcenco O, Grumezescu AM, Teleanu DM. Neurotransmitters—Key factors in neurological and neurodegenerative disorders of the central nervous system. Int J Mol Sci. 2022;23(11):5954. https://doi.org/10.3390/ijms23115954.
25. Straughan DW. Neurotransmitters and the hippocampus. In: Isaacson RL, Pribram KH, editors. The hippocampus. Boston: Springer; 1975. https://doi.org/10.1007/978-1-4684-2976-3_10.
26. Nedergaard M, Takano T, Hansen AJ. Beyond the role of glutamate as a neurotransmitter. Nat Rev Neurosci. 2002;3(9):748–55. https://doi.org/10.1038/nrn916.
27. Zhou Y, Danbolt NC. Glutamate as a neurotransmitter in the healthy brain. J Neural Transm (Vienna). 2014;121(8):799–817. https://doi.org/10.1007/s00702-014-1180-8.
28. Traynelis SF, Wollmuth LP, McBain CJ, Menniti FS, Vance KM, Ogden KK, Hansen KB, Yuan H, Myers SJ, Dingledine R. Glutamate receptor ion channels: structure, regulation, and function. Pharmacol Rev. 2010;62(3):405–96. https://doi.org/10.1124/pr.109.002451.
29. Lüscher C, Malenka RC. NMDA receptor-dependent long-term potentiation and long-term depression (LTP/LTD). Cold Spring Harb Perspect Biol. 2012;4(6):a005710. https://doi.org/10.1101/cshperspect.a005710.
30. Jewett BE, Sharma S. Physiology, GABA. [Updated 2023 Jul 24]. In: StatPearls [internet]. Treasure Island: StatPearls Publishing; 2024. https://www.ncbi.nlm.nih.gov/books/NBK513311/.
31. Akyuz E, Celik BR, Aslan FS, Sahin H, Angelopoulou E. Exploring the role of neurotransmitters in multiple sclerosis: an expanded review. ACS Chem Neurosci. 2023;14(4):527–53. https://doi.org/10.1021/acschemneuro.2c00589.
32. Vizi ES, Kiss JP. Neurochemistry and pharmacology of the major hippocampal transmitter systems: synaptic and nonsynaptic interactions. Hippocampus. 1998;8(6):566–607. https://doi.org/10.1002/(SICI)1098-1063(1998)8:6<566::AID-HIPO2>3.0.CO;2-W.
33. Song NN, Huang Y, Yu X, Lang B, Ding YQ, Zhang L. Divergent roles of central serotonin in adult hippocampal neurogenesis. Front Cell

Neurosci. 2017;11:185. https://doi.org/10.3389/fncel.2017.00185.
34. Bossong MG, Antoniades M, Azis M, Samson C, Quinn B, Bonoldi I, Modinos G, Perez J, Howes OD, Stone JM, Allen P, McGuire P. Association of hippocampal glutamate levels with adverse outcomes in individuals at clinical high risk for psychosis. JAMA Psychiatry. 2019;76(2):199–207. https://doi.org/10.1001/jamapsychiatry.2018.3252.
35. Jiménez-Balado J, Ycaza Herrera A, Igwe K, Klem L, Buyukturkoglu K, Irimia A, Chen L, Guo J, Brickman AM, Eich TS. Reduced hippocampal GABA+ is associated with poorer episodic memory in healthy older women: a pilot study. Front Behav Neurosci. 2021;15:695416. https://doi.org/10.3389/fnbeh.2021.695416.
36. Schmitz TW, Correia MM, Ferreira CS, Prescot AP, Anderson MC. Hippocampal GABA enables inhibitory control over unwanted thoughts. Nat Commun. 2017;8(1):1311. https://doi.org/10.1038/s41467-017-00956-z.
37. Teles-Grilo Ruivo LM, Baker KL, Conway MW, Kinsley PJ, Gilmour G, Phillips KG, Isaac JTR, Lowry JP, Mellor JR. Coordinated acetylcholine release in prefrontal cortex and hippocampus is associated with arousal and reward on distinct timescales. Cell Rep. 2017;18(4):905–17. https://doi.org/10.1016/j.celrep.2016.12.085.
38. Palacios-Filardo J, Udakis M, Brown GA, Tehan BG, Congreve MS, Nathan PJ, Brown AJH, Mellor JR. Acetylcholine prioritises direct synaptic inputs from entorhinal cortex to CA1 by differential modulation of feedforward inhibitory circuits. Nat Commun. 2021;12(1):5475. https://doi.org/10.1038/s41467-021-25280-5.
39. Takács VT, Cserép C, Schlingloff D, Pósfai B, Szőnyi A, Sos KE, Környei Z, Dénes Á, Gulyás AI, Freund TF, Nyiri G. Co-transmission of acetylcholine and GABA regulates hippocampal states. Nat Commun. 2018;9(1):2848. https://doi.org/10.1038/s41467-018-05136-1.
40. Garrido Zinn C, Clairis N, Silva Cavalcante LE, Furini CR, de Carvalho Myskiw J, Izquierdo I. Major neurotransmitter systems in dorsal hippocampus and basolateral amygdala control social recognition memory. Proc Natl Acad Sci U S A. 2016;113(33):E4914–9. https://doi.org/10.1073/pnas.1609883113.
41. Karayol R, Medrihan L, Warner-Schmidt JL, Fait BW, Rao MN, Holzner EB, Greengard P, Heintz N, Schmidt EF. Serotonin receptor 4 in the hippocampus modulates mood and anxiety. Mol Psychiatry. 2021;26(6):2334–49. https://doi.org/10.1038/s41380-020-00994-y.
42. Bliss TV, Collingridge GL. A synaptic model of memory: long-term potentiation in the hippocampus. Nature. 1993;361(6407):31–9. https://doi.org/10.1038/361031a0.
43. Lynch MA. Long-term potentiation and memory. Physiol Rev. 2004;84(1):87–136. https://doi.org/10.1152/physrev.00014.2003.
44. Hayashi Y. Molecular mechanism of hippocampal long-term potentiation—Towards multiscale understanding of learning and memory. Neurosci Res. 2022;175:3–15.
45. Kumar A. Long-term potentiation at CA3-CA1 hippocampal synapses with special emphasis on aging, disease, and stress. Front Aging Neurosci. 2011;3:7. https://doi.org/10.3389/fnagi.2011.00007.
46. Romero-Barragán MT, Gruart A, Delgado-García JM. Transsynaptic long-term potentiation in the hippocampus of behaving mice. Front Synaptic Neurosci. 2022;13:811806. https://doi.org/10.3389/fnsyn.2021.811806.
47. Schreurs A, Sabanov V, Balschun D. Distinct properties of long-term potentiation in the dentate gyrus along the dorsoventral axis: influence of age and inhibition. Sci Rep. 2017;7(1):5157. https://doi.org/10.1038/s41598-017-05358-1.
48. Radahmadi M, Hosseini N, Alaei H. Effect of exercise, exercise withdrawal, and continued regular exercise on excitability and long-term potentiation in the dentate gyrus of hippocampus. Brain Res. 2016;1653:8–13. https://doi.org/10.1016/j.brainres.2016.09.045.
49. Wearick-Silva LE, Sebben AD, Costa-Ferro ZSM, Marinowic DR, Nunes ML. Undernourishment and recurrent seizures early in life impair long-term potentiation and alter NMDAR and AMPAR expression in rat hippocampus. Int J Dev Neurosci. 2019;75:13–8. https://doi.org/10.1016/j.ijdevneu.2019.03.005.

Structural and Synaptic Plasticity in the Hippocampus

6

A. Cihangir Uğuz and Züleyha Doğanyiğit

6.1 Introduction

For the majority of the twentieth century, the prevailing view in the neuroscience community was that the mammalian central nervous system achieved structural stability shortly after birth, thereby discounting the possibility of neural remodeling in adulthood. The hippocampal formation is characterized as a brain region that develops relatively late. Unlike the static and unchanging portrayal in earlier anatomical and physiological discussions, the hippocampus is recognized for undergoing various structural changes throughout life, commonly termed as "developmental." However, advances in technology over recent decades have conclusively demonstrated that the brain exhibits persistent plasticity across the life. Although the initial discovery of adult neurogenesis in the 1960s was largely overlooked and evidence in birds dismissed as irrelevant to mammals, subsequent research has increasingly substantiated the occurrence of adult neurogenesis in numerous mammalian species. The hippocampus is widely recognized as a brain region where the continuous production of new granule neurons persists throughout life. In neuroscience, "plasticity" describes the brain's capacity to modify and adapt due to experiences, environmental factors, and different internal and external stimuli. The integration of new neurons into existing circuits triggers a series of structural changes that enhance the region's structural plasticity. These new neurons develop axons and dendrites and participate in synaptogenesis [1]. Adult neurogenesis enhances hippocampal plasticity through various mechanisms, as the generation of new granule neurons includes the development of their axons, dendrites, and synapses. These advancements are often followed by a decline, characterized by cell death and the loss of neural connections [2]. As adult neurogenesis has gained recognition as a vital process in the mammalian brain, research has identified various factors and conditions that impact its development. This increasing body of evidence suggests potential functions for neurons produced during adulthood. This adaptability encompasses changes at multiple levels, including molecular, cellular, and structural, and it is crucial for processes such as learning, memory, development, and recovery from injury. This phenomenon can be analyzed through two primary categories: structural and synaptic plasticity.

A. C. Uğuz (✉)
Department of Biophysics, Karamanoğlu Mehmetbey University Faculty of Medicine, Karaman, Türkiye
e-mail: cihangiruguz@gmail.com

Z. Doğanyiğit
Department of Histology and Embryology, Yozgat Bozok University Faculty of Medicine, Yozgat, Türkiye

M. Turgut et al. (eds.), *The Human Hippocampus*, https://doi.org/10.1007/978-3-031-88907-3_6

Structural plasticity in the hippocampus denotes the brain's capacity to alter its physical structure in response to learning, experiences, and environmental factors. Unlike synaptic plasticity, which focuses on the strength of synaptic connections, structural plasticity involves changes in the number, shape, and organization of neurons and their connections. Structural plasticity provides the physical substrate for synaptic plasticity by creating and modifying the architecture of neural circuits.

Synaptic plasticity, in turn, fine-tunes these connections to optimize information processing and storage. Together, they enable the hippocampus to support learning, memory, and cognitive flexibility. Various forms of structural plasticity have been associated with a wide range of psychiatric and neurological disorders, highlighting the potential significance of these processes in understanding, and potentially treating such conditions. Several studies have shown that the hippocampal formation can undergo significant reorganization following damage and undergo substantial structural changes even when intact. Abovementioned region seems to experience ongoing dynamic changes, such as the extension and retraction of dendrites, as well as the formation and elimination of synapses.

Synaptic plasticity in the hippocampus allows synapses (the connections between neurons) to strengthen or weaken over time, reflecting their activity levels. Synaptic plasticity involves changes in the strength and effectiveness of synaptic connections, primarily through mechanisms like long-term potentiation (LTP) and long-term depression (LTD). LTP enhances synaptic strength following high-frequency stimulation, while LTD weakens synaptic connections in response to low-frequency stimulation. Synaptogenesis and dendritic spine morphology changes are also involved in this process. These dynamic processes are regulated by a complex interplay of molecular signals, including neurotransmitters, receptors, and intracellular signaling cascades. In the hippocampus, these plastic changes are crucial for the encoding, consolidation, and retrieval of memories, highlighting the region's role as a critical hub in the brain's cognitive architecture. The properties of LTP can be significantly affected by various factors, including the specific cellular pathways under investigation. Additionally, the age of the animal plays a crucial role.

The interaction between structural and synaptic plasticity enables the hippocampus to perform its role in cognitive functions, including learning and memory consolidation. Advances in understanding these mechanisms not only illuminate the cellular basis of cognition but also provide potential therapeutic targets for neurological disorders characterized by cognitive impairments. This chapter delves into the intricate processes of structural and synaptic plasticity in the hippocampus, exploring their roles in neural adaptation and cognitive function. This chapter also highlights the vital role of the N-methyl-D-aspartate (NMDA) receptor in the initiation of LTP and its connection to hippocampus-dependent learning. Glutamate interacts with two primary neurotransmitter receptors on the postsynaptic neuron: NMDA and α-amino-3-hydroxy-5-methyl-4-isoxazolepropionic acid (AMPA) receptors. These ligand-gated ionotropic receptors are crucial for synaptic transmission and plasticity. AMPA receptors mainly facilitate Na^+ influx, leading to the depolarization of the postsynaptic neuron, while NMDA receptors regulate the influx of calcium ions (Ca^{2+}), essential for signal transduction. It has been established that the conditions needed to induce LTP at most hippocampal synapses are the same as those required to activate the NMDA receptor. The Ca^{2+} influx through the activated NMDA receptor is the cellular event that triggers LTP induction, highlighting the receptor's essential function in synaptic plasticity (Fig. 6.1).

For over a century, neuroanatomists have proposed the adult brain's potential for structural changes. In recent decades, increasing evidence has highlighted the hippocampus as a region with significant structural plasticity. These results collectively suggest that the hippocampal formation is a dynamic region, constantly undergoing structural changes, both at the microscopic and macroscopic levels.

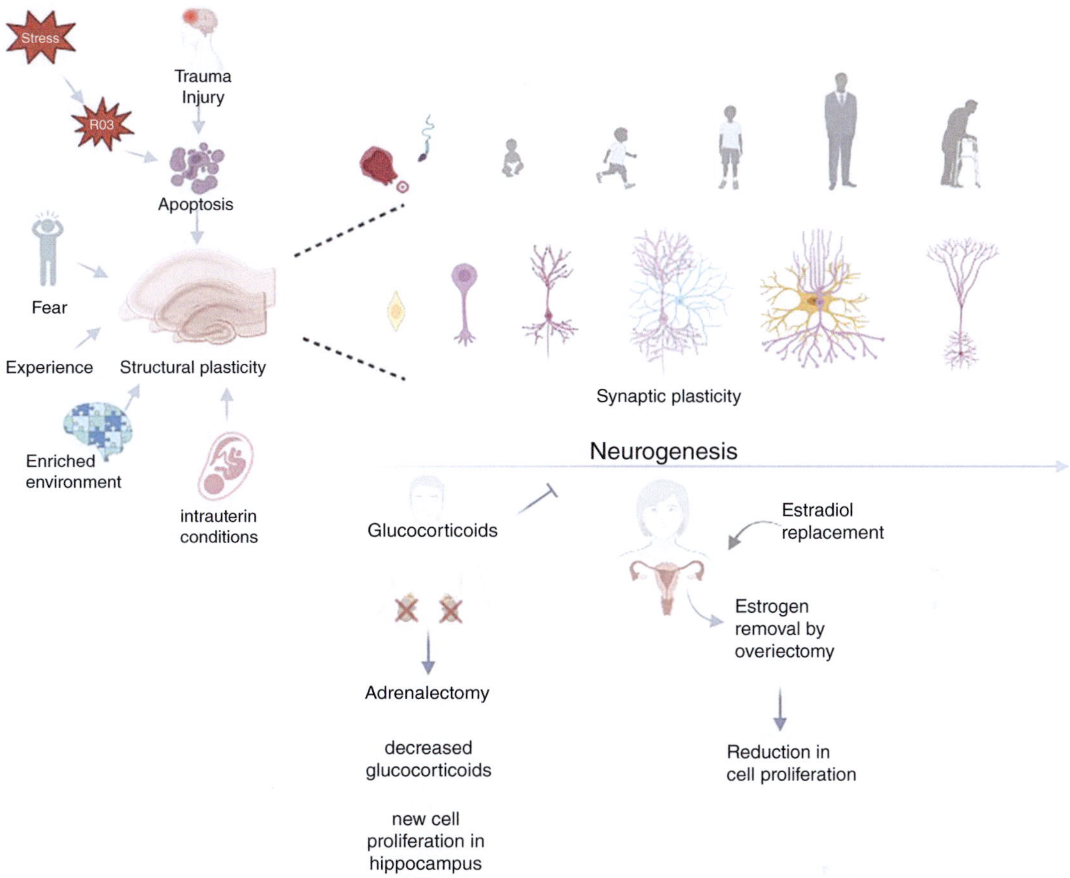

Fig. 6.1 Modulators of structural plasticity. This figure illustrates the multifactorial influences on the structural plasticity of the hippocampus, highlighting the complex interplay among oxidative stress, enriched environments, fear, trauma, intrauterine conditions, and age. Oxidative stress is shown as a detrimental factor that induces cellular damage and reduces dendritic branching and synaptic density through the production of ROS. An enriched environment, depicted through increased cognitive and physical stimulation, enhances neurogenesis, synaptic plasticity, and dendritic complexity. The figure also details the adverse effects of fear and trauma, where chronic exposure activates stress hormone pathways, leading to hippocampal atrophy and decreased neurogenesis. Intrauterine conditions, influenced by maternal health, nutrition, and stress during pregnancy, impact fetal hippocampal development and can predispose individuals to neurodevelopmental disorders. Additionally, the figure incorporates the influence of age, illustrating how aging affects hippocampal plasticity by reducing neurogenesis and synaptic plasticity, while potentially exacerbating the negative impacts of oxidative stress, fear, and trauma. The crosstalk between these factors is emphasized, showing how an enriched environment can mitigate oxidative stress and age-related decline, whereas fear and trauma can amplify these detrimental effects, underscoring the intricate regulatory mechanisms of hippocampal plasticity

6.2 Factors Driving Structural Plasticity in the Brain

6.2.1 Stress

Understanding the intricate interplay between stress and hippocampal synaptic plasticity offers profound insights into the mechanisms through which external challenges influence cognitive function and emotional resilience. Stress conditions can significantly impact dendritic architecture in the hippocampus. Chronic stress typically has detrimental effects on dendritic structure in this region. Studies on adult male rats and tree shrews subjected to repeated restraint stress have shown a reduction in the dendritic tree size of

CA3 pyramidal neurons, similar to observations following chronic glucocorticoid treatment [3, 4]. Stress- and recovery-induced changes in the dendritic tree align with hippocampus-dependent task performance. Acute stress impacts dendritic spine numbers on CA1 pyramidal cells in adult rats, with variations based on sex and, in females, the estrous stage. In males, brief, intermittent tail shocks increase the number of dendritic spines on pyramidal cells [5, 6]. In female rats, tail shocks during diestrus decrease dendritic spines on pyramidal cells, whereas shocks during estrus do not have this effect [7]. The functional significance of these changes is unclear, but stress has been reported to have opposite effects on learning in males and females, paralleling the changes in dendritic spines and suggesting a functional relationship [5, 6].

Experience is a crucial regulatory factor in neurogenesis, encompassing sensory and motor activities, cognitive demands, and stress [8]. The catabolic effects of stress and stress hormones (such as cortisol and corticosterone) have long been acknowledged. The term "fight-or-flight" is commonly used to describe the response to dangerous situations. This response suppresses energy-intensive processes, such as growth, that are future oriented. As a result, stress hinders the generation of new granule cells both during early development and in adulthood [9]. Exposure to various stressors, such as predator odors, electric shocks, or restraint, has been found to decrease cell proliferation in the dentate gyrus of three mammalian species, including postnatal and adult rats. Stress has been shown to reduce the proliferation of granule cell precursors and the formation of early-stage neurons in the dentate gyrus. This effect seems to be localized to this region, as the number of Brd-U labeled cells in the subventricular zone, another neurogenic site, remained unchanged following stress [10]. Prenatal stress has a lasting impact on granule cell production. In both monkeys and rodents, it persistently reduces the proliferation of granule cell precursors into adulthood, highlighting the enduring effects of early stress on hippocampal neurogenesis [11]. Postnatal stress, such as maternal separation, leads to lasting suppression of cell proliferation and the production of early-stage neurons in adult rats, emphasizing the long-term impact of early stress on neurogenesis [12]. Prenatal and early postnatal stress result in lasting changes to the hypothalamic–pituitary–adrenal (HPA) axis, causing inefficient glucocorticoid regulation [13]. Consequently, animals exposed to early life stress may experience persistently inhibited cell production due to elevated glucocorticoid levels or increased sensitivity to these hormones. Lowering glucocorticoid levels enhances the proliferation of new cells in the hippocampus. In ovariectomized animals, estradiol replacement restores cell proliferation rates, with natural fluctuations seen across the estrous cycle—peaking during proestrus when estrogen levels are highest. Estrogen's stimulation of cell proliferation leads to a higher production of immature neurons in females. However, these estrogen-induced increases are transient, disappearing within weeks after BrdU injection [14]. Housing conditions of laboratory animals may also affect estrogen-related differences in neuron production. Additionally, estrogen may boost new neuron production by promoting their survival.

Elevated glucocorticoid levels have been linked to dendritic shrinkage in the CA3 area. Prolonged corticosterone exposure in rodents leads to a reduction in the complexity and size of CA3 pyramidal neuron apical dendrites [15]. While these changes could indicate impending cell death, which has been linked to chronic high glucocorticoid levels [16, 17], some suggest that dendritic retraction may serve an adaptive function by shielding the hippocampus from excessive glutamate [18, 19].

In adult female rats and monkeys, the number of dendritic spines in the CA1 region undergoes constant changes over hours and days, driven by fluctuating ovarian steroid levels [20]. Elevated levels of ovarian steroids correspond to increased dendritic spine numbers, while lower hormone levels lead to reduced spine numbers. The fluctuations in spine numbers are accompanied by corresponding changes in the number of excitatory synapses in the same area [21, 22]. Research findings regarding the influence of estrogen on hippocampus-related behaviors are inconsistent. Some studies propose that estrogen improves performance in tasks reliant on the hippocampus, correlating with changes in dendritic spines [23]. However, other studies have not observed these improvements with estrogen treatment or throughout the estrous cycle [24, 25]. Therefore, the relationship between behavior and the number of dendritic spines in the hippocampus remains inconclusive.

The ovarian steroid estradiol has a beneficial effect on the production of new cells in the adult rat dentate gyrus [26, 27]. Ovariectomy, which removes estrogen, reduces cell proliferation, while estradiol replacement restores it [26]. Additionally, natural fluctuations in cell proliferation correspond with the estrous cycle, peaking during proestrus when estrogen levels are highest. This estrogen-induced stimulation results in a higher production of immature neurons in females, highlighting a significant sex difference in hippocampal neurogenesis [28]. Estrogen may additionally augment the production of new neurons by promoting their survival [29].

6.2.2 Environment

Enriched environments, whether experienced during development or adulthood, have been consistently lead to larger brains in animals compared to standard conditions [30]. Although most studies concentrate on the neocortex, similar effects are observed in the hippocampus. Enriched environments promote the growth and complexity of dendritic trees, leading to an increase in the number of dendritic spines and synapses [31]. Enriched laboratory environments typically include stimuli that promote stress, physical activity, and learning. Many studies have investigated the specific factors driving brain changes in these environments [32, 33]. Research suggests that stress is not the primary mediator of hippocampal changes, as evidenced by comparable stress hormone levels between groups and varying effects on brain structure after specific stressors. Some studies suggest that physical activity contributes to the hippocampal structural changes associated with enriched environments [34]. It is unclear whether a single factor or a combination of factors drives hippocampal structural changes in animals raised in complex environments [35]. The roles of learning and synaptic plasticity in these structural changes have also been examined [36]. Numerous studies have shown that learning and LTP, a type of synaptic plasticity linked to learning, induce structural changes in the hippocampus [37, 38]. Among the many properties of hippocampal synapses, their capacity for long-lasting changes in synaptic efficacy in response to specific activation patterns is particularly significant and extensively studied. Although synaptic plasticity occurs at many excitatory synapses in the brain, most of the understanding of LTP and LTD comes from research in the hippocampus, where these phenomena were initially discovered. The hippocampus remains central to our understanding of synaptic plasticity. These modifications encompass changes in dendritic spine shape, an increase in spine numbers, and variations in synapse quantity and distribution along dendrites. These effects are similar to those observed in enriched environments. Nevertheless, the precise role of these structural changes in the learning process remains uncertain.

6.2.3 Injury/Trauma

Structural plasticity is essential for the brain's response to injury, facilitating recovery and functional restoration through the reorganization of neural circuits. When an injury occurs, such as a traumatic brain injury or stroke, the brain initiates several adaptive processes to compensate for the damage. One key aspect of this response is dendritic remodeling, where dendrites may retract or degenerate but also sprout new branches to reestablish lost connections. Axonal sprouting, where axons grow new branches, further aids in forming new synaptic connections, essential for neural network reorganization [39]. Synaptogenesis, the formation of new synapses, is another critical component of structural plasticity. It allows neurons to create new connections, enhancing communication within neural networks [40]. Neurogenesis, particularly in the hippocampus, can also be stimulated by injury, leading to the production of new neurons that integrate into existing circuits and contribute to neural repair [41]. Glial cells, including astrocytes and microglia, play supportive roles by stabilizing the extracellular environment, promoting synapse formation, and modulating inflammation to aid in the repair process [42]. These adaptive mechanisms underscore the brain's remarkable capacity for self-repair, highlighting the importance of structural plasticity in recovery from injury. Studies of disease models, such as epilepsy and stroke, have identified other instances of structural changes in the hippocampal formation [35, 43]. These conditions result in significant cell death within the hippocampus. Over time, following a seizure or stroke, substantial structural reorganization occurs, including axon sprouting, which can be considered a regenerative process. Research on the brains of individuals with epilepsy and Alzheimer's disease has revealed evidence of structural reorganization [44, 45]. The brain is protected from physical damage by the skull and chemical harm by the blood–brain barrier. Consequently, it is not immediately evident that the hippocampus's regenerative abilities evolved specifically to repair brain damage, particularly in older age. However, adult neurogenesis has been observed in the hippocampus of almost every mammalian species studied. Nevertheless, axonal sprouting and synaptogenesis might be developmental mechanisms reinitiated by certain environmental stimuli. Alternatively, these processes may occur at a minimal level under normal circumstances, with their activity intensifying in response to injury. Considering the significant evidence supporting the hippocampus's structural plasticity in typical conditions, the latter hypothesis seems especially plausible. The absence of reported structural changes in the hippocampus by early neuroanatomists, despite the substantial addition of new neurons in the dentate gyrus, is puzzling. Approximately 9000 new cells are added daily, with many extending axons, developing dendrites, and forming synapses. Yet, traditional histological methods likely overlooked these changes because they did not alter the overall size, shape, or number of cells and synapses. The prevailing belief that structural changes were solely a developmental phenomenon may have led to misinterpretations of adult changes. For example, variations in mossy fiber terminals in the CA3 region could suggest different stages of development for connections formed by newly generated granule cells, highlighting the dynamic nature of the adult hippocampus. This oversight underscores the importance of reconsidering traditional views on neurogenesis and structural plasticity in the adult brain.

Head trauma can significantly disrupt synaptic plasticity, a crucial mechanism for learning and memory. Traumatic brain injury (TBI) often results in disruptions to neural circuits and synaptic functions, leading to both immediate and long-term cognitive deficits. The initial mechanical insult from TBI can cause widespread neuronal depolarization and excessive release of neurotransmitters, notably glutamate, leading to excitotoxicity and subsequent neuronal damage. This acute phase can be followed by alterations in synaptic plasticity, including impaired LTP and enhanced LTD. These alterations are mediated by the disruption of calcium homeostasis, oxidative stress, and inflammatory responses. Moreover, TBI can alter the expression and function of key receptors and signaling pathways involved in synaptic plasticity, such as NMDA and AMPA receptors, and the downstream activation of kinases like CaMKII and ERK. Comprehending the complex relationship between head trauma and synaptic plasticity is crucial for creating therapeutic strategies to alleviate cognitive impairments linked to TBI and to enhance neural repair and recovery. This is significant because accumulating evidence suggests that TBI, especially repetitive injuries common in sports like boxing and American football, can contribute to a neurodegenerative condition known as chronic traumatic encephalopathy (CTE) [46]. CTE pathology is marked by the presence of hyperphosphorylated tau protein forming neurofibrillary tangles, especially in the depths of the sulci. In cases of TBI, increased glutamate levels result in abnormal intracellular Ca^{2+} concentrations, which contribute to neuronal death. AMPA receptors are made up of four subunits (GluR1–4), and research has demonstrated that TBI-related increases in glutamate can induce alterations in AMPA receptors. These changes exacerbate glutamate toxicity by intensifying Ca^{2+} overload [46]. Intracellular Ca^{2+} release triggers the expression of AMPA receptors without the GluR2 subunit, resulting in receptors with high Ca^{2+} permeability. This alteration in the AMPA receptor subtype plays a crucial role in amplifying excitotoxic neuronal death shortly after injury. The NMDA receptor binds to PSD95, initiating a cascade of interactions that ultimately leads to the phosphorylation of the GluR2 subunit of the AMPA receptor by PKC-alpha. This interaction results in the induction of GluR2-deficient AMPA receptors, thereby exacerbating excitotoxicity in TBI.

6.3 Neurogenesis

An important aspect of neurogenesis involves its regulation, potentially through endocrine mechanisms. The hippocampal formation is densely populated with hormone receptors, affecting various cell types. As previously discussed, hormonal manipulations in adulthood can induce significant structural changes in this region, underscoring the regulatory role of hormones in hippocampal neurogenesis [47, 48]. Numerous studies have demonstrated that new neuron production in the dentate gyrus is influenced by circulating steroid hormone levels [9, 49]. This regulation can either decrease the number of immature granule neurons or increase the number of new neurons, depending on hormone levels. This highlights the crucial role of steroid hormones in modulating neurogenesis in the hippocampus. Glucocorticoids inhibit neurogenesis by limiting the growth of granule cell precursors. In rat pups, corticosterone administration during development and adulthood has been shown to decrease the production of new granule cells. This underscores the inhibitory effect of glucocorticoids on hippocampal neurogenesis [50, 51]. Removing adrenal steroids in adults through adrenalectomy has a positive impact on neurogenesis, increasing the proliferation of granule cell precursors and the generation of new neurons. This effect is also observed in aged animals, where the removal of adrenal steroids restores cell proliferation in the dentate gyrus to levels seen in young adults. This highlights the significant role of adrenal steroids in regulating neurogenesis throughout life. New neurons can be visualized through electron microscopy, retrograde tracing, immunohistochemistry for neuronal markers, and electro physiological techniques. Techniques such as ^{3}H-thymidine autoradiography, BrdU labeling, and GFP-tagged retrovirus can be utilized in conjunction with the methods. Fluctuations in circulating glucocorticoid levels during development and with experience suggest that the inhibition of cell proliferation by glucocorticoids may be a natural process important for hippocampal function [52]. An illustrative example is found in the dentate gyrus of Meadow voles. As seasonal breeders, female meadow voles exhibit higher glucocorticoid levels during the breeding season, resulting in suppressed cell proliferation. This highlights the potential impact of glucocorticoids on hippocampal neurogenesis in response to environmental and physiological changes (Figs. 6.1 and 6.2) [53].

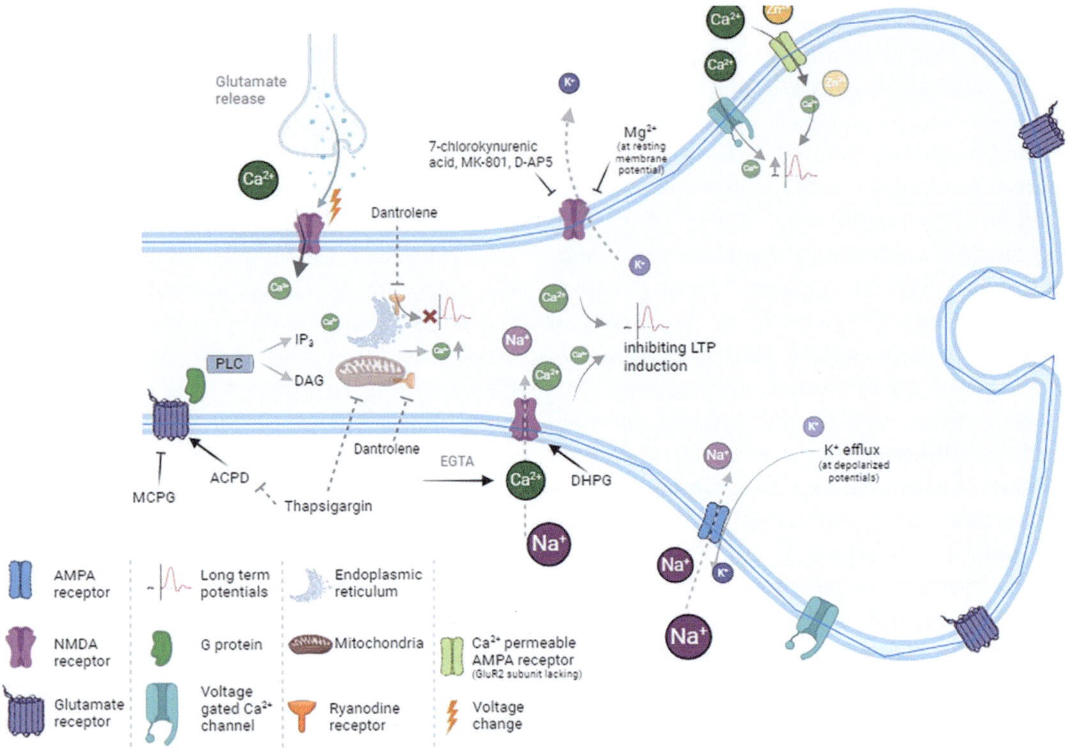

Fig. 6.2 Modulators of synaptic plasticity. The figure illustrates the mechanisms underlying the synaptic plasticity. The specific NMDA receptor antagonist D-2-amino-5-phosphonopentanoic acid (DAP5) blocks LTP induction, as does the introduction of the Ca^{2+} chelator EGTA into CA1 pyramidal cells, highlighting the essential role of NMDA receptor-mediated Ca^{2+} influx. At resting membrane potential, NMDA receptors are blocked by magnesium ions, requiring membrane depolarization to relieve the Mg^{2+} block and allow bound glutamate to open the receptors. This results in an inward current primarily carried by Na^+ and Ca^{2+} as Mg^{2+} is displaced with increasing depolarization. Additionally, at resting potentials, Na^+ influx dominates, but the channel also permits K^+ efflux at depolarized potentials. Another significant source of Ca^{2+} in neurons is the release from intracellular stores, and Ca^{2+} entry via voltage-gated Ca^{2+} channels enables NMDA receptor-independent LTP. The inhibitor of ryanodine receptors, dantrolene, also blocks LTP induction at these synapses. NMDA antagonists, such as the use-dependent noncompetitive antagonist MK-801 and the glycine site antagonist 7-chlorokynurenic acid, further demonstrate the necessity of NMDA receptor activity for LTP. AMPA receptors lacking GluR2 subunits are permeable to Ca^{2+} and Zn^{2+}, while the presence of the GluA2 subunit prevents the influx of these cations. Additionally, the mGluR agonist dihydroxyphenylglycine (DHPG) modulates synaptic plasticity, illustrating the complex interplay of multiple receptors and ionic movements in LTP induction

6.4 Synaptic Plasticity

Synaptic plasticity, a term defined by Polish psychologist Jerzy Konorski, describes enduring changes in synaptic strength that are influenced by neuronal activity. This concept, aligned with the views of neuroscientists since Santiago Ramon y Cajal, is believed to underpin information storage in the brain. Canadian psychologist, Donald Hebb, advanced a formal hypothesis embodying this idea. Hebb's postulate suggests that if an axon of cell A repeatedly or persistently stimulates cell B, a growth process or metabolic change occurs in one or both cells, increasing the efficiency of A in firing B. This principle is fundamental to our understanding of synaptic plasticity and learning [54]. Nearly 25 years after Hebb's hypothesis, Bliss and Lømo discovered a Hebb-like synapse in the mammalian brain [55]. Subsequently, they identified enduring changes in synaptic strength at hippocampal perforant path-granule cell synapses after brief tetanic stimulation, introducing the term long-term potentiation (LTP) [55]. This discovery emerged from their research on frequency potentiation in excitatory hippocampal pathways, where high-frequency stimulation resulted in sustained enhanced responses. In evaluating synaptic efficacy, comparisons are made before and after various learning experiences. It is anticipated that specific types of learning will be linked to a lasting enhancement in synaptic efficacy at relevant synapses. Subsequent studies by Bliss, Lømo, and Gardner-Medwin verified that tetanus-induced potentiation could last for days and demonstrated input specificity [56]. The advent of hippocampal slice preparation further clarified LTP mechanisms, revealing its presence across various hippocampal pathways and underscoring its importance in synaptic plasticity. Unlike in vivo studies, which find stable recording conditions in the dentate gyrus, in vitro studies typically concentrate on the CA1 region. This preference stems from the fact that inducing LTP in the dentate gyrus in vitro is challenging without reducing tonic inhibition using a $GABA_A$ receptor antagonist. A limitation of slice preparations is that perforant path fibers are disconnected from their parent cell bodies. It is crucial to remember that hippocampal slices are highly simplified models [57]. While they enable various physiological and pharmacological experiments that are otherwise impossible, a comprehensive understanding of hippocampal synaptic plasticity requires both in vitro and in vivo approaches. In 1983, two key discoveries provided important insights into the mechanisms underlying LTP induction. Graham Collingridge and his team showed that D-2-amino-5-phosphonopentanoic acid (DAP5), a specific antagonist of NMDA receptors, prevents the induction of LTP in the CA1 region without altering synaptic responses at low stimulation rates [58]. Additionally, Gary Lynch's team found that injecting the Ca^{2+} chelator EGTA into CA1 pyramidal cells also blocks LTP induction [59]. These findings highlighted the critical role of the postsynaptic cell and established calcium's essential role in the LTP induction process [60]. Excitatory synapses possess up to four types of postsynaptic glutamate receptors: AMPA, kainate, NMDA, and metabotropic glutamate (mGlu) receptors. Low-rate stimulation mainly activates AMPA receptors, while NMDA receptors contribute minimally to these excitatory postsynaptic currents (EPSCs) but are crucial for synaptic plasticity. In the 1980s, two key properties of NMDA receptors were discovered. First, NMDA channels are both voltage- and ligand-activated; at resting membrane potentials, magnesium (Mg^{2+}) blocks the channel, requiring substantial depolarization for activation [61]. This dual dependence makes the NMDA receptor a molecular coincidence detector, essential for cooperativity, input specificity, and associativity. Second, NMDA receptors are permeable to Ca^{2+}, suggesting that Ca^{2+} entry through these channels triggers the cellular processes leading to LTP by activating Ca^{2+}-dependent enzymes in dendritic spines [62]. Conversely, intracellular injection of BAPTA did not produce the same effect, indicating a Ca^{2+}-insensitive induction process. Learning activities that involve the hippocampal formation are expected to result in synaptic potentiation, while other learning types might not. However, it remains uncertain if these synaptic changes can be easily observed through global measures like field excitatory postsynaptic potentials (fEPSPs).

By the mid-1980s, research had advanced to provide a compelling molecular explanation for LTP's key properties: cooperativity, input specificity, and associativity. The hypothesis that LTP induction conditions align with NMDA receptor activation was further supported in 1986 when Holger Wigström and Bengt Gustafsson demonstrated that low-frequency stimuli could induce LTP in CA1 pyramidal cells if paired with strong depolarizing pulses [63]. This finding showed that LTP induction does not strictly require high-frequency trains, validating its physiological relevance. Concurrently, Richard Morris and colleagues showed that blocking NMDA receptors impaired spatial learning and memory in animals, reinforcing LTP's role in cognitive functions. These developments solidified LTP's acceptance as a crucial physiological pathway for learning and memory, marking a significant milestone in neuroscience research [60, 64]. Synaptic transmission at the single synapse level is inherently stochastic and can be affected by prior activity in both presynaptic and postsynaptic neurons, along with the influence of various neuromodulators on both sites. In the hippocampal formation, activity-dependent changes in synaptic efficacy, including facilitation, depression, augmentation, and potentiation, are generally short-lived. These forms of plasticity, despite their brief duration, play a crucial role in synaptic function and modulation.

The degree of facilitation is inversely proportional to the initial release probability, indicating a predominantly presynaptic mechanism. In addition to the presynaptic component, a postsynaptic contribution to facilitation has been observed in cortical and CA1 pyramidal neurons [65]. McNaughton introduced a test to distinguish presynaptic from postsynaptic mechanisms in LTP, based on the principle that higher release probability limits presynaptic facilitation, resulting in reduced paired-pulse facilitation [66]. A reduction in paired-pulse facilitation after LTP expression suggests an increase in release probability. Early studies showed that paired-pulse facilitation decreased during post-tetanic potentiation (PTP) but remained unchanged in LTP, indicating a postsynaptic mechanism. However, the paired-pulse facilitation test is not definitive; no change would occur if LTP involved recruiting previously silent boutons that then adopted similar release probabilities to the active population.

The temporary reduction in synaptic efficacy with the second of paired stimuli is primarily caused by vesicle depletion, but receptor desensitization also plays a significant role. Evidence for transmitter depletion is seen in observations where depression diminishes when transmitter release is reduced, such as by lowering extracellular calcium levels, thereby revealing the complete facilitation time course [67]. Additionally, transmitter release can reduce responses to subsequent stimuli through negative feedback via presynaptic glutamate receptors. Similarly, presynaptic $GABA_B$ autoreceptors on inhibitory terminals may contribute to reduced inhibition, thereby increasing the effectiveness of primed burst stimulation in inducing LTP [68]. Facilitation and depression closely interact at synapses. The arrival of an action potential at the presynaptic terminal initiates facilitation, regardless of neurotransmitter release. Depression, on the other hand, depends on neurotransmitter release. Therefore, if two action potentials occur within milliseconds, the second response is shaped by both facilitation and depression if transmitter release happened, but by facilitation alone if it did not. When the release probability is high, depression tends to dominate, as observed in vitro at medial, but not lateral, perforant path synapses. Grasping this interaction is essential for understanding synaptic dynamics. Repeated synaptic stimulation triggers a short-term increase in synaptic strength known as post-tetanic potentiation (PTP). At perforant path synapses, a mild high-frequency train significantly enhances test responses, which return to baseline levels within minutes. Longer stimulus trains overlay PTP on LTP, complicating the analysis. Controlled experiments reveal that PTP comprises two exponential processes: augmentation and potentiation, with time constants of 7 s and 2–3 min, respectively. An increase in Ca^{2+} concentration in terminal boutons during PTP indicates a presynaptic process, akin to facilitation. Understanding PTP is crucial for analyzing synaptic plasticity mechanisms [60]. In a typical LTP experiment, tetanic stimulation significantly amplifies the response. LTP is easily elicited and produces immediate, notable results. Synaptic potentiation is typically referred to as LTP if the enhanced response is sustained for over 30–60 min without significant decline. This timeframe only addresses the initial phase of LTP and does not account for later phases. In field potential studies, LTP is categorized into three phases based on drug susceptibility. When LTP induction occurs with gene transcription or protein synthesis inhibitors, it reverts to baseline within 3–5 h.

Glutamate is the primary excitatory neurotransmitter in both the hippocampus and broader mammalian central nervous system.

Glutamate receptors can be divided into two main groups such as ionotropic and metabotropic receptors. Ionotropic receptors include AMPA, kainate, and NMDA receptors. Metabotropic receptors have eight subtypes. The three primary classes of ionotropic glutamate receptors are identified by the agonists that selectively bind to them. At the majority of excitatory synapses within the hippocampus, EPSCs are mediated by NMDA and AMPA receptors, which exhibit markedly distinct pharmacological and biophysical characteristics. Glutamate depolarizes hippocampal neurons. In addition to this information, electrical stimulation triggers glutamate release. Glutamate absorbed by astrocytes is decarboxylated through the action of glutamine synthetase. Glutamate uptake is driven by the coupled transport of three Na+ ions down their electrochemical gradient. This process maintains a low intracellular concentration of glutamate within astrocytes, a condition essential for the energetically favorable continued uptake of glutamate. While glutamate is widely recognized as the primary mediator of fast excitatory transmission at "glutamatergic" synapses, there is ongoing evidence suggesting that aspartate may also be released in an activity-dependent manner from hippocampal slices. Aspartate immunoreactivity has been observed in specific presynaptic terminals that release glutamate, and this signal diminishes following periods of intense synaptic activity. Aspartate is one of the weak NMDA agonists. In the presence of protein synthesis inhibitors like emetine, anisomycin, or transcription inhibitors like actinomycin, the amplified response in the CA1 region of hippocampal slices is reduced to baseline within 5–6 h [69]. STP relies on NMDA receptors but does not depend on protein synthesis or protein kinase activity. Weak stimulus trains can induce E-LTP without leading to L-LTP. However, there are no known stimulus protocols that can trigger L-LTP without first inducing E-LTP. Associativity, which enhances co-active inputs, is central to the role LTP is likely to play in hippocampal memory functions. Cooperativity, associativity, and input specificity can be accounted for by the necessity for LTP induction, which requires presynaptic activity coupled with concurrent activation of the postsynaptic neuron. This concept closely aligns with Hebb's criterion for strengthening synaptic connections in neural circuits involved in associative conditioning [70]. This theory suggests that "neurons that fire together, wire together." Simultaneously active neuronal connections become strengthened, which ensures that these neurons are activated together during memory recall. Hebb's postulate asserts that for synaptic weight changes to happen, the postsynaptic cell must fire. While some forms of plasticity deviate from the Hebbian rule, postsynaptic firing may not be required for LTP, even in the traditional Hebbian synapse between Schaffer commissural fibers and CA1 pyramidal cells. Although it is experimentally feasible to bypass the postsynaptic action potential, in a physiological context, achieving sufficient depolarization might necessitate postsynaptic firing, as Hebb suggested. Even if sufficient depolarization is the primary factor for NMDA receptor activation, the strict Hebbian coincidence rule can be somewhat relaxed without compromising its core principle. The essential requirement for LTP induction remains the simultaneous firing of the presynaptic cell with adequate depolarization of the postsynaptic cell within a narrow time window. One of the most significant advancements in understanding the cellular mechanisms underlying LTP is the elucidation of the NMDA receptor's role in its induction. The importance of this glutamate receptor subtype was established in 1983 when the NMDA receptor antagonist D-aminophosphopentanoic acid (D-AP5) was shown to prevent LTP induction in the CA1 region of hippocampal slices [58]. AP5 does not alter the response to single stimuli, indicating that the NMDA receptor does not play a direct role in synaptic transmission at low frequencies. Furthermore, AP5 has no effect on established LTP, indicating that its influence is primarily on the induction phase rather than the expression phase. This finding, that NMDA receptor blockade suppresses LTP induction, has been corroborated using other competitive NMDA antagonists, the use-dependent noncompetitive antagonist MK-801 [71], and the glycine site antagonist 7-chlorokynurenic acid [72]. Tsien and colleagues generated mice with a deletion of NMDAR1, the essential subunit of the NMDA receptor, in the CA1 region but not in the dentate gyrus [73]. Slices from these mice exhibited no LTP in CA1 but displayed normal LTP in the dentate gyrus. In subsequent site-directed mutagenesis experiments, researchers created mice with a deletion of the intracellular carboxy terminus of the NMDA receptor [74]. Despite having unaltered Ca^{2+} currents through the mutant receptor, these mice could not exhibit LTP, indicating that the NMDA receptor has an additional role in LTP formation beyond its function as a Ca^{2+}-permeable ionophore. Following the discovery of NMDA receptors' critical role in LTP induction, an explanation of their mechanism emerged. It was already established that NMDA receptor-mediated responses are significantly inhibited by Mg^{2+} in spinal cord neurons. Unlike all other known ligand-gated ion channels, NMDA receptor-mediated responses exhibit strong voltage dependence. The crucial insight was the realization that this voltage dependence is due to the blockade by Mg^{2+} [61, 75]. The NMDA receptor channel, when activated by glutamate or other ligands, exhibits its minimal inward current at resting membrane potential. Depolarization leads to the expulsion of Mg^{2+} ions from the channel, enabling Na^+ and Ca^{2+} ions to flow inward. An NMDA receptor-mediated response necessitates both presynaptic glutamate release and adequate postsynaptic depolarization to unblock the NMDA receptor channel. Unlike most ionotropic AMPA receptor channels, activated NMDA receptor channels show high permeability to Ca^{2+}. Thus, it is reasonable to conclude that the influxion of Ca^{2+} through the NMDA receptor-associated channel serves as the trigger for LTP induction. While NMDA receptor activation is essential for inducing a major component of LTP, it is not always sufficient on its own. NMDA alone can induce short-term potentiation, but neither NMDA nor glutamate alone can reliably induce long-lasting potentiation in hippocampal slices. This finding challenges a strictly postsynaptic model of LTP induction and expression, instead supporting models where presynaptic mechanisms also play a role in LTP expression.

Evidence indicates that particular subtypes of NMDA receptors contribute to LTP. Experiments involving the knockout of the NR2A subunit have suggested a role for NR2A-containing NMDA receptors in LTP [76, 77]. Although animals lacking the NR2A subunit still showed LTP, its magnitude was about half that observed in wild-type animals. This reduction may be due to developmental effects of the knockout model. Later pharmacological experiments also suggested that various NR2 subtypes play a role in LTP. For instance, LTP showed greater sensitivity to NR2A/2B selective antagonists, while LTD was more sensitive to NR2C/D selective antagonists [78]. This varying sensitivity suggests a role for NR2A and/or NR2B receptors in LTP.

6.5 Role of Calcium Signaling in LTP

Calcium signaling plays a pivotal role in LTP, a process critical for synaptic plasticity and memory formation. The induction of LTP primarily hinges on the influx of Ca^{2+} ions into the postsynaptic neuron, chiefly through NMDA receptors. These receptors, which are highly permeable to calcium, require both presynaptic glutamate release and postsynaptic depolarization to open, ensuring that Ca^{2+} entry is activity dependent. Once inside the neuron, calcium ions activate key intracellular signaling pathways, including calcium/calmodulin-dependent protein kinase II (CaMKII), protein kinase A (PKA), and the MAPK pathway. These pathways facilitate the phosphorylation of substrates, gene transcription, and protein synthesis necessary for synaptic strengthening. Furthermore, calcium-induced calcium release (CICR) from intracellular stores amplifies the Ca^{2+} signal, promoting robust signaling cascade activation. Calcium signaling also plays a role in synaptic tagging and capture, which ensures that potentiated synapses maintain long-term changes by capturing newly synthesized plasticity-related proteins. Additionally, under strong synaptic activation, voltage-gated calcium channels (VGCCs) can contribute to calcium influx, activating similar pathways as NMDA receptor-mediated entry. Thus, calcium signaling is essential for the induction, maintenance, and specificity of LTP, underpinning the cellular mechanisms of learning and memory. A primary source of Ca^{2+} in neurons is the release from intracellular reserves. Evidence suggests that Ca^{2+} stores play a crucial role in synaptic plasticity, as thapsigargin, which blocks their refilling, can prevent the induction of LTP at CA1 synapses without affecting its expression [79]. Intracellular Ca^{2+} release can be initiated by the generation of inositol trisphosphate (IP3) or through a Ca^{2+}-induced Ca^{2+} release mechanism involving IP3 and ryanodine receptors. The importance of Ca^{2+}-induced Ca^{2+} release is supported by evidence showing that dantrolene, a ryanodine receptor inhibitor, blocks the induction of LTP at these synapses [80].

While it is widely accepted that Ca^{2+} entry through the NMDA receptor channel triggers NMDAR-dependent LTP, this is not the sole pathway for increasing intracellular Ca^{2+} levels. Specifically, depolarization during summated EPSPs can result in Ca^{2+} influx via VGCCs. Since NMDA receptors contribute significantly to the depolarization during high-frequency synaptic responses, this source of Ca^{2+} is primarily dependent on NMDAR activity [81]. The logical conclusion is that LTP can be induced if sufficient Ca^{2+} enters through VGCCs, as this Ca^{2+} can reach the sites typically accessed by Ca^{2+}-permeating NMDARs [81]. Consequently, VGCCs are generally not required for the induction of NMDAR-dependent LTP. However, stronger induction protocols may activate VGCCs, causing depolarization to spread to other regions of the cell. Initially, activation of NMDA receptors requires occupancy of a second agonist-binding site in addition to glutamate. The glycine-binding site is often occupied, based on estimates of the tonic extracellular glycine concentration in the brain. D-serine can also bind to the glycine site on the NMDA receptor, suggesting a potential physiological role in regulating NMDA receptor activity. This might cause a partial loss of input specificity and lead to Ca^{2+} entry at the soma, where it could function as a nuclear signal. Secondly, NMDA receptors exhibit significant permeability to both Ca^{2+} ions and monovalent cations. The influx of Ca^{2+} through NMDA receptors is pivotal in various types of long-term synaptic plasticity. Ca^{2+} ions serve as widespread secondary messengers, and the activation of NMDA receptors has been demonstrated to induce additional release of Ca^{2+} from intracellular stores. Thirdly, the ionophore of NMDA receptors is blocked by Mg^{2+} ions in a voltage-dependent manner. Consequently, at resting membrane potentials (approximately −50 mV), NMDA receptors cannot generate an excitatory postsynaptic current (EPSC) even in the presence of glutamate and glycine or D-serine. They facilitate ion flux only upon membrane depolarization. The Ca^{2+} permeability and Mg^{2+} blockade of NMDA receptors elucidate their function as synaptic coincidence detectors: Ca^{2+} influx occurs exclusively upon the conjunction of presynaptic glutamate release and postsynaptic depolarization, which occurs when both pre- and postsynaptic activities coincide. AMPA receptors, found at most excitatory synapses in the hippocampus, regulate a channel permeable to cations. The channel is primarily permeable to Na^+ ions at resting membrane potential, but it also allows the passage of other small monovalent cations. This enables K^+ efflux at depolarized potentials. The predominant AMPA receptor configuration found in pyramidal neurons of the adult hippocampus consists of tetramers comprising GluR1–2 or GluR2–3 subunits. Depolarization through AMPA receptors alone might allow Ca^{2+} entry and LTP that appears independent of NMDARs but engages the same downstream pathways as NMDAR-dependent LTP. AMPA receptors facilitate rapid synaptic transmission at excitatory synapses in the central nervous system and play essential roles in neuronal development, synaptic plasticity, and structural remodeling. The Ca^{2+} permeability of the AMPAR complex is influenced by the presence of the GluA2 subunit and the R residue in GluA2-containing AMPARs. AMPA receptors lacking the GluR2 subunits are permeable to Ca^{2+} and Zn^{2+} ions. The presence of GluA2 subunit in

AMPARs prevents the influx of cations such as Ca^{2+} and Zn^{2+}. CP-AMPARs can enhance biological signal amplification. The positively charged R residue in GluA2-containing AMPARs has multiple effects, including blocking Ca^{2+} influx, slowing channel kinetics, reducing channel conductance, increasing synaptic event amplitude and paired-pulse facilitation, enhancing neuronal excitability, and regulating AMPAR trafficking and anchoring. However, Noh et al. likely conclude that blocking calcium-permeable AMPA receptors can provide neuroprotection to hippocampal neurons during events of global ischemia. By preventing excessive calcium influx through these receptors, neuronal death and damage in the hippocampus can be mitigated, suggesting a potential therapeutic target for conditions involving global cerebral ischemia. This approach might reduce the extent of brain injury and improve outcomes after ischemic events such as stroke [82]. Moreover, Guo and Ma conclude that calcium-permeable AMPA receptors play a significant role in excitotoxicity, which is a pathological process that contributes to neuronal damage in various neurological disorders. The article likely emphasizes that excessive activation of these receptors leads to an overload of calcium ions in neurons, triggering a cascade of deleterious cellular processes that result in cell death. The study might suggest that targeting calcium-permeable AMPA receptors could be a promising strategy for developing therapeutic interventions to mitigate neuronal damage in diseases such as stroke, epilepsy, and neurodegenerative disorders [83]. Thus, calcium-permeable AMPA receptors (CP-AMPARs) are vital for synaptic plasticity, the brain's ability to modify synaptic strength in response to activity, which underpins learning and memory. By facilitating and allowing calcium ions to enter the postsynaptic neuron, CP-AMPARs trigger intracellular signaling pathways essential for LTP and LTD, processes that respectively strengthen and weaken synapses. This calcium influx is crucial for activating protein kinases and phosphatases that modulate synaptic efficacy. Additionally, CP-AMPARs play a significant role in homeostatic plasticity, helping neurons maintain stable activity levels by adjusting synaptic receptor composition. They also influence the morphology of dendritic spines, affecting their size and shape, which is integral to the structural basis of synaptic changes. Thus, CP-AMPARs are central to the dynamic regulation of synaptic connections, facilitating the adaptability and functionality of neural networks. Park et al. mentioned that CP-AMPARs are crucial for the induction and maintenance of LTP in principal neurons of the rodent hippocampus. It likely emphasizes that the Ca^{2+} influx through CP-AMPARs triggers intracellular signaling cascades necessary for synaptic strengthening, highlighting their essential role in LTP and, consequently, in learning and memory processes. The study may also suggest that modulating CP-AMPAR activity could influence hippocampal function and plasticity [84].

Metabotropic glutamate receptors (mGluRs) influence various intracellular signaling pathways potentially involved in LTP through their coupling to G proteins. Direct evidence supporting mGluR activation during LTP induction emerged with the discovery of the first selective mGluR antagonist, alpha-methyl-4-carboxyphenylglycine (MCPG). MCPG selectively inhibited LTP induction without affecting basal synaptic transmission or the expression of LTP [85]. The slow-onset potentiation induced by aminocyclopentane-1,3-dicarboxylic acid (ACPD) was blocked by thapsigargin [86]. This finding suggests a potential mechanism by which ACPD might facilitate NMDAR-dependent LTP, with intracellular Ca^{2+} stores as the convergence point. Typically, Ca^{2+} release from these stores is initiated by Ca^{2+} entry through NMDA receptors. However, under certain conditions, Ca^{2+} release may be triggered by activating mGluRs without NMDAR activation [87]. Group I mGluRs are linked to phospholipase C, which generates IP_3 that subsequently acts on IP_3 receptors to release Ca^{2+} from intracellular stores. Group I mGluR agonist dihydroxyphenylglycine (DHPG) exposure consistently triggers strong LTD in adult slice preparations, particularly in the CA1 region. One of the most perplexing discoveries was that LY341495, a compound capable of inhibiting all known mGluRs (mGluR1–8) at high doses, did not affect LTP [88]. The observed differences between MCPG and LY341495 within the same pathway suggest that MCPG may have a target other than mGlu1–8, or that the broad inhibition of mGluRs produces effects that differ from the selective inhibition of specific subunits [89]. DHPG-induced LTD has been extensively utilized as a model to investigate the signaling mechanisms underlying mGluR-dependent LTD. mGluRs are linked to G proteins, and as anticipated, DHPG-induced LTD necessitates G protein activation. Given that group I mGluRs are positively linked to PLC, leading to the generation of diacylglycerol and IP^3, one might expect that inhibitors of PKC or blockers of Ca^{2+} release from stores would inhibit mGluR-dependent LTD. However, numerous potent PKC inhibitors and Ca^{2+} store blockers, whether applied individually or together, did not affect DHPG-induced LTD. Moreover, DHPG-induced LTD is unaffected by PKA inhibitors, whereas CaMKII inhibitors slightly enhance the effect. Although there is a consensus on the involvement of MAPK cascades, there is debate over whether the mechanism includes Ras-activated ERKs or Rap-activated p38 MAPK [90]. Evidence suggests that the PI3K-Akt-mTOR signaling pathway plays a role in DHPG-induced LTD [91]. Lao-Peregrin et al. discovered that hippocampal synaptic plasticity facilitated by BDNF relies on effective interaction with mGluR5. Detailed investigations uncovered a universal mechanism of cross talk between receptor tyrosine kinases (RTKs) and G protein-coupled receptors (GPCRs), driven by synergistic G-protein activity [92].

The primary inhibitory neurotransmitter GABA is synthesized through the decarboxylation of glutamate. It has long been established that inhibiting GABA can significantly enhance the induction of NMDAR-dependent LTP. The biphasic inhibitory postsynaptic potential (IPSP) following EPSPs in hippocampal slices consists of fast $GABA_A$ and slower $GABA_B$ components. The hyperpolarizing effect of IPSPs strengthens the Mg^{2+} block of NMDARs, reducing their contribution to synaptic responses. Blocking GABA inhibition reveals NMDAR-mediated responses during low-frequency stimulation. Priming, a technique that temporarily reduces GABA inhibition, is crucial for LTP induction. Theta activity patterns and $GABA^B$ antagonists can facilitate LTP induction by manipulating GABA inhibition [93]. This precise system suppresses GABA inhibition during theta activity, indicating that hippocampal synapses are especially susceptible to potentiation during behaviors like exploration and locomotion that generate theta rhythms [94]. Many neurotransmitters and neuromodulators can theoretically influence the LTP induction process by affecting GABA inhibition. For instance, it has been demonstrated that endocannabinoids released from pyramidal neurons can suppress GABA inhibition, thereby facilitating LTP. One set of experiments revealed that LTP is connected to endocannabinoid-mediated LTD of GABA inhibition in a specific region near the potentiated synapses. This LTD of synaptic inhibition subsequently facilitates the induction of LTP in the surrounding area. They highlight the necessity for an optimal level of NMDAR activation to induce LTP; persistently low levels of activation can counteract LTP induction through metaplasticity [95]. mGluR receptor activation seems to enhance the induction of subsequent LTP and can alter the future requirement for mGluR activation in LTP induction. In freely moving animals, the induction of LTP in the dentate gyrus results in elevated levels of NR1 and NR2B, which remain increased for at least 48 hours. In expression systems, phosphorylation of NR2B enhances NMDAR-mediated currents by extending the channel's open time. While this has minimal impact on predominantly AMPA receptor-mediated synaptic responses, it significantly influences subsequent efforts to induce plasticity. However, both LTD and LTP are initiated by rises in intracellular Ca^{2+}, LTD requires a lower Ca^{2+} concentration than LTP. The increased Ca^{2+} influx through NMDARs supports both types of plasticity. The activation of NMDA receptors involves several crucial steps: removing the Mg^{2+} block, allowing Ca^{2+} influx through the receptor, and activating Ca^{2+}-dependent protein kinases, all essential for the induction process. Mechanisms underlying LTP expression include enhanced neurotransmitter release, AMPA receptor phosphorylation, and their incorporation into the postsynaptic density [96]. An early experiment investigating postsynaptic sensitivity changes found that LTP was linked to a postponed increase in response to AMPAR agonists. This indicates that while E-LTP coincides with changes in postsynaptic action, although STP occurs without these alterations. The easiest explanation for this finding is that STP is presynaptic [97]. Additionally, a discovered issue is that the magnitude of STP is affected by the afferent stimulation frequency, with higher frequencies leading to greater STP. Specifically, a brief high-frequency tetanus after low-frequency test stimulation, commonly used in field potential recordings, tends to generate large-magnitude STP with a slow decay. There is substantial confirmation indicating that the kinetic and amplitude of the synaptically triggered Ca^{2+} transients are crucial in determining the direction of synaptic change. A decreased Ca^{2+} transfer through the NMDAR protein is linked to LTD, while an elevated transfer is associated with LTP. LTD is considered NMDAR-dependent if it can be inhibited by D-AP5 or other NMDA receptor antagonists. The induction of NMDAR-dependent LTD is dependent on an increase in postsynaptic Ca^{2+}, with the amplitude and duration of the Ca^{2+} transient determining the extent of synaptic plasticity. [98]. The generation of LTD can be inhibited by introducing a Ca^{2+} chelator into CA1 pyramidal cells. Reducing external Ca^{2+} levels transforms a procedure that typically induces LTP into one that results in LTD. Ca^{2+} entry into the postsynaptic cell can be influenced in several ways, such as by applying low doses of AP5 to limit Ca^{2+} influx during tetanus, which converts LTP to LTD. In brief, low levels of Ca^{2+} elevation activate protein phosphatases, leading to LTD, while higher concentrations activate protein kinases, resulting in LTP. The phosphorylation of AMPA receptors is known to increase AMPA receptor-mediated currents, contributing to the enhanced synaptic efficacy observed in LTP [99]. Phosphorylated CaMKII is a substrate for the protein phosphatase PP1, but PP1 is not Ca^{2+}-sensitive. A phosphatase cascade is initiated by the activation of the Ca^{2+}-sensitive protein phosphatase calcineurin. Calcineurin, once activated, dephosphorylates a phosphoprotein known as inhibitor 1 (I1), which, when activated, inhibits PP1. Subsequently calcineurin activates PP1 through disinhibition, enabling the dephosphorylation of CaMKII. An essential aspect of this mechanism is that calcineurin has a higher sensitivity than CaMKII to Ca^{2+}, supporting the hypothesis that slight increases in $[Ca^{2+}]_i$ result in LTD, while significant increases lead to LTP [100]. Calcineurin, a Ca^{2+}/calmodulin-dependent serine/threonine phosphatase, can be inhibited by several agents. LTD was inhibited by the injection of specific PP1 and calcineurin inhibitors into CA1 neurons. Some key inhibitors of calcineurin can be listed as follows: Cyclosporine A (CsA) (immunosuppressive drug forms a complex with cyclophilin), Tacrolimus (FK506) (immunosuppressive agent, tacrolimus binds to FK506-binding protein (FKBP)), Ascomycin (binds to FKBP and inhibits calcineurin activity), Okadaic Acid (while primarily an inhibitor of protein phosphatases PP1 and PP2A, it can indirectly affect calcineurin activity in some certain contexts), Deltamethrin (a synthetic pyrethroid insecticide), and Pyridone 6 (P6). These inhibitors are often used in research to study the role of calcineurin in various cellular processes, as well as in clinical settings to suppress immune responses, especially in organ transplantation. Moreover, the involvement of NR2B-containing NMDARs was indicated by experiments using a knockout of this subunit. These animals exhibited an absence of LTD, experienced significant developmental issues, and had a shortened lifespan. In some reports, NR2B inhibitors have been found to enhance LTD [101]. Studies suggest that NR2D receptors contribute to the slower phase of the synaptic response in the hippocampus [102]. The NR2D component of the NMDAR-mediated EPSC, with its slow kinetics and low Mg^{2+} sensitivity, effectively summates at the low stimulation frequencies typically used to induce LTD [103].

The mechanism of LTD is influenced by whether the AMPA receptors at the synapse are calcium-permeable or calcium-impermeable. For both synapse types, induction occurs postsynaptically; however, evidence suggests presynaptic expression in the case of calcium-permeable receptors and postsynaptic expression for calcium-impermeable receptors [62]. Interneurons and pyramidal cells have distinct AMPA and NMDA receptor subunit compositions, with NR2B being the primary NMDA subunit in interneurons, leading to prolonged NMDAR-mediated currents. AMPA receptors on inhibitory interneurons typically lack the GluR2 subunit, resulting in increased calcium permeability and higher AMPAR-mediated calcium fluxes. The unique composition of glutamate receptor subtypes at interneuron synaptic membranes within the hippocampus results in varying types of plasticity. High-frequency stimulation can lead to no change, LTP, or LTD, depending on the specific protocol or the type of interneuron. This diversity underscores the complexity and specificity of synaptic plasticity mechanisms within the hippocampus. The induction of LTP at excitatory synapses is dependent on mGluR1 and necessitates postsynaptic Ca^{2+} entry, with the expression of LTP appearing to be presynaptic. Tetanus-triggered LTP was detected in stratum lucidum interneurons that express Ca^{2+}-impermeable GluR2-containing AMPA receptors [104]. It seems that a significant contributor to age-related changes in LTP is the modification of conditions influencing NMDA receptor function and its associated Ca^{2+} signaling. The changes in NMDA receptor-dependent plasticity appear to explain many of the modifications in the dynamics of neuronal assemblies associated with aging [105].

The first kinase identified as playing a role in LTP was protein kinase C (PKC) [106]. Several studies have reported that potent and more specific PKC inhibitors have no effect on LTP. Conversely, a role for PKC has been suggested based on the intracellularly application of peptide inhibitors. The most plausible description is that PKC is involved in certain, but not all, aspects of E-LTP. The complexity of PKC's role in LTP is due to the existence of multiple isoforms, including the Ca^{2+} and diacylglycerol-activated conventional PKCs, and the Ca^{2+}-independent, diacylglycerol independent atypical PKCs. The induction of PKC-dependent potential may be triggered by Ca^{2+} entry through NMDAR. Alternatively, glutamate binding to group I mGluRs can lead to the production of diacylglycerol and the release of Ca^{2+} from intracellular stores, which are the two activators of cPKCs. Given that information, potent PKC inhibitors can effectively prevent mGluR-mediated metaplasticity [107]. CaMKII is a significant constituent of the postsynaptic density. When calcium-loaded calmodulin attaches to the calmodulin-binding region in the molecule's C-terminal region, it induces a structural transformation/conformational change. This alteration leads to a conformational change that separates the autoinhibitory domain from the ATP binding site, activating the enzyme. In the adult hippocampus, two isoforms of calcium/calmodulin-dependent protein kinase II are present: alpha CaMKII and beta CaMKII. The predominant isoform in the CA1 region is alpha CaMKII. Substantial evidence indicates that alpha CaMKII is essential for LTP in the CA1 region of the adult hippocampus and can induce LTP when activated. This underscores the critical role of alpha CaMKII in synaptic plasticity and memory formation [108]. Specific peptide inhibitors of alpha CaMKII, which are derived from its autoinhibitory domain, suppress both calcium-dependent and calcium-independent activities, thereby preventing the induction of LTP.

1. This finding underscores the critical role of alpha CaMKII in the mechanism underlying LTP induction Is it possible to frame these three lines as ''Highlights''?
2. The membrane-permeable CaMKII inhibitors KN-62 and KN-93 reversibly inhibit the induction of LTP.
3. LTP is impaired in alpha CaMKII knockout mice (alpha-CaMKII−/−), although a background effect may also contribute to this deficit.

Nitric oxide (NO) is a more persistent candidate for a retrograde messenger than arachidonic acid (AA), although the evidence remains inconclusive. NO's tiny, membrane-permeable nature and short semi-life make it suitable for this role. In neurons, NO is produced from L-arginine by nitric oxide synthase (NOS), with the isoforms nNOS and eNOS present in the dendrites of hippocampal neurons [109]. NO targets soluble guanylyl cyclase, leading to the activation of cyclic guanosine monophosphate (cGMP). One of cGMP's substrates is protein kinase G (PKG). The effector molecules associated with this second messenger system in hippocampal synaptic terminals are still largely unknown. Inhibiting NO synthase prevented LTP induction. In hippocampal slice experiments, NO synthesis inhibitors blocked LTP at room temperature. Research suggests that NO acts specifically on stimulated terminals, selectively affecting neighboring active terminals and the terminal responsible for its release from the postsynaptic cell. NMDA receptor antagonists blocked these changes, indicating that NO is produced postsynaptically and functions as a diffusible retrograde messenger [110]. Additionally, evidence shows that NO influences vesicle recycling in LTD, suggesting that its interaction with presynaptic factors determines the direction of change.

Cell adhesion molecules (CAMs) are membrane-spanning proteins. Their ligand-binding domains are located in the extracellular part. These molecules play a vital role in cell–cell recognition during neural development and are potential candidates for anterograde and retrograde transsynaptic signaling [111]. Neuronal CAMs (NCAMs) are classified into four types: integrins, cadherins, immunoglobulin superfamily CAMs, and neuroligins with their ligands, neurexins. They can bind to identical or different family members via their extracellular domains. Many CAMs' C-terminal regions associate with the actin cytoskeleton, directly or through accessory proteins, facilitating potential changes in spine or bouton morphology. Mice treated with NCAM antibodies or having an NCAM deletion show reduced LTP in both the CA1 area and the mossy fiber projection to the CA3 area [112]. The enzymatic removal of PSA from NCAM hinders LTP induction in the CA1 region. However, LTP remained unaffected in the CA1 region of Thy-1 knockout mice. However, it was impaired in the dentate gyrus in vivo, with no impact on performance in the Morris water maze test [113].

Cadherins are a diverse family of cell adhesion molecules with cytoplasmic signaling domains connected to the actin cytoskeleton through beta-catenin, a protein that binds to cadherin. Alpha-catenin beta-catenin, and actin. Cadherin dimers form strong homophilic interactions across the synapse. Both E- and N-cadherins are widely expressed in the hippocampus, but N-cadherins are specifically localized to excitatory synapses. Peptide blockers of N-cadherin reduce tetanus-induced LTP in the CA1 region. Late-onset LTP induced by a membrane-permeable cAMP analog in hippocampal slices is linked to an increase in synaptic puncta containing N-cadherins and the presynaptic marker synaptophysin [114]. N-cadherin function is modulated by synaptic activity, with strong activation triggering changes in N-cadherin conformation and the NMDAR-dependent translocation of catenin from dendrite to spine [115]. The translocation of β-catenin into the spine is facilitated by tyrosine dephosphorylation, providing a potential regulatory mechanism for synaptic activity via a calcium-dependent tyrosine phosphatase [116].

Integrins are glycoproteins consisting of two different subunits, each with a single membrane-spanning domain that connects the extracellular ligand-binding region to the cytoplasmic tail. This tail is linked to actin via the accessory proteins vinculin, talin, and beta-actinin [117]. Broad-spectrum peptide inhibitors of integrins reduce the magnitude and duration of LTP and can reverse LTP if administered shortly after induction [118]. Neuroligins and their ligand proteins, neurexins, are essential for cell adhesion in the brain and exist in various isoforms produced by alternative splicing. Neurexins, located presynaptically, form heterophilic connections with postsynaptic neuroligins. Intracellular C-termini of neuroligins bind to the PDZ domain of PSD-95, while the C-termini of neurexins interact with the PDZ-containing protein CASK [119]. Disrupting intercellular interactions minimally affects baseline transmission but significantly impairs the magnitude and duration of LTP. A key area of uncertainty is how CAMs detect and respond to synaptic activity. CAMs likely play a crucial role in mediating structural changes, ensuring synchronized modifications between pre- and postsynaptic regions. Calcium influx through NMDA receptors triggers NMDAR-mediated LTP. Thus, it is unsurprising that calcium, especially when bound to calmodulin, plays a crucial role in signaling to the nucleus. There is ongoing debate about the mechanisms involved. One theory suggests that Ca^{2+} is released from the endoplasmic reticulum in a wave-like manner, traveling from the dendrite to the soma and activating nuclear calmodulin-dependent kinases. Another theory posits that calcium/calmodulin translocates from the synapse to the nucleus. A third possibility is that Ca^{2+} enters the nucleus through voltage-gated Ca^{2+} channels in the plasma membrane. A key target for nuclear calcium/calmodulin is CaMKIV, a member of the calcium/calmodulin-dependent kinase family predominantly found in the nucleus. Phosphorylation at ser133 activates CaMKIV, enabling it to bind to the accessory CRE-binding protein (CBP) and initiate transcription. Additionally, other Ca^{2+}-activated signaling pathways, such as the Ras/MAP kinase and PKA pathways, are also engaged [120]. The MAP kinase signaling cascade can be activated by various stimuli, including extracellular ligands binding to receptor tyrosine kinases and G protein-coupled receptors. Additionally, synaptic activity resulting in calcium entry through ionotropic glutamate receptors or voltage-gated Ca^{2+} channels can also trigger this pathway. Synaptic activity triggers the phosphorylation of ERK1/2 in both dendrites and the soma. However, the mechanism by which the activated ERK signal is transported to the nucleus remains unclear [87]. Unlike growth factors or neuromodulators, neurotransmitters elevate intracellular calcium levels by either directly activating calcium-permeable receptor channels, such as NMDA receptors, or indirectly depolarizing the membrane to open L-type voltage-sensitive calcium channels [121]. Elevated Ca^{2+} levels can result in the phosphorylation of both calcium response factor (CRF) and CREB at specific serine residues (ser141 and ser143). This phosphorylation reduces the ability of ser133 phospho-CREB to bind to CBP [122].

PKMζ, a PKC family member, is a potential key player in the long-term persistence of LTP. Activated 30 min after NMDAR-dependent LTP induction, PKMζ seems to be both necessary and sufficient, as it can mimic and occlude LTP. PKMζ inhibition, using either chelerythrine or the dominant-negative PKMζ-K281W, prevents LTP [123]. The mechanism underlying the maintenance of PKMζ-dependent L-LTP seems to involve an increase in AMPAR numbers. PKMζ-induced potentiation of synaptic transmission is characterized by an increase in the amplitude, but not the frequency, of AMPAR-mediated mEPSCs, without affecting their kinetics or mean single-channel conductance. Tetanic stimulation resulted in the activation of presynaptic kainate receptors containing GluR5 subunits (LY382884 target), which subsequently led to a persistent increase in transmitter release via a Ca^{2+}-dependent pathway.

6.5.1 Modulating Factors of Long-Term Potentiation

Previous studies suggest that septal cholinergic input plays a facilitating role in LTP induction in the hippocampus. Some evidences support this; first, the inhibition of LTP by wide-range muscarinic antagonists, and second, the lowering of the LTP threshold by muscarinic agonists or septal stimulation. Pretreatment with the muscarinic antagonist scopolamine lowered LTP to near-baseline levels, as did the injection of the neurotoxin immunoglobulin G (IgG)-saporin, targeting cholinergic neurons, into the medial septum [124]. These findings suggest that LTP can be modulated by endogenous, behaviorally driven cholinergic input from the septum. This effect is likely mediated by high-affinity postsynaptic M2 receptors. NMDA receptor conductance is directly potentiated by ACh at lower concentrations than those required to inactivate K^+ channels [62]. Muscarinic receptor activation in the hippocampus results in a G protein-mediated blockade of several K^+ channels [125]. These effects could play a role in the positive modulation of LTP induction by ACh via muscarinic receptors.

Monoamines like serotonin, dopamine, norepinephrine, and brain-derived neurotrophic factors also contribute to the LTP induction process. Since β-adrenergic receptors are positively associated with adenylate cyclase, norepinephrine likely operates similarly to dopamine. This process increases cAMP levels, leading to the activation of PKA [126]. Selective serotonin depletion reduced the magnitude of LTP in the dentate gyrus in vivo study. Evidence supporting the idea that 5-HT primarily suppresses LTP includes studies demonstrating that the 5-HT uptake inhibitor fluvoxamine reduces LTP by activating 5-HT1A receptors. But this molecule needs to be largely explored. Compelling evidence indicates that dopamine significantly modulates the induction of protein synthesis-dependent L-LTP in area CA1. Mice with a knockout of the D1 dopamine receptor exhibit a lack of L-LTP but maintain normal E-LTP [127]. Prolongation of the cAMP action by phosphodiesterase inhibitors and dopaminergic antagonists enhances the activity of protein kinase A, the primary downstream effector of dopamine can convert E-LTP into L-LTP. Dopamine influences cellular processes by activating PKA, which plays a crucial role in dendrite-to-nucleus signaling that initiates gene transcription, as well as in local protein synthesis within dendrites. Dopamine, through cAMP and PKA, is directly involved in the synthesis of proteins related to plasticity. Research has indicated that synaptic activity prompts the release of the neurotrophic peptide brain-derived neurotrophic factor (BDNF) at hippocampal synapses [128]. BDNF is likely released from both pre- and postsynaptic sites at hippocampal synapses and acts on populations of pre- and/or postsynaptic tropomyosin receptor kinase B (TrkB) receptors, whose numbers are regulated by synaptic activity. BDNF seems to be crucial for the induction of LTP, as BDNF-null mice exhibit impaired LTP. BDNF is also capable of inducing LTP. BDNF-induced late LTP occludes with late tetanus-induced LTP; however, unlike rapid-onset LTP observed in the dentate gyrus in vitro, it does not rely on NMDARs. Despite this, BDNF causes CREB phosphorylation by activating ERK1/2 and upregulating immediate early genes in granule cells. BDNF can partially activate the normal late LTP inducing machinery in the postsynaptic cell, downstream of the NMDA receptor. Interleukins, including IL-18, IL-8, IL-6, and IL-2, also inhibit LTP induction. Interferons may also inhibit LTP. TNF-alpha inhibits early LTP, but not late LTP, through the activation of a p38MAPK cascade and may exert its effects by influencing AMPA receptor trafficking [90, 129].

Stress is typically linked to elevated plasma concentrations of the corticosterone. The hippocampus is rich in two types of adrenal hormone (corticosteroid) receptors: the high-affinity mineralocorticoid receptor (type 1) and the low-affinity glucocorticoid receptor (type II). Type 1 receptor activation enhances LTP, while glucocorticoid receptors, activated by elevated corticosterone levels, predominantly mediate the stress-induced deficits in LTP and spatial learning. Therefore, the relationship between plasma corticosterone levels and their impact on LTP and learning follows a significant pattern. At low corticosterone concentrations, such as after adrenalectomy, mineralocorticoid receptor occupancy is minimal. Conversely, during severe stress, high corticosterone levels fully occupy glucocorticoid receptors. Both scenarios are linked to reduced LTP, increased LTD, and impaired learning. LTP and learning peak at intermediate circulating corticosterone levels [130]. However, it is a generalization to assume that corticosterone is the sole mediator of stress effects on LTP and behavior. For instance, stress can still inhibit LTP in rats that have undergone adrenalectomy to remove corticosterone, and lesions in the amygdala can prevent stress-induced LTP impairment despite elevated corticosterone levels in experimental animals. Evidence suggests that stress activates certain signaling pathways that are also implicated in synaptic plasticity. [131]. Also, stress leads to a significant increase in the phosphorylation of the MAP kinase signaling molecules ERK1/2. Blocking ERK1/2 phosphorylation can counteract the detrimental effects of stress on learning and LTP [132]. The LTP impairment caused by stress is likely due to a metaplastic effect mediated by NMDA receptor activation, as NMDA receptor antagonists can block this reduction [62].

6.5.2 Periodic Factors Affecting LTP Induction

Estrogen levels significantly influence synaptic plasticity, particularly in the hippocampus. Elevated estrogen levels, such as those observed during proestrus in female rats, increase the number of dendritic spines on CA1 pyramidal cells, enhancing synaptic connectivity and plasticity. Estrogen can modulate the induction of LTP, with higher levels generally promoting LTP, while the absence or low levels of estrogen, as seen in ovariectomized animals, may not always lead to a predicted reduction in LTP. This suggests that estrogen plays a crucial role in regulating synaptic plasticity and associated cognitive functions. There are several cyclic modulators, such as hormones, that periodically affect LTP induction. For instance, during proestrus in female rats, estrogen levels peak, leading to a significant increase in the number of spines on CA1 pyramidal cells compared to during estrus [133]. Studies in ovariectomized animals have shown that the complete absence of estrogen does not consistently result in the expected reduction in LTP [134, 135].

Sleep and the circadian cycle have profound effects on synaptic plasticity, influencing learning, memory consolidation, and overall brain function. Sleep promotes memory consolidation and synaptic homeostasis, while the circadian cycle regulates gene expression, hormonal levels, and neuronal activity patterns, all of which are crucial for maintaining synaptic plasticity and cognitive function. Sleep, especially REM and slow-wave sleep, is crucial for consolidating memories. During sleep, neural activity patterns from wakefulness are replayed, strengthening synaptic connections. The synaptic homeostasis hypothesis suggests that sleep helps balance synaptic strength by downscaling synapses that were potentiated during wakefulness, thus preventing saturation and maintaining cellular homeostasis. Both LTP and LTD are influenced by sleep. Studies have shown that sleep promotes LTP in hippocampal circuits, while sleep deprivation impairs LTP induction and enhances LTD, negatively impacting learning and memory [136]. The circadian cycle regulates the expression of genes involved in synaptic plasticity, including those encoding for receptors, signaling molecules, and synaptic proteins. The circadian cycle regulates the expression of genes involved in synaptic plasticity, including those encoding for receptors, signaling molecules, and synaptic proteins [137]. Neuronal activity follows a circadian rhythm, with periods of heightened plasticity corresponding to wakefulness when learning and memory processes are most active.

Hibernation profoundly impacts synaptic plasticity, reflecting the brain's adaptive strategies to extreme physiological conditions. During hibernation, the brain undergoes significant synaptic pruning and reduction in neurotransmitter release, markedly lowering synaptic activity. This downscaling conserves energy and minimizes metabolic demands, offering neuroprotection against excitotoxic damage. Despite this temporary suppression of synaptic plasticity, hibernation facilitates remarkable resilience and adaptability. Upon arousal, synaptic reorganization and regeneration occur swiftly, restoring neurotransmitter levels and synaptic connections critical for cognitive and motor functions. LTP, essential for learning and memory, rebounds rapidly, coinciding with the reactivation of metabolic processes and gene expression profiles that support synaptic repair and plasticity. Additionally, the protective downregulation of synaptic activity during hibernation helps preserve neuronal integrity, highlighting hibernation's role in safeguarding brain function [138]. Thus, hibernation underscores a unique balance between energy conservation and neuroprotection, illustrating the brain's capacity to dynamically regulate synaptic plasticity in response to extreme environmental and physiological challenges.

Both LTP and learning experiments have shown a reported increase in glutamate release. Increased glutamate release was observed in association with learning. Tissue from the hemisphere where LTP was induced exhibited a more significant rise in glutamate release in rats during the early stages of training compared to the later stages. The changes in evoked responses due to learning have been noted in various brain regions. For instance, increased fEPSPs have been recorded in the piriform cortex during olfactory learning. Additionally, both the enhancement of LTP and facilitation of LTD have been observed [139]. A multitude of studies have demonstrated that competitive NMDA antagonists hinder hippocampus-dependent learning. The study utilized a diverse range of learning paradigms, including spatial learning, specific olfactory learning tasks, T-maze alternation, contextual fear conditioning, social transmission of food preferences, flavor-place paired associate learning, and various operant tasks. However, during watermaze training, diffuse NMDA receptor blockade can sometimes lead to sensorimotor disturbances. These disturbances may include falling off the platform, displaying thigmotaxis, and failing to climb onto the platform. When administered in high doses, AP5-treated animals show stereotypic behaviors. Although these doses are substantially greater than those needed for in vivo LTP blockade with region-specific infusion, acute intrahippocampal infusion can significantly address these concerns. Inhibiting NMDA receptors distinguishes various elements of spatial learning, hindering the animal's capacity to learn the required strategy without affecting its knowledge of the room's landmarks where the watermaze is located. Despite intraperitoneal administration of the NMDA antagonist CGS19755 at a dose that effectively blocks LTP in freely moving animals, "strategy learning" in the watermaze remains unaffected during non-spatial pretraining. In experienced animals, the reference memory task may demand minimal hippocampal synaptic plasticity. Despite extensive training, NMDAR-dependent plasticity could still play a significant role in one-trial encoding. In line with this, several genetic interventions have shown selective impairment in rapid learning. Mice lacking the GluR1 (GluR-A) subunit of the AMPA receptor exhibit significant deficits in CA1 LTP but can still acquire a water-maze reference memory task. However, they struggle with spatial working memory tasks, such as T-maze nonmatching to place. In this chapter, we have discussed the most significant advancements in the field of synaptic plasticity that have emerged from neuroscientific research. The significant progress in understanding the mechanisms and functions of synaptic plasticity provides confidence that we now possess the conceptual and instrumental tools to address these challenges. While both the study of consciousness and LTP have their easy and hard problems, the challenging aspect of LTP—the connection between plasticity and memory—is considered solvable.

References

1. Kaplan MS, Hinds JW. Neurogenesis in the adult rat: electron microscopic analysis of light radioautographs. Science. 1977;197:1092–4.
2. Lattanzi D, Savelli D, Pagliarini M, Cuppini R, Ambrogini P. Short-term, voluntary exercise affects morpho-functional maturation of adult-generated neurons in rat hippocampus. Int J Mol Sci. 2022;23:6866.
3. Schnell C, Janc OA, Kempkes B, et al. Restraint stress intensifies interstitial K(+) accumulation during severe hypoxia. Front Pharmacol. 2012;3:53.
4. Rico AM, Mendoza AL, Durán DA, de la Luz Torres H, Mendoza GA, Gómez AB. The effects of chronic restraint on the morphology of ventral CA1 neurons in female long Evans rats. Stress. 2015;18:67–75.
5. Dalla C, Whetstone AS, Hodes GE, Shors TJ. Stressful experience has opposite effects on dendritic spines in the hippocampus of cycling versus masculinized females. Neurosci Lett. 2009;449:52–6.
6. Shors TJ, Chua C, Falduto J. Sex differences and opposite effects of stress on dendritic spine density in the male versus female hippocampus. J Neurosci. 2001;21:6292–7.
7. Shors TJ, Falduto J, Leuner B. The opposite effects of stress on dendritic spines in male vs. female rats are NMDA receptor-dependent. Eur J Neurosci. 2004;19:145–50.
8. Kumar J, Patel T, Sugandh F, et al. Innovative approaches and therapies to enhance neuroplasticity and promote recovery in patients with neu-

rological disorders: a narrative review. Cureus. 2023;15:e41914.
9. Schoenfeld TJ, Gould E. Stress, stress hormones, and adult neurogenesis. Exp Neurol. 2012;233:12–21.
10. Gould E, Gross CG. Neurogenesis in adult mammals: some progress and problems. J Neurosci. 2002;22:619–23.
11. Gould E, Tanapat P. Stress and hippocampal neurogenesis. Biol Psychiatry. 1999;46:1472–9.
12. Lucassen PJ, Oomen CA, Naninck EF, et al. Regulation of adult neurogenesis and plasticity by (early) stress, glucocorticoids, and inflammation. Cold Spring Harb Perspect Biol. 2015;7:a021303.
13. Duthie L, Reynolds RM. Changes in the maternal hypothalamic-pituitary-adrenal axis in pregnancy and postpartum: influences on maternal and fetal outcomes. Neuroendocrinology. 2013;98:106–15.
14. Yagi S, Splinter JEJ, Tai D, Wong S, Wen Y, Galea LAM. Sex differences in maturation and attrition of adult neurogenesis in the hippocampus. eNeuro. 2020;7.
15. Yau SY, Li A, Tong JB, et al. Chronic corticosterone administration reduces dendritic complexity in mature, but not young granule cells in the rat dentate gyrus. Restor Neurol Neurosci. 2016;34:849–57.
16. Tata DA, Anderson BJ. The effects of chronic glucocorticoid exposure on dendritic length, synapse numbers and glial volume in animal models: implications for hippocampal volume reductions in depression. Physiol Behav. 2010;99:186–93.
17. Schmidt S, Rainer J, Ploner C, Presul E, Riml S, Kofler R. Glucocorticoid-induced apoptosis and glucocorticoid resistance: molecular mechanisms and clinical relevance. Cell Death Differ. 2004;11(Suppl 1):S45–55.
18. Christian KM, Miracle AD, Wellman CL, Nakazawa K. Chronic stress-induced hippocampal dendritic retraction requires CA3 NMDA receptors. Neuroscience. 2011;174:26–36.
19. Conrad CD. What is the functional significance of chronic stress-induced CA3 dendritic retraction within the hippocampus? Behav Cogn Neurosci Rev. 2006;5:41–60.
20. Sheppard PAS, Choleris E, Galea LAM. Structural plasticity of the hippocampus in response to estrogens in female rodents. Mol Brain. 2019;12:22.
21. Erli F, Palmos AB, Raval P, et al. Estradiol reverses excitatory synapse loss in a cellular model of neuropsychiatric disorders. Transl Psychiatry. 2020;10:16.
22. Hara Y, Waters EM, McEwen BS, Morrison JH. Estrogen effects on cognitive and synaptic health over the lifecourse. Physiol Rev. 2015;95:785–807.
23. Sandstrom NJ, Williams CL. Spatial memory retention is enhanced by acute and continuous estradiol replacement. Horm Behav. 2004;45:128–35.
24. Chesler EJ, Juraska JM. Acute administration of estrogen and progesterone impairs the acquisition of the spatial Morris water maze in ovariectomized rats. Horm Behav. 2000;38:234–42.
25. Dubal DB, Wise PM. Estrogen and neuroprotection: from clinical observations to molecular mechanisms. Dialogues Clin Neurosci. 2002;4:149–61.
26. Tanapat P, Hastings NB, Gould E. Ovarian steroids influence cell proliferation in the dentate gyrus of the adult female rat in a dose- and time-dependent manner. J Comp Neurol. 2005;481:252–65.
27. Hajszan T, Milner TA, Leranth C. Sex steroids and the dentate gyrus. Prog Brain Res. 2007;163:399–415.
28. Tanapat P, Hastings NB, Reeves AJ, Gould E. Estrogen stimulates a transient increase in the number of new neurons in the dentate gyrus of the adult female rat. J Neurosci. 1999;19:5792–801.
29. Bustamante-Barrientos FA, Méndez-Ruette M, Ortloff A, et al. The impact of estrogen and estrogen-like molecules in neurogenesis and neurodegeneration: beneficial or harmful? Front Cell Neurosci. 2021;15:636176.
30. Zentall TR. Effect of environmental enrichment on the brain and on learning and cognition by animals. Animals (Basel). 2021;11:973.
31. Risher WC, Patel S, Kim IH, et al. Astrocytes refine cortical connectivity at dendritic spines. Elife. 2014;3:e04047.
32. Tooley UA, Bassett DS, Mackey AP. Environmental influences on the pace of brain development. Nat Rev Neurosci. 2021;22:372–84.
33. Smail MA, Smith BL, Nawreen N, Herman JP. Differential impact of stress and environmental enrichment on corticolimbic circuits. Pharmacol Biochem Behav. 2020;197:172993.
34. Mandolesi L, Polverino A, Montuori S, et al. Effects of physical exercise on cognitive functioning and wellbeing: biological and psychological benefits. Front Psychol. 2018;9:509.
35. Leuner B, Gould E. Structural plasticity and hippocampal function. Annu Rev Psychol. 2010;61:111–40, c1–3.
36. Abraham WC, Jones OD, Glanzman DL. Is plasticity of synapses the mechanism of long-term memory storage? npj Sci Learn. 2019;4:9.
37. Hao L, Yang Z, Lei J. Underlying mechanisms of cooperativity, input specificity, and associativity of long-term potentiation through a positive feedback of local protein synthesis. Front Comput Neurosci. 2018;12:25.
38. Chen HX, Jiang M, Akakin D, Roper SN. Long-term potentiation of excitatory synapses on neocortical somatostatin-expressing interneurons. J Neurophysiol. 2009;102:3251–9.
39. Ming GL, Song H. Adult neurogenesis in the mammalian brain: significant answers and significant questions. Neuron. 2011;70:687–702.
40. Murphy TH, Corbett D. Plasticity during stroke recovery: from synapse to behaviour. Nat Rev Neurosci. 2009;10:861–72.
41. Obernier K, Alvarez-Buylla A. Neural stem cells: origin, heterogeneity and regulation in the adult mammalian brain. Development. 2019;146:dev156059.

42. Burda JE, Sofroniew MV. Reactive gliosis and the multicellular response to CNS damage and disease. Neuron. 2014;81:229–48.
43. Scharfman HE. Epilepsy as an example of neural plasticity. Neuroscientist. 2002;8:154–73.
44. Lee DA, Lee HJ, Kim SE, Park KM. Brain networks and epilepsy development in patients with Alzheimer disease. Brain Behav. 2023;13:e3152.
45. Zhang D, Chen S, Xu S, et al. The clinical correlation between Alzheimer's disease and epilepsy. Front Neurol. 2022;13:922535.
46. Jamjoom AAB, Rhodes J, Andrews PJD, Grant SGN. The synapse in traumatic brain injury. Brain. 2021;144:18–31.
47. Balu DT, Lucki I. Adult hippocampal neurogenesis: regulation, functional implications, and contribution to disease pathology. Neurosci Biobehav Rev. 2009;33:232–52.
48. Jorgensen C, Wang Z. Hormonal regulation of mammalian adult neurogenesis: a multifaceted mechanism. Biomolecules. 2020;10:1151.
49. Spritzer MD, Roy EA. Testosterone and adult neurogenesis. Biomolecules. 2020;10:225.
50. McEwen BS, Nasca C, Gray JD. Stress effects on neuronal structure: hippocampus, amygdala, and prefrontal cortex. Neuropsychopharmacology. 2016;41:3–23.
51. Saaltink DJ, Vreugdenhil E. Stress, glucocorticoid receptors, and adult neurogenesis: a balance between excitation and inhibition? Cell Mol Life Sci. 2014;71:2499–515.
52. De Alcubierre D, Ferrari D, Mauro G, Isidori AM, Tomlinson JW, Pofi R. Glucocorticoids and cognitive function: a walkthrough in endogenous and exogenous alterations. J Endocrinol Invest. 2023;46:1961–82.
53. Galea LA, McEwen BS. Sex and seasonal differences in the rate of cell proliferation in the dentate gyrus of adult wild meadow voles. Neuroscience. 1999;89:955–64.
54. Morris RG, Hebb DO. The organization of behavior, Wiley: New York; 1949. Brain Res Bull. 1999;50:437.
55. Bliss TV, Lomo T. Long-lasting potentiation of synaptic transmission in the dentate area of the anaesthetized rabbit following stimulation of the perforant path. J Physiol. 1973;232:331–56.
56. Bliss TV, Gardner-Medwin AR. Long-lasting potentiation of synaptic transmission in the dentate area of the unanaesthetized rabbit following stimulation of the perforant path. J Physiol. 1973;232:357–74.
57. Wigström H, Gustafsson B. Facilitated induction of hippocampal long-lasting potentiation during blockade of inhibition. Nature. 1983;301:603–4.
58. Collingridge GL, Kehl SJ, McLennan H. Excitatory amino acids in synaptic transmission in the Schaffer collateral-commissural pathway of the rat hippocampus. J Physiol. 1983;334:33–46.
59. Lynch G, Larson J, Kelso S, Barrionuevo G, Schottler F. Intracellular injections of EGTA block induction of hippocampal long-term potentiation. Nature. 1983;305:719–21.
60. Andersen P, Morris R, Amaral D, Bliss T, O'Keefe J. The hippocampus book. Oxford: Oxford University Press; 2006.
61. Mayer ML, Westbrook GL, Guthrie PB. Voltage-dependent block by Mg2+ of NMDA responses in spinal cord neurones. Nature. 1984;309:261–3.
62. Lüscher C, Malenka RC. NMDA receptor-dependent long-term potentiation and long-term depression (LTP/LTD). Cold Spring Harb Perspect Biol. 2012;4:a005710.
63. Wigström H, Gustafsson B, Huang YY, Abraham WC. Hippocampal long-term potentiation is induced by pairing single afferent volleys with intracellularly injected depolarizing current pulses. Acta Physiol Scand. 1986;126:317–9.
64. Morris RG, Anderson E, Lynch GS, Baudry M. Selective impairment of learning and blockade of long-term potentiation by an N-methyl-D-aspartate receptor antagonist, AP5. Nature. 1986;319:774–6.
65. Bagal AA, Kao JP, Tang CM, Thompson SM. Long-term potentiation of exogenous glutamate responses at single dendritic spines. Proc Natl Acad Sci U S A. 2005;102:14434–9.
66. McNaughton BL. Long-term synaptic enhancement and short-term potentiation in rat fascia dentata act through different mechanisms. J Physiol. 1982;324:249–62.
67. Holohean AM, Magleby KL. The number of components of enhancement contributing to short-term synaptic plasticity at the neuromuscular synapse during patterned nerve stimulation progressively decreases as basal release probability is increased from low to normal levels by changing extracellular Ca2+. J Neurosci. 2011;31:7060–72.
68. Sanchez-Vives MV, Barbero-Castillo A, Perez-Zabalza M, Reig R. GABA(B) receptors: modulation of thalamocortical dynamics and synaptic plasticity. Neuroscience. 2021;456:131–42.
69. Karpova A, Mikhaylova M, Thomas U, Knöpfel T, Behnisch T. Involvement of protein synthesis and degradation in long-term potentiation of Schaffer collateral CA1 synapses. J Neurosci. 2006;26:4949–55.
70. Hebb DO. The organization of behavior: a neuropsychological theory. New York: Wiley; 1949.
71. Gilbert ME, Mack CM. The NMDA antagonist, MK-801, suppresses long-term potentiation, kindling, and kindling-induced potentiation in the perforant path of the unanesthetized rat. Brain Res. 1990;519:89–96.
72. Kemp JA, Foster AC, Leeson PD, et al. 7-Chlorokynurenic acid is a selective antagonist at the glycine modulatory site of the N-methyl-D-aspartate receptor complex. Proc Natl Acad Sci U S A. 1988;85:6547–50.
73. Tsien JZ, Chen DF, Gerber D, et al. Subregion- and cell type-restricted gene knockout in mouse brain. Cell. 1996;87:1317–26.

74. Sprengel R, Suchanek B, Amico C, et al. Importance of the intracellular domain of NR2 subunits for NMDA receptor function in vivo. Cell. 1998;92:279–89.
75. MacDonald JF, Wojtowicz JM. Two conductance mechanisms activated by applications of L-glutamic, L-aspartic, DL-homocysteic, N-methyl-D-aspartic, and DL-kainic acids to cultured mammalian central neurones. Can J Physiol Pharmacol. 1980;58:1393–7.
76. Jin SX, Feig LA. Long-term potentiation in the CA1 hippocampus induced by NR2A subunit-containing NMDA glutamate receptors is mediated by Ras-GRF2/Erk map kinase signaling. PLoS One. 2010;5:e11732.
77. Philpot BD, Cho KK, Bear MF. Obligatory role of NR2A for metaplasticity in visual cortex. Neuron. 2007;53:495–502.
78. Hrabetova S, Serrano P, Blace N, et al. Distinct NMDA receptor subpopulations contribute to long-term potentiation and long-term depression induction. J Neurosci. 2000;20:RC81.
79. Lauri SE, Bortolotto ZA, Nistico R, et al. A role for Ca2+ stores in kainate receptor-dependent synaptic facilitation and LTP at mossy fiber synapses in the hippocampus. Neuron. 2003;39:327–41.
80. Foskett JK, White C, Cheung KH, Mak DO. Inositol trisphosphate receptor Ca2+ release channels. Physiol Rev. 2007;87:593–658.
81. Lau CG, Takeuchi K, Rodenas-Ruano A, et al. Regulation of NMDA receptor Ca2+ signalling and synaptic plasticity. Biochem Soc Trans. 2009;37:1369–74.
82. Noh KM, Yokota H, Mashiko T, Castillo PE, Zukin RS, Bennett MV. Blockade of calcium-permeable AMPA receptors protects hippocampal neurons against global ischemia-induced death. Proc Natl Acad Sci U S A. 2005;102:12230–5.
83. Guo C, Ma YY. Calcium permeable-AMPA receptors and excitotoxicity in neurological disorders. Front Neural Circuits. 2021;15:711564.
84. Park P, Kang H, Sanderson TM, et al. The role of calcium-permeable AMPARs in long-term potentiation at principal neurons in the rodent hippocampus. Front Synaptic Neurosci. 2018;10:42.
85. Latif-Hernandez A, Faldini E, Ahmed T, Balschun D. Separate ionotropic and metabotropic glutamate receptor functions in depotentiation vs. LTP: a distinct role for Group1 mGluR subtypes and NMDARs. Front Cell Neurosci. 2016;10:252.
86. Bortolotto ZA, Collingridge GL. On the mechanism of long-term potentiation induced by (1S,3R)-1-aminocyclopentane-1,3-dicarboxylic acid (ACPD) in rat hippocampal slices. Neuropharmacology. 1995;34:1003–14.
87. Hagenston AM, Bading H. Calcium signaling in synapse-to-nucleus communication. Cold Spring Harb Perspect Biol. 2011;3:a004564.
88. Fitzjohn SM, Bortolotto ZA, Palmer MJ, et al. The potent mGlu receptor antagonist LY341495 identifies roles for both cloned and novel mGlu receptors in hippocampal synaptic plasticity. Neuropharmacology. 1998;37:1445–58.
89. Hovelsø N, Sotty F, Montezinho LP, Pinheiro PS, Herrik KF, Mørk A. Therapeutic potential of metabotropic glutamate receptor modulators. Curr Neuropharmacol. 2012;10:12–48.
90. Falcicchia C, Tozzi F, Arancio O, Watterson DM, Origlia N. Involvement of p38 MAPK in synaptic function and dysfunction. Int J Mol Sci. 2020;21:5624.
91. Jong YI, Izumi Y, Harmon SK, Zorumski CF, ÓMalley KL. Striatal mGlu(5)-mediated synaptic plasticity is independently regulated by location-specific receptor pools and divergent signaling pathways. J Biol Chem. 2023;299:104949.
92. Lao-Peregrin C, Xiang G, Kim J, et al. Synaptic plasticity via receptor tyrosine kinase/G-protein-coupled receptor crosstalk. Cell Rep. 2024;43:113595.
93. Davies CH, Collingridge GL. Regulation of EPSPs by the synaptic activation of GABAB autoreceptors in rat hippocampus. J Physiol. 1996;496(Pt 2):451–70.
94. Davies CH, Collingridge GL. The physiological regulation of synaptic inhibition by GABAB autoreceptors in rat hippocampus. J Physiol. 1993;472:245–65.
95. Zorumski CF, Izumi Y. NMDA receptors and metaplasticity: mechanisms and possible roles in neuropsychiatric disorders. Neurosci Biobehav Rev. 2012;36:989–1000.
96. Granger AJ, Nicoll RA. Expression mechanisms underlying long-term potentiation: a postsynaptic view, 10 years on. Philos Trans R Soc Lond Ser B Biol Sci. 2014;369:20130136.
97. Volianskis A, Jensen MS. Transient and sustained types of long-term potentiation in the CA1 area of the rat hippocampus. J Physiol. 2003;550:459–92.
98. Neubauer FB, Min R, Nevian T. Presynaptic NMDA receptors influence Ca(2+) dynamics by interacting with voltage-dependent calcium channels during the induction of long-term depression. Neural Plast. 2022;2022:2900875.
99. Diering GH, Huganir RL. The AMPA receptor code of synaptic plasticity. Neuron. 2018;100:314–29.
100. Creamer TP. Calcineurin. Cell Commun Signal. 2020;18:137.
101. Cui Z, Feng R, Jacobs S, et al. Increased NR2A:NR2B ratio compresses long-term depression range and constrains long-term memory. Sci Rep. 2013;3:1036.
102. Pampaloni NP, Riva I, Carbone AL, Plested AJR. Slow AMPA receptors in hippocampal principal cells. Cell Rep. 2021;36:109496.
103. Harney SC, Jane DE, Anwyl R. Extrasynaptic NR2D-containing NMDARs are recruited to the synapse during LTP of NMDAR-EPSCs. J Neurosci. 2008;28:11685–94.
104. Polepalli JS, Sullivan RK, Yanagawa Y, Sah P. A specific class of interneuron mediates inhibi-

tory plasticity in the lateral amygdala. J Neurosci. 2010;30:14619–29.
105. Billard JM. Changes in serine racemase-dependent modulation of NMDA receptor: impact on physiological and pathological brain aging. Front Mol Biosci. 2018;5:106.
106. Lovinger DM, Wong KL, Murakami K, Routtenberg A. Protein kinase C inhibitors eliminate hippocampal long-term potentiation. Brain Res. 1987;436:177–83.
107. Wang H, Zhuo M. Group I metabotropic glutamate receptor-mediated gene transcription and implications for synaptic plasticity and diseases. Front Pharmacol. 2012;3:189.
108. Pi HJ, Otmakhov N, Lemelin D, De Koninck P, Lisman J. Autonomous CaMKII can promote either long-term potentiation or long-term depression, depending on the state of T305/T306 phosphorylation. J Neurosci. 2010;30:8704–9.
109. Umar S, van der Laarse A. Nitric oxide and nitric oxide synthase isoforms in the normal, hypertrophic, and failing heart. Mol Cell Biochem. 2010;333:191–201.
110. Xu L, Mabuchi T, Katano T, et al. Nitric oxide (NO) serves as a retrograde messenger to activate neuronal NO synthase in the spinal cord via NMDA receptors. Nitric Oxide. 2007;17:18–24.
111. Gorlewicz A, Kaczmarek L. Pathophysiology of trans-synaptic adhesion molecules: implications for epilepsy. Front Cell Dev Biol. 2018;6:119.
112. Benson DL, Schnapp LM, Shapiro L, Huntley GW. Making memories stick: cell-adhesion molecules in synaptic plasticity. Trends Cell Biol. 2000;10:473–82.
113. Errington ML, Bliss TV, Morris RJ, Laroche S, Davis S. Long-term potentiation in awake mutant mice. Nature. 1997;387:666–7.
114. Bozdagi O, Valcin M, Poskanzer K, Tanaka H, Benson DL. Temporally distinct demands for classic cadherins in synapse formation and maturation. Mol Cell Neurosci. 2004;27:509–21.
115. Thoumine O, Lambert M, Mège RM, Choquet D. Regulation of N-cadherin dynamics at neuronal contacts by ligand binding and cytoskeletal coupling. Mol Biol Cell. 2006;17:862–75.
116. Murase T, Nagasawa A, Suzuki J, Hase T, Tokimitsu I. Beneficial effects of tea catechins on diet-induced obesity: stimulation of lipid catabolism in the liver. Int J Obes Relat Metab Disord. 2002;26:1459–64.
117. Campbell ID, Humphries MJ. Integrin structure, activation, and interactions. Cold Spring Harb Perspect Biol. 2011;3:a004994.
118. Sandkühler J, Gruber-Schoffnegger D. Hyperalgesia by synaptic long-term potentiation (LTP): an update. Curr Opin Pharmacol. 2012;12:18–27.
119. Südhof TC. Neuroligins and neurexins link synaptic function to cognitive disease. Nature. 2008;455:903–11.
120. Naqvi S, Martin KJ, Arthur JS. CREB phosphorylation at Ser133 regulates transcription via distinct mechanisms downstream of cAMP and MAPK signalling. Biochem J. 2014;458:469–79.
121. Teleanu RI, Niculescu AG, Roza E, Vladâcenco O, Grumezescu AM, Teleanu DM. Neurotransmitters—Key factors in neurological and neurodegenerative disorders of the central nervous system. Int J Mol Sci. 2022;23:5954.
122. Pulimood NS, Contreras M, Pruitt ME, Tarasiewicz A, Medina AE. Phosphorylation of CREB at serine 142 and 143 is essential for visual cortex plasticity. eNeuro. 2021;8.
123. Sacktor TC. Memory maintenance by PKMζ—an evolutionary perspective. Mol Brain. 2012;5:31.
124. Winger G, Jutkiewicz EM, Woods JH. Comparison of the muscarinic antagonist effects of scopolamine and L-687,306. Behav Pharmacol. 2020;31:359–67.
125. Brown DA. Regulation of neural ion channels by muscarinic receptors. Neuropharmacology. 2018;136:383–400.
126. Xing B, Li YC, Gao WJ. Norepinephrine versus dopamine and their interaction in modulating synaptic function in the prefrontal cortex. Brain Res. 2016;1641:217–33.
127. Granado N, Ortiz O, Suárez LM, et al. D1 but not D5 dopamine receptors are critical for LTP, spatial learning, and LTP-induced arc and zif268 expression in the hippocampus. Cereb Cortex. 2008;18:1–12.
128. Solinas SMG, Edelmann E, Leßmann V, Migliore M. A kinetic model for brain-derived neurotrophic factor mediated spike timing-dependent LTP. PLoS Comput Biol. 2019;15:e1006975.
129. Wall AM, Mukandala G, Greig NH, O'Connor JJ. Tumor necrosis factor-α potentiates long-term potentiation in the rat dentate gyrus after acute hypoxia. J Neurosci Res. 2015;93:815–29.
130. de Kloet ER, Meijer OC, de Nicola AF, de Rijk RH, Joëls M. Importance of the brain corticosteroid receptor balance in metaplasticity, cognitive performance and neuro-inflammation. Front Neuroendocrinol. 2018;49:124–45.
131. Wang A, Zou X, Wu J, et al. Early-life stress alters synaptic plasticity and mTOR signaling: correlation with anxiety-like and cognition-related behavior. Front Genet. 2020;11:590068.
132. Albert-Gascó H, Ros-Bernal F, Castillo-Gómez E, Olucha-Bordonau FE. MAP/ERK signaling in developing cognitive and emotional function and its effect on pathological and neurodegenerative processes. Int J Mol Sci. 2020;21:4471.
133. Rocks D, Kundakovic M. Hippocampus-based behavioral, structural, and molecular dynamics across the estrous cycle. J Neuroendocrinol. 2023;35:e13216.
134. Zhang K, Yang Q, Yang L, et al. CB1 agonism prolongs therapeutic window for hormone replacement in ovariectomized mice. J Clin Invest. 2019;129:2333–50.
135. Pollard KJ, Wartman HD, Daniel JM. Previous estradiol treatment in ovariectomized mice provides

lasting enhancement of memory and brain estrogen receptor activity. Horm Behav. 2018;102:76–84.
136. Peters A, Reisch C, Langemann D. LTP or LTD? Modeling the influence of stress on synaptic plasticity. eNeuro. 2018;5.
137. Frank MG. Circadian regulation of synaptic plasticity. Biology (Basel). 2016;5:31.
138. Sonntag M, Arendt T. Neuronal activity in the hibernating brain. Front Neuroanat. 2019;13:71.
139. Trieu BH, Kramár EA, Cox CD, et al. Pronounced differences in signal processing and synaptic plasticity between piriform-hippocampal network stages: a prominent role for adenosine. J Physiol. 2015;593:2889–907.

3D Visualization of the Human Hippocampus Using a Different Imaging Technique

7

Nihal Gürlek Çelik and Sümeyye Uçar

7.1 Introduction

The hippocampus consists of the archicortex [1]. Anatomically, it is located in the inner parts of the temporal lobe and has functions related to memory, learning, and processing emotions. The hippocampus consists of several parts, including the cornu ammonis (CA1–CA4), dentate gyrus, subiculum, presubiculum, and fimbria [2]. The hippocampus, dentate gyrus, and subiculum form the hippocampal formation [3]. The hippocampus solves spatial problems by transforming experiences from short-term to long-term memory throughout a person's life and is critical for remembering past location experiences [4].

Because of the hippocampus's location and functional interaction with many brain structures, inferences can be made about the symptoms of certain diseases [5]. Synapse loss in Alzheimer's disease can affect the transmission of excitatory signals in both the hippocampus and the cortex, leading to memory loss [6]. It is important to evaluate the hippocampus in people with systemic neurological disorders [7]. In the past, manual segmentation was frequently used in the evaluation of the hippocampus, but automatic segmentation techniques have recently replaced it. Automatic segmentation techniques are advantageous in that they are objective, reliable, and repeatable [8]. The data must be three-dimensional to comprehensively analyze the structure of neurodegenerative diseases [9]. Thus, automatic segmentation methods are being developed to obtain more detailed results regarding the hippocampus and its subsections [10].

Different imaging methods can be used to evaluate the hippocampus. Hippocampal morphology has mostly been examined by ultrasound (US), magnetic resonance imaging (MRI), and computed tomography (CT).

7.1.1 Ultrasound (US)

Three-dimensional ultrasound is of great importance in evaluating the fetus's central nervous system (CNS) during pregnancy [11, 12]. The combined use of transvaginal sonography and 3D ultrasound is a suitable method for analyzing anomalies and vascular problems that occur in structures related to the CNS [13, 14]. In their study, Gindes et al. evaluated the fetal hippocampus and fornix using 3D sonography at 14–37 weeks [15].

N. G. Çelik (✉)
Department of Anatomy, Amasya University Faculty of Medicine, Amasya, Türkiye
e-mail: nihal.g.celik@gmail.com

S. Uçar
Department of Anatomy, Erciyes University Faculty of Medicine, Kayseri, Türkiye

M. Turgut et al. (eds.), *The Human Hippocampus*, https://doi.org/10.1007/978-3-031-88907-3_7

7.1.2 Magnetic Resonance Imaging

MRI offers advantages such as no radiation [16] and no invasive procedures [17]. It is an important imaging method that provides information about the structure and function of the healthy and pathological human body [17]. A study evaluated the hippocampus of Down syndrome patients and healthy individuals using 7 Tesla MRI. Using T1-weighted MP2RAGE images, gray and white matter volume calculations of the whole brain were made with the FreeSurfer 6.0 software program, and the volumes of the subdivisions of the hippocampus were evaluated with the Automated Segmentation of Hippocampal Subfields (ASHS) software [18]. In a human cadaver study by Gross et al., the hippocampus was imaged with T2-weighted MRI along the caudal-rostral axis using the curved multiplanar reformatting (CMPR) method [19].

7.1.3 Computed Tomography

CT is one of the first methods used for the analysis of the brain and related structures [20]. The reasons for preferring CT are that it provides thin sections, the area of interest can be viewed in different planes, and it has the ability to visualize hippocampal calcification [21, 22]. Postmortem human hippocampus [9] and hippocampal segmentation [23] have been evaluated with CT. Eckermann et al., in their study, examined the three-dimensional histological structure of the hippocampal tissues of people with and without Alzheimer's disease based on phase-contrast X-ray CT. The histological evaluation stated that Ilastik software was used for semi-automatic segmentation of the nuclei of granular cells, and Avizo Lite was used for another parameter [9].

7.1.4 Positron Emission Tomography

Positron emission tomography (PET) is a noninvasive molecular imaging tool that evaluates the development of the brain and related diseases [24]. Some PET data are 2D, while others can be collected in 3D mode [25].

7.2 Discussion

Due to its location and connections, the hippocampus is always an area worth investigating. The hippocampus and its subsections are measured using different software programs. Morphometric results are very important in aging and many neurological/psychiatric diseases.

In Gurlek Celik et al.'s study, volume analysis of the hippocampus and its subdivisions was performed in healthy individuals. Using T1-weighted MRI and supported by HIPS, a tab of the vol2Brain software program, the study reported that hippocampus, Cornu Ammonis1 (CA1), Cornu Ammonis2–3 (CA2–CA3), and Cornu Ammonis4-Dentate Gyrus (CA4-DG) values of both sexes were higher on the right side. When hippocampus and CA2–CA3 values were evaluated according to sex and age, it was reported that there was no difference in the age group of 50 and above. In contrast, in individuals under the age of 50, the average values were higher in males [26].

In their study, Mak et al. examined the hippocampus of patients with mild cognitive impairment (MCI), Alzheimer's disease (AD), and dementia with Lewy bodies (DLB). FreeSurfer 6 software program was used to segment the hippocampus and its subdivisions of 84 individuals participating in the study on 3T MRI and diffusion tensor imaging (DTI) data. It was stated that the hippocampus volume of the healthy group and DLB was higher than AD and MCI. It was reported that high hippocampal mean diffusivity in MCI was associated with atrophy of CA1, one of its subdivisions, in the data obtained from DTI images [27]. In another study conducted with the FreeSurfer 6 software program, the relationship between hippocampus volume and cognitive function of patients with schizophrenia and mood disorders was evaluated [28].

The brain structure of Parkinson's disease (PD) patients was evaluated with the FreeSurfer software program in high-resolution T1-weighted 3D brain imaging. In the study that continued for 1 year, it was reported that there was no volumetric change in the brain structure examined between the PD and healthy groups. However, it was stated that there was a decrease in the posterior cingulate cortex volume in PD patients with motor impairment and in the amygdala and hippocampus volume in PD patients with psychobehavioral impairment [29].

Gorno-Tempini et al. evaluated the neuroimaging of 31 individuals with primary progressive aphasia. When analyzed with voxel-based morphometry on MRI, atrophies were observed in the left anterior hippocampus in the logopenic progressive aphasia group [30].

In another study, the region to be evaluated in Frontotemporal Dementia (FTD) was compared with hippocampal volumetry, tensor-based volumetry and voxel-based morphometry using MRI. Voxel-based morphometry was reported to have high accuracy in distinguishing FTD and AD patients, while hippocampal volumetry and tensor-based volumetry had low accuracy [31].

Zhu et al. used the 3D Zernike transform (3DZMs) to determine hippocampal changes in 428 adults with normal cognition, MCI, and clinical diagnoses of AD dementia. A significant correlation was observed between hippocampus volume values and shape measurements at 24 months [32].

A functional MRI (fMRI) study used resting-state fMRI and T2 MRI to analyze sex differences in cognitive connectivity in the hippocampus in AD patients and healthy individuals. As a result, it was reported that the connectivity between the left and right hippocampus was stronger in males than in females [33]. Uchida et al. used ultra-high-field diffusion tensor imaging and tractography to study postmortem brain tissues of individuals with and without AD. In this study, they detected neurodegenerative changes in the entorhinal-hippocampal pathway of AD patients [34]. Meysami et al. evaluated brain measurements with T1-weighted MRI in early- and late-onset AD patients. The brain volumes of AD patients were calculated using a volumetric program, and it was reported that the left putamen and hippocampus were smaller in the late-onset group [35].

The following table shows the imaging methods used in different studies and their results (Table 7.1).

Table 7.1 Measurement parameters and results of hippocampus

Study	How many samples	Methods	Measurements parameters	Result
Carlson et al. (2021) [36]	Twenty-four subjects (five amyloid-positive probable AD, six amyloid-positive mild cognitive impairment (MCI), and 13 healthy control subjects)	3T PET/MRI	CA1, dentate gyrus and CA2–4 as one combined subfield (DG), subiculum (SUB), entorhinal and perirhinal cortices (ERC/PRC), and white matter within the parahippocampal gyrus (PHG)	Parahippocampal gyral fractional anisotropy was lower in AD and MCI compared to the healthy group
Wisse et al. (2015) [37]	Twenty-five patients with amnestic mild cognitive impairment (aMCI) ($n = 15$) or early AD ($n = 10$) and 17 controls	7T MRI	7 Tesla MRI to obtain ERC and hippocampal subfield volumes	CA1, dentate gyrus, and CA4 volumes were significantly higher in healthy individuals than in AD and MCI patients
Sone et al. (2017) [38]	Eighteen patients with early AD and 18 healthy control subjects	PET and high-resolution T2 MRI	Segmentation of hippocampal subfields	While subunits of the hippocampus did not show a significant correlation with tau accumulation, right CA3 volumes showed a correlation with amyloid accumulation
Wisse et al. (2014) [39]	Sixteen patients with amnestic MCI, 9 with AD, and 29 controls	7T MRI	ERC, subiculum, CA1, CA2, CA3, and DG&CA4 volumes	The lower parts of the hippocampus in healthy groups have higher volumes than in disease groups
Bullich et al. (2022) [40]	MCI/AD: 74	PET	Temporal cortex, fusiform gyrus, inferior temporal cortex, superior temporal, occipital, and parietal regions	An inverse relationship was initially observed between increasing tau load and hippocampus volume
Wu et al. (2018) [41]	Aβ positive AD ($n = 151$) and Aβ negative cognitively unimpaired subjects ($n = 271$)	PET, MRI	Hippocampal segmentation	Results have been reported as a biomarker for AD
Özdemir et al. (2019) [42]	302 healthy individuals	MRI 1.5T	Hippocampus volume with volBrain	The hippocampus volume is higher in males, with the right hippocampus mean being higher than the left
Gurlek Celik et al. (2023) [26]	90 healthy individuals	1.5T T1-weighted volumetric MRI	Pallidum, hippocampus, CA1, CA2–CA3, and CA4-DG with Vol2Brain	Pallidum, hippocampus, CA1, CA2–CA3, and CA4-DG are higher in the right region in both sexes

7.3 Conclusion

The hippocampus, which is different from the cerebral cortex in structure and function, establishes direct or indirect connections with brain regions. We accumulate many memories throughout our lives, but sometimes, we may not remember these memories regarding time and place. The hippocampus has a special and important role in episodic memory. It stores our lives like a movie frame.

When the structure of neuropsychiatric diseases is examined in the literature, there is information that there are differences in the hippocampus depending on the gender and age of the person. It is thought that three-dimensional imaging methods that clinicians use are very valuable in identifying these differences, following the disease's prognosis, and better managing the treatment process.

References

1. Young PA, Young PH. Basic clinical neuroanatomy. Baltimore: Lippincott, Williams & Wilkins; 1997.
2. Small SA, Schobel SA, Buxton RB, Witter MP, Barnes CA. A pathophysiological framework of hippocampal dysfunction in ageing and disease. Nat Rev Neurosci. 2011;12:585–601.
3. Zola S. Amnesia: neuroanatomic and clinical aspects. In: Feinberg TE, Farah MJ, editors. Behavioral neurology and neuropsychology. New York: McGraw-Hill; 1997.
4. Chauhan P, Jethwa K, Rathawa A, Chauhan G, Mehra S. Chapter 2. The anatomy of the hippocampus. In: Pluta R, editor. Cerebral ischemia [internet]. Brisbane: Exon Publications; 2021. https://www.ncbi.nlm.nih.gov/books/NBK575732/.
5. Mueller SG, Weiner MW. Selective effect of age, Apo e4, and Alzheimer's disease on hippocampal subfields. Hippocampus. 2009;19:558–64.
6. Li N, Li Y, Li LJ, Zhu K, Zheng Y, Wang XM. Glutamate receptor delocalization in postsynaptic membrane and reduced hippocampal synaptic plasticity in the early stage of Alzheimer's disease. Neural Regen Res. 2019;14(6):1037–45.
7. Yang Q, Wang C, Pan K, Xia B, Xie R, Shi J. An improved 3D-UNet-based brain hippocampus segmentation model based on MR images. BMC Med Imaging. 2024;24(1):166. https://doi.org/10.1186/s12880-024-01346-w.
8. Maglietta R, Amoroso N, Boccardi M, Bruno S, Chincarini A, Frisoni GB, Inglese P, Redolfi A, Tangaro S, Tateo A, Bellotti R, Alzheimer's Disease Neuroimaging Initiative. Automated hippocampal segmentation in 3D MRI using random undersampling with boosting algorithm. Pattern Anal Appl. 2016;19:579–91.
9. Eckermann M, Schmitzer B, van der Meer F, Franz J, Hansen O, Stadelmann C, Salditt T. Three-dimensional virtual histology of the human hippocampus based on phase-contrast computed tomography. Proc Natl Acad Sci U S A. 2021;118(48):e2113835118.
10. Haukvik UK, Tamnes CK, Söderman E, Agartz I. Neuroimaging hippocampal subfields in schizophrenia and bipolar disorder: a systematic review and meta-analysis. J Psychiatr Res. 2018;104:217–26.
11. The nervous system. In: Moore KL, Persaud TVN, Shiota K, editors. Color atlas of clinical embryology. Philadelphia: WB Saunders; 1994. pp. 209–220.
12. Pooh RK, Pooh K. Assessment of fetal central nervous system. Donald School J Ultrasound Obstet Gynecol. 2013;7(4):369–84.
13. Pooh RK, Pooh K, Nakagawa Y, Nishida S, Ohno Y. Clinical application of three-dimensional ultrasound in fetal brain assessment. Croat Med J. 2000;41(3):245–51.
14. Pooh RK, Pooh KH. Fetal neuroimaging. Fetal Matern Med Rev. 2008;19(1):1–31.
15. Gindes L, Weissmann-Brenner A, Weisz B, Zajicek M, Geffen KT, Achiron R. Identification of the fetal hippocampus and fornix and role of 3-dimensional sonography. J Ultrasound Med. 2011;30:1613–8.
16. Tavlayan E, Olacak N, Yavuz A. Beyin tümörlerinin radyoterapi planlamasında, BT ve MRG görüntü eşleştirilmesinin hedef volüm belirlenmesine ve doz dağılımına etkisi. Türk Onkoloji Dergisi. 2011;26(2):67–75. https://doi.org/10.5505/tjoncol.2011.188.
17. Vijayalaxmi M, Fatahi O. Speck. Magnetic resonance imaging (MRI): A review of genetic damage investigations. Mutation Res Rev Mutation Res, vol 764. Elsevier; 2015. pp. 51–63. https://doi.org/10.1016/j.mrrev.2015.02.002.
18. Koenig KA, Oh SH, Stasko MR, Roth EC, Taylor HG, Ruedrich S, Wang ZI, Leverenz JB, Costa ACS. High resolution structural and functional MRI of the hippocampus in young adults with Down syndrome. Brain Commun. 2021;3(2):fcab088.
19. Gross DW, Misaghi E, Steve TA, Wilman AH, Beaulieu C. Curved multiplanar reformatting provides improved visualization of hippocampal anatomy. Hippocampus. 2020;30(2):156–61.
20. Buchlak QD, Tang CHM, Seah JCY, et al. Effects of a comprehensive brain computed tomography deep learning model on radiologist detection accuracy. Eur Radiol. 2024;34:810–22. https://doi.org/10.1007/s00330-023-10074-8.
21. Chew APT, Gupta G, Alatakis S, Schneider-Kolsky M, Stuckey SL. Hippocampal calcification prevalence at CT: a retrospective review. Radiology. 2012;265(2):504–10. https://doi.org/10.1148/radiol.12110636.
22. Kockelkoren R, De Vis JB, Mali WP, Hendrikse J, de Jong PA, Rozemuller AM, Koek HL. Hippocampal calcification on computed tomography in relation to cognitive decline in memory clinic patients: a case-control study. PLoS One. 2016;11(11):e0167444. https://doi.org/10.1371/journal.pone.0167444.
23. Porter E, Fuentes P, Siddiqui Z, Thompson A, Levitin R, Solis D, et al. Hippocampus segmentation on noncontrast CT using deep learning. Med Phys. 2020;47(7):2950–61.
24. Lameka K, Farwell MD, Ichise M. Positron emission tomography. Handb Clin Neurol. 2016;135:209–27.
25. Fahey FH. Data acquisition in PET imaging. J Nucl Med Technol. 2002;30(2):39–49.
26. Gurlek Celik N, Tiryaki S. Changes in the volumes and asymmetry of subcortical structures in healthy individuals according to gender. Anat Sci Int.

2023;98(4):506–19. https://doi.org/10.1007/s12565-023-00714-w. Epub 2023 Mar 22. Erratum in: Anat Sci Int. 2023 Sep;98(4):520.

27. Mak E, Gabel S, Su L, et al. Multi-modal MRI investigation of volumetric and microstructural changes in the hippocampus and its subfields in mild cognitive impairment, Alzheimer's disease, and dementia with Lewy bodies. Int Psychogeriatr. 2017;29(4):545–55. https://doi.org/10.1017/S1041610216002143.
28. Yasuda K, Yamada S, Uenishi S, Ikeda N, Tamaki A, Ohoshi Y, Tsuji T, Takahashi S. Hippocampal subfield volumes and cognitive function in schizophrenia and mood disorders. Neuropsychobiology. 2022;81(3):204–14.
29. Marques A, Macias E, Pereira B, Durand E, Chassain C, Vidal T, Defebvre L, Carriere N, Fraix V, Moro E, Thobois S, Metereau E, Mangone G, Vidailhet M, Corvol JC, Lehéricy S, Menjot de Champfleur N, Geny C, Spampinato U, Meissner WG, Frismand S, Schmitt E, Doé de Maindreville A, Portefaix C, Remy P, Fénelon G, Houeto JL, Colin O, Rascol O, Peran P, Bonny JM, Fantini ML, Durif F, R2* Study Group. Volumetric changes and clinical trajectories in Parkinson's disease: a prospective multicentric study. J Neurol. 2023;270(12):6033–43.
30. Gorno-Tempini ML, Dronkers NF, Rankin KP, Ogar JM, Phengrasamy L, Rosen HJ, Johnson JK, Weiner MW, Miller BL. Cognition and anatomy in three variants of primary progressive aphasia. Ann Neurol. 2004;55(3):335–46.
31. Muñoz-Ruiz MÁ, Hartikainen P, Koikkalainen J, Wolz R, Julkunen V, Niskanen E, Herukka SK, Kivipelto M, Vanninen R, Rueckert D, Liu Y, Lötjönen J, Soininen H. Structural MRI in frontotemporal dementia: comparisons between hippocampal volumetry, tensor-based morphometry and voxel-based morphometry. PLoS One. 2012;7(12):e52531.
32. Zhu DC, Gwo CY, Deng AW, Scheel N, Dowling MA, Zhang R, Alzheimer's Disease Neuroimaging Initiative. Hippocampus shape characterization with 3D Zernike transformation in clinical Alzheimer's disease progression. Hum Brain Mapp. 2023;44(4):1432–44.
33. Williamson J, James SA, Mukli P, Yabluchanskiy A, Wu DH, Sonntag W, Alzheimer's Disease Neuroimaging Initiative Consortium, Yang Y. Sex difference in brain functional connectivity of hippocampus in Alzheimer's disease. Geroscience. 2024;46(1):563–72.
34. Uchida Y, Onda K, Hou Z, Troncoso JC, Mori S, Oishi K. Microstructural neurodegeneration of the entorhinal-hippocampus pathway along the Alzheimer's disease continuum. J Alzheimers Dis. 2023;95(3):1107–17.
35. Meysami S, Raji CA, Merrill DA, Porter VR, Mendez MF. Quantitative MRI differences between early versus late onset Alzheimer's disease. Am J Alzheimers Dis Other Dement. 2021;36:15333175211055325.
36. Carlson ML, Toueg TN, Khalighi MM, Castillo J, Shen B, Azevedo EC, DiGiacomo P, Mouchawar N, Chau G, Zaharchuk G, James ML, Mormino EC, Zeineh MM. Hippocampal subfield imaging and fractional anisotropy show parallel changes in Alzheimer's disease tau progression using simultaneous tau-PET/MRI at 3T. Alzheimers Dement (Amst). 2021;13(1):e12218.
37. Wisse LE, Reijmer YD, ter Telgte A, Kuijf HJ, Leemans A, Luijten PR, Koek HL, Geerlings MI, Biessels GJ, Utrecht Vascular Cognitive Impairment (VCI) Study Group. Hippocampal disconnection in early Alzheimer's disease: a 7 tesla MRI study. J Alzheimers Dis. 2015;45(4):1247–56. https://doi.org/10.3233/JAD-142994.
38. Sone D, Imabayashi E, Maikusa N, Okamura N, Furumoto S, Kudo Y, Ogawa M, Takano H, Yokoi Y, Sakata M, Tsukamoto T, Kato K, Matsuda H. Regional tau deposition and subregion atrophy of medial temporal structures in early Alzheimer's disease: a combined positron emission tomography/magnetic resonance imaging study. Alzheimers Dement (Amst). 2017;9:35–40.
39. Wisse LE, Biessels GJ, Heringa SM, Kuijf HJ, Koek DH, Luijten PR, Geerlings MI, Utrecht Vascular Cognitive Impairment (VCI) Study Group. Hippocampal subfield volumes at 7T in early Azheimer's disease and normal aging. Neurobiol Aging. 2014;35(9):2039–45.
40. Bullich S, Mueller A, De Santi S, Koglin N, Krause S, Kaplow J, Kanekiyo M, Roé-Vellvé N, Perrotin A, Jovalekic A, Scott D, Gee M, Stephens A, Irizarry M. Evaluation of tau deposition using 18F-PI-2620 PET in MCI and early AD subjects—a MissionAD tau sub-study. Alzheimers Res Ther. 2022;14(1):105. https://doi.org/10.1186/s13195-022-01048-x.
41. Wu J, Zhang J, Shi J, Chen K, Caselli RJ, Reiman EM, Wang Y. Hippocampus morphometry study on pathology-confirmed Alzheimer's disease patients with surface multivariate morphometry statistics. Proc IEEE Int Symp Biomed Imaging. 2018;2018:1555–9.
42. Özdemir M, Soysal H, Eraslan Ö, Dilli A. Normative hippocampal volumetric measurements using magnetic resonance imaging. Turk J Med Sci. 2019;49(5):1464–70.

Neuropathology of the Human Hippocampus

8

Afife Uğuz

8.1 Introduction

The hippocampus can be affected by numerous pathologies that affect the brain. In the hippocampus, various clinical conditions such as epilepsy and dementia may arise due to functional loss associated with any pathology [1].

The term "neuropathology of the human hippocampus" refers to the study of diseases and disorders that affect the structure and function of the hippocampus, a critical region of the brain involved in memory and spatial navigation. This field encompasses the examination of various pathological changes, including neurodegenerative diseases such as Alzheimer's disease, where significant neuronal loss, amyloid plaque accumulation, and neurofibrillary tangles in the hippocampus lead to cognitive decline and memory impairment. The hippocampus, particularly the CA1 region, is highly vulnerable to ischemic injury due to its high metabolic demand, resulting in selective neuronal loss and impaired cognitive function in conditions like stroke or cardiac arrest. Temporal lobe epilepsy often involves the hippocampus, characterized by hippocampal sclerosis, which includes neuronal loss and gliosis, contributing to seizure development and progression of hippocampal damage [2]. Traumatic brain injuries can lead to diffuse axonal injury, neuronal loss, and subsequent cognitive and emotional impairments in the hippocampus. Infections such as herpes simplex encephalitis specifically target the hippocampus, causing acute inflammation, neuronal loss, and long-term cognitive deficits, while chronic inflammatory conditions like autoimmune encephalitis affect hippocampal function. Exposure to neurotoxic substances, such as heavy metals (e.g., lead, iron), drugs (e.g., trimethyltin), and other environmental toxins, can lead to hippocampal damage, resulting in neuronal death, synaptic dysfunction, and cognitive impairments. In addition to proinflammatory chemokines and cytokines such as TNFα, activated microglia release various toxic molecules that can harm structural components of the central nervous system. These molecules include reactive nitrogen intermediates (RNIs), reactive oxygen species (ROS), and derivatives of arachidonic acid. Furthermore, activated microglia enhance the expression of MHC class I and II molecules. A prevalent experimental approach involves examining microglial responses to ischemic, traumatic, or seizure-related brain injuries in animal models, frequently with an emphasis on the hippocampus [3]. Abnormalities in hippocampal development can result in conditions such as autism spectrum disorders and intellectual disabilities, where hippocampal structure and connectivity are often altered.

A. Uğuz (✉)
Department of Pathology, Karamanoğlu Mehmetbey University Faculty of Medicine, Karaman, Türkiye
e-mail: afifeuguz@kmu.edu.tr

M. Turgut et al. (eds.), *The Human Hippocampus*, https://doi.org/10.1007/978-3-031-88907-3_8

Changes in hippocampal volume and function are associated with psychiatric conditions like depression, schizophrenia, and post-traumatic stress disorder (PTSD), often contributing to the clinical symptoms observed in these disorders. By studying the neuropathology of the hippocampus, researchers aim to understand the underlying mechanisms of these diseases, identify potential biomarkers for early diagnosis, and develop targeted therapeutic strategies to mitigate hippocampal damage and preserve cognitive function [4].

There are some types of neuronal cell death mechanisms. These "programmed" mechanisms, which involve specific signaling pathways, have distinct molecular signatures. However, they are not easily distinguishable using routine histopathological examination. In addition to necrosis, various "non-accidental" pathways of neuronal cell death have been identified. These pathways include delayed neuronal cell death, necroptosis, autophagy, and apoptosis [5, 6]. Necrotic tissue damage is typically categorized into two types: selective neuronal necrosis, which affects neurons while sparing glial cells, and pan-necrosis, where all tissue elements, including neurons, glial cells, and blood vessels, eventually die, leading to cavitation over time [7].

When the etiology of pan-necrosis is ischemia, it is referred to as infarction. Both selective neuronal necrosis and pan-necrosis can impact nearly any region of the brain. That pathway should not be mistaken for selective vulnerability. Selective vulnerability refers to the macro brain insults that lead to focal lesions that are concentrated in specific brain regions, with the affected location depending on the nature of the insult. As previously discussed, there are various mechanisms of neuronal cell death. A comparison of the patterns of neuronal death in ischemia, hypoglycemia, and epilepsy shows that all three types of insults can lead to selective neuronal necrosis [8, 9]. This happens because neurons are more prone to dying from overstimulation compared to astrocytes. This vulnerability is not linked to their higher metabolic rates or larger size, but rather to the excessive release of excitatory amino acids [10]. It has been known for a long time that neurons can be harmed by overstimulation from glutamate or other excitatory amino acids, such as aspartate [11]. The neurotoxic potential of acidic amino acids is directly associated with their neuroexcitatory activity [12].

Glutamate, a widespread neurotransmitter, interacts with four distinct receptor types: (1) N-methyl-D-aspartate (NMDA) receptors, (2) kainate receptors, which preferentially bind to kainate (though the marine toxin domoic acid, associated with shellfish poisoning, exhibits an even higher affinity), (3) α-amino-3-hydroxy-5--methyl-4-isoxazole propionic acid (AMPA) receptors, and (4) metabotropic receptors, which are involved in initiating intracellular signaling pathways without necessarily inducing ion flow across the membrane [13]. The first three types of glutamate receptors facilitate ion fluxes, which are accompanied by the movement of water across the dendritic cell membrane. These receptors are involved in the morphological changes seen in excitotoxicity, leading to swelling in dendrites but not affecting the axons. NMDA receptors can trigger lasting changes in synaptic efficiency when subjected to different stimulation patterns. These changes may persist for hours, days, or even longer, and they have a crucial role in memory processes. The prolonged increase in synaptic transmission is called long-term potentiation (LTP), whereas the persistent decrease in synaptic transmission is termed long-term depression (LTD) [14]. LTP is linked to changes in the postsynaptic spines and dendrites [15, 16]. NMDA receptors are highly expressed in the CA1 pyramidal neurons of the hippocampus, which might explain why Sommer's sector is particularly vulnerable to ischemia, hypoglycemia, and epilepsy. The cell death associated with these conditions is likely due to shared underlying mechanisms [17].

The phrases "delayed neuronal death" or "maturation phenomenon" were initially coined to describe findings in the hippocampus of rodents exposed to overall ischemia [18]. The crucial observation is that neurons do not die right after the 5- to 10-min episode of global ischemia but rather begin to deteriorate and die in the following hours to days. Some experimental models and human cases [19] delayed neuronal death pertains to the gradual neuronal degeneration that occurs days after a brief episode of complete ischemia, as opposed to the degeneration resulting from chronic, partial, or incomplete ischemia [20, 21]. After injury, cell death in the central nervous system (CNS) can happen through a combination of apoptosis, necrosis, and hybrid forms along the apoptosis-necrosis continuum, commonly referred to as 'necroptosis'. Apoptosis is the most well-characterized form of programmed cell death in many types of cells and represents the primary type of delayed type of neuronal cell death following ischemia. Mild ischemic injury tends to induce cell death through an apoptotic-like mechanism. That pathway can be initiated by factors like DNA damage, oxygen free radicals, protease activation, ionic imbalances, and binding of death receptor ligands. After cerebral ischemia, two main apoptotic signaling pathways are activated. The intrinsic pathway is initiated by internal events involving mitochondrial disruption and the release of cytochrome [22].

The extrinsic pathway, on the other hand, is triggered by the binding of ligands to cell surface death receptors, such as Fas and tumor necrosis factor (TNF) receptors [23]. Both pathways culminate in the activation of a series of downstream caspases, which ultimately result in DNA fragmentation and cell death [24]. During ischemic injury, programmed necrosis (necroptosis) can be initiated by the stimulation of death receptors with various ligands, such as TNFα, FasL, and Trail, which are also capable of triggering apoptosis [5]. Autophagy, derived from the Greek term meaning "self-eating," is a regulated cellular process activated to facilitate the bulk degradation and removal of organelles and cellular proteins. Autophagy's role in cell death can differ, acting as either cytoprotective or cytotoxic, depending on the particular conditions. During autophagic degradation, proteins are directed to and enclosed within membrane-bound vesicles. These vesicles then fuse with lysosomes, where their contents are broken down by lysosomal enzymes such as cathepsin B and D. Autophagy can be activated in both neurons and glial cells following ischemic injury [25]. The autophagy pathway involves the participation of up to 50 autophagy-related genes and their associated proteins, such as ATG5, ATG12, and microtubule-associated protein light chain 3 (LC3). Many insults that affect the entire brain result in damage confined to specific regions. Nearly all neuropathological diseases exhibit selective vulnerability, yet the underlying mechanisms remain poorly understood. The pattern of selective vulnerability acts as a diagnostic "signature" for specific diseases or groups of disease processes, giving it important diagnostic value. Moreover, this pattern can provide important insights into the pathogenesis of brain injuries resulting from deficiencies, genetic mutations, or exposure to toxins. Several factors contribute to selective vulnerability, and significant progress has been made in understanding certain aspects of this phenomenon. GABA, calcium ions (Ca^{2+}) NMDA receptors contribute to selective vulnerability. Experimental studies highlight that transient global cerebral ischemia (GCI) causes glutamate to accumulate in the brain's extracellular fluid, resulting in excitotoxic neuronal death. This process involves overstimulation of NMDA receptors, excessive Ca^{2+} influx into neurons, activation of various enzymes such as neuronal nitric oxide synthase, phospholipases, and calpain, as well as oxidative and enzymatic damage to macromolecules, culminating in cell death. The NMDA receptor is particularly abundant in the dendritic fields of vulnerable neurons within the CA1 region of the hippocampus [15].

8.2 Developmental and Perinatal Brain Diseases

8.2.1 Hypoxia-Ischemia

The term hypoxia is frequently paired with ischemia (hypoxia-ischemia), but the casual use of this expression is generally discouraged. While tissue hypoxia results from ischemia, hypoxia also has an anti-ischemic effect, as it stimulates an increase in cerebral blood flow (CBF). In many forms of hypoxia, there is an increase in CBF, which helps maintain a steady supply of nutrients and facilitates the removal of potentially harmful byproducts. This elevated CBF is crucial in sustaining tissue health under low oxygen conditions by compensating for the reduced oxygen availability. During ischemia, not only is the blood supply reduced, but the clearance of harmful metabolites is also compromised. This dual impact exacerbates tissue damage and contributes to the progression of injury within the affected area. Brain ischemia can occur alongside systemic hypoxia in certain situations, such as strangulation and cardiorespiratory arrest. In these instances, it is advisable to use the term specific to the clinical event, unless there is precise knowledge of the CBF or the partial oxygen pressure (pO_2). The pO_2 refers to the pressure oxygen would exert in a liquid or gas if it occupied the entire volume alone, regardless of the presence of different molecules and the overall pressure. Hypoxia and ischemia are distinct pathophysiological conditions. Ischemia is invariably pathological, whereas hypoxia is not necessarily so. Hypoxia is experienced in varying degrees with increasing altitude, transitioning seamlessly from normal physiology without an immediate link to pathology. In contrast, ischemia is more directly associated with tissue damage and pathological processes. Low oxygen tension in the blood alone does not cause cerebral necrosis. However, ischemia lasting even as short as 2 min can lead to necrosis in selectively vulnerable brain regions such as the hippocampus. In clinical cases of brain ischemia, systemic hypoxemia is typically not present. Conversely, hypoxic conditions generally occur without concurrent ischemia. Therefore, hypoxia must be differentiated from ischemia, recognizing that ischemia invariably leads to reduced tissue oxygenation (tissue hypoxia) at the cellular level [26, 27].

Severe brain hypoxemia may cause a prolonged coma lasting up to 2 weeks, with the potential for a full and remarkable recovery. In ischemic injury, the range of perfusion between the level where electrophysiological responses or functions are impaired and the level where irreversible damage occurs is known as the penumbra. The term "stroke" refers to an acute disruption or loss of brain function caused by either brain ischemia or hemorrhage. Within the hippocampus, the neurons most sensitive to even mild ischemic insults are those in the CA1 area of Lorente de Nó, known in humans as the H1 cells of Rose. Experimental studies have shown that these neurons can undergo necrosis after just 2 min of global ischemia [28].

Within the first 8–24 h of the acute (non-inflammatory) phase, hyper-eosinophilic and/or apoptotic neurons in the pyramidal layer of the hippocampus and subiculum begin to appear as single cells or small clusters. These neurons are observed in the deep gray matter, as well as in the middle layers or throughout all segments of the cerebral cortex. The recent understanding of the association between gray matter lesions and periventricular leukomalacia (PVL) has led to the emergence of the concept of "encephalopathy of prematurity." In this paradigm, subtle injuries in the cerebral cortex, hippocampus, and thalamus accompany the more prominent lesions of PVL, contributing to later cognitive impairments in premature survivors [29].

Hippocampal neuron loss has been reported in cases of placental insufficiency and in infants suffering from congenital heart defects or respiratory distress. Following experimental ischemic or excitotoxic damage to the hippocampus, microglial activation and proliferation commence on the first day, becoming well established by the fourth day, particularly in the dentate hilus [30]. During mid-pregnancy, hippocampal neurons become particularly susceptible to hypoxic-ischemic brain injury. Apoptotic neurons with karyorrhectic nuclei can increasingly be observed in the Ammon's horn, subiculum, and dentate gyrus towards term and in the early postnatal period, potentially forming a component of ponto-subicular necrosis. Sommer's sector neurons may remain immature until the age of two. A hallmark histological feature of mild hypoxic-ischemic brain injury seen around term and up to 9 months of age is the aggregation of microglia along the inner margin of the dentate gyrus [31]. Typically, selective neuronal necrosis is observed rather than pan-necrosis. However, if the global ischemic insult is severe, infarction of the entire hippocampus can occur. Selective neuronal necrosis is a characteristic of the developing brain exposed to hypoxia-ischemia, referring to the necrosis of individual neurons or groups of neurons in specific gray matter locations such as the cerebral cortex, hippocampus, thalamus, basal ganglia, or brainstem. This condition differs from the typical injury pattern observed in adults [32]. In fact, the entire brain can become necrotic following severe insults of high magnitude. The hippocampus and cerebellum contribute to the discussion that these regions are highly susceptible to chronic hypoxia during periods of accelerated growth [33].

8.2.2 Fetal Growth Restriction

Morphological changes in the gray matter, including reductions in the volume of the motor and visual cortex, hippocampus, basal ganglia, and cerebellum, are commonly observed in animal models of fetal growth restriction (FGR). In the FGR brain, neuronal loss occurs in multiple regions, including the hippocampus, and the surviving neurons exhibit selective morphological changes in the dendrites of hippocampal cells. The hippocampus is notably vulnerable to damage in growth-restricted offspring, exhibiting decreased volume and a reduced total number of neurons. The neurons that do survive often display irregularities in axonal and dendritic structure, along with changes in connectivity [33].

8.2.3 Preterm Brain Damage

Reductions in caudate, hippocampal, and cortical volumes are associated with lower intelligence quotients in prematurely born children aged 7–8 years. Additionally, a reduction in hippocampal volume is associated with particular memory impairments in children as they enter early adolescence [34, 35].

Effects on specific brain structures, including the reduction in hippocampal volume, have been reported, yet it remains unclear whether these alterations predict neurological function independently of white matter pathology [35]. This model aligns with histological evidence from primate studies on preterm birth and neonatal intensive care, showing reduced white matter volume in the temporal, frontal, and parietal lobes, relative preservation in the occipital lobe, and tissue loss in the deep gray cortex [36].

8.2.4 Hydrocephaly

Deformation in the hippocampus may also be associated with hydrocephalus [31].

8.2.5 Holoprosencephaly

The hippocampus is almost always present, even if it exhibits incomplete or abnormal development. Depending on the type of holoprosencephaly (HPE), other structures may be more or less recognizable [31].

8.2.6 Thanatophoric Dysplasia

Thanatophoric dysplasia is a severe congenital skeletal disorder classified under the umbrella of short-limbed dwarfism, caused by mutations in the fibroblast growth factor receptor 3 (FGFR3) gene. This condition is characterized by disproportionately short limbs, a narrow thoracic cavity, and a unique cloverleaf-shaped skull deformity, often resulting in respiratory failure and death shortly after birth. The significance of thanatophoric dysplasia in the neuropathology of the human hippocampus stems from its impact on brain development. Mutations in FGFR3 can lead to various brain abnormalities such as ventriculomegaly, hydrocephalus, and disrupted cortical architecture, including malformations of the hippocampus. In thanatophoric dysplasia, severe disorganization of the medial temporal lobe is a well-documented phenomenon. This disorganization often involves hypoplasia or even aplasia of the dentate gyrus. As observed in thanatophoric dysplasia, the olive may exhibit hyperconvolution [37]. These hippocampal malformations can affect memory, learning, and spatial navigation due to altered neuronal circuits. The disorder highlights the critical role of FGFR3 signaling in neurodevelopment and offers insights into how aberrant signaling pathways can lead to profound neurological and developmental abnormalities. Mutations in FGFR2 have been identified in Apert syndrome, while FGFR3 mutations are associated with thanatophoric dysplasia. Cortical malformations, ventriculomegaly, reduced hippocampal volume, gliosis, disrupted FGFR3 signaling, and hydrocephalus are the main findings of the human hippocampus of diagnosed individuals. Regarding the dentate nucleus, it can become hyperconvoluted in cases of thanatophoric dysplasia and may appear simplified or fragmented into islands or segments in instances of cerebellar hypoplasia, as well as in some cases of trisomy 13 [38] and 18 [39, 40]. Abnormal development of the occipitotemporal regions, deep transverse sulci on the inferomedial temporal surface, and hippocampal dysplasia have been observed [39]. Thanatophoric dwarfism, also known as lethal congenital chondrodysplasia, is a skeletal disorder characterized by severely shortened limbs (micromelia), a narrow thoracic cavity, and a disproportionately large head, often with a trilobed or cloverleaf skull shape [41].

8.2.7 Lissencephaly

Lissencephaly, meaning "smooth brain," is a rare brain malformation characterized by the absence or reduction of normal brain folds and grooves (gyri and sulci), leading to a smooth cortical surface. This condition results from abnormal neuronal migration during brain development, typically between the 12th and 24th weeks of gestation. Autosomal recessive lissencephaly in humans results from mutations in the RELN gene, responsible for producing reelin, the human equivalent. Studies on the genes involved in neuronal migration disorders, such as lissencephaly and heterotopias, have underscored the significance of molecular control over microtubule dynamics in migrating neuroblasts. Similarly, mutations in the ARX gene lead to a spectrum of neurodevelopmental anomalies, including lissencephaly [42]. Lissencephaly is furthermore associated with genetic mutations such as those in the LIS1 or DCX genes and can lead to a spectrum of neurological impairments, including intellectual disability, developmental delays, seizures, and muscle spasticity. Children who survive in utero CMV infection may experience severe neurological impairments and brain malformations, such as microcephaly, lissencephaly, polymicrogyria, and cerebellar hypoplasia [43–45]. For instance, individuals with co-deletion of the closely located genes LIS1 and 14-3-3 display Miller–Dieker lissencephaly syndrome [46]. These proteins interact with other molecules, including Ndel1, to regulate the activity of the microtubule motor protein dynein. This regulation plays a vital role in controlling the orientation of the mitotic spindle during cell division and in facilitating nuclear movement during neuroblast migration. Neu-Laxova syndrome is an uncommon and lethal autosomal recessive condition marked by severe intrauterine growth restriction, microcephaly, distinct facial abnormalities, limb flexion deformities, skin dysplasia, and underdevelopment of various organs. Central nervous system anomalies are prominent, including microcephaly, lissencephaly, agenesis of the corpus callosum, atrophy of the brain, cerebellum, and pons, as well as the absence of olfactory bulbs [42]. This syndrome highlights the impact of genetic mutations on fetal development and emphasizes the critical role of neural structures in shaping phenotypic outcomes. The severe neurological impairments associated with Neu-Laxova syndrome underscore the importance of early developmental processes in determining the integrity and functionality of the central nervous system. Understanding this syndrome offers valuable insights into the pathophysiology of neural development disorders and the complex interactions between genetics and developmental biology [42]. In these individuals, the brain is 10 to 20 times smaller than average. The cerebral hemispheres often appear smooth or cobblestone-like, with a granular texture. The corpus callosum is typically absent, and the lateral ventricles are dilated. The hippocampus is either undeveloped or not identifiable [47].

In type I lissencephaly, neuronal migration to the cerebral cortex is significantly disrupted, with several genetic variants of this condition being recognized. Isolated lissencephaly sequence (ILS) is found in patients with deletions of the LIS1 gene, which encodes for the non-catalytic subunit of platelet-activating factor acetylhydrolase (PAFAH) [48, 49]. Miller–Dieker lissencephaly is another condition that arises from the deletion of the LIS1 gene [46, 49]. Recent research has uncovered the genetic origins of other forms of type I lissencephaly: one is an autosomal recessive type due to mutations in the RELN gene [50], and another is lissencephaly associated with ambiguous genitalia, which is caused by mutations in the ARX gene [51]. In type II lissencephaly, the brain's surface appears cobblestone-like, caused by the accumulation of heterotopic neurons in the marginal zone. The pathogenesis of type II lissencephaly appears to involve defects in O-glycosylation, specifically affecting dystroglycan, based on the functions of all the genes identified as being associated with this condition. Walker–Warburg syndrome, associated with mutations in the POMT1 and POMT2 genes, is one of the disorders within this group. Other mutations in three genes are POMGnT1, FCMD and LARGE. The product of the LARGE gene is also necessary for O- glycosylation [52].

8.2.8 Apert Syndrome

Apert syndrome is an uncommon genetic condition marked by the early fusion of specific skull bones, resulting in craniofacial deformities, syndactyly, and different levels of intellectual impairment. This condition is caused by mutations in the FGFR2 (fibroblast growth factor receptor) gene, which plays a crucial role in cell growth, differentiation, and signaling during embryonic development. The importance of Apert syndrome in the neuropathology of the human hippocampus lies in its impact on brain development and cognitive function. Apert syndrome is characterized by the absence of olfactory bulbs and tracts, fusion of the olfactory tubercles, and hippocampal dysplasia [53]. The early fusion of cranial sutures can lead to increased intracranial pressure, potentially affecting the growth and development of brain structures, including the hippocampus. The hippocampus is vital for memory and learning, and any disruption in its development can lead to cognitive impairments commonly observed in individuals with Apert syndrome. In a study examining the amygdala and hippocampal region in Apert syndrome, bilateral verticalized and malrotated hippocampi were identified in two patients (5.4%). In one patient, persistent epilepsy and a resected amygdala tumor were diagnosed as ganglioglioma [54]. It has been suggested that heightened FGFR3 and FGFR2 activity at the cortical margin, a signaling center near the dorsal midline and choroid plexus, leads to the excessive growth of the temporal lobe. This overgrowth is suggested to lead to hippocampal dysplasia, polymicrogyria, and subependymal neuronal heterotopia [55]. Gain-of-function mutations provide a new function to the mutant gene product. This is exemplified in Apert syndrome, where mutations give FGFR2 a constitutive or "always on" activity. Mutations in FGFR2 have been linked to Apert syndrome, whereas mutations in FGFR3 are associated with thanatophoric dysplasia.

8.2.9 Hippocampal Hypoplasia

Hippocampal hypoplasia may impair neurogenesis and synaptogenesis, potentially influencing functional abilities and playing a role in the factors that lead to intellectual disability (ID) in Down syndrome [56]. Genetically defined syndromes encompass autosomal recessive lissencephaly with cerebellar and hippocampal hypoplasia, which is caused by mutations in the RELN gene. Hippocampal malformations can vary from mild hippocampal malrotations to severe hypoplasia with dentate gyrus aplasia, as observed in thanatophoric dysplasia [57].

Smith-Lemli-Opitz syndrome (SLOS) is a multifaceted genetic condition marked by unique facial characteristics and microcephaly, along with neurodevelopmental and cognitive difficulties. Neuropathological assessment in Smith-Lemli-Opitz syndrome (SLOS) has demonstrated a degree of severe hippocampal hypoplasia. Key findings include delayed rotation of the hippocampus, a shallow hippocampal fissure, absence of the inner and outer blades of the dentate gyrus, a mild dispersion of immature cells in the pyramidal and polymorphic layers of the dentate gyrus, and stunted, thicker CA3–CA1 sections [58].

8.2.10 Rotational Anomaly

Studies on epilepsy have reported rotational malformations, known as 'hippocampal malrotation' (HIMAL) or 'incomplete hippocampal inversion' (IHI), which are developmental anomalies where the hippocampus does not achieve its typical mature form during brain development. When examined using MR imaging, HIMAL is typically characterized by an abnormal round appearance of the hippocampus, with blurred internal structures and a vertical orientation of the collateral sulcus. There may also be mild dilation of the temporal horn and deviation of the fornix. However, unlike hippocampal sclerosis, there is no associated volume loss or signal change. Some imaging studies suggest that hippocampal malrotation is more prevalent among epilepsy patients, while others indicate that its frequency is comparable to that found in healthy individuals. HIMAL is often unilateral and isolated, but it can also be bilateral or accompanied by other developmental abnormalities. These can include agenesis of the corpus callosum, heterotopia, polymicrogyria, or increased complexity of temporal cortex folding [57]. HIMAL can sometimes be confined to a specific section of the hippocampus along its longitudinal axis. Hippocampal rotation and shape abnormalities, although more pronounced in patients with periventricular heterotopia, are observed in various types of malformations of cortical development (MCD) [59]. Patients with various types of MCD typically exhibit a small hippocampus. In cases of widespread MCD due to abnormal neuron migration (i.e., lissencephaly-agyria-pachygyria and subcortical band heterotopia), abnormally enlarged hippocampi are present [59]. There are relatively few pathological studies of HIMAL. Cases of concurrent hippocampal sclerosis are rare, and the pyramidal cell layer in the CA1/subiculum region is usually maintained. However, it frequently shows signs of excessive convolution and a "heaped" appearance [60].

8.2.11 Focal Cortical Dysplasia

Over 40 years ago, Taylor and colleagues first coined the term "focal cortical dysplasia" (FCD) to describe a lesion associated with epilepsy that they believed was distinct from the cortical lesions observed in tuberous sclerosis [61]. FCD is best defined as a developmental malformation of the cortical plate that is presumed to cause clinical seizures. It is characterized by abnormal cytoarchitecture that is limited in scope while maintaining the preservation of the gyral pattern. FCD has significant implications for the neuropathology of the human hippocampus due to its role in the development of epilepsy and its potential impact on hippocampal function. In the following decades, a broader range of localized lesions was grouped under the term "cortical dysplasia." This included various abnormalities identified with the growing use of MRI in assessing epilepsy patients and more subtle histological abnormalities in cortical layering. Lesions associated with FCD exhibit distinct morphological indications of disrupted cytodifferentiation. These lesions exhibit atypical expression of markers linked to the differentiation of neurons and glial cells. Dysplastic lesions may arise when neuroepithelial cells become halted at different stages during their progression toward increased cell specialization. The point at which cell lineage is determined, such as when neuronal and glial cells differentiate in the ventricular zone, suggests that dysplastic lesions with cells displaying mixed neuronal and glial characteristics probably arise in the early stages of prenatal development [62]. FCD, the second most commonly resected pathology, has slightly lower reported seizure-free rates. Over the years, several classification schemes have been developed to categorize dysplasia into subtypes based on histological features. These classifications aim to enhance consistency in reporting across different medical centers. The latest ILAE classification from 2011 introduces a third subtype, FCD III, which distinguishes between dysplasias located adjacent to lesions and those that are isolated [63]. FCD is the third most common structural abnormality found in adult epilepsy surgical cases, following HS and tumors. The relative incidence of FCD subtypes varies across different studies. FCD I is reported more frequently, but there is significant variation in its incidence compared to FCD II [64].

ILAE: FCD III describes irregularities in cortical layering that are linked to a primary lesion located near or impacting the same cortical region or lobe. It is important to distinguish between four unique variants: Type IIIa (FCD) associated with HS; Type IIIb FCD associated with tumors; Type IIIc FCD linked to vascular malformations; and Type IIId FCD connected to primary lesions that develop early in life [63, 65]. In rare instances, Type IIa or IIb (FCD) occurs concurrently with other principal lesions such as HS, cavernomas, or tumors. Apoptotic neurons have been identified in type III of FCD. In the advanced stages, widespread neuronal degeneration, astrocytic gliosis, and cortical spongiosis characterize large regions of pan-laminar or patchy cortical necrosis, while the inflammatory response becomes less noticeable [42].

In this variant, patients with hippocampal sclerosis exhibit alterations in the temporal cortex, either in architectural organization (cortical dislamination) or cytoarchitectural composition (i.e., hypertrophic neurons outside of layer 5). Dual pathology refers to patients with hippocampal sclerosis who also present with a second primary brain lesion, potentially located outside the ipsilateral temporal lobe. This includes tumors, vascular malformations, glial scars, limbic/Rasmussen encephalitis, or malformations of cortical development (including type IIa/IIb FCD) [63, 65]. This form of hemimegalencephaly shares many cytological characteristics with the more localized lesions seen in focal cortical dysplasia and tuberous sclerosis, both of which now have identified associated genes [61, 66]. One case report describes a patient with HS hippocampal sclerosis who also presented with FCD, type IIIa, and polymicrogyria [67]. The pathology of a patient who underwent epilepsy surgery reportedly exhibited a prominent micro-columnar architecture, which is indicative of FCD type Ia [68].

8.3 Metabolic and Neurodegenerative Diseases of Childhood: Neuronal Disorders

8.3.1 Rett Syndrome

In females, X chromosome inactivation during development can result in mosaic expression of the non-functional gene in about half of the neurons. Rett syndrome is a neurodevelopmental disorder that typically emerges between 6 and 18 months of age, predominantly affecting girls (X-linked dominant disorder). It is characterized by a halt and regression in development, loss of speech, and repetitive, stereotypical hand movements. This syndrome is mainly triggered by de novo mutations in the methyl-CpG binding protein 2 (MECP2) gene, which is found on the X chromosome. [69, 70]. The MECP2 gene produces a methyl-CpG-binding protein that plays a role in recruiting histone deacetylase, which suppresses gene transcription [71, 72]. Girls with Rett Syndrome seem to develop typically during the first 12–18 months of life, but they later experience regression, losing much of their previously acquired language, cognitive, social, and motor abilities [73, 74]. Mutations in the CDKL5 gene, which is located on the X chromosome, are responsible for a variety of epileptic syndromes. These range from atypical Rett syndrome and early onset epileptic encephalopathies to milder forms of the disorder [75, 76]. A proven CDKL5 mutation can lead to a syndrome characterized by infantile spasms and Rett-like symptoms, which can include the risk of sudden unexpected death in epilepsy (SUDEP). FOXG1 mutations can also cause a congenital form of Rett syndrome. The underlying mechanisms of Rett syndrome and the complex relationship between genotype and phenotype are still not well understood. However, these areas have been significantly clarified through the development of several transgenic models. MECP2 is essential for adult neural function, as evidenced by the development of a Rett syndrome-like phenotype when the gene is selectively knocked out in mature mice [77]. MECP2 plays a crucial role in neuronal regulation and is at least partially activated in response to neuronal activity [78]. Additionally, mutations in MECP2 elevate the expression of retrotransposons in developing neural progenitors. This raises the intriguing possibility that MECP2 mutations may lead to secondary somatic mutations during neural development [79]. Furthermore, MECP2 mutations result in increased expression of retrotransposons within developing neural progenitors. This suggests the intriguing possibility that these mutations might contribute to secondary somatic mutations during the process of neural development. After a few months, this phase progresses to rapid regression characterized by increased irritability, loss of purposeful hand movements, the emergence of stereotypical hand motions, and a decline in communication abilities. After a few weeks of regression, girls transition into a more stable phase that may persist for several years. This phase is characterized by significant learning disabilities, slowed head growth (or obvious microcephaly), repetitive hand movements, ataxia or apraxia, and irregular breathing patterns. Patients often develop epilepsy and scoliosis. The final stage, usually occurring after age 10, is marked by signs of both upper and lower motor neuron involvement, reduced mobility, worsening scoliosis, and a general decline in seizure activity. The most prominent changes are observed in body and brain growth. Diminished head growth is noticed within the first few months of life, followed by a reduction in overall body growth, impacting both height and weight. In the initial neuropathological examination, Jellinger and Seitelberger observed decreased melanin levels in the pars compacta of the substantia nigra. This finding, later confirmed by subsequent studies, was linked to a slight reduction in tyrosine hydroxylase [73]. Research on the cerebral cortex has shown several cytoarchitectural alterations, including smaller neuron size and a higher packing density of neurons [80]. A significant decrease in dendritic branching is evident in the frontal motor and temporal cortex, and this reduction is more severe than what is seen in the brains of individuals with Down syndrome. Additional findings include variable pathological changes in the basal ganglia and hippocampus, progressive loss of Purkinje cells in the cerebellum, and abnormalities of the olive. Degenerative alterations have also been observed in both the white and gray matter of the spinal cord [73].

8.4 Lysosomal Diseases

Lysosomal diseases, also referred to as lysosomal storage disorders (LSD), are a group of rare, inherited metabolic conditions caused by impaired lysosomal function. Lysosomes are cellular organelles enclosed by membranes and filled with primary acid hydrolases and other proteins essential for breaking down both naturally occurring and phagocytized intracellular macromolecules that digest and recycle materials within the cell. When lysosomes fail to function properly due to enzyme deficiencies, substrates accumulate within the cell, leading to a variety of clinical symptoms depending on the specific disorder. Lysosomal diseases play a significant role in the neuropathology of the hippocampus, primarily due to their impact on cellular waste processing and recycling mechanisms. The hippocampus, crucial for memory and learning, is particularly sensitive to disruptions in cellular homeostasis. The complexity starts with the nomenclature of lysosomal storage diseases, which is historically based partly on the type of primary storage material identified as accumulating within the lysosomes. In LSDs, mutations in lysosomal enzyme genes lead to deficient enzymatic activity, causing an accumulation of substrates such as lipids, glycoproteins, and mucopolysaccharides. This accumulation disrupts neuronal function, impairs synaptic integrity, and promotes cell death within the hippocampus. Additionally, the buildup of waste products can induce inflammation and oxidative stress, further damaging hippocampal neurons and contributing to neurodegeneration. The result is often manifest in cognitive deficits, memory loss, and learning disabilities, highlighting the critical need for targeted therapeutic strategies to address lysosomal dysfunction in the hippocampus. This area of research not only advances our understanding of LSDs but also provides insights into more common neurodegenerative diseases involving lysosomal pathways. A significant characteristic of most lysosomal diseases is their progressive nature, often leading to fatal outcomes. However, individuals affected by these conditions may appear completely normal early in life. These conditions are typically caused by mutations in the genes responsible for encoding lysosomal enzymes, resulting in deficient or non-functional enzymes. Notable examples of lysosomal storage disorders include Gaucher's disease, Fabry disease, Hunter syndrome, and Tay-Sachs disease. These disorders can impact various body systems, such as the skeleton, brain, skin, heart, and central nervous system, frequently leading to severe and progressively worsening symptoms. Available treatment options may involve enzyme replacement therapy, substrate reduction therapy, and supportive care to alleviate symptoms and enhance quality of life [81].

Sphingolipidoses are a category of LSD lysosomal storage disorders resulting from insufficient activity of lysosomal enzymes or defects in proteins essential for the breakdown and recycling of sphingolipids. These protein defects lead to blockages in degradation and salvage pathways, resulting in the excessive accumulation of one or more sphingolipids. A common accompanying feature of these disorders is the intraneuronal accumulation of unesterified cholesterol [42].

8.4.1 Fabry Disease

Fabry disease (FD) is an X-linked recessive inherited metabolic disorder characterized by the accumulation of glycolipids, primarily globotriaosylceramide (Gb3), due to a deficiency of alpha-galactosidase A. It was firstly described as a dermatological disease in 1898 by two different dermatologists. In hemizygous males with Fabry disease, common clinical features include angiokeratomas on the skin, painful peripheral neuropathy, transient ischemic attacks or cerebral infarcts, myocardial infarctions, and renal failure. Corneal and lenticular opacities are usually detected using a slit-lamp examination. These symptoms arise from the gradual buildup of glycolipids in vascular endothelial cells, which results in ischemia. These symptoms usually emerge in childhood or adolescence. While heterozygous females are generally asymptomatic, some symptoms may appear as they age. In the majority of patients with FD, cerebrovascular damage is believed to result from vascular stenosis caused by mural thickening due to glycolipid deposits within the vessel walls or from luminal blockage or thrombosis associated with intraendothelial storage [82]. Even with these earlier observations, atrophy in certain brain areas, especially the thalamus and hippocampus, has been documented in FD patients when compared to healthy controls, after accounting for cerebrovascular events [83]. The hippocampus, a structure linked to psychiatric disorders, has shown volume abnormalities in conditions like depression and schizophrenia, and it has also been reported to be affected in FD. Postmortem studies have revealed the accumulation of Gb3 in neurons within specific brain regions, such as the hippocampus [84]. The accumulation of Gb3 in FD is critically important for the neurophysiology of the hippocampus due to its multifaceted impact on neuronal health and function. Gb3 buildup disrupts normal cellular processes, leading to neuronal dysfunction and compromised synaptic transmission, both of which are vital for learning and memory. This accumulation induces cellular stress and oxidative damage, which can trigger apoptosis and result in the loss of hippocampal neurons. Additionally, Gb3 deposits in the endothelial cells of hippocampal blood vessels can cause vascular stenosis and occlusion, reducing blood flow and exacerbating neuronal injury through ischemia. The inflammatory response activated by Gb3 further contributes to neurodegeneration, as chronic inflammation damages neurons and synapses, impairing cognitive functions. Ultimately, these pathological changes can lead to hippocampal atrophy, a condition associated with significant cognitive deficits, underscoring the critical role of Gb3 in the neuropathology of FD Fabry disease. In the central nervous system, enlarged storage neurons have been identified in areas such as the amygdala, hypothalamic nuclei, hippocampus, entorhinal cortex, and brainstem [42].

8.4.2 Mucopolysaccharidoses

MPSs are a category of lysosomal storage disorders resulting from enzyme deficiencies that impair the progressive breakdown of glycosaminoglycans. MPSs are primarily autosomal recessive disorders, except for Hunter disease (MPS type II), which is X-linked. Each disorder is caused by a deficiency in an enzyme required for a particular step in the glycosaminoglycan (GAG) degradation process, resulting in the excessive buildup of partially degraded GAGs in almost all tissues. To identify common brain pathologies associated with neurological MPS, brain cortex tissues from autopsy materials of patients affected by MPS I, II, IIIA, IIIC, and IIID have been examined [85].

Eleven identified enzyme deficiencies are responsible for the development of seven distinct types of MPSs). There are clinical similarities among different enzyme deficiencies, yet within each deficiency, there is also a wide spectrum of clinical severity. All MPSs exhibit numerous common clinical characteristics, including a chronic and progressive nature, involvement of multiple systems, organ enlargement, dysostosis multiplex, distinctive facial features, hearing impairment, joint issues, and cardiovascular corneal clouding. The specific enzymatic defect dictates the pattern of the primary storage materials [86].

Abnormal ganglioside accumulation has been noted in multiple brain regions of patients with MPS-I and MPS-III, including the cerebral cortex and hippocampus [87]. Neuronal accumulation was confined to the hippocampus. Localized neuronal accumulation in the hippocampus has been documented in MPS IVA cases [88]. Unlike MPS I, MPS II, and MPS III, the extent of neuronal accumulation in the CNS varied, with the most significant storage found in the hippocampus, subiculum, presubiculum, substantia nigra, inferior olive, and neurons in the spinal anterior horn and intermediolateral columns. In patients with MPS III, the accumulation of GAGs and lamellar residues has been observed in the endothelial cells and pericytes of the blood-brain barrier (BBB). In patients with MPS IIIA and IIID, swelling and degeneration of cells, perivascular edema, and collagen buildup have also been observed in the hippocampus, putamen, and cerebellar capillaries, accompanied by significant reductions in tight junction proteins [89].

8.4.3 Mucolipidosis

Mucolipidosis, particularly mucolipidosis IV (ML IV), is of significant importance to the neurophysiology of the hippocampus due to its profound impact on neuronal function and brain architecture. ML-IV is passed down through an autosomal recessive inheritance pattern. Various lipids and acid mucopolysaccharides, including gangliosides, have been identified as the substances that accumulate in storage. The abnormal accumulation is probably due to a dysfunctional endocytic process or a disrupted membrane fusion event caused by the functional absence of the ML-IV protein [90]. In normal cells, endocytosed membrane components or degradation products eventually make their way to the Golgi apparatus. However, in ML-IV disease, this final pathway seems to be obstructed, causing endocytosed membrane components to remain trapped within storage vesicles [91]. ML-IV, resulting from mutations in the MCOLN1 gene, causes the buildup of lipids and other materials in lysosomes, interfering with normal cellular functions. The MCOLN1 gene codes for mucolipin 1, a membrane protein that shares similarities with a class of calcium channels belonging to the transient receptor potential (TRP/TRPL) family. In the hippocampus, this manifests as the formation of smaller punctate inclusions within pyramidal neurons, notably in the CA3 region. Significant neuronal loss and gliosis were detected in the hippocampal CA3 region. These disruptions hinder synaptic function and plasticity, both of which are crucial for learning and memory. Moreover, the accumulation of specific gangliosides, such as GM3, in the hippocampal neurons further exacerbates cellular stress and neuronal dysfunction. These pathological changes contribute to the severe neurological deficits observed in ML-IV patients, highlighting the critical role of lysosomal integrity in maintaining hippocampal health and cognitive functions. Patients with ML-IV experience severe neurological impairments, and the brain pathology linked to this lysosomal disorder remains poorly understood [92]. Neuropathology in lysosomal storage disorder models is generally marked by atrophy and neuronal loss in particular brain regions. To assess the progression of anatomical changes in the Mcoln1−/− brain, volume reductions were noted in four brain regions (cortex, hippocampus, thalamus, and striatum) in mice at 2, 3, and 7 months of age [92].

MLIV: Additionally, smaller punctate inclusions have been observed in the pyramidal neurons of the hippocampal CA3 region in mice. In contrast to GM2 accumulation, GM3 is present in a smaller group of cortical pyramidal neurons, mainly in the infragranular neocortex, entorhinal cortex, and hippocampal CA3 region in mice [93].

8.4.4 Neuronal Ceroid Lipofuscinosis - CLN1 Disease

The term "neuronal ceroid lipofuscinosis" (NCLs) was introduced to categorize a clinically diverse group of diseases that were previously grouped with Tay-Sachs disease and other conditions referred to as "amaurotic family idiocies." Recently, research has identified up to 14 genetically distinct loci associated with different types of (NCL), now designated as CLN1 through CLN14. It is highly probable that this number will continue to increase in the future. NCLs are a group of inherited lysosomal storage disorders marked by the buildup of autofluorescent lipopigments, including ceroid and lipofuscin, within lysosomes. The foundation of this new classification lies in the identification of distinct neuropathological characteristics, particularly the build-up of autofluorescent ceroid lipofuscin-like inclusions within the cytoplasm of neurons and various other cell types. This accumulation occurs due to deficiencies in specific lysosomal enzymes or proteins, disrupting the normal degradation and recycling processes. In NCLs, the inability to adequately break down lipids and proteins leads to their buildup within neurons and other cell types, resulting in progressive cellular dysfunction and death. The accumulation of undigested substrates in hippocampal neurons causes significant neuronal degeneration, disrupting the intricate neuronal circuits and impairing cognitive functions. Different NCL subtypes exhibit distinct patterns of neuronal loss within the hippocampus, with regions such as CA2–CA4 being particularly vulnerable. This selective vulnerability highlights the importance of these regions in hippocampal function. NCLs also trigger glial activation, with microglial activation being most prominent in areas of substantial neuronal loss, while astrocyte activation is observed in regions with less severe neuronal loss [94].

NCLs are lysosomal storage disorders inherited in a recessive manner and are currently categorized into eight types (CLN1-CLN8). CLN1 is/was observed in Finland. In the final stage of CLN1, the hippocampus is nearly entirely destroyed [86]. Individuals affected by CLN1 disease exhibit reduced activity of the lysosomal enzyme palmitoyl protein thioesterase I (PPT1). This enzyme is crucial for the proper breakdown of certain lipids within the lysosomes. Its deficiency leads to the accumulation of lipid-modified proteins, resulting in the progressive neurodegeneration observed in CLN1. The impairment of PPT1 disrupts normal cellular processes, leading to the build-up of autofluorescent lipopigments, primarily in the neurons, which is a hallmark of the disease. Understanding the role of PPT1 in lipid metabolism is essential for developing targeted therapies aimed at alleviating the symptoms and slowing the progression of this debilitating condition.

PPT1 is an essential housekeeping enzyme that is ubiquitously expressed across various tissues. It is especially prevalent in the spleen, brain, and testes, where it plays a critical role in maintaining cellular homeostasis. In the hippocampus, neuronal degeneration in various subtypes of NCL predominantly affects the CA2-CA4 regions. A similar pattern of selective degeneration is seen in immunohistochemically identified interneurons, with the extent of loss varying by phenotype, whereas calretinin-positive interneurons remain largely intact.

NCLs induce more selective pathological changes in the hippocampus, including glial activation. Microglial activation is most prominent in regions with substantial neuronal loss, while astrocyte activation is more noticeable in areas with milder neuronal loss. As a result, NCLs exhibit a distinctive pattern of selective hippocampal pathology, which contrasts with that seen in most temporal lobe epilepsies [95]. Mutations in several other genes have been identified to produce phenotypes resembling NCL, although they have not been officially classified as CLN genes. One of the most significant findings involves mutations in chloride channels, specifically those functioning as Cl^-/H^+ exchangers, which are part of a highly conserved gene family. Mutations in the CLCN7 gene, which encodes a protein located in late endosomes and lysosomes, have been linked to bone disorders in children as well as various central nervous system effects, including cortical atrophy. These findings emphasize the vital role that chloride channels play in maintaining cellular homeostasis and highlight their potential impact on neurological health. CLCN3, which is localized in both endosomes and synaptic vesicles, plays a critical role in cellular function. Its deficiency results in the accumulation of subunit c of mitochondrial ATP synthase (SCMAS) and is associated with neuronal loss in the hippocampus. This loss highlights the importance of CLCN3 in maintaining neuronal integrity and function [96].

8.5 Mitochondrial Diseases

Mitochondrial diseases encompass a wide range of disorders stemming from dysfunctional mitochondria, the cell's powerhouse responsible for ATP production through oxidative phosphorylation. These disorders are caused by mutations in either mitochondrial DNA (mtDNA) or nuclear DNA (nDNA), which impair mitochondrial function. The hippocampus, a brain region integral to learning and memory, is particularly susceptible to the impacts of mitochondrial dysfunction due to its high energy demands. In the hippocampus, impaired mitochondrial function can lead to decreased ATP production, elevated oxidative stress, and disrupted calcium homeostasis, all of which contribute to neuronal damage and death. These pathological processes can manifest as neuronal degeneration, synaptic dysfunction, and impaired neurogenesis, particularly in the dentate gyrus. Consequently, mitochondrial diseases can result in significant cognitive deficits and memory impairments [97].

The most frequently observed neuropathological pattern is seen in mitochondrial encephalomyopathy, lactic acidosis, and stroke-like episodes (MELAS) syndrome. Brain frequently exhibits a broad pattern of nonspecific atrophy along with enlarged ventricular spaces. The defining neuropathological characteristics of MELAS are the presence of infarct or "infarct-like" lesions within the cerebral cortex and subcortical white matter [98, 99]. In a MELAS patient, there was little correlation between threshold levels in different neuronal populations and the pattern of neuropathological changes. Interestingly, the threshold for the m.3243A>G mutation was surprisingly low in the hippocampus. In COX-positive neurons of the occipital cortex and cerebellum, mutation levels surpassed 70%, indicating that the threshold for these neurons was higher than 70% [100].

8.6 Excitotoxicity

Excitotoxicity is a pathological process where neurons are damaged or killed as a result of overstimulation by excitatory neurotransmitters like glutamate. This happens when glutamate receptors, especially NMDA and AMPA receptors, become overactivated, causing an excessive influx of calcium ions into the neuron. This calcium overload triggers a cascade of destructive intracellular processes, including the activation of enzymes that degrade proteins, lipids, and nucleic acids, and the generation of ROS, which cause oxidative stress. Excitotoxicity is especially harmful in the hippocampus, a key brain region involved in learning and memory. The hippocampus has a high density of glutamate receptors, making it especially vulnerable to excitotoxic damage. This susceptibility plays a major role in several neuropathological conditions, including ischemic stroke, traumatic brain injury, and neurodegenerative disorders like Alzheimer's disease. Excitotoxicity in the hippocampus can lead to neuronal death, synaptic dysfunction, and impaired neurogenesis, contributing to cognitive deficits and memory loss [101].

It is now broadly recognized that the abnormal activation of excitatory amino acid receptors is a crucial factor in triggering cellular responses to metabolic disturbances, ultimately leading to cell death. The NMDA and AMPA glutamate receptors, which are most closely linked to excitotoxicity, are most densely concentrated in the hippocampus, cortex, and striatum. This distribution of receptors parallels, to some extent, the regions of necrotic lesions observed after general ischemia in mitochondrial disease, though the correlation is weak [102]. Histological examination revealed relatively preserved neurons and proliferated capillaries in very small, recent, and old necrotic foci within the cerebral cortex, amygdala, hippocampus, and cerebellar cortex [103].

8.7 Nutritional and Toxic Neuropathologies of the Hippocampus

8.7.1 Wernicke-Korsakoff Syndrome

Wernicke and Korsakoff syndromes share the same pathology, though they exhibit some clinical differences. Both are reported to result from thiamine deficiency, with possible contributions from direct ethanol neurotoxicity. A triad of symptoms characterizes fully developed Wernicke syndrome: abnormal mental state, ocular motor dysfunction, and gait disturbances. Korsakoff syndrome (KS), qualitatively different from Wernicke syndrome, is entirely a mental disorder [104]. While it is well established that thiamine deficiency can cause Wernicke encephalopathy in both alcoholic and non-alcoholic individuals, with similar underlying neuropathology in both groups, Korsakoff syndrome is rarely seen in those without a history of alcoholism [105, 106].

On rare occasions, a gray discoloration suggesting necrosis is seen in the regions around the third and fourth ventricles. The most notable microscopic characteristic is gliosis, partly attributed to the increased cell packing density caused by the atrophy in these areas. In most instances, neurons appear largely preserved; however, quantitative studies show that neuron counts are lower in most regions when compared to controls. Evidence of previous hemorrhage, indicated by the presence of hemosiderin-laden macrophages, and endothelial hypertrophy, a hallmark of Wernicke encephalopathy, are also observed. Numerous studies have shown that the hippocampus is selectively affected by alcohol. It has been demonstrated that hippocampal volume decreases, undergoes atrophy, and affects memory. Epilepsy is also a common clinical condition among alcoholics, with hippocampal sclerosis observed in many of these patients [107].

8.7.2 Carbon Monoxide

The pathophysiological mechanism of carbon monoxide (CO) poisoning involves CO binding to hemoglobin, resulting in the formation of carboxyhemoglobin (COHb). The pathological changes in brain tissue from patients with CO poisoning are similar to those seen in hypoxic encephalopathy, featuring brain edema and varying levels of necrosis. Cerebral edema and ischemia can disrupt cerebral circulation, potentially causing ischemic brain necrosis and worsening cerebral hypoxia [108].

Acute CO poisoning involving the cerebral cortex focally has been reported much less frequently. There is a preference for the temporal lobe and hippocampus. Injury may present as transient vasogenic edema or overt necrosis (infarction) without occlusion of cerebral arteries [109]. Hypoxic-ischemic leukoencephalopathy can also occur due to anesthesia mishaps, drug intoxications, and acute CO poisoning. During prolonged hypoxic brain injury, the hippocampus, cerebral cortex, caudate, or cerebellum may suffer diffuse or periarterial damage, which can be followed by localized or widespread cessation of cerebral blood flow [110].

CO poisoning-induced ischemia or hypoxia can result in neuronal necrosis in the hippocampal CA1 region, cerebral cortex, cerebellum, and basal ganglia. Significant edema and hemorrhage were observed in the head of the caudate nucleus. At autopsy, the basal ganglia, cerebral white matter, and hippocampus were found to have a soft texture. Histological findings consistent with edema indicated that necrosis occurred in the cerebral cortex, hippocampus, and globus pallidus. While it is widely recognized that the hippocampal CA region is highly vulnerable to ischemia or hypoxia, it is unexpected that apoptotic cells are more prevalent in the CA2 region than in the CA1 region [111].

8.7.3 Lead

Lead (Pb) exposure is of significant concern for the neuropathology of the hippocampus due to its profound and detrimental effects on neuronal function and development. Lead can disrupt synaptic plasticity, reduce dendritic branching, and impair neurogenesis, all of which are essential processes for cognitive function. The toxic effects of lead include oxidative stress, mitochondrial dysfunction, and inflammation, leading to neuronal cell death and synaptic loss. These changes can result in significant cognitive deficits, particularly affecting memory and learning abilities. Additionally, lead exposure during critical periods of brain development can cause long-lasting structural and functional abnormalities in the hippocampus. The vital impact of lead on the hippocampus is crucial for devising strategies to mitigate its neurotoxic effects and safeguard cognitive health, particularly in vulnerable populations such as children. Lead can be directly ingested from contaminated water, air, and soil, or indirectly through the consumption of contaminated animal meat, fruits, vegetables, and their derivatives. Environmental or occupational lead exposure can be worsened by insufficient protective practices, personal habits, and socio-economic factors. Lead can be found in batteries, solder, plastics, household paints, and gasoline, as well as in items like glass bottles, toys, glazed pottery, granite countertops, calcium supplements, herbal medicines, wild game, face makeup, and cigars. Lead affects the brain, kidneys, liver, blood, and testes, and it has detrimental effects on learning, attention, IQ, memory, hearing, and sociability [112].

The disruption of hippocampal structure and function in rats exposed to lead acetate from adolescence to adulthood underscores the sensitivity of the hippocampus to this metal. Lead exposure induces alterations in the expression of critical proteins involved in the physiological functions of neural tissue and leads to increased levels of lipid peroxidation. This disrupts the redox balance of the hippocampus and results in significant neuronal loss. The combination of these factors leads to functional impairments in mnemonic behavior. These findings highlight the profound impact of lead on hippocampal integrity and cognitive functions, emphasizing the need for strategies to mitigate lead exposure and protect brain health [111].

In summary, research shows that intermittent lead exposure leads to changes in the interaction between astrocytes and microglia in the hippocampus, contributing to neuroinflammation. Evidence from the literature indicates that chronic neuroinflammation caused by lead exposure may heighten the risk of adverse events in individuals with preexisting cardiovascular conditions and/or the elderly. One of the other longitudinal studies tracks the progression of low-level intermittent lead exposure from the fetal stage through adulthood [113].

8.7.4 Domoic Acid

Postmortem examinations have revealed brain injuries characterized by neuronal necrosis and astrogliosis, with the hippocampus and amygdaloid nucleus being particularly affected [114]. Histological analysis of the hippocampus has shown total neuronal loss in the CA1 and CA3 regions, nearly total loss in the CA4 region, and moderate loss in the CA2 region. Irregular neuronal loss has been observed in the medial and basal parts of the amygdala, accompanied by reports of neuronal loss and gliosis in the cortex. Mild to moderate gliosis and neuronal loss have been observed in the dorsal and ventral septal nuclei, the nucleus accumbens, and secondary olfactory regions. Reactive astrocytes have been identified in the sixth cortical layer and the adjacent white matter of the orbital and lateral basal regions, as well as in the first and second temporal gyri, fusiform gyrus, parietal parasagittal cortex, and the insula [115].

Studies have shown that administering domoic acid (DA) systemically to rats at a dose of 2.5 mg/kg leads to the presence of the toxin in the cytoplasm of pyramidal neurons within the hippocampus, as detected by immunohistochemistry 6 and 10 h after treatment. This observation reinforces the theory that this brain region functions as an epileptogenic center as a result of excitotoxin-triggered seizure activity. Neurological symptoms have been reported in animals that survived a 2.5 mg/kg dose of DA, with brain damage becoming evident by the fifth day post-administration. This damage consists of delayed selective neuronal death, large necrotic areas, and significant reactive gliosis. Pathological assessments conducted 54 days post-DA exposure showed ongoing neuronal death, calcified lesions in the thalamus, and persistent gliosis. The presence of nitric oxide synthase (NOS) at 5 and 54 days after exposure indicates that nitric oxide (NO) may play a role in the development of DA-induced neuronal degeneration and ongoing brain inflammation [116].

8.7.5 Trimethyltin

Trimethyltin-induced neurotoxicity was defined by significant acute multifocal neuronal degeneration and necrosis, leading to widespread cell loss primarily in the auditory cortex and hippocampus. Neuronal degeneration in the auditory cortex was confined to the pyramidal neurons, which were distributed extensively across the auditory cortex. In the hippocampus, the pyramidal neurons in the CA1 region of Ammon's horn were the most significantly impacted [117]. Trimethyltin (TMT) is significant for the neuropathology of the human hippocampus due to its potent neurotoxic effects, which provide insights into the mechanisms of neuronal damage and degeneration. TMT exposure causes significant neuronal degeneration and necrosis, especially in the hippocampus, with the CA1 region of Ammon's horn being the most heavily impacted. This selective vulnerability mirrors certain pathological conditions in humans, such as Alzheimer's disease and other neurodegenerative disorders, making TMT a valuable model for studying hippocampal damage. Understanding the impact of TMT on the hippocampus is crucial for several reasons: TMT-induced neurotoxicity helps elucidate the molecular and cellular mechanisms underlying hippocampal neuron degeneration, including excitotoxicity, oxidative stress, and inflammatory responses. The acute multifocal neuronal degeneration and necrosis observed with TMT exposure resemble the histopathological changes seen in various human neurological conditions, aiding in the identification of biomarkers and potential therapeutic targets. TMT serves as a research model for testing neuroprotective agents and strategies aimed at mitigating hippocampal damage, thereby contributing to the development of interventions for hippocampal-related cognitive deficits and diseases. The selective loss of hippocampal neurons due to TMT provides a basis for studying the functional consequences of hippocampal damage, particularly in relation to memory and learning impairments, which are hallmarks of many neurodegenerative conditions. Trimethyltin-induced neurotoxicity is marked by severe acute multifocal neuronal degeneration and necrosis, leading to significant cell loss concentrated in the auditory cortex and hippocampus. Neuronal degeneration in the auditory cortex has been observed to be restricted to pyramidal neurons, which are diffusely distributed across the auditory cortex. In the hippocampus, pyramidal neurons in the CA1 region of Ammon's horn were the most significantly impacted [117].

In summary, at the given doses, TMT toxicity specifically targets the pyramidal cell region within the hippocampus. The dorsal hippocampus has been identified as the main area affected. In the dorsal hippocampus, the CA4 region is the most vulnerable to TMT toxicity, followed by the CA1 and CA3b regions. The dorsal CA3a pyramidal cells and granule cells have been reported to remain unaffected, suggesting possible functional differences in these regions [118].

A distinct group of 14 cases involving sudden-onset amnesia, characterized by acute, complete, and bilateral hippocampal ischemia, was identified in Massachusetts between 2012 and 2016. While a clear etiology was not evident, it was noted that 13 of the 14 individuals tested positive for opioids or had a recorded history of opioid use at the initial assessment [119].

Following excessive iron exposure in the newborn, the principal neurons in the adult hippocampal CA1, CA3, and dentate gyrus (DG) regions exhibit neurodegeneration and neuronal loss [120].

The substantial negative impact of neonatal iron intake on the morphological and biochemical outcomes in particular brain regions, including the substantia nigra, basal ganglia, and hippocampus, has been demonstrated [121].

8.8 Epilepsy

Houser and colleagues reported varying levels of dispersion in the granular layer of the dentate gyrus, including cases of bilaminar organization, in a group of patients with severe temporal lobe epilepsy and a history of early-life febrile seizures.

8.8.1 Hippocampal Sclerosis

Hippocampal sclerosis (HS), also referred to as mesial temporal sclerosis (MTS) or Ammon's horn sclerosis (AHS), is a pathological condition involving the shrinkage and scarring of the hippocampus. This condition involves a stereotypical pattern of neuron loss and gliosis (i.e., an increase in glial cells). The CA1 region of the hippocampus is particularly affected, although other regions such as CA3 and the dentate gyrus may also show signs of damage. HS is most commonly associated with temporal lobe epilepsy (TLE) and is a leading cause of drug-resistant epilepsy. The condition can result from various factors, including prolonged seizures (status epilepticus), head trauma, infections, or developmental abnormalities. Patients with HS often present with chronic seizures, memory impairments, and other cognitive deficits due to the critical role of the hippocampus in memory formation and spatial navigation. Diagnosing HS typically involves imaging techniques such as MRI, which can reveal hippocampal atrophy and increased signal intensity in the affected regions. Histopathological examination post-surgery or post-mortem confirms the diagnosis by showing the characteristic neuronal loss and gliosis. HS appears as a decrease in the volume of the hippocampus, observable through neuroimaging or macroscopic examination. It is closely linked to the clinical syndrome of mesial temporal lobe epilepsy (MTLE), though it may also be present in other epilepsy syndromes and in aging individuals without epilepsy [122].

The histological features of hippocampal sclerosis (HS) include significant loss of pyramidal cells in the CA1 (Sommer's sector), CA3, and CA4 regions of the hippocampus, with the CA2 pyramidal cells and the granular cells of the dentate gyrus being relatively spared [123].

In most of the examined hippocampal resections, morphological changes such as dispersed granular cells in the dentate gyrus are observed [124]. In limbic encephalitis, histological examination reveals significant astrogliosis in the temporal neocortex and hippocampus, a tendency for microglial activation to form nodules, and infiltration of T-lymphocytes into the brain parenchyma [124].

In Rasmussen's encephalitis, there may be patchy inflammation and underlying white matter demyelination, along with involvement of the deep gray nuclei. Inflammation may extend to the hippocampus, potentially leading to the presence of hippocampal sclerosis (HS). When postmortem tissue is available, true bilateral disease accompanied by inflammatory changes is likely quite rare. In this model, contralateral hippocampal sclerosis, along with tissue lesion severity, neuronal loss, and inflammation, was unique and strongly associated with the onset and severity of post-traumatic epilepsy (PTE), as well as accompanying behavioral and cognitive deficits. Therefore, this mouse model could be valuable in evaluating the pathophysiology and novel treatments for PTE [125].

8.9 Neurodegenerative Diseases

8.9.1 Parkinson's Disease

Parkinson's disease (PD) is a widespread neurodegenerative condition with an unknown origin. From a neuropathological perspective, PD is marked by the misfolding and accumulation of alpha-synuclein within cells (α-syn), leading to the formation of Lewy bodies and neurites [126, 127]. Reports indicate that no cell death was detected in the studied cell populations of patients with Parkinson's disease dementia (PDD) and non-demented Parkinson's disease (NPD) [127]. The gross findings in PDD are notable for cortical atrophy, particularly affecting the medial temporal lobe, as well as atrophy of the hippocampus and the rostral brainstem tegmentum, which overlap with the atrophy observed in Alzheimer's disease [128]. Both Parkinson's disease (PD) and dementia with Lewy bodies exhibit substantial neuritic degeneration. These abnormal swollen neuronal processes are not detectable with H&E or standard silver stains but are clearly visible through ubiquitin immunohistochemistry. They are most concentrated in the CA2 subregion of the hippocampus but can also be found in the amygdala, nucleus basalis, and several brainstem nuclei [129].

Following exposure to misfolded species material, neurons develop alpha-synuclein (α-Syn) pathology in a time-dependent manner after a latency period. This pathology begins as neuritic inclusions that progressively transform into larger inclusions within the cell body. The formation of pathology occurs more rapidly in vulnerable neurons, such as Math2+ cells in the hippocampus, leading to accelerated cell death. The formation and accumulation of (α-Syn) pathology in resilient Prox1+ neurons are relatively slower, partly due to their lower levels of aSyn expression. In vulnerable Math2+ neurons, degenerative mechanisms are triggered when the accumulation of α-Syn reaches a critical threshold. In the case of Math2+ neurons, the loss of this key transcription factor precedes cell death. Math2+ neurons reach this threshold more rapidly than adjacent Prox1+ neurons due to their higher endogenous levels of α-Syn protein, which facilitates the faster spread of pathological α-Syn [130].

Notably, a marked decrease in pathology over time is observed in the CA2/3 subfield, aligning with the consolidation of α-Syn pathology into dense somatic inclusions, which is followed by neurodegeneration. Silver-positive degenerating neurites were detected prior to neuronal loss, indicating that this may be an early sign of fibril-induced neurotoxicity and a precursor to neurodegeneration. Importantly, mice injected with α-Syn fibrils exhibited worsening impairments in spatial learning and memory over time. These findings support that the formation of α-Syn aggregates in the forebrain of mice accelerates neurodegenerative changes characteristic of Lewy-associated cognitive dysfunction [131]. The key neuropathological features are Lewy bodies (LB) and Lewy neurites (LN), which are immunopositive neuronal inclusions for alpha-synuclein protein. This analysis revealed a crowded membrane environment, including vesicular structures and dysmorphic organelles. Filaments distributed among membranes and organelles are detectable in most, but not all, alpha-synuclein inclusions [132]. Lewy bodies (LBs) have been identified as the primary pathological hallmark of Parkinson's disease (PD) and dementia with Lewy bodies (DLB) since their microscopic discovery more than a century ago. The protein α-Syn has been identified as a key component of Lewy bodies and is believed to be crucial in their formation, as well as in the development of other Lewy-related pathologies, such as pale bodies and Lewy neurites [133]. McGeer and colleagues were the first to report an increase in the number of HLA-DR+ microglia in the substantia nigra of PD patients compared to healthy brains. In a follow-up study, they also observed an increased presence of microglia in the hippocampus, transentorhinal region, cingulate cortex, and temporal cortex of individuals with Parkinson's disease [134]. In a postmortem study, digital measurements following immunohistochemical staining revealed a significant increase in the severity of alpha-synuclein pathology in the hippocampus, entorhinal cortex, and occipitotemporal cortex of (PDD) cases compared to Parkinson's disease without dementia (PDND) cases. Additionally, there was evidence of an enhanced immune response in limbic and cortical brain regions, including increased microglial activation, T lymphocyte infiltration, upregulation of pro-inflammatory cytokines, and TLR gene expression [134].

8.9.2 Corticobasal Degeneration

Corticobasal degeneration (CBD) holds significant importance for the neuropathology of the human hippocampus due to its role in elucidating the broader mechanisms of neurodegenerative diseases. Although CBD primarily affects the frontoparietal cortex, it also impacts the hippocampus, contributing to cognitive deficits such as memory loss and impaired spatial navigation. The characteristic accumulation of hyperphosphorylated tau protein in CBD provides insights into similar tau pathologies in other neurodegenerative disorders. CBD is a neurodegenerative disorder that advances over time and is defined by specific pathological characteristics. Focal cortical atrophy is present, primarily impacting the parasagittal superior frontoparietal regions, while the temporal and occipital lobes remain relatively unaffected [128].

In cases of corticobasal degeneration-corticobasal syndrome (CBD-CBS), tau pathology has been reported to be prominent in forebrain structures, including the motor and somatosensory cortices and the putamen. In patients with corticobasal degeneration, Richardson syndrome (CBD-RS), a greater presence of tau pathology has been observed in limbic structures compared to CBD-CBS patients. These structures include hippocampal subfields, the dentate gyrus, the anterior nucleus of the thalamus, as well as the medullary tegmentum and cerebellar white matter [135].

8.9.3 Huntington Disease

Huntington's disease (HD) is a neurodegenerative disorder known to cause motor, cognitive, and neuropsychiatric impairment, starting in the neostriatum and leading to more pronounced degeneration in other parts of the brain [136]. HD is a progressive autosomal dominant disorder, which is characterized by a combination of progressive motor and cognitive impairments [137]. HD is marked by degeneration of the striatum and cortical atrophy, but other regions of the brain, including the hippocampus, thalamus, globus pallidus, and substantia nigra, are also impacted [138]. Evidence from human and animal models of HD suggests that various mechanisms disrupt short- and long-term functional neuronal plasticity, potentially affecting learning and memory [137]. Additionally, dendritic atrophy and impaired neurogenesis in the dentate gyrus further compromise hippocampal connectivity and adaptability. These pathological changes collectively elucidate the hippocampus's role in the cognitive decline associated with HD. In conclusion, hippocampal volume changes in HD are minimal or localized, while shape analyses reveal specific alterations in certain regions of the hippocampus.

8.9.4 Aging

Brain regions that are especially vulnerable to aging are involved in functions like learning, planning, memory, and motor control. These include (1) the cortex, particularly the frontal cortex as well as the temporal and other cortical lobes, (2) the striatum (caudate, putamen, nucleus accumbens), (3) the substantia nigra, and (4) the hippocampus. As we age, there is a decline in homeostatic functions, neuromodulation, and neuroprotection involving adenosine and oxytocin receptors. In contrast, receptors associated with inflammation, oxidative stress, and excitotoxicity show an increase in density as aging progresses.

8.9.5 Alzheimer's Disease

The basic pathological findings of the human hippocampus in Alzheimer's disease (AD) are characterized by several key features. Neurofibrillary tangles (NFTs), composed of hyperphosphorylated tau protein, accumulate within neurons, particularly in the CA1 region, leading to neuronal dysfunction and death. Extracellular amyloid plaques, formed from beta-amyloid peptides, deposit throughout the hippocampus and contribute to neurotoxicity and inflammation. There is a significant loss of neurons, especially in the CA1 area, resulting in cognitive impairments, including severe memory loss. Synaptic dysfunction is evident, with a reduction in synaptic density and alterations in synaptic proteins, impairing the synaptic transmission and plasticity essential for learning and memory. Additionally, gliosis, marked by an increased presence of reactive astrocytes and microglia, indicates an inflammatory response to neuronal injury. Granulovacuolar degeneration (GVD) and dystrophic neurites further reflect the extensive cellular stress and degeneration occurring within the hippocampus in AD. These pathological changes collectively underlie the profound cognitive deficits observed in AD.

Nevertheless, Alzheimer's disease (AD) remains the most widespread and well-researched neurodegenerative condition. From a neuropathological perspective, the distribution of two proteinopathic lesions characterizes Alzheimer's disease neuropathological change (ADNC). Amyloid beta (Aβ) plaques gradually develop in the neocortex, then spread to the hippocampus and limbic structures, and eventually reach the brainstem and cerebellum [139].

The neuropathology of Alzheimer's disease (AD) is characterized by the buildup of beta-amyloid plaques and hyperphosphorylated tau in neurofibrillary tangles. These features have been identified in multiple brain areas, including the nucleus basalis of Meynert (NBM) in the basal forebrain, the frontal lobe, hippocampus, cingulate gyrus, amygdala, substantia nigra, various brainstem nuclei, and the cerebral cortex [140]. AD causes long-term suffering in patients and leads to problems in society. Amyloid-beta (Aβ) deposits, neurofibrillary tangles (NFTs), and neuritic plaques (NPs) are prominent pathological findings in the brains of AD patients. Microglia form a barrier that prevents the neurotoxic protofibrillar Aβ42 hotspots around plaques.

In Alzheimer's disease, neuropathological changes are linked to localized rather than widespread Aβ plaques in the hippocampus. Notably, focal Aβ plaques in the hippocampus, rather than all plaque types, show a positive correlation with recognition impairments. Additionally, compared to widespread Aβ plaques, focal Aβ plaques are more strongly associated with neuroinflammation, regardless of whether the response is innate or acquired [141]. The fornix has been confirmed to exhibit p-MAPT neuropathology, indicative of neurofibrillary degeneration at NIA/AA stages B2 and B3. As AD progresses, it is proposed that the fornix disseminates p-MAPT neuropathology from the hippocampal formation to the basal forebrain [142]. In the hippocampus of Alzheimer's patients, this pathology can impact any subregion, though research indicates that the entorhinal cortex and CA1 are the most significantly affected. Pathology in the CA2 region generally begins to appear at Braak stage IV. This distribution can be assessed by examining the presence and extent of Aβ accumulation in the hippocampus [143].

8.9.6 Frontotemporal Dementia with Parkinsonism Linked to Chromosome 17

Frontotemporal dementia with parkinsonism linked to chromosome 17 (FTDP-17) is a hereditary neurodegenerative disorder, with the majority of cases caused by mutations in the microtubule-associated protein tau (MAPT) gene. A patient with FTDP-17 carrying the MAPT R406W mutation was clinically diagnosed with Alzheimer's disease (AD) and showed significant neuronal loss and gliosis in the anteromedial temporal cortex and the pyramidal cell layer of the hippocampus. Moderate neurodegeneration was noted in the amygdaloid nucleus, parahippocampal gyrus, and dorsal vagal nucleus. Additionally, uneven distribution of neurodegeneration was noted in the frontal and occipital cortices [144].

8.9.7 α-Synuclein

The accumulation of alpha-synuclein within the cytoplasm is a shared characteristic of a group of neurodegenerative diseases referred to as synucleinopathies, which include Parkinson's disease, dementia with Lewy bodies (DLB), and multiple system atrophy (MSA). In these synucleinopathies, abnormal α-synuclein aggregates appear as neuronal perikaryal inclusions (NPIs), neurites, or glial cytoplasmic inclusions (GCIs) [145]. Furthermore, the induced α-synuclein pathology aligned with progression from the specific brain injection sites, such as the hippocampus, cortex, or striatum. The observed pathology in the brainstem and spinal cord likely indicated the selective regional vulnerability of these transgenic mice [146].

It is common for the three pathologies—pTau, Aβ, and pSyn—to co-occur in the hippocampus of patients with Alzheimer's disease (AD) and dementia with Lewy bodies (DLB), although the severity is markedly increased in AD cases. Each pathology—pTau, Aβ, and pSyn—appears to follow a specific subregional pattern that is predominantly preserved in both AD and dementia with Lewy bodies (DLB). The combinations and severity of subregional misfolded protein pathologies in the hippocampus transform the local microglial phenotypic composition. High pTau and pSyn loads, associated with increased microglial changes, may exacerbate the vulnerability of the CA1 region in (AD) [147]. Pathological comorbidity is common in neurodegenerative diseases. For example, co-occurrence of a beta- and α-syn at autopsy is found in more than 50% of patients diagnosed with (AD) [148].

8.10 Schizophrenia

Schizophrenia is a complex neuropsychiatric condition with risk factors strongly linked to brain development and maturation [149]. Hippocampal pathology is present in schizophrenia and includes morphological, molecular, and functional components. The fundamental alteration likely involves changes in the precise organization and functioning of neural circuits within the hippocampus and its connections to other structures, particularly the prefrontal cortex. Glutamatergic neurons and synapses are significantly impacted, with a portion of the pathology likely stemming from developmental origins and influenced by a combination of genetic and environmental factors [150].

In schizophrenia, hippocampal size is reduced, resulting in hippocampal shape deformation. Numerous studies have identified a bilateral reduction in hippocampal volume and an enlargement of the lateral ventricles. The molecular profile exhibits changes in the expression levels of various proteins. Neuronal evidence points to dendritic and synaptic abnormalities, including simplified dendritic arbors, a higher occurrence of basal dendrites on granule cells, lower dendritic spine density, variable expression of synaptic proteins, a reduced number of dentate mossy fiber terminal synapses, and changes in nicotinic, glutamatergic, GABA, and serotonin receptors or their subunits.

In schizophrenia, the primary morphological and molecular neuroanatomical findings in the hippocampal/parahippocampal region include entorhinal cytoarchitectural disorganization, reduced neuron size, abnormal cytoskeletal protein expression, aberrant axonal innervation, and alterations in synapse-associated proteins [151, 152]. In chronic schizophrenia, there is evidence of disrupted hippocampal connectivity, which is subsequently followed by impaired neurogenesis. This disruption likely contributes to cognitive deficits [153].

8.11 Infection

Many infectious agents affect the hippocampus. Reduced hippocampal neurogenesis may play a role in the cognitive and emotional symptoms seen in COVID-19 patients, which include impairments in verbal memory and learning, working memory and executive functions, set-shifting, attention, as well as symptoms of anxiety, depression, and PTSD [154].

In summary, COVID-19 patients exhibited perivascular infiltrations along with an increase in microglial cells. A notable histological observation is the substantial loss of neurons in the granular layer of the dentate gyrus, along with neuropil loosening and edema. In vivo analyses have revealed that patients with post-COVID syndrome (PCS) exhibit hippocampal gray matter atrophy, altered microstructural integrity, hypoperfusion, and changes in functional connectivity. These hippocampal structural and functional changes are particularly associated with cognitive dysfunctions, including impairments in attention and memory. The biomarkers GFAP, MOG, CCL11, and NfL have revealed changes in PCS and shown correlations with hippocampal volume alterations in selective hippocampal subfields. Additionally, acute phase samples demonstrated increased presence of GFAP and CCL11 in the hippocampus, along with decreased concentrations of MOG [155].

In a study on a mouse model of leprosy, it was observed that the disease does not directly affect the nervous system but induces changes in working memory, the morphology of the dorsal region of the hippocampus, and neurotransmitter levels [156]. Hippocampal sclerosis (HS) observed in paracoccidioidomycosis can be described as dual pathology. This is because HS is a common pathological finding in temporal lobe epilepsy. The etiology of HS is debated, whether it is acquired or developmental. However, factors that may trigger HS include status epilepticus, febrile seizures, infections, and vascular diseases or malformations [157].

Since the neuroanatomical locations of calcified neurocysticercosis (NCC) lesions do not always correlate with areas of abnormality on electroencephalograms (EEGs), some researchers suggest that the calcifications may be incidental findings in patients with epilepsy. Recent studies have proposed that calcifications may be linked to the hippocampus, which could potentially explain the discrepancy between NCC lesions and EEG abnormalities. HS is defined by the presence of both hippocampal atrophy and visible sclerosis on MRI. NCC may lead to HS through two possible mechanisms: (1) NCC lesions may cause chronic interictal or ictal discharges that injure and exacerbate the hippocampus, or (2) inflammation associated with NCC may directly damage the hippocampus [158].

The reduction in hippocampal neurogenesis may be a factor in the cognitive and emotional symptoms seen in COVID-19 patients, such as difficulties with verbal memory and learning, working memory, executive functions, set-shifting, attention, along with anxiety, depression, and PTSD symptoms [154].

Overall, perivascular infiltrations and an increase in microglial cells were observed in COVID-19 patients. An intriguing histological finding is the significant neuronal loss observed in the granular neurons of the dentate gyrus, accompanied by neuropil loosening and edema. An interesting histological finding is the significant neuronal loss in the granular neurons of the dentate gyrus, with observations of loosened and edematous neutrophil. In vivo analyses have revealed that patients with post-COVID syndrome (PCS) exhibit hippocampal gray matter atrophy, altered microstructural integrity, hypoperfusion, and changes in functional connectivity. These hippocampal structural and functional changes are particularly associated with cognitive dysfunctions, including impairments in attention and memory. The biomarkers GFAP, MOG, CCL11, and NfL have revealed changes in PCS and shown correlations with hippocampal volume alterations in selective hippocampal subfields. Additionally, acute phase samples demonstrated increased presence of GFAP and CCL11 in the hippocampus, along with decreased concentrations of MOG [155].

In a study on a mouse model of leprosy, it was observed that the disease does not directly affect the nervous system but induces changes in working memory, the morphology of the dorsal region of the hippocampus, and neurotransmitter levels [156]. HS observed in paracoccidioidomycosis can be described as dual pathology. This is because HS is a common pathological finding in temporal lobe epilepsy. The etiology of HS is debated, whether it is acquired or developmental. However, factors that may trigger HS include status epilepticus, febrile seizures, infections, and vascular diseases or malformations [157].

Since the neuroanatomical locations of calcified NCC lesions do not always correlate with areas of abnormality on electroencephalograms (EEGs), some researchers suggest that the calcifications may be incidental findings in patients with epilepsy. It has recently been proposed that calcifications may be linked to the hippocampus, which might help explain the discrepancy between NCC lesions and EEG irregularities. HS is defined by the presence of both hippocampal atrophy and visible sclerosis on MRI. NCC may lead to HS through two possible mechanisms: (1) NCC lesions may cause chronic interictal or ictal discharges that injure and exacerbate the hippocampus, or (2) inflammation associated with NCC may directly damage the hippocampus [158].

8.12 Multiple Sclerosis

Multiple sclerosis (MS) is a long-term, inflammatory, demyelinating, and neurodegenerative disorder affecting the CNS. The disease progression is often marked by physical disabilities, including motor and sensory symptoms, typically linked to recurring inflammatory episodes. Alongside these symptoms, 40–70% of individuals with MS also suffer from cognitive impairments [159]. Postmortem examinations of MS patients have documented hippocampal demyelination, nerve cell loss, and diminished expression of neuronal proteins, leading to tissue shrinkage [160].

Experimental data from mouse studies support the role of hippocampal damage in MS-related epileptogenesis. In mice treated with cuprizone, a model of toxic demyelination, extensive hippocampal demyelination has been observed alongside noticeable seizure activity. Cuprizone-induced degeneration of CA1 interneurons in the hippocampus triggers seizures, further disrupting the integrity of intrinsic hippocampal networks. This, in turn, increases susceptibility to neuroinflammatory and neurodegenerative processes [161]. Hippocampal demyelination is frequent and widespread in MS). An autopsy study of patients with chronic multiple sclerosis revealed large demyelinating lesions within and around the hippocampus, independent of cognitive impairment. These lesions were not reported in any of the control individuals. In MS) patients, small, isolated lesions within the hippocampus are commonly observed and show a distinct anatomical pattern, affecting both the CA2 subregion and the hilus of the dentate gyrus [162].

8.13 Paraneoplastic

A peripheral tumor can cause changes such as elevated levels of glucocorticoids and pro-inflammatory cytokines, which induce neuroinflammation, reduce neurogenesis, and alter synaptic plasticity in the hippocampus, impacting the immune and endocrine systems. The presence of a peripheral tumor is thought to cause behavioral dysfunctions, such as cognitive impairment and mood disorders, likely through alterations in the molecular and cellular parameters of the hippocampus [163].

References

1. Anand KS, Dhikav V. Hippocampus in health and disease: an overview. Ann Indian Acad Neurol. 2012;15:239–46.
2. You J, Huang H, Chan CTY, Li L. Pathological targets for treating temporal lobe epilepsy: discoveries from microscale to macroscale. Front Neurol. 2021;12:779558.

3. Wojnarowicz MW, Fisher AM, Minaeva O, Goldstein LE. Considerations for experimental animal models of concussion, traumatic brain injury, and chronic traumatic encephalopathy—these matters matter. Front Neurol. 2017;8:240.
4. Nasb M, Tao W, Chen N. Alzheimer's disease puzzle: delving into pathogenesis hypotheses. Aging Dis. 2024;15:43–73.
5. Christofferson DE, Yuan J. Necroptosis as an alternative form of programmed cell death. Curr Opin Cell Biol. 2010;22:263–8.
6. Vandenabeele P, Galluzzi L, Vanden Berghe T, Kroemer G. Molecular mechanisms of necroptosis: an ordered cellular explosion. Nat Rev Mol Cell Biol. 2010;11:700–14.
7. Baron JC, Yamauchi H, Fujioka M, Endres M. Selective neuronal loss in ischemic stroke and cerebrovascular disease. J Cereb Blood Flow Metab. 2014;34:2–18.
8. Bartolini E, Ferrari AR, Fiori S, Della Vecchia S. Glycaemic imbalances in seizures and epilepsy of Paediatric age: a literature review. J Clin Med. 2023;12:2580.
9. Auer RN, Siesjö BK. Hypoglycaemia: brain neurochemistry and neuropathology. Baillieres Clin Endocrinol Metab. 1993;7:611–25.
10. Murayama S, Bouldin TW, Suzuki K. Selective sparing of Betz cells in primary motor area in hypoxic-ischemic encephalopathy. Acta Neuropathol. 1990;80:560–2.
11. Olney JW. Brain lesions, obesity, and other disturbances in mice treated with monosodium glutamate. Science. 1969;164:719–21.
12. Frandsen A, Schousboe A, Griffiths R. Cytotoxic actions and effects on intracellular Ca2+ and cGMP concentrations of sulphur-containing excitatory amino acids in cultured cerebral cortical neurons. J Neurosci Res. 1993;34:331–9.
13. Nakanishi S. Molecular diversity of glutamate receptors and implications for brain function. Science. 1992;258:597–603.
14. Bliss TV, Gardner-Medwin AR. Long-lasting potentiation of synaptic transmission in the dentate area of the unanaesthetized rabbit following stimulation of the perforant path. J Physiol. 1973;232:357–74.
15. Lüscher C, Malenka RC. NMDA receptor-dependent long-term potentiation and long-term depression (LTP/LTD). Cold Spring Harb Perspect Biol. 2012;4:a005710.
16. Lüscher C, Nicoll RA, Malenka RC, Muller D. Synaptic plasticity and dynamic modulation of the postsynaptic membrane. Nat Neurosci. 2000;3:545–50.
17. Ribe EM, Serrano-Saiz E, Akpan N, Troy CM. Mechanisms of neuronal death in disease: defining the models and the players. Biochem J. 2008;415:165–82.
18. Iwai T, Niwa M, Hara A, Mori H, Uematsu T, Sakai N. DNA fragmentation in the CA2 sector of gerbil hippocampus following transient forebrain ischemia. Brain Res. 2000;857:275–8.
19. Neumann JT, Cohan CH, Dave KR, Wright CB, Perez-Pinzon MA. Global cerebral ischemia: synaptic and cognitive dysfunction. Curr Drug Targets. 2013;14:20–35.
20. Rots ML, de Borst GJ, van der Toorn A, et al. Effect of bilateral carotid occlusion on cerebral hemodynamics and perivascular innervation: an experimental rat model. J Comp Neurol. 2019;527:2263–72.
21. de la Torre JC, Fortin T. Partial or global rat brain ischemia: the SCOT model. Brain Res Bull. 1991;26:365–72.
22. Jan R, Chaudhry GE. Understanding apoptosis and apoptotic pathways targeted cancer therapeutics. Adv Pharm Bull. 2019;9:205–18.
23. Broughton BR, Reutens DC, Sobey CG. Apoptotic mechanisms after cerebral ischemia. Stroke. 2009;40:e331–9.
24. Radak D, Katsiki N, Resanovic I, et al. Apoptosis and acute brain ischemia in ischemic stroke. Curr Vasc Pharmacol. 2017;15:115–22.
25. Corti O, Blomgren K, Poletti A, Beart PM. Autophagy in neurodegeneration: new insights underpinning therapy for neurological diseases. J Neurochem. 2020;154:354–71.
26. Claassen J, Park S. Spontaneous subarachnoid haemorrhage. Lancet. 2022;400:846–62.
27. Wijdicks EFM. Historical appreciation of brain vulnerability from pure hypoxemia. Neurocrit Care. 2023;39:522–6.
28. Fricker M, Tolkovsky AM, Borutaite V, Coleman M, Brown GC. Neuronal cell death. Physiol Rev. 2018;98:813–80.
29. Folkerth RD. The neuropathology of acquired pre- and perinatal brain injuries. Semin Diagn Pathol. 2007;24:48–57.
30. Del Bigio MR, Becker LE. Microglial aggregation in the dentate gyrus: a marker of mild hypoxic-ischaemic brain insult in human infants. Neuropathol Appl Neurobiol. 1994;20:144–51.
31. Adle-Biassette H, Golden JA, Harding B. Developmental and perinatal brain diseases. Handb Clin Neurol. 2017;145:51–78.
32. Folkerth RD. Neuropathologic substrate of cerebral palsy. J Child Neurol. 2005;20:940–9.
33. Miller SL, Huppi PS, Mallard C. The consequences of fetal growth restriction on brain structure and neurodevelopmental outcome. J Physiol. 2016;594:807–23.
34. Dean JM, Bennet L, Back SA, McClendon E, Riddle A, Gunn AJ. What brakes the preterm brain? An arresting story. Pediatr Res. 2014;75:227–33.
35. Thompson DK, Wood SJ, Doyle LW, et al. Neonate hippocampal volumes: prematurity, perinatal predictors, and 2-year outcome. Ann Neurol. 2008;63:642–51.
36. Ball G, Boardman JP, Rueckert D, et al. The effect of preterm birth on thalamic and cortical development. Cereb Cortex. 2012;22:1016–24.

37. Wainwright H. Thanatophoric dysplasia: a review. S Afr Med J. 2016;106:S50–3.
38. Hammou HA, Sennaoui M, Bouzid F, et al. Phenotypic and cytogenetic variability of patau syndrome in Morocco. Afr Health Sci. 2023;23:575–81.
39. Hevner RF. The cerebral cortex malformation in thanatophoric dysplasia: neuropathology and pathogenesis. Acta Neuropathol. 2005;110:208–21.
40. Cereda A, Carey JC. The trisomy 18 syndrome. Orphanet J Rare Dis. 2012;7:81.
41. Meyer AN, Modaff P, Wang CG, et al. Typical achondroplasia secondary to a unique insertional variant of FGFR3 with in vitro demonstration of its effect on FGFR3 function. Am J Med Genet A. 2021;185:798–805.
42. Love S, Perry A, Ironside J, Budka H. Greenfield's neuropathology, 2 vol set. CRC Press; 2018.
43. Jiménez-Caballero PE. Crossed cerebellar atrophy: update. J Neurosci Rural Pract. 2012;3:235–6.
44. Bahari H, Hajaj H, Ayyad A, Messaoudi S, Amrani R. Trisomy 13 with bilateral congenital anophthalmia: a case report. Cureus. 2024;16:e60264.
45. Williams GM, Brady R. Patau syndrome. In: StatPearls. edn. Treasure Island ineligible companies. Disclosure: Robert Brady declares no relevant financial relationships with ineligible companies.: StatPearls Publishing Copyright © 2024, StatPearls Publishing LLC.; 2024.
46. Toyo-oka K, Shionoya A, Gambello MJ, et al. 14-3-3epsilon is important for neuronal migration by binding to NUDEL: a molecular explanation for Miller-Dieker syndrome. Nat Genet. 2003;34:274–85.
47. Allias F, Buenerd A, Bouvier R, et al. The spectrum of type III lissencephaly: a clinicopathological update. Fetal Pediatr Pathol. 2004;23:305–17.
48. Chandrasekaran PK, Jambunathan ST, Zainal NZ. Characteristics of patients with organic brain syndromes: a cross-sectional 2-year follow-up study in Kuala Lumpur, Malaysia. Ann Gen Psychiatry. 2005;4:9.
49. Hattori M, Adachi H, Tsujimoto M, Arai H, Inoue K. Miller-Dieker lissencephaly gene encodes a subunit of brain platelet-activating factor acetylhydrolase [corrected]. Nature. 1994;370:216–8.
50. Hong SE, Shugart YY, Huang DT, et al. Autosomal recessive lissencephaly with cerebellar hypoplasia is associated with human RELN mutations. Nat Genet. 2000;26:93–6.
51. Kitamura K, Yanazawa M, Sugiyama N, et al. Mutation of ARX causes abnormal development of forebrain and testes in mice and X-linked lissencephaly with abnormal genitalia in humans. Nat Genet. 2002;32:359–69.
52. van Tol W, Wessels H, Lefeber DJ. O-glycosylation disorders pave the road for understanding the complex human O-glycosylation machinery. Curr Opin Struct Biol. 2019;56:107–18.
53. Breik O, Mahindu A, Moore MH, Molloy CJ, Santoreneos S, David DJ. Central nervous system and cervical spine abnormalities in Apert syndrome. Childs Nerv Syst. 2016;32:833–8.
54. Munarriz PM, Pascual B, Castaño-Leon AM, et al. Apert syndrome: cranial procedures and brain malformations in a series of patients. Surg Neurol Int. 2020;11:361.
55. Stark Z, McGillivray G, Sampson A, et al. Apert syndrome: temporal lobe abnormalities on fetal brain imaging. Prenat Diagn. 2015;35:179–82.
56. Hasina Z, Wang N, Wang CC. Developmental neuropathology and neurodegeneration of Down syndrome: current knowledge in humans. Front Cell Dev Biol. 2022;10:877711.
57. Sen A, Thom M, Martinian L, Dawodu S, Sisodiya SM. Hippocampal malformations do not necessarily evolve into hippocampal sclerosis. Epilepsia. 2005;46:939–43.
58. Grynspan D, Michaud J, Nikkel SM, Creede E, Staines WA. Hippocampal hypoplasia in Smith-Lemli-Opitz syndrome. Pediatr Dev Pathol. 2013;16:318–20.
59. Montenegro MA, Kinay D, Cendes F, et al. Patterns of hippocampal abnormalities in malformations of cortical development. J Neurol Neurosurg Psychiatry. 2006;77:367–71.
60. Hokkanen SRK, Hunter S, Polvikoski TM, et al. Hippocampal sclerosis, hippocampal neuron loss patterns and TDP-43 in the aged population. Brain Pathol. 2018;28:548–59.
61. McGonigal A. Frontal lobe seizures: overview and update. J Neurol. 2022;269:3363–71.
62. Grove EA, Williams BP, Li DQ, Hajihosseini M, Friedrich A, Price J. Multiple restricted lineages in the embryonic rat cerebral cortex. Development. 1993;117:553–61.
63. Blümcke I, Thom M, Aronica E, et al. The clinicopathologic spectrum of focal cortical dysplasias: a consensus classification proposed by an Ad Hoc Task Force of the ILAE Diagnostic Methods Commission. Epilepsia. 2011;52:158–74.
64. Mohsin SN, Grezenko H, Khan S, et al. Bridging development and disruption: comprehensive insights into focal cortical dysplasia and epileptic management. Cureus. 2023;15:e45996.
65. Najm I, Lal D, Alonso Vanegas M, et al. The ILAE consensus classification of focal cortical dysplasia: an update proposed by an Ad Hoc Task Force of the ILAE Diagnostic Methods Commission. Epilepsia. 2022;63:1899–919.
66. Jayalakshmi S, Nanda SK, Vooturi S, et al. Focal cortical dysplasia and refractory epilepsy: role of multimodality imaging and outcome of surgery. AJNR Am J Neuroradiol. 2019;40:892–8.
67. Fasulo L, Saucedo S, Cáceres L, Solis S, Caraballo R. Migrating focal seizures during infancy: a case report and pathologic study. Pediatr Neurol. 2012;46:182–4.
68. Weckhuysen S, Holmgren P, Hendrickx R, et al. Reduction of seizure frequency after epilepsy surgery in a patient with STXBP1 encephalopathy and

clinical description of six novel mutation carriers. Epilepsia. 2013;54:e74–80.
69. Lioy DT, Wu WW, Bissonnette JM. Autonomic dysfunction with mutations in the gene that encodes methyl-CpG-binding protein 2: insights into Rett syndrome. Auton Neurosci. 2011;161:55–62.
70. Amir RE, Van den Veyver IB, Wan M, Tran CQ, Francke U, Zoghbi HY. Rett syndrome is caused by mutations in X-linked MECP2, encoding methyl-CpG-binding protein 2. Nat Genet. 1999;23:185–8.
71. Damen D, Heumann R. MeCP2 phosphorylation in the brain: from transcription to behavior. Biol Chem. 2013;394:1595–605.
72. Zimmermann CA, Hoffmann A, Raabe F, Spengler D. Role of mecp2 in experience-dependent epigenetic programming. Genes (Basel). 2015;6:60–86.
73. Armstrong DD. Neuropathology of Rett syndrome. J Child Neurol. 2005;20:747–53.
74. He L, Caudill MS, Jing J, et al. A weakened recurrent circuit in the hippocampus of Rett syndrome mice disrupts long-term memory representations. Neuron. 2022;110:1689–99.e6.
75. MacKay CI, Wong K, Demarest ST, Benke TA, Downs J, Leonard H. Exploring genotype-phenotype relationships in the CDKL5 deficiency disorder using an international dataset. Clin Genet. 2021;99:157–65.
76. Weaving LS, Christodoulou J, Williamson SL, et al. Mutations of CDKL5 cause a severe neurodevelopmental disorder with infantile spasms and mental retardation. Am J Hum Genet. 2004;75:1079–93.
77. McGraw CM, Samaco RC, Zoghbi HY. Adult neural function requires MeCP2. Science. 2011;333:186.
78. Ebert DH, Gabel HW, Robinson ND, et al. Activity-dependent phosphorylation of MeCP2 threonine 308 regulates interaction with NCoR. Nature. 2013;499:341–5.
79. Muotri AR, Marchetto MC, Coufal NG, et al. L1 retrotransposition in neurons is modulated by MeCP2. Nature. 2010;468:443–6.
80. Bauman ML, Kemper TL, Arin DM. Pervasive neuroanatomic abnormalities of the brain in three cases of Rett's syndrome. Neurology. 1995;45:1581–6.
81. Picache JA, Zheng W, Chen CZ. Therapeutic strategies for Tay-Sachs disease. Front Pharmacol. 2022;13:906647.
82. Okeda R, Nisihara M. An autopsy case of Fabry disease with neuropathological investigation of the pathogenesis of associated dementia. Neuropathology. 2008;28:532–40.
83. Cortés-Saladelafont E, Fernández-Martín J, Ortolano S. Fabry disease and central nervous system involvement: from big to small, from brain to synapse. Int J Mol Sci. 2023;24
84. Varela P, Carvalho G, Martin RP, Pesquero JB. Fabry disease: GLA deletion alters a canonical splice site in a family with neuropsychiatric manifestations. Metab Brain Dis. 2021;36:265–72.
85. Viana GM, Priestman DA, Platt FM, Khan S, Tomatsu S, Pshezhetsky AV. Brain pathology in mucopolysaccharidoses (MPS) patients with neurological forms. J Clin Med. 2020:9.
86. Platt FM, d'Azzo A, Davidson BL, Neufeld EF, Tifft CJ. Lysosomal storage diseases. Nat Rev Dis Primers. 2018;4:27.
87. Khan SA, Peracha H, Ballhausen D, et al. Epidemiology of mucopolysaccharidoses. Mol Genet Metab. 2017;121:227–40.
88. Koto A, Horwitz AL, Suzuki K, Tiffany CW, Suzuki K. The Morquio syndrome: neuropathology and biochemistry. Ann Neurol. 1978;4:26–36.
89. Bigger BW, Begley DJ, Virgintino D, Pshezhetsky AV. Anatomical changes and pathophysiology of the brain in mucopolysaccharidosis disorders. Mol Genet Metab. 2018;125:322–31.
90. Wakabayashi K, Gustafson AM, Sidransky E, Goldin E. Mucolipidosis type IV: an update. Mol Genet Metab. 2011;104:206–13.
91. Chen CS, Bach G, Pagano RE. Abnormal transport along the lysosomal pathway in mucolipidosis, type IV disease. Proc Natl Acad Sci U S A. 1998;95:6373–8.
92. Grishchuk Y, Sri S, Rudinskiy N, et al. Behavioral deficits, early gliosis, dysmyelination and synaptic dysfunction in a mouse model of mucolipidosis IV. Acta Neuropathol Commun. 2014;2:133.
93. Micsenyi MC, Dobrenis K, Stephney G, et al. Neuropathology of the Mcoln1(−/−) knockout mouse model of mucolipidosis type IV. J Neuropathol Exp Neurol. 2009;68:125–35.
94. Lawrence JM, Schardien K, Wigdahl B, Nonnemacher MR. Roles of neuropathology-associated reactive astrocytes: a systematic review. Acta Neuropathol Commun. 2023;11:42.
95. Tyynelä J, Cooper JD, Khan MN, Shemilts SJ, Haltia M. Hippocampal pathology in the human neuronal ceroid-lipofuscinoses: distinct patterns of storage deposition, neurodegeneration and glial activation. Brain Pathol. 2004;14:349–57.
96. Nita DA, Mole SE, Minassian BA. Neuronal ceroid lipofuscinoses. Epileptic Disord. 2016;18:73–88.
97. Klein IL, van de Loo KFE, Smeitink JAM, et al. Cognitive functioning and mental health in mitochondrial disease: a systematic scoping review. Neurosci Biobehav Rev. 2021;125:57–77.
98. Tetsuka S, Ogawa T, Hashimoto R, Kato H. Clinical features, pathogenesis, and management of stroke-like episodes due to MELAS. Metab Brain Dis. 2021;36:2181–93.
99. Kuriyama M, Umezaki H, Fukuda Y, et al. Mitochondrial encephalomyopathy with lactate-pyruvate elevation and brain infarctions. Neurology. 1984;34:72–7.
100. Ryytty S, Hämäläinen RH. The mitochondrial m.3243A>G mutation on the dish, lessons from in vitro models. Int J Mol Sci. 2023;24:13478.
101. Baracaldo-Santamaría D, Ariza-Salamanca DF, Corrales-Hernández MG, Pachón-Londoño MJ, Hernandez-Duarte I, Calderon-Ospina CA. Revisiting excitotoxicity in traumatic brain

injury: from bench to bedside. Pharmaceutics. 2022;14:152.
102. Brown GK, Squier MV. Neuropathology and pathogenesis of mitochondrial diseases. J Inherit Metab Dis. 1996;19:553–72.
103. Pietronigro DD, Seligman ML, Jones BG, Demopoulos HB. Retarding effects of DNA on the autoxidation of liposomal suspensions. Lipids. 1976;11:808–13.
104. Brust JCM. Chapter 117—Neurology of drug and alcohol addictions. In: Schapira AHV, Byrne E, DiMauro S, Frackowiak RSJ, Johnson RT, Mizuno Y, Samuels MA, Silberstein SD, Wszolek ZK, editors. Neurology and clinical neuroscience. Philadelphia: Mosby; 2007. p. 1537–43.
105. Kopelman MD. The Korsakoff syndrome. Br J Psychiatry. 1995;166:154–73.
106. Homewood J, Bond NW. Thiamin deficiency and Korsakoff's syndrome: failure to find memory impairments following nonalcoholic Wernicke's encephalopathy. Alcohol. 1999;19:75–84.
107. Harper C. The neuropathology of alcohol-specific brain damage, or does alcohol damage the brain? J Neuropathol Exp Neurol. 1998;57:101–10.
108. Li W, Meng J, Lei J, Li C, Yue W. Carbon monoxide poisoning with hippocampi lesions on MRI: cases report and literature review. BMC Neurol. 2024;24:159.
109. Lo CP, Chen SY, Lee KW, et al. Brain injury after acute carbon monoxide poisoning: early and late complications. AJR Am J Roentgenol. 2007;189:W205–11.
110. Sato H, Tanaka T, Kasai K, Tanaka N. An autopsy case of acute carbon monoxide poisoning after a long-term vegetative state. Am J Forensic Med Pathol. 2012;33:341–3.
111. Alves Oliveira AC, Dionizio A, Teixeira FB, et al. Hippocampal impairment triggered by long-term Lead exposure from adolescence to adulthood in rats: insights from molecular to functional levels. Int J Mol Sci. 2020;21:6937.
112. Olympio KP, Gonçalves C, Günther WM, Bechara EJ. Neurotoxicity and aggressiveness triggered by low-level lead in children: a review. Rev Panam Salud Publica. 2009;26:266–75.
113. Shvachiy L, Amaro-Leal Â, Outeiro TF, Rocha I, Geraldes V. Intermittent lead exposure induces behavioral and cardiovascular alterations associated with neuroinflammation. Cells. 2023;12:818.
114. Pulido OM. Domoic acid toxicologic pathology: a review. Mar Drugs. 2008;6:180–219.
115. Cendes F, Andermann F, Carpenter S, Zatorre RJ, Cashman NR. Temporal lobe epilepsy caused by domoic acid intoxication: evidence for glutamate receptor-mediated excitotoxicity in humans. Ann Neurol. 1995;37:123–6.
116. Vieira AC, Alemañ N, Cifuentes JM, Bermúdez R, Peña ML, Botana LM. Brain pathology in adult rats treated with domoic acid. Vet Pathol. 2015;52:1077–86.
117. Vuorimaa M, Kareinen I, Toivanen P, Karlsson S, Ruohonen S. Deep learning-based segmentation of morphologically distinct rat hippocampal reactive astrocytes after trimethyltin exposure. Toxicol Pathol. 2022;50:754–62.
118. Robertson DG, Gray RH, de la Iglesia FA. Quantitative assessment of trimethyltin induced pathology of the hippocampus. Toxicol Pathol. 1987;15:7–17.
119. Barash JA, Somerville N, DeMaria A Jr. Cluster of an unusual amnestic syndrome—Massachusetts, 2012-2016. MMWR Morb Mortal Wkly Rep. 2017;66:76–9.
120. Li J, Ding Y, Zhang J, et al. Iron overload suppresses hippocampal neurogenesis in adult mice: implication for iron dysregulation-linked neurological diseases. CNS Neurosci Ther. 2024;30:e14394.
121. Agrawal S, Berggren KL, Marks E, Fox JH. Impact of high iron intake on cognition and neurodegeneration in humans and in animal models: a systematic review. Nutr Rev. 2017;75:456–70.
122. Bandopadhyay R, Liu JY, Sisodiya SM, Thom M. A comparative study of the dentate gyrus in hippocampal sclerosis in epilepsy and dementia. Neuropathol Appl Neurobiol. 2014;40:177–90.
123. Blümcke I, Coras R, Miyata H, Ozkara C. Defining clinico-neuropathological subtypes of mesial temporal lobe epilepsy with hippocampal sclerosis. Brain Pathol. 2012;22:402–11.
124. Metodiev D, Minkin K, Ruseva M, Ganeva R, Parvanov D, Nachev S. Pathomorphological diagnostic criteria for focal cortical dysplasias and other common epileptogenic lesions-review of the literature. Diagnostics (Basel). 2023;13:1311.
125. Golub VM, Reddy DS. Contusion brain damage in mice for modelling of post-traumatic epilepsy with contralateral hippocampus sclerosis: comprehensive and longitudinal characterization of spontaneous seizures, neuropathology, and neuropsychiatric comorbidities. Exp Neurol. 2022;348:113946.
126. Dickson DW. Parkinson's disease and parkinsonism: neuropathology. Cold Spring Harb Perspect Med. 2012;2:a009258.
127. Villar-Conde S, Astillero-Lopez V, Gonzalez-Rodriguez M, et al. The human hippocampus in Parkinson's disease: an integrative stereological and proteomic study. J Parkinsons Dis. 2021;11:1345–65.
128. Dickson DW, Braak H, Duda JE, et al. Neuropathological assessment of Parkinson's disease: refining the diagnostic criteria. Lancet Neurol. 2009;8:1150–7.
129. Montanez-Miranda C, Bramlett SN, Hepler JR. RGS14 expression in CA2 hippocampus, amygdala, and basal ganglia: implications for human brain physiology and disease. Hippocampus. 2023;33:166–81.
130. Luna E, Decker SC, Riddle DM, et al. Differential α-synuclein expression contributes to selective vulnerability of hippocampal neuron subpopula-

tions to fibril-induced toxicity. Acta Neuropathol. 2018;135:855–75.
131. Dues DJ, Nguyen APT, Becker K, Ma J, Moore DJ. Hippocampal subfield vulnerability to α-synuclein pathology precedes neurodegeneration and cognitive dysfunction. npj Parkinsons Dis. 2023;9:125.
132. Shahmoradian SH, Lewis AJ, Genoud C, et al. Lewy pathology in Parkinson's disease consists of crowded organelles and lipid membranes. Nat Neurosci. 2019;22:1099–109.
133. Goedert M, Spillantini MG, Del Tredici K, Braak H. 100 years of Lewy pathology. Nat Rev Neurol. 2013;9:13–24.
134. Kouli A, Camacho M, Allinson K, Williams-Gray CH. Neuroinflammation and protein pathology in Parkinson's disease dementia. Acta Neuropathol Commun. 2020;8:211.
135. Kouri N, Murray ME, Hassan A, et al. Neuropathological features of corticobasal degeneration presenting as corticobasal syndrome or Richardson syndrome. Brain. 2011;134:3264–75.
136. Wilkes FA, Jakabek D, Walterfang M, et al. Hippocampal morphology in Huntington's disease, implications for plasticity and pathogenesis: the IMAGE-HD study. Psychiatry Res Neuroimaging. 2023;335:111694.
137. Weerasinghe-Mudiyanselage PDE, Ang MJ, Kang S, Kim JS, Moon C. Structural plasticity of the hippocampus in neurodegenerative diseases. Int J Mol Sci. 2022;23:3349.
138. Puigdellívol M, Saavedra A, Pérez-Navarro E. Cognitive dysfunction in Huntington's disease: mechanisms and therapeutic strategies beyond BDNF. Brain Pathol. 2016;26:752–71.
139. Thal DR, Rüb U, Orantes M, Braak H. Phases of a beta-deposition in the human brain and its relevance for the development of AD. Neurology. 2002;58:1791–800.
140. Wenk GL. Neuropathologic changes in Alzheimer's disease. J Clin Psychiatry. 2003;64(Suppl 9):7–10.
141. Liu F, Sun J, Wang X, et al. Focal-type, but not diffuse-type, amyloid beta plaques are correlated with Alzheimer's neuropathology, cognitive dysfunction, and neuroinflammation in the human hippocampus. Neurosci Bull. 2022;38:1125–38.
142. Plowey ED, Ziskin JL. Hippocampal phospho-tau/MAPT neuropathology in the fornix in Alzheimer disease: an immunohistochemical autopsy study. Acta Neuropathol Commun. 2016;4:114.
143. Walker JM, Orr ME, Orr TC, et al. Spatial proteomics of hippocampal subfield-specific pathology in Alzheimer's disease and primary age-related tauopathy. Alzheimers Dement. 2024;20:783–97.
144. Ishida C, Kobayashi K, Kitamura T, Ujike H, Iwasa K, Yamada M. Frontotemporal dementia with Parkinsonism linked to chromosome 17 with the MAPT R406W mutation presenting with a broad distribution of abundant senile plaques. Neuropathology. 2015;35:75–82.
145. Alafuzoff I, Parkkinen L, Al-Sarraj S, et al. Assessment of alpha-synuclein pathology: a study of the BrainNet Europe consortium. J Neuropathol Exp Neurol. 2008;67:125–43.
146. Uchihara T, Giasson BI. Propagation of alpha-synuclein pathology: hypotheses, discoveries, and yet unresolved questions from experimental and human brain studies. Acta Neuropathol. 2016;131:49–73.
147. Fixemer S, Ameli C, Hammer G, et al. Microglia phenotypes are associated with subregional patterns of concomitant tau, amyloid-β and α-synuclein pathologies in the hippocampus of patients with Alzheimer's disease and dementia with Lewy bodies. Acta Neuropathol Commun. 2022;10:36.
148. Hamilton RL. Lewy bodies in Alzheimer's disease: a neuropathological review of 145 cases using alpha-synuclein immunohistochemistry. Brain Pathol. 2000;10:378–84.
149. Birnbaum R, Weinberger DR. The genesis of schizophrenia: an origin story. Am J Psychiatry. 2024;181:482–92.
150. Schmitt A, Falkai P, Papiol S. Neurodevelopmental disturbances in schizophrenia: evidence from genetic and environmental factors. J Neural Transm (Vienna). 2023;130:195–205.
151. Arnold SE. Cellular and molecular neuropathology of the parahippocampal region in schizophrenia. Ann N Y Acad Sci. 2000;911:275–92.
152. Harrison PJ. The hippocampus in schizophrenia: a review of the neuropathological evidence and its pathophysiological implications. Psychopharmacology (Berl). 2004;174:151–62.
153. Falkai P, Malchow B, Wetzestein K, et al. Decreased oligodendrocyte and neuron number in anterior hippocampal areas and the entire hippocampus in schizophrenia: a stereological postmortem study. Schizophr Bull. 2016;42(Suppl 1):S4–s12.
154. Méndez R, Balanzá-Martínez V, Luperdi SC, et al. Long-term neuropsychiatric outcomes in COVID-19 survivors: a 1-year longitudinal study. J Intern Med. 2022;291:247–51.
155. Díez-Cirarda M, Yus-Fuertes M, Sanchez-Sanchez R, et al. Hippocampal subfield abnormalities and biomarkers of pathologic brain changes: from SARS-CoV-2 acute infection to post-COVID syndrome. EBioMedicine. 2023;94:104711.
156. Becerril-Villanueva E, Ponce-Regalado MD, Pérez-Sánchez G, et al. Chronic infection with Mycobacterium lepraemurium induces alterations in the hippocampus associated with memory loss. Sci Rep. 2018;8:9063.
157. Rosa-JÚnior M, Grenfell MLR, PeÇanha PM. Hippocampal sclerosis in paracoccidioidomycosis. Arq Neuropsiquiatr. 2020;78:384.
158. Herrick JA, Bustos JA, Clapham P, Garcia HH, Loeb JA. Unique characteristics of epilepsy development in neurocysticercosis. Am J Trop Med Hyg. 2020;103:639–45.

159. Rocca MA, Amato MP, De Stefano N, et al. Clinical and imaging assessment of cognitive dysfunction in multiple sclerosis. Lancet Neurol. 2015;14:302–17.
160. Dutta R, Chang A, Doud MK, et al. Demyelination causes synaptic alterations in hippocampi from multiple sclerosis patients. Ann Neurol. 2011;69:445–54.
161. Lapato AS, Szu JI, Hasselmann JPC, Khalaj AJ, Binder DK, Tiwari-Woodruff SK. Chronic demyelination-induced seizures. Neuroscience. 2017;346:409–22.
162. Klaver R, De Vries HE, Schenk GJ, Geurts JJ. Grey matter damage in multiple sclerosis: a pathology perspective. Prion. 2013;7:66–75.
163. Yang M, Kim J, Kim JS, et al. Hippocampal dysfunctions in tumor-bearing mice. Brain Behav Immun. 2014;36:147–55.

Neuroimaging Techniques for the Investigation of Hippocampus

9

Gökhan Kahraman, Pınar Yazgan, and Ahmet Tuncay Turgut

9.1 Introduction

The hippocampus, a vital part of the limbic system, is a bilateral structure found in the medial temporal lobes [1].Because of its critical role in cognitive processes and participation in various neurological and psychiatric diseases, studying the function and structure of the hippocampus has been a major focus of neuroscience research. As a result, imaging of the hippocampus has become an important component of clinical and research neuroimaging. Neuroimaging methods have transformed the assessment of the hippocampus, allowing researchers to study its morphology, connectivity, and functional features. This chapter provides an overview of the primary neuroimaging methods used to study the human hippocampus, detailing their principles and applications, followed by a brief discussion of normal hippocampal anatomy and lesions.

G. Kahraman (✉)
Department of Radiology, Hacettepe University Faculty of Medicine, Ankara, Türkiye
e-mail: gokhankahraman1@outlook.com

P. Yazgan
Department of Radiology, Ankara Bilkent City Hospital, Ankara, Türkiye

A. T. Turgut
Department of Radiology, Ankara Medipol University Faculty of Medicine, Ankara, Türkiye

9.2 Imaging Modalities

9.2.1 Ultrasound

High-frequency sound waves are used in ultrasound imaging to produce visualizations of the body's internal structures. Lower-frequency transducers are required for transcranial ultrasound to penetrate through the dense skull bone; alternatively, other approaches, like the temporal bone window, can be used where the bone is thinner. Ultrasound imaging is extremely operator-dependent, necessitating expertise and experience to provide accurate and repeatable images [2].

The fetal fornix and hippocampus can be visualized with ultrasonography during the second-trimester abnormality scanning [3]. Nevertheless, the effectiveness of ultrasound imaging of the hippocampus is limited, particularly in adults. The hippocampus is located deep inside the temporal lobe, making it difficult to image with ultrasonography. The presence of acoustic artifacts, like reverberations and shadowing, can deteriorate image quality and obscure important anatomical details. One of the key disadvantages of hippocampal ultrasound imaging is its inferior resolution and image quality compared to magnetic resonance imaging (MRI). These limitations can restrict the applicability of this method. Multi-modal imaging approaches can provide a more comprehensive understanding of the hippocampus.

M. Turgut et al. (eds.), *The Human Hippocampus*, https://doi.org/10.1007/978-3-031-88907-3_9

9.2.2 Computed Tomography (CT)

CT utilizes X-rays to create detailed cross-sectional images of the body. The fundamental components include an X-ray tube as the source, a detector array, and a computer system for reconstructing images. The X-ray tube rotates around the patient, acquiring numerous projections at various angles. The attenuation of the body part that was scanned is measured by detectors. To generate a 3D image of the area that is scanned, algorithms are used to reconstruct these projections [4]. The various densities of air, fat, water, and bone can be measured in Hounsfield units. In situations where MRI is not an appropriate modality, an intravenous iodinated contrast agent may be administered during the scan to obtain angiograms of the cerebral arteries or to assess tumors. While MRI is more commonly used in neuroimaging due to its superior soft tissue contrast, CT has unique advantages. Calcification, hemorrhage, and gross structural abnormalities can all be detected with CT imaging [5].

Hippocampal calcification is frequently observed as an incidental finding in older patients, but its clinical significance is unknown. In a group of patients with acute ischemic stroke, hippocampal calcification was found in 34% of cases and was significantly associated with age, hyperlipidemia, and diabetes mellitus, indicating an atherosclerotic cause [6]. CT is the preferred imaging technique for evaluating traumatic brain injuries and acute intracranial hemorrhage, and it is highly effective in detecting cerebral contusions, including those involving the hippocampal region [7, 8]. Because its soft tissue contrast is inferior relative to MRI, CT is not as useful for revealing subtle pathological changes or precise hippocampus anatomy. MRI is the most efficient technique to evaluate conditions like hippocampal sclerosis or inflammatory changes [9].

Recent developments in CT technology, including dual-energy and photon-counting CT, show potential for improving the quality of images while decreasing radiation exposure [10]. These advancements could increase the diagnostic capabilities of CT for hippocampal imaging.

9.2.3 Nuclear Medicine

Nuclear medicine methods, particularly brain perfusion single-photon emission computed tomography (SPECT) and metabolic positron emission tomography (PET) using fluorine-18 fluorodeoxyglucose (FDG), are often used in the assessment of hippocampal function and related pathways [11]. For brain perfusion SPECT, patients should rest in a quiet, low-lit room before and after radiopharmaceutical injection. Epilepsy evaluations can only be performed with SPECT and require electroencephalography monitoring [12]. FDG PET is more accurate than perfusion SPECT for hippocampus-related disorders and superior to SPECT for the diagnosis of neurodegenerative diseases [13].

9.2.4 Magnetic Resonance Imaging

MRI generates detailed cross-sectional images of the brain by utilizing strong magnetic fields and radio waves. Images are created according to the magnetic characteristics of hydrogen atoms in water molecules, and this modality does not require radiation. MRI is the ideal method for brain imaging owing to its higher resolution of soft tissue contrast [5]. Thus, the preferred-imaging method for hippocampus assessment is MRI. It is widely used for the assessment of hippocampal volume and morphology [11]. However, in some cases, such as evaluating intracranial hemorrhage, calcifications, and bone structures (e.g., fractures), CT is often preferred. Additionally, contraindications of MRI are critical for ensuring patient safety and the effective use of this imaging modality. One primary contraindication is the presence of MRI-incompatible internal metallic devices or implants, such as various aneurysm clips, cochlear implants, pacemakers, and some prosthetic devices. These items can malfunction or become displaced in the magnetic field, posing significant risks. MRI-compatible metal objects, like spinal fixation hardware or dental braces, can create susceptibility artifacts that may disrupt the quality of the images. Anxiety and claustrophobia are also notable contraindications, as the enclosed nature of MRI machines can cause significant distress. In such cases, sedation or the use of open MRI systems might be considered. Other factors limiting the use of MRI include the inability to remain physically still, and morbid obesity [14].

Several MRI pulse sequences have been developed to improve tissue characterization. T1- and T2-weighted imaging are fundamental techniques in MRI that provide critical information about tissue characteristics and are essential for accurate diagnosis and assessment in neuroimaging. T1-weighted imaging (T1WI) highlights differences in longitudinal relaxation times of tissues, producing images where fluids appear. T1WI is commonly used to visualize normal anatomical structures and to detect abnormalities such as fat, hemorrhage, and gadolinium-enhanced lesions with high spatial resolution. On the other hand, T2-weighted imaging (T2WI) emphasizes differences in transverse relaxation times, which makes it highly sensitive to fluid content and edema within tissues. T2WI is particularly valuable for identifying pathological changes, such as inflammation, edema, tumors, and certain lesions that exhibit higher water content. Fluid-attenuated inversion recovery (FLAIR) sequence is an essential technique designed to enhance the visibility of brain and spinal cord lesions, providing T2-weighted images, making cerebrospinal fluid (CSF) appear dark by nullifying the signals from CSF. This method involves the use of an inversion recovery pulse sequence with a long inversion time to suppress the bright signal of CSF, thereby improving the contrast between pathological tissues and normal brain structures. This technique enhances the visibility of T2-bright lesions within the brain parenchyma, like masses, ischemia, edema, and demyelinating diseases. Fat suppression methods are advanced imaging techniques designed to selectively nullify the signal from adipose tissue, enhancing the visibility of underlying anatomical structures and pathological conditions. Fat suppression methods can be used in T1WI as well as T2WI. T1WI with fat suppression is superior for contrast enhancement, whereas T2WI with fat suppression is more effective for detecting edema.

Susceptibility-weighted imaging (SWI) is a sophisticated MRI method that uses differences in magnetic susceptibility to enhance the visibility of various tissues and substances, such as calcifications, iron deposits, and deoxygenated blood. SWI is particularly useful in the detection and evaluation of small veins, microhemorrhages, and other pathologies that are not easily detectable with conventional MRI sequences.

Diffusion MRI measures the random motion of water molecules within tissues, known as Brownian motion. The technique primarily includes diffusion-weighted imaging (DWI) and diffusion tensor imaging (DTI). DWI generates contrast in images through detecting changes in water molecule diffusion. Areas with restricted water diffusion, such as acute ischemic stroke, appear hyperintense on DWI, making this approach sensitive for early stroke diagnosis. DWI is also useful in detecting and characterizing tumors, abscesses, and other lesions in which cellular density influences diffusion characteristics. DTI provides information by measuring the directional dependence of water diffusion, known as anisotropy. DTI is a valuable technique for mapping the white matter tracts in the brain through tractography. This enables the visualization of neuronal pathways. DTI plays an essential role in studying brain connections, neural development, and various diseases. White matter tracts can be qualitatively evaluated using color maps that represent different orientations with distinct colors, as well as three-dimensional tractography images. The quantitative metrics derived from diffusion MRI, such as apparent diffusion coefficient (ADC) and fractional anisotropy (FA), contribute to its diagnostic utility. To distinguish between vasogenic edema, which does not exhibit restricted diffusion, and cytotoxic edema, which does, ADC maps are crucial. FA values obtained from DTI provide information about white matter integrity and are used to examine various neurological conditions in both clinical settings and research.

Magnetic resonance spectroscopy (MRS) is an advanced imaging method that enhances the capabilities of conventional MRI by providing metabolic and biochemical information. MRS analyzes the chemical composition of tissues, enabling the identification and measurement of various metabolites. This method can be applied to both single and multi-voxels, depending on the size, homogeneity, and location of the lesion. MRS provides information about cellular function, energy metabolism, and biochemical changes associated with various disorders by detecting relative metabolite concentrations, including choline (Cho), creatine (Cr), N-acetylaspartate (NAA), lactate, and myoinositol. MRS is frequently used for the assessment of tumors, infections, demyelinating diseases, and neurodegenerative diseases. For instance, a decrease in NAA levels, an indicator of neuronal health, can suggest neuronal loss or dysfunction, whereas elevated Cho levels can indicate increased cellular turnover associated with tumor growth or inflammation. Increased lactate levels may indicate anaerobic metabolism, which is commonly seen in tumors or ischemia.

Perfusion MRI, an advanced imaging technique, provides crucial information about tissue blood flow, volume, and vascular permeability. This method is essential in both diagnosing and managing a variety of neurological diseases, such as stroke, tumors, and neurodegenerative diseases. Perfusion MRI is useful in the early detection and monitoring of stroke because it identifies areas of reduced blood flow and tissue at risk of infarction (penumbra). Perfusion MRI helps characterize tumors, guide biopsy, plan treatment, and monitor therapeutic response by evaluating changes in tumor vascularity and perfusion. It also detects regional perfusion deficits that correlate with cognitive decline in neurodegenerative diseases such as Alzheimer's disease. Perfusion MRI can be performed using various techniques, including dynamic susceptibility contrast (DSC), dynamic contrast-enhanced (DCE), and arterial spin labeling (ASL). DSC MRI involves the intravenous injection of a gadolinium-based contrast agent, followed by image acquisition to monitor the agent's transit through the cerebral blood vessels. This technique provides quantitative maps of relative cerebral blood flow (rCBF), relative cerebral blood volume (rCBV), time to peak (TTP), and mean transit time (MTT), which play a crucial role in determining the extent and severity of ischemic stroke, identifying penumbral tissue, and classifying tumor grades based on their perfusion characteristics. In ASL, gadolinium contrast agents are not needed; instead, magnetically labeled arterial blood water serves as an endogenous tracer for measuring CBF. ASL is especially beneficial for patients who cannot be given contrast agents due to allergies or renal dysfunction. DCE perfusion imaging is a valuable technique that involves the injection of a gadolinium contrast agent intravenously, followed by rapid and continuous imaging. DCE perfusion provides detailed information about blood flow, volume, and capillary permeability within tissues. Parameters of capillary permeability, including the transfer constant ($\boldsymbol{K^{trans}}$), fractional plasma volume ($\boldsymbol{v_p}$), and fractional volume of the tissue extracellular space ($\boldsymbol{v_e}$), can be measured.

Functional MRI (fMRI) is a technique that measures and maps brain activity by detecting changes in blood flow and oxygenation. fMRI relies on the theoretical concept of neurovascular coupling, which suggests that there is a strong relationship between neural activity and changes in blood flow in the brain. fMRI can detect changes in blood oxygenation levels caused by increased oxygen use in a specific area of the brain. The most common fMRI technique is blood oxygen level-dependent (BOLD) imaging, which uses differences in the magnetic properties of oxygenated and deoxygenated blood. When neural activity increases in a specific brain area, oxygenated blood flow changes the BOLD signal, allowing researchers to detect brain activation patterns associated with particular tasks or stimuli. In clinical practice, fMRI is used to map eloquent brain areas before surgery, such as tumor removal or epilepsy treatment. This mapping helps surgeons avoid critical areas, preserving essential brain functions and reducing the risk of postoperative deficits. fMRI has revolutionized cognitive neuroscience by providing insights into the functional organization of the brain. Furthermore, fMRI has important applications in psychiatric and neurological research, aiding in the understanding of disorders such as depression and neurodegenerative diseases.

As previously mentioned, MRI is the preferred method for evaluating the hippocampus. For certain clinical conditions, such as epilepsy, dementia, and transient global amnesia (TGA), specialized MRI sequences focused on the hippocampus are used alongside standard brain MRI protocols. In epilepsy and dementia, T2-weighted sequences with coronal views perpendicular to the hippocampal body are essential. For TGA, coronal and DWI with thin 3 mm sections and a high b value of 2000 s/mm^2 are recommended. The volumetric quantitative assessment continues to be a subject of discussion [11].Techniques such as voxel-based morphometry (VBM) and manual segmentation can help provide accurate measurements of hippocampal subfields, which are essential for the study of conditions like depression, Alzheimer's disease, and epilepsy [15–17]. In the clinical management of dementia, these techniques can provide a crucial and effective complement. In contrast, for epilepsy, visual MRI analysis appears to be more sensitive than automated volumetry, since structural changes in the hippocampal subfields usually occur before any noticeable volumetric changes [11].

9.3 Anatomy, Normal Imaging Properties, and Lesions of the Hippocampus

9.3.1 Anatomy and Normal Imaging Properties

The hippocampus, situated in the hippocampal sulcus below the temporal horn of the lateral ventricle and above the parahippocampal gyrus, is part of the hippocampal formation, which includes the dentate gyrus and subicular plexus. The hippocampus itself is divided into three parts: head, body, and tail, and consists of four regions: cornu ammonis 1–4 (CA1, CA2, CA3, and CA4) (Fig. 9.1). In MRI evaluations, the hippocampal internal architecture (HIA) is assessed by observing the laminar appearance of gray and white matter on coronal sections, identifying all segments of the cornu ammonis, and observing the continuity of the stratum radiatum, lacunosum, and moleculare (SRLM) [18].

The hippocampus interacts with the amygdala, parahippocampal gyrus, subiculum, and entorhinal cortex. The alveus, along with the fimbria, extends into the temporal horn, while the uncus projects from the parahippocampal gyrus [11].

9.3.2 Hippocampal Lesions

Diagnosing hippocampal lesions can be challenging and requires careful clinical and laboratory correlation since similar imaging characteristics can appear in different diseases. Alves et al. proposed an algorithm to assist in the imaging diagnosis of hippocampal lesions and related structures [11] (Table 9.1).

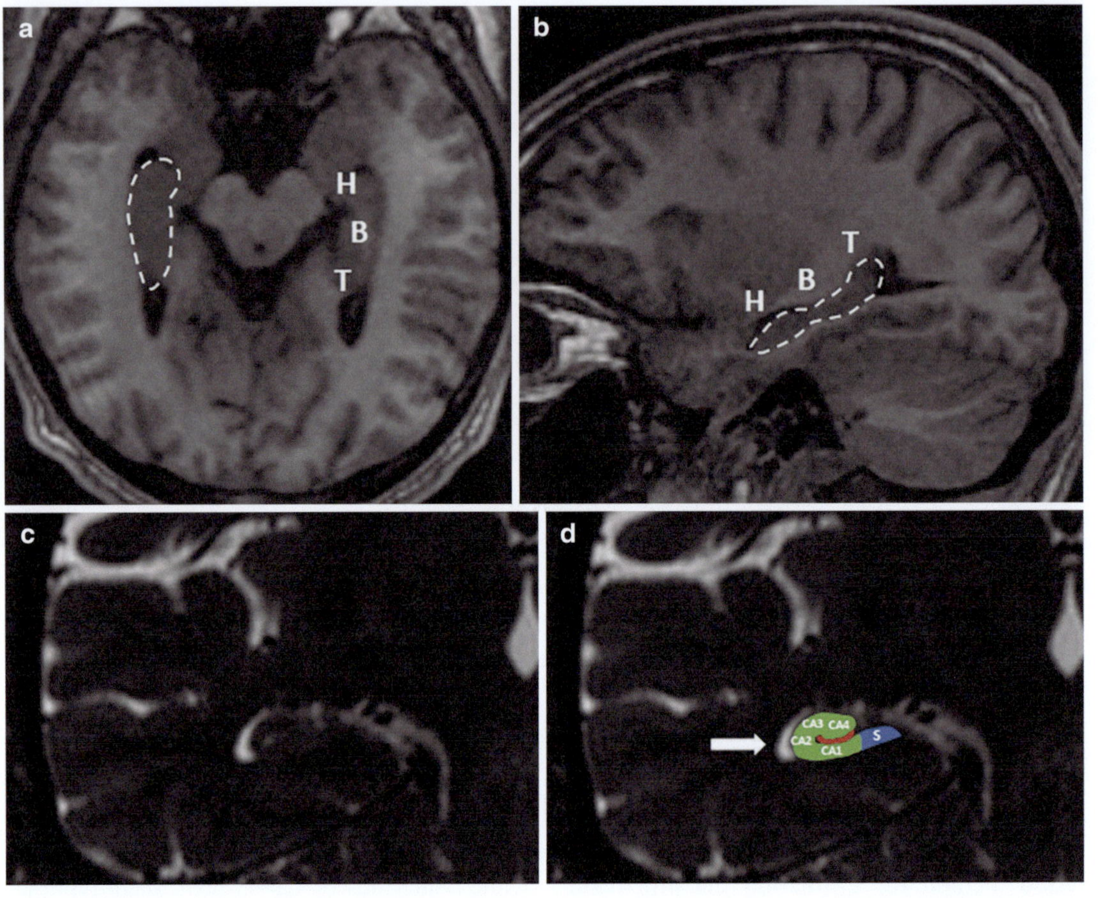

Fig. 9.1 Normal anatomy of the hippocampus (marked with a dotted line in **a**, **b**). Axial (**a**) and sagittal (**b**) T1-weighted images showing head (H), body (B), and tail (T) of the hippocampus. Coronal T2-weighted images at the level of the hippocampal body (**c**, **d**) show the hippocampal formation, which consists of cornu ammonis 1 (CA1), cornu ammonis 2 (CA2), cornu ammonis 3 (CA3), and cornu ammonis 4 (CA4), dentate gyrus (red-colored), and subicular plexus (S, blue-colored). The temporal horn of the lateral ventricle is marked with a white arrow

Table 9.1 Hippocampal lesions

Non-nodular	Unilateral	Volumetric changes	– **Reduction** Degenerative MTLE+HS (Fig. 9.2) Sequelae – **Increase** Neoplasia Infectious encephalitis (Fig. 9.3) Autoimmune encephalitis Ictal changes
		DWI changes	– Autoimmune encephalitis – Infectious encephalitis – Ictal changes – Toxic metabolic – Transient global amnesia (TGA) – Ischemia
Non-nodular	Bilateral	Volumetric changes	– **Reduction** Degenerative MTLE+HS – **Increase** NF-1 Infectious encephalitis Autoimmune encephalitis
		DWI changes	– Autoimmune encephalitis – Infectious encephalitis – Toxic metabolic – TGA (Less frequently than unilateral)
Nodular	Cystic		– Choroidal fissure cyst – Hippocampal remnant cyst (Fig. 9.4) – Multinodular and vacuolating tumor – Tumefactive perivascular space – Neoplasia
	Non-cystic		– Vascular malformation – Malformation of cortical development NF 1 – Calcifications – Neoplasia
	Hyperintensity T1		– Neurocutaneous melanosis – Vascular malformations – Angiocentric glioma

MTLE + HS mesial temporal lobe epilepsy plus hippocampal sclerosis, *NF-1* neurofibromatosis type 1

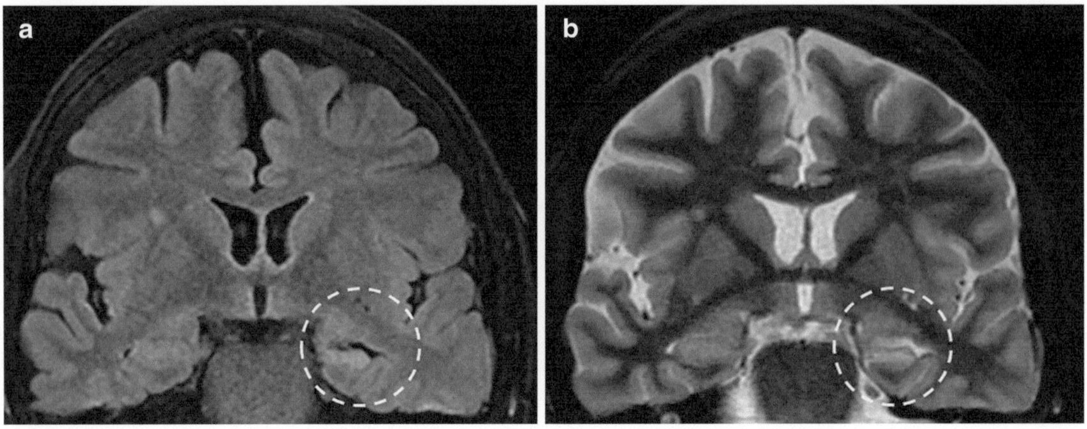

Fig. 9.2 Mesial temporal lobe epilepsy plus hippocampal sclerosis affecting the left temporal lobe and hippocampus (dotted circles). Coronal FLAIR (**a**) and T2-weighted (**b**) images reveal volume loss and increased T2 signal compared to the normal side

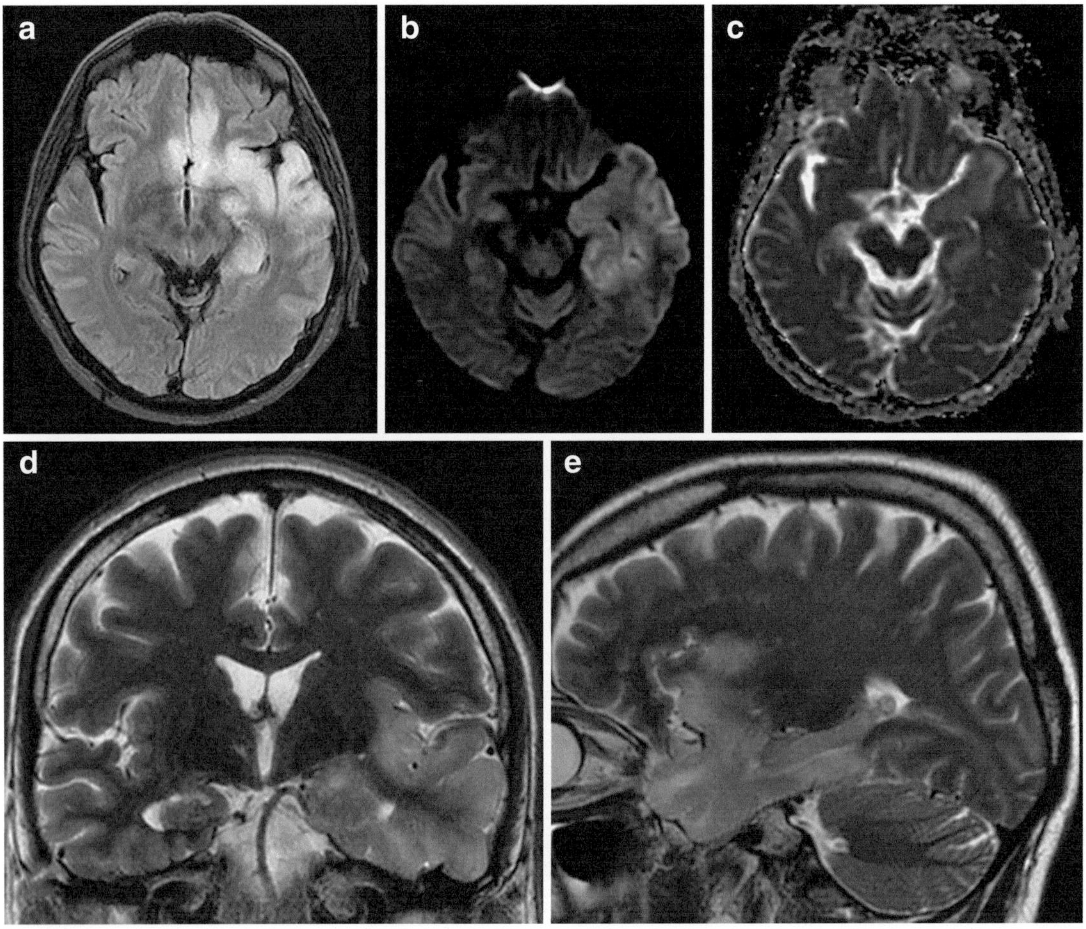

Fig. 9.3 Herpes simplex encephalitis affecting the left anterior and medial temporal lobe, hippocampus, insula, and partially frontal lobes. Axial FLAIR (**a**), diffusion-weighted (**b**), ADC map (**c**), coronal T2-weighted (**d**), and sagittal T2-weighted (**e**) images show severe edema and increased T2 signal accompanied by weak diffusion restriction at the cortical level

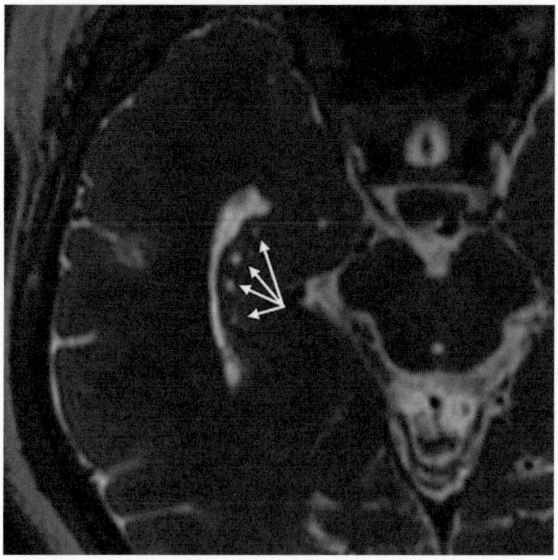

Fig. 9.4 Hippocampal remnant cysts located at the body of the hippocampus marked with white arrows on an axial T2-weighted image

9.4 Conclusion

The hippocampus is a complex and vital structure within the brain that plays an important role in memory as well as various neurological and psychiatric disorders. Accurate imaging of the hippocampus is essential for understanding its anatomy, function, and involvement in disease. Among the imaging modalities, MRI is the preferred method for evaluating the hippocampus due to its superior soft tissue contrast and resolution. Techniques like fMRI, DWI, and MRS have further enhanced our ability to study the anatomy and connectivity of the hippocampus, as well as diagnose hippocampal lesions. Although CT provides important findings, especially in specific clinical settings such as trauma, MRI remains the gold standard for in-depth hippocampal evaluation. Continuous improvements in imaging technology and the development of specialized protocols will improve our diagnostic abilities and understanding of hippocampal diseases.

References

1. Witter MP, Amaral DG. Hippocampal formation. In: Paxinos G, editor. The rat nervous system. 3rd ed. Amsterdam: Elsevier/Academic Press; 2004.
2. Lupetin AR, Davis DA, Beckman I, Dash N. Transcranial Doppler sonography. Part 1. Principles, technique, and normal appearances. Radiographics. 1995;15(1):179–91.
3. Toprak E, Sayal HB. Ultrasonographic imaging of the fetal hippocampus. Arch Gynecol Obstet. 2024;309(5):1943–9.
4. Calhoun PS, Kuszyk BS, Heath DG, Carley JC, Fishman EK. Three-dimensional volume rendering of spiral CT data: theory and method. Radiographics. 1999;19(3):745–64.
5. Nadgir RL, Yousem DM. Neuroradiology: the core requisites, 5th ed. Elsevier; 2024.
6. Kockelkoren R, De Vis JB, Stavenga M, Mali W, Hendrikse J, Rozemuller AM, et al. Hippocampal calcification on brain CT: prevalence and risk factors in a cerebrovascular cohort. Eur Radiol. 2018;28(9):3811–8.
7. Davis PC, Drayer BP, Anderson RE, Braffman B, Deck MD, Hasso AN, et al. Head trauma. American College of Radiology. ACR appropriateness criteria. Radiology. 2000;215(Suppl):507–24.
8. Salmela MB, Mortazavi S, Jagadeesan BD, Broderick DF, Burns J, Deshmukh TK, et al. ACR appropriateness criteria(®) cerebrovascular disease. J Am Coll Radiol. 2017;14(5s):S34–s61.
9. Dekeyzer S, De Kock I, Nikoubashman O, Vanden Bossche S, Van Eetvelde R, De Groote J, et al. "Unforgettable"—A pictorial essay on anatomy and pathology of the hippocampus. Insights Imaging. 2017;8(2):199–212.
10. Srinivas-Rao S, Cao J, Marin D, Kambadakone A. Dual-energy computed tomography to photon counting computed tomography: emerging technological innovations. Radiol Clin North Am. 2023;61(6):933–44.
11. Alves IS, Coutinho AMN, Vieira APF, Rocha BP, Passos UL, Gonçalves VT, et al. Imaging aspects of the hippocampus. Radiographics. 2022;42(3):822–40.
12. Juni JE, Waxman AD, Devous MD Sr, Tikofsky RS, Ichise M, Van Heertum RL, et al. Procedure guideline for brain perfusion SPECT using (99m) Tc radiopharmaceuticals 3.0. J Nucl Med Technol. 2009;37(3):191–5.
13. Nobili F, Arbizu J, Bouwman F, Drzezga A, Agosta F, Nestor P, et al. European Association of Nuclear Medicine and European Academy of Neurology recommendations for the use of brain (18) F-fluorodeoxyglucose positron emission tomog-

raphy in neurodegenerative cognitive impairment and dementia: Delphi consensus. Eur J Neurol. 2018;25(10):1201–17.
14. Ghadimi M, Sapra A. Magnetic resonance imaging contraindications. StatPearls. Treasure Island ineligible companies. Disclosure: Amit Sapra declares no relevant financial relationships with ineligible companies: StatPearls Publishing. Copyright © 2024, StatPearls Publishing LLC.; 2024.
15. Bergouignan L, Chupin M, Czechowska Y, Kinkingnéhun S, Lemogne C, Le Bastard G, et al. Can voxel based morphometry, manual segmentation and automated segmentation equally detect hippocampal volume differences in acute depression? Neuroimage. 2009;45(1):29–37.
16. Huang H, Zheng S, Yang Z, Wu Y, Li Y, Qiu J, et al. Voxel-based morphometry and a deep learning model for the diagnosis of early Alzheimer's disease based on cerebral gray matter changes. Cereb Cortex. 2023;33(3):754–63.
17. Martin P, Winston GP, Bartlett P, de Tisi J, Duncan JS, Focke NK. Voxel-based magnetic resonance image postprocessing in epilepsy. Epilepsia. 2017;58(9):1653–64.
18. Su L, Hayes L, Soteriades S, Williams G, Brain SAE, Firbank MJ, et al. Hippocampal stratum radiatum, lacunosum, and moleculare sparing in mild cognitive impairment. J Alzheimers Dis. 2018;61(1):415–24.

10 Volume of the Human Hippocampus Using MRI

Handan Soysal, Seher Yılmaz, and Niyazi Acer

10.1 Introduction

The hippocampus is a part of the brain known as the memory center of the brain and is involved in transferring information from short-term memory to long-term memory [1, 2]. It has a major role in learning and spatial reasoning [3–6]. The hippocampus and its parts are known to change in volume with age [7]. Due to its clinical importance, it is frequently the subject of research in neuroscience. In the literature, there are reports on the neuroanatomical structure of the hippocampus. Some of these reports are based on the potential effect of race, age, lateralization, or gender on hippocampal subfield volumes using various methodologies (histometric, radiological, etc.) [8, 9]. It has also been reported that the volume of the hippocampus decreases in neurodegenerative diseases, epilepsy, Alzheimer's disease (AD), and dementia [10]. Reduction of hippocampal volume is one of the most consistent findings found on Magnetic Resonance Imaging (MRI) of AD patients. Hippocampal atrophy has been used as both a diagnostic and prognostic feature in clinical studies of AD [11].

There are studies in the literature in which the volumes of the brain and subcortical structures have been calculated using digital segmentation methods [12, 13]. There are some software programs such as ITK-SNAP and 3D Slicer that manually measure the hippocampus volume in MR images [14, 15]. By using ITK-SNAP 4.2.0 (www.itksnap.org) and 3D Slicer 5.6.2 (https://www.slicer.org/), the boundary of the hippocampus can be determined manually.

Some studies use VolBrain, while others use different software programs such as FreeSurfer, SPM12, and MRICloud [16–18]. VolBrain processes three-dimensional MRI images and provides voxel-based morphometric data regarding not only the hippocampus but also the whole brain. Hippocampal subfield segmentation and volumetric evaluation have already been research fields as well as cerebral or cerebellar lobular segmentation. In the literature, there are studies in which the volumes of hippocampus subfields are calculated using different software programs [19–21], but some studies have used HIPS, one of the VolBrain pipelines [19, 22]. A comparative evaluation has also been reported on FreeSurfer and VolBrain, i.e., HIPS [23]. FreeSurfer used FreeSurfer 7.0 version for hippocampal and amygdala volume calculation [24]. Hammer atlas

H. Soysal
Department of Anatomy, Yıldırım Beyazıt University Faculty of Dentistry, Ankara, Türkiye

S. Yılmaz
Department of Anatomy, Yozgat Bozok University Faculty of Medicine, Yozgat, Türkiye

N. Acer (✉)
Department of Anatomy, Sanko University Faculty of Medicine, Gaziantep, Türkiye
e-mail: niyazi.acer@sanko.edu.tr

M. Turgut et al. (eds.), *The Human Hippocampus*, https://doi.org/10.1007/978-3-031-88907-3_10

was used by SPM12 for hippocampus volume calculation [14]. Finally, all different semi- or fully automated techniques are based on various operational environments, constant thresholds, and anatomical brain maps, among others. Volume images are determined by image recording conditions and the characteristics of MRI post-processing software [15].

Volumetric adjustment (relative volume data) based on an external anatomical structure like total intracranial volume is an analytical method as well as the analysis of the direct (absolute) volumetric measurements of the hippocampus and hippocampal subfields [25].

The hippocampus can be subdivided into the following regions: Dentate Gyrus, critical for pattern separation; Cornu Ammonis (CA), central to pattern completion; CA1, important for input integration from CA3 and the entorhinal cortex; and Subiculum, relevant for memory retrieval. Most past structural neuroimaging work has combined all these regions, using measures of whole hippocampal volume [24]. Neuroanatomic data retrieved by the methods mentioned above may help describe the normal anatomical structure of the hippocampus. Besides, these data may help researchers conduct a comparative analysis of neuropathological anatomy and normal anatomy [26].

Because the racial properties of neuroanatomical structures, especially the hippocampus, are not well understood, local data may help clinicians improve their clinical view of their population, based on normal anatomy and normal morphometric data. In the literature, we have accessed some publications that partially report the volume properties of the hippocampus as well as other brain parts in the normal, healthy Turkish population [27].

10.2 Automatic Segmentation Methods

There are several techniques, such as FreeSurfer, FSL (FMRIB Software Library), and SPM (Statistical Parametric Mapping), that are commonly used for hippocampal volume calculations [28–31].

Each technique has its own advantages and disadvantages, and researchers usually choose the one that best suits their research goals and data. Some techniques, such as FreeSurfer, can be very time-consuming and have disadvantages such as requiring some coding, while others do not provide all the structures related to the brain. FSL has disadvantages such as being Mac-based. SPM also has disadvantages, such as working with a program such as MATLAB, and CAT12 takes too long. Various automated segmentation tools and algorithms have been developed over the years, in particular machine learning models. Those using deep convolutional neural network (CNN) architecture are increasingly being applied to improve the accuracy of automated methods. The most widely used ones are AssemblyNet, BrainSegNet, and Bayesian QuickNAT. There are some algorithms, such as hippoDeep and CAST, that are becoming widespread for hippocampal and subfield volume using deep CNN [31].

10.2.1 Obtaining Hippocampus Volume Using VolBrain

The T1-weighted images were uploaded to the MRI brain volumetry system, VolBrain, for brain volumetric analysis. The volume of hippocampal segments and the ratio of total hippocampal volume to intracranial volume were calculated using VolBrain.

VolBrain (https://www.volBrain.net) is an online fully automated segmentation technique whose algorithm is based on multi-atlas patch-based tag merge segmentation technology. It processes the MRIs to calculate the intensity differences between the tissues and regions of interest by voxel-based morphometry [18]. A specialized tool of the VolBrain system, the HIPS pipeline, segments the hippocampal subfields and hippocampus and processes the NIFTI (Neuroimaging Informatics Technology Initiative) format of MRIs. Sagittal 3D T1 DICOM format image files were converted to NIFTI format and uploaded to VolBrain. Anonymized subject files were transferred to the VolBrain system. The volumetric structure of the hippocampus and three subfields [(CA1, CA2, CA3, DG-CA4, and subiculum)

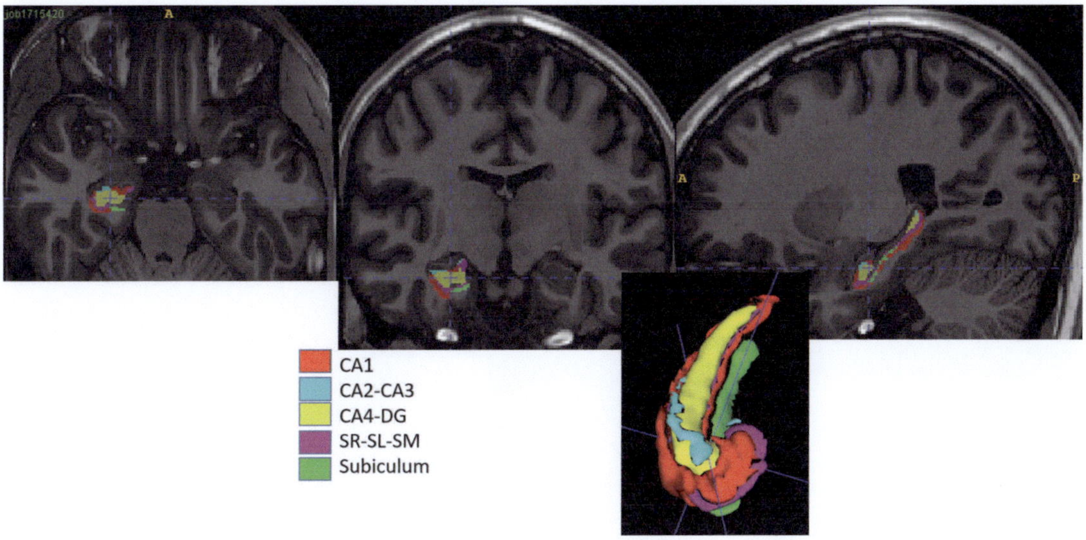

Fig. 10.1 Winterburn protocol for hippocampal subfield segmentation using the VolBrain-HIPS pipeline

(first three ones together abbreviated as CA1–3)] were processed and analyzed using the Kulaga-Yoskovitz protocol during the HIPS process [32]. VolBrain automatically calculates the volumetric data (absolute volume measurements in cm^3 and relative volume data adjusted to total intracranial volume (TIV) in % for each subject. Then, it generates the output files, including the visuals of the hippocampus and hippocampal subfields and the quantitative data as a report of the segmentation process. Besides, the segmentation files for reproducibility of the structures are also produced.

HIPS-monospectral and multispectral provide hippocampus subfields using monomodal MR images. Monospectral uses T1w MR images, but multispectral uses T2w MR images. These processes can carry out the HIPS segmentation over two different protocols: Winterburn and Kulaga-Yoskovitz [18]. In the Winterburn protocol, hippocampus segmentation is analyzed over five subfields (CA1, CA2/CA3, CA4/DG, SR/SL/SM, and subiculum) (Fig. 10.1) [33], while the Kulaga-Yoskovitz protocol, which is preferred in this research report, provides three subfields (Fig. 10.2) [32]. In the Kulaga-Yoskovitz protocol, the tag subiculum represents the presubiculum, the parasubiculum, and the subiculum; CA1–3 represents CA1, 2, and 3; CA4-DG represents CA4 and DG. If you enter gender and age, the expected bounds for each structure are included in the report (Figs. 10.3 and 10.4).

10.2.2 Automated Segmentation Using FreeSurfer

FreeSurfer represents a comprehensive open-source software suite tailored for the analysis of neuroimaging data. Among its array of functionalities, FreeSurfer incorporates tools specifically designed for segmenting the hippocampus into subfields. This capability proves invaluable for investigating the intricate role of the hippocampus across diverse neurological and psychiatric disorders [30].

FreeSurfer is an advanced open-source software suite designed for the comprehensive analysis of neuroimaging data. Its workflow encompasses several crucial steps aimed at preprocessing brain images to ensure accurate and reliable results. These steps include [28–31]:

1. Motion Correction: This initial step corrects for any motion artifacts present in the acquired images. Motion artifacts can distort the images and affect subsequent analyses, so correcting for them is essential for accurate results.

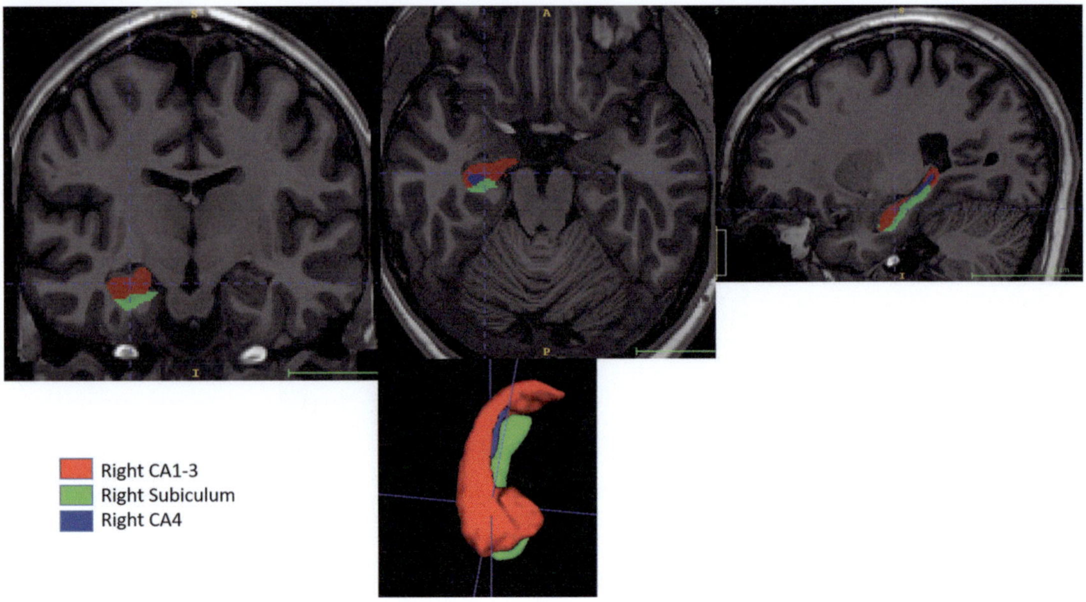

Fig. 10.2 Kulaga-Yoskovitz protocol for hippocampal subfield segmentation using the VolBrain-HIPS pipeline

Segmentation protocol: Winterburn[2]

Volumes[3]	Total ($cm^3/\%$)	Right ($cm^3/\%$)	Left ($cm^3/\%$)	Asym.($\%$)[4]
Hippocampus	5.34 (0.3539)	2.72 (0.1801)	2.62 (0.1738)	3.5292
	[0.28 - 0.42][5]	[0.14 - 0.22]	[0.14 - 0.21]	[-19.21 - 14.42]
CA1	1.89 (0.1256)	0.95 (0.0630)	0.94 (0.0626)	0.6377
	[0.09 - 0.15]	[0.05 - 0.08]	[0.05 - 0.08]	[-24.84 - 20.08]
CA2-CA3	0.31 (0.0203)	0.19 (0.0123)	0.12 (0.0080)	42.2460
	[0.02 - 0.03]	[0.01 - 0.02]	[0.01 - 0.02]	[-58.33 - 22.24]
CA4-DG	1.28 (0.0846)	0.68 (0.0453)	0.59 (0.0393)	14.1103
	[0.07 - 0.11]	[0.04 - 0.06]	[0.03 - 0.06]	[-33.74 - 21.61]
SR-SL-SM	1.19 (0.0787)	0.59 (0.0389)	0.60 (0.0399)	-2.4483
	[0.05 - 0.09]	[0.03 - 0.05]	[0.03 - 0.04]	[-28.42 - 31.23]
Subiculum	0.67 (0.0447)	0.31 (0.0206)	0.36 (0.0241)	-15.4360
	[0.03 - 0.05]	[0.01 - 0.03]	[0.02 - 0.03]	[-16.97 - 32.45]

Fig. 10.3 VolBrain results

2. Affine Transformation to Talairach Image Space: FreeSurfer aligns the images with a standardized reference space, typically the Talairach atlas. This alignment facilitates spatial normalization, allowing for consistent analysis across different subjects and studies.
3. Nonuniform Intensity Normalization: In this step, FreeSurfer corrects for nonuniform intensity variations within the images. Intensity inhomogeneities can arise due to variations in scanner sensitivity or imaging artifacts. Correcting for these variations ensures uniform intensity across the images, improving the accuracy of subsequent analyses.
4. Removal of Nonbrain Tissues: FreeSurfer identifies and removes nonbrain tissues from the images, such as the skull, scalp, and soft tissues. This step is crucial for focusing the analysis specifically on brain structures and

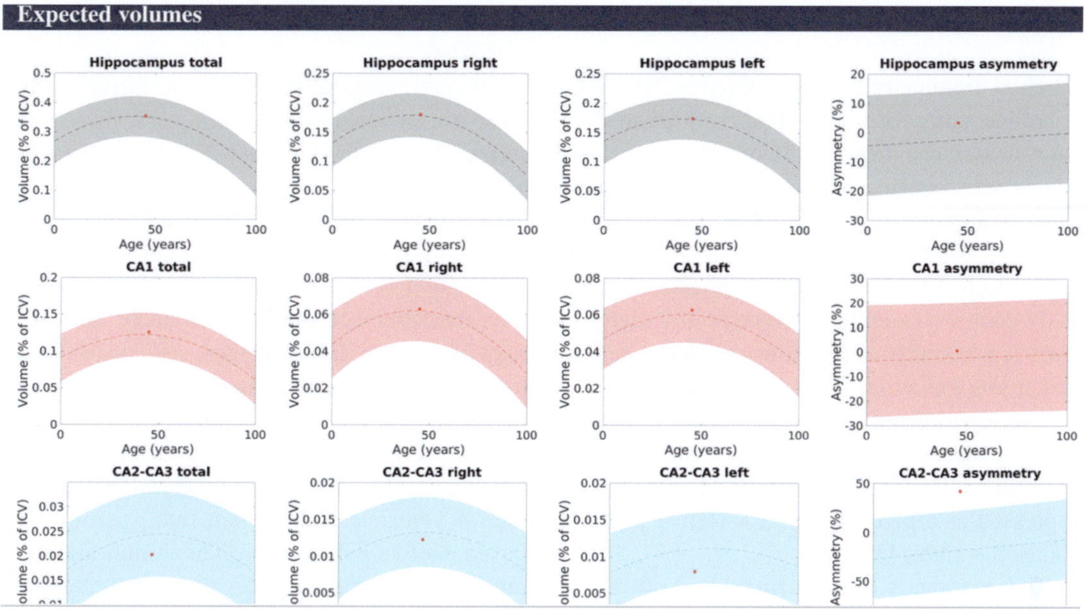

Fig. 10.4 VolBrain expected volume

removing extraneous information that could interfere with subsequent processing steps.

By performing these preprocessing steps, FreeSurfer prepares the neuroimaging data for more advanced analyses, such as segmentation of brain structures, cortical surface reconstruction, and measurement of cortical thickness. This comprehensive preprocessing pipeline enhances the quality and reliability of the data, enabling researchers to extract meaningful insights from neuroimaging studies. FreeSurfer's hippocampal subfield segmentation algorithm is based on a probabilistic model that was trained on manually segmented hippocampi. The algorithm assigns each voxel in the hippocampus to one of 18 subfields, based on its intensity and location relative to other voxels [17].

10.2.3 Voxel-Based Morphometry (VBM) Analysis with SPM12

Volumetric brain analysis was performed using CAT12 (Computational Anatomy Toolbox) within Statistical Parametric Mapping (SPM12) in MATLAB. Sagittal T1 DICOM (Digital Imaging and Communications in Medicine) images were converted to NIFTI-1 (Neuroimaging Informatics Technology Initiative) format. TIV and overall tissue volume (TTV) were determined using white matter (WM), gray matter (GM), and cerebrospinal fluid (CSF). The Automated Anatomical Labeling Atlas was used alongside one of the SPM toolboxes to calculate lobar GM volumes.

In the CAT12 toolbox, the final stages of segmentation, normalization, and modulation were seamlessly executed in conjunction. Through CAT12, regional tissue volumes were estimated across various regions utilizing probabilistic atlases. These volumes were approximated in their native space following high-dimensional spatial registration, preceding any spatial normalization. This integrated approach ensures accurate and efficient processing of MRI data, facilitating comprehensive analysis of regional tissue volumes while preserving the integrity of the native spatial information [31]. The Hammer atlas is commonly used for hippocampal segmentation in neuroimaging studies. It provides accurate and reliable segmentation of the hippocampus and its subregions [34].

10.2.4 Hippocampal Asymmetry

Cerebral dominance may be defined as "the functional inequality of the cerebral hemispheres" in the neuroscience literature. In our research, dominance was considered only in the context of being volumetrically large. The word "dominance" is used for descriptive purposes specific to this study. We would like to emphasize that our study does not examine asymmetry as a factor in brain functionality as it is often discussed in neuroanatomy literature.

The Asymmetry Index (AI) is determined by dividing the mean hippocampus volumes of the right and left sides by their respective means, expressed as a percentage. AI = (left − right) × 100/(left + right) [35].

The asymmetry data generated by VolBrain in individual outputs is frequently treated as an arithmetic average in the existing literature [9, 36–40]. In certain studies, this asymmetry is aggregated and presented by means, with findings often separated by within-study groups, such as gender (e.g., women and men) [9, 36]. Consequently, comparisons between women and men in terms of asymmetry are made, yielding specific findings and prompting further discussion.

The asymmetry index provided by VolBrain serves as an indicator aiding the assessment of the magnitude of size differences between brain hemispheres within an individual. Essentially, the asymmetry data from VolBrain offers insights into the larger side of the brain and the extent of the disparity when examined on an individual basis. It's worth noting that no mathematical or logical errors are anticipated in the individual evaluation of the asymmetry index, ensuring the reliability of these measurements for assessing brain structural asymmetry.

10.3 Discussion

10.3.1 Different Techniques for Hippocampus Volume

Some studies have examined manual tracing methods and automatic and semi-automatic methods to evaluate hippocampal volume using MRI, such as FSL, AFNI, FreeSurfer, VolBrain, and SPM.

Manual tracking is usually performed using three-dimensional (3D) MRI in each slice of the series. Manual tracking is the gold standard for segmentation in volumetric research studies.

Automated techniques have advantages and disadvantages. Most of the automated techniques are freely available to perform more specific volume measurement, quickly and easily. However, they have some disadvantages, such as the structures being small and the intensity of spatial relations being nonuniform and non-contrasted.

The whole brain can be segmented into most structures, including the cerebrum, cerebellum, lateral ventricle, brain stem, etc., using web-based software VolBrain. The individual volumes of the hippocampus and subregions can be parcellated automatically using VolBrain [38–40].

VolBrain has the advantages of both providing fast results and offering hippocampal subfield volumes. No difference was found between the techniques in the hippocampal volume calculations using these techniques [41].

FreeSurfer stands out as a widely accessible and extensively automated tool, characterized by thorough documentation, active development, and ease of implementation. The current iteration, FreeSurfer 7.4.1, maintains these qualities and notably retains the hippocampal subfield tool without significant alterations from version 6.0. This consistency ensures continuity and reliability for users familiar with earlier versions while still providing robust functionality for the segmentation and analysis of hippocampal subfields. The biggest advantage of FreeSurfer is its unbiased and uniform segmentation. FreeSurfer is updated with newer versions, and human error is minimized.

SPM12 (Statistical Parametric Mapping 12) is a software package for neuroimaging data analysis. CAT12 provides a comprehensive set of tools for analyzing brain structures, including the hippocampus. It offers automated segmentation and volumetry, allowing researchers to accurately quantify hippocampal volume changes [42]. CAT12 gives results using the Hammers and Neuromorphometrics atlases.

10.3.2 Clinical Importance for Hippocampal Volumetry

Hippocampal volume segmentation can be used to study the relationship between hippocampal volume and various neurological and psychiatric disorders. For example, studies have shown that hippocampal volume is reduced in people with Alzheimer's disease, schizophrenia, and depression [30].

With the VolBrain tool, not only was the normal anatomical structure of the hippocampus investigated, but also the anatomical structural differences between patient groups and the healthy population were examined. It has been reported that the volumes of the subiculum, cornu ammonis (CA1–3), gyrus dentatus, and CA4 subfields in Alzheimer's disease are smaller than those in the healthy control group [37, 38]. In another study, in which VolBrain-HIPS segmentation was preferred as in our methodology, it was reported that the hippocampal volumes of Alzheimer's patients were smaller than those in the control group [39]. In epilepsy research, hippocampal volume was also examined with VolBrain. In epilepsy patients, characteristic structural changes, such as heterotrophy or atrophy in the total hippocampal volume, were found [13, 26, 40].

Tae et al. (2019) observed a bilateral reduction in hippocampal volume among individuals diagnosed with Major Depressive Disorder (MDD), with predominant left-sided hippocampal atrophy evident in both absolute measurements and volumes corrected for age and intracranial volume (ICV) [43]. This finding aligns with numerous studies reporting hippocampal atrophy in MDD patients, particularly in those with a history of multiple depressive episodes. These consistent observations underscore the potential role of hippocampal volume reduction as a neurobiological marker associated with the recurrent nature of depression, shedding light on the structural alterations within the brain linked to the pathophysiology of MDD [44].

Zilioli et al. (2023) found a statistically significant increase in the absolute mean hippocampal volume, as well as in the ipsilateral and contralateral hippocampal volumes, in the group of glioblastoma patients [45].

We saw that the volume of the hippocampus was the most reduced structure in AD according to the literature [46, 47]. Bartos et al. (2019) found that the most atrophied structures were both hippocampi and the amygdala using FreeSurfer [48].

10.3.3 Age-Related and Gender Volumetric Changes of the Hippocampus

Studies have shown that there is gender- and age-related volume differences in the hippocampus [9, 27, 49, 50]. Hippocampal volume decreases with age, with a more significant decline in middle age and a steeper decline in men than women [51].

Total volume of the hippocampus is reported to be between 4.48 ± 0.66 cm^3 in females and 4.70 ± 0.79 cm^3 in males [9]. Another study reports results of 3.94 ± 0.49 cm^3 in males and 3.74 ± 0.42 cm^3 in females [27]. Some research has reported the volumes of the right and left hippocampi as well as gender differences [27]. The volume of CA1, CA2–3, and subiculum was found to be higher in men under 50 years of age [9]. CA4-DG volume was found to be higher in both genders under the age of 50 than in those over the age of 50 [9]. Another study reported that the volume of the CA2–3 and CA4-DG subfields was higher in young people than in the over-50 age group [50].

Bartos et al. (2019) found that age was significantly correlated with right hippocampal volume [48]. The right hippocampus is mostly larger than the left hippocampus in normal subjects [46].

Gender, age, and side may have a natural effect on the size of an anatomical structure. According to the literature, all absolute volumetric values for hippocampal volume are higher in men than in women. The fact that direct measurements (absolute volumetric data) were larger in men was found to be compatible with biological reality [52].

Taxi drivers in London are trained and tested before they start work. During this test, they are asked to recognize many locations and the shortest routes between them. Taxi drivers undergo an MRI scan, and their hippocampus is compared with a control group. In this study conducted at University College London, it was found that the right posterior gray matter volume of the hippocampus in taxi drivers was larger and that this size increased as the drivers' experience increased [53].

10.4 Conclusion

Knowledge of how the volume of the hippocampus in healthy individuals varies depending on age and gender may contribute to understanding brain functions. Hippocampus-adjusted volume information should also be taken into account when analyzing VolBrain, FreeSurfer, and SPM12 and making clinical or anatomical decisions.

The asymmetry data produced by VolBrain should be considered a magnitude scale of the asymmetry in individuals with a right- or left-dominant hippocampus or hippocampal subfield. Clear and accurate definitions regarding asymmetry in individuals are needed. It may help to correctly interpret the findings when applying individual data to the population.

The analysis of human hippocampal volume using various automated software packages for MRI-based measurements has yielded valuable insights into the consistency and discrepancies among different methodologies. Our chapter compared results obtained from multiple software tools, revealing both commonalities and variations in volumetric measurements.

While the automated software packages generally agreed on the gross morphology and spatial localization of the hippocampus, notable differences were observed in the specific volume measurements. These variations may stem from differences in algorithms, image processing techniques, segmentation protocols, or anatomical definitions employed by each software.

Despite these disparities, overarching trends and patterns remained consistent across different software platforms. This suggests that while absolute volume measurements may differ, the relative differences between individuals or groups are likely to be preserved across various software packages.

Our chapter emphasizes the importance of carefully considering the selection of software and validation procedures when conducting MRI-based volumetric analysis of the hippocampus. Researchers and clinicians should be mindful of the potential influence of software choice on study outcomes and interpretation.

Moving forward, efforts should focus on standardizing protocols, improving algorithmic accuracy, and enhancing inter-software agreement to ensure greater consistency and reliability in hippocampal volumetric measurements. Additionally, future investigations could explore the clinical implications of software discrepancies and delve deeper into the sources of variability.

In conclusion, while automated software tools offer efficient means for hippocampal volumetric analysis, cautious interpretation and validation of results are necessary, given the inherent variability associated with different methodologies. Addressing these challenges will advance our understanding of hippocampal structure and function, facilitating more accurate diagnoses and treatments of neurological and psychiatric disorders.

Volumetric changes in the hippocampus provide important information for understanding hippocampal abnormalities in patients with neuropsychiatric and neurodegenerative diseases.

Finally, intracranial, especially hippocampal, volumes may be useful in diagnosing various brain diseases [54, 55]. A method for obtaining higher contrast through changes in imaging acquisition factors and the development of software for accurate measurement is required.

References

1. Hartley T, Bird CM, Chan D, Cipolotti L, Husain M, Vargha-Khadem F, Burgess N. The hippocampus is required for short-term topographical memory in humans. Hippocampus. 2007;17(1):34–48. https://doi.org/10.1002/hipo.20240.
2. Marek J, Maj E, Przybyla OK, Skrzynski W, Pasicz K, Fabiszewska E, et al. The impact of studying on

the hippocampal volume in medical students and its correlation with the results of the final medical examination: a single-centre, prospective observational cohort study. Pol J Radiol. 2023;88:e22–30. https://doi.org/10.5114/pjr.2023.124433.
3. Anand KS, Dhikav V. Hippocampus in health and disease: an overview. Ann Indian Acad Neurol. 2012;15(4):239–46. https://doi.org/10.4103/0972-2327.104323.
4. Hafting T, Fyhn M, Molden S, Moser M-B, Moser EI. Microstructure of a spatial map in the entorhinal cortex. Nature. 2005;436(7052):801–6. https://doi.org/10.1038/nature03721.
5. Romero JE, Coupe P, Manjon JV. HIPS: a new hippocampus subfield segmentation method. Neuroimage. 2017;163:286–95. https://doi.org/10.1016/j.neuroimage.2017.09.049.
6. Schmajuk NA. Role of the hippocampus in temporal and spatial navigation: an adaptive neural network. Behav Brain Res. 1990;39(3):205–29. https://doi.org/10.1016/0166-4328(90)90028-D.
7. O'Shea A, Cohen RA, Porges EC, Nissim NR, Woods AJ. Cognitive aging and the hippocampus in older adults. Front Aging Neurosci. 2016;8:298. https://doi.org/10.3389/fnagi.2016.00298.
8. Genç H. A Comparison of the right and left hippocampus in the rabbit brain in terms of cell density. Turk J Biol. 1998;22(3):263–70. https://journals.tubitak.gov.tr/biology/vol22/iss3/2/.
9. Gurlek Celik N, Tiryaki S. Changes in the volumes and asymmetry of subcortical structures in healthy individuals according to gender. Anat Sci Int. 2023;98(4):506–19. https://doi.org/10.1007/s12565-023-00714-w.
10. Braak H, Braak E. Neuropathological stageing of Alzheimer-related changes. Acta Neuropathol. 1991;82(4):239–59. https://doi.org/10.1007/BF00308809.
11. Mueller SG, Schuff N, Yaffe K, Madison C, Miller B, Weiner MW. Hippocampal atrophy patterns in mild cognitive impairment and Alzheimer's disease. Hum Brain Mapp. 2010;31(9):1339–47. https://doi.org/10.1002/hbm.20934.
12. Acer N, Kamaşak B, Karapınar BO, Akkus Yetkin E, Ipekten F, Değirmencioğlu L, Ilica AT. A comparison of automated segmentation and manual tracing of magnetic resonance imaging to quantify lateral ventricle volumes. Erciyes Med J. 2022;44(2):148–55. https://doi.org/10.14744/etd.2021.73920.
13. Wozniak MM, Zbroja M, Matuszek M, Pustelniak O, Cyranka W, Drelich K, et al. Epilepsy in pediatric patients—evaluation of brain structures' volume using VolBrain software. J Clin Med. 2022;11(16):4657. https://doi.org/10.3390/jcm11164657.
14. Khlif MS, Egorova N, Werden E, Redolfi A, Boccardi M, DeCarli CS, Fletcher E, Singh B, Li Q, Bird L, Brodtmann A. A comparison of automated segmentation and manual tracing in estimating hippocampal volume in ischemic stroke and healthy control participants. Neuroimage Clin. 2019;21:101581. https://doi.org/10.1016/j.nicl.2018.10.019. Epub 2018 Oct 22. PMID: 30606656; PMCID: PMC6411582.
15. Kim J, Kim S, Shin H, Seo H, Choi H, Son S, Tae W, Kim S. Evaluation of hippocampal volume based on various inversion time in normal adults by manual tracing and automated segmentation methods. Investig Magn Reson Imaging. 2015;19:67. https://doi.org/10.13104/imri.2015.19.2.67.
16. Sone D, Sato N, Maikusa N, Ota M, Sumida K, Yokoyama K, Kimura Y, Imabayashi E, Watanabe Y, Watanabe M, Okazaki M, Onuma T, Matsuda H. Automated subfield volumetric analysis of hippocampus in temporal lobe epilepsy using high-resolution T2-weighed MR imaging. Neuroimage Clin. 2016;12:57–64. https://doi.org/10.1016/j.nicl.2016.06.008. PMID: 27489767; PMCID: PMC4960104.
17. Schmidt MF, Storrs JM, Freeman KB, Jack CR Jr, Turner ST, Griswold ME, Mosley TH Jr. A comparison of manual tracing and FreeSurfer for estimating hippocampal volume over the adult lifespan. Hum Brain Mapp. 2018;39(6):2500–13. https://doi.org/10.1002/hbm.24017. Epub 2018 Feb 21. PMID: 29468773; PMCID: PMC5951757.
18. Manjón JV, Romero JE, Coupe P. A novel deep learning based hippocampus subfield segmentation method. Sci Rep. 2022;12(1):1333. https://doi.org/10.1038/s41598-022-05287-8.
19. Bussy A, Plitman E, Patel R, Tullo S, Salaciak A, Bedford SA, et al. Hippocampal subfield volumes across the healthy lifespan and the effects of MR sequence on estimates. Neuroimage. 2021;233:117931. https://doi.org/10.1016/j.neuroimage.2021.117931.
20. Menzler K, Hamer HM, Mross P, Rosenow F, Deichmann R, Wagner M, et al. Validation of automatic MRI hippocampal subfield segmentation by histopathological evaluation in patients with temporal lobe epilepsy. Seizure. 2021;87:94–102. https://doi.org/10.1016/j.seizure.2021.03.007.
21. Wisse LEM, Chételat G, Daugherty AM, de Flores R, la Joie R, Mueller SG, et al. Hippocampal subfield volumetry from structural isotropic 1 mm(3) MRI scans: a note of caution. Hum Brain Mapp. 2021;42(2):539–50. https://doi.org/10.1002/hbm.25234.
22. Baek MS, Lee N, Kim JW, Hong JY. Association of hippocampal subfield volumes with amyloid-beta deposition in Alzheimer's disease. J Clin Med. 2022;11(6):1526. https://doi.org/10.3390/jcm11061526.
23. Samara A, Raji CA, Li Z, Hershey T. Comparison of hippocampal subfield segmentation agreement between 2 automated protocols across the adult life span. AJNR Am J Neuroradiol. 2021;42(10):1783–9. https://doi.org/10.3174/ajnr.A7244.
24. Kahhale I, Buser NJ, Madan CR, et al. Quantifying numerical and spatial reliability of hippocampal and amygdala subdivisions in FreeSurfer. Brain Inf. 2023;10:9. https://doi.org/10.1186/s40708-023-00189-5.

25. Comino Garcia-Munoz A, Alemán-Gómez Y, Toledano R, Poch C, García-Morales I, Aledo-Serrano Á, et al. Morphometric and microstructural characteristics of hippocampal subfields in mesial temporal lobe epilepsy and their correlates with mnemonic discrimination. Front Neurol. 2023;14:1096873. https://doi.org/10.3389/fneur.2023.1096873.
26. Moghaddam HS, Aarabi MH, Mehvari-Habibabadi J, Sharifpour R, Mohajer B, Mohammadi-Mobarakeh N, et al. Distinct patterns of hippocampal subfield volume loss in left and right mesial temporal lobe epilepsy. Neurol Sci. 2021;42(4):1411–21. https://doi.org/10.1007/s10072-020-04653-6.
27. Ozdemir M, Soysal H, Eraslan O, Dilli A. Normative hippocampal volumetric measurements using magnetic resonance imaging. Turk J Med Sci. 2019;49(5):1464–70. https://doi.org/10.3906/sag-1903-233.
28. Brown EM, Pierce ME, Clark DC, Fischl BR, Iglesias JE, Milberg WP, et al. Test-retest reliability of FreeSurfer automated hippocampal subfield segmentation within and across scanners. Neuroimage. 2020;210:116563. https://doi.org/10.1016/j.neuroimage.2020.116563.
29. Liu D, Chen J, Ge H, Yan Z, Luo B, Hu X, et al. Structural plasticity of the contralesional hippocampus and its subfields in patients with glioma. Eur Radiol. 2023;33(9):6107–15. https://doi.org/10.1007/s00330-023-09582-4.
30. Sämann PG, Iglesias JE, Gutman B, Grotegerd D, Leenings R, Flint C, et al. FreeSurfer-based segmentation of hippocampal subfields: a review of methods and applications, with a novel quality control procedure for ENIGMA studies and other collaborative efforts. Hum Brain Mapp. 2022;43(1):207–33. https://doi.org/10.1002/hbm.25326.
31. Seyedi S, Jafari R, Talaei A, et al. Comparing VBM and ROI analyses for detection of gray matter abnormalities in patients with bipolar disorder using MRI. Middle East Curr Psychiatry. 2020;27:69. https://doi.org/10.1186/s43045-020-00076-3.
32. Kulaga-Yoskovitz J, Bernhardt BC, Hong SJ, Mansi T, Liang KE, van der Kouwe AJ, et al. Multi-contrast submillimetric 3 Tesla hippocampal subfield segmentation protocol and dataset. Sci Data. 2015;2:150059. https://doi.org/10.1038/sdata.2015.59.
33. Winterburn JL, Pruessner JC, Chavez S, Schira MM, Lobaugh NJ, Voineskos AN, Chakravarty MM. A novel in vivo atlas of human hippocampal subfields using high-resolution 3 T magnetic resonance imaging. Neuroimage. 2013;74:254–65. https://doi.org/10.1016/j.neuroimage.2013.02.003.
34. Rodionov R, Chupin M, Williams E, Hammers A, Kesavadas C, Lemieux L. Evaluation of atlas-based segmentation of hippocampi in healthy humans. Magn Reson Imaging. 2009;27(8):1104–9. https://doi.org/10.1016/j.mri.2009.01.008. Epub 2009 Mar 4. PMID: 19261422.
35. Manjon JV, Coupe P. volBrain: an online MRI brain volumetry system. Front Neuroinform. 2016;10:30.
36. Isiklar S, Demir I, Ozdemir ST, Ozpar R. Examination of the development and asymmetry of the cerebellum and its lobules in individuals aged 1-18 years: a retrospective MRI study. Brain Topogr. 2023;36:901. https://doi.org/10.1007/s10548-023-00997-2.
37. Öncel Ç, Özdemir B, Erdoğan Ç, Kıroğlu Y, Şahiner T. MRI measures of hippocampus in mild cognitive impairment and Alzheimer patients. Turk J Neurol. 2011;17(4):181–4. https://jag.journalagent.com/z4/vi.asp?pdir=tjn&plng=eng&un=TJN-35582.
38. Wisse LE, Biessels GJ, Heringa SM, Kuijf HJ, Koek DH, Luijten PR, Geerlings MI. Hippocampal subfield volumes at 7T in early Alzheimer's disease and normal aging. Neurobiol Aging. 2014;35(9):2039–45. https://doi.org/10.1016/j.neurobiolaging.2014.02.021.
39. Zamani J, Sadr A, Javadi AH. A large-scale comparison of cortical and subcortical structural segmentation methods in Alzheimer's disease: a statistical approach. bioRxiv. 2020. 2020.2008.2018.256321. https://doi.org/10.1101/2020.08.18.256321.
40. Princich JP, Donnelly-Kehoe PA, Deleglise A, Vallejo-Azar MN, Pascariello GO, Seoane P, et al. Diagnostic performance of MRI volumetry in epilepsy patients with hippocampal sclerosis supported through a random Forest automatic classification algorithm. Front Neurol. 2021;12:613967. https://doi.org/10.3389/fneur.2021.613967.
41. Droby A, Thaler A, Giladi N, Hutchison RM, Mirelman A, Ben Bashat D, Artzi M. Whole brain and deep gray matter structure segmentation: quantitative comparison between MPRAGE and MP2RAGE sequences. PLoS One. 2021;16(8):e0254597. https://doi.org/10.1371/journal.pone.0254597.
42. Farokhian F, Beheshti I, Sone D, Matsuda H. Comparing CAT12 and VBM8 for detecting brain morphological abnormalities in temporal lobe epilepsy. Front Neurol. 2017;24(8):428. https://doi.org/10.3389/fneur.2017.00428. PMID: 28883807; PMCID: PMC5573734.
43. Tae WS, Kim SS, Lee KU, Nam EC, Choi JW, Park JI. Hippocampal shape deformation in female patients with unremitting major depressive disorder. AJNR Am J Neuroradiol. 2011;32(4):671–6. https://doi.org/10.3174/ajnr.A2367. Epub 2011 Mar 3. PMID: 21372170; PMCID: PMC7965897.
44. Campbell S, Marriott M, Nahmias C, et al. Lower hippocampal volume in patients suffering from depression: a meta-analysis. Am J Psychiatry. 2004;161:598–607.
45. Zilioli A, Misirocchi F, Mutti C, et al. Volumetric hippocampal changes in glioblastoma: a biomarker for neuroplasticity? J Neurooncol. 2023;163:261–7. https://doi.org/10.1007/s11060-023-04315-5.
46. Shi F, Liu B, Zhou Y, Yu C, Jiang T. Hippocampal volume and asymmetry in mild cognitive impairment and Alzheimer's disease: meta-analyses of MRI studies. Hippocampus. 2009;19:1055–64. https://doi.org/10.1002/hipo.20573.
47. da Silva Filho SRB, Barbosa JHO, Rondinoni C, Dos Santos AC, Salmon CEG, da Costa Lima NK,

Ferriolli E, Moriguti JC. Neuro-degeneration profile of Alzheimer's patients: a brain morphometry study. Neuroimage Clin. 2017;15:15–24. https://doi.org/10.1016/j.nicl.2017.04.001. PMID: 28459000; PMCID: PMC5397580.
48. Bartos A, Gregus D, Ibrahim I, Tintěra J. Brain volumes and their ratios in Alzheimer's disease on magnetic resonance imaging segmented using FreeSurfer 6.0. Psychiatry Res Neuroimaging. 2019;287:70–4.
49. Duvernoy HM, Cattin F, Risold PY. The human hippocampus: functional anatomy, vascularization and serial sections with MRI. Springer Berlin Heidelberg. 2013.
50. Zheng F, Liu Y, Yuan Z, Gao X, He Y, Liu X, et al. Age-related changes in cortical and subcortical structures of healthy adult brains: a surface-based morphometry study. J Magn Reson Imaging. 2019;49(1):152–63. https://doi.org/10.1002/jmri.26037.
51. Mark AF, Erin I, Walsh M, Shaw E, Abhayaratna WP, Anstey KJ, Sachdev PS, Cherbuin N. Longitudinal trajectories of hippocampal volume in middle to older age community dwelling individuals. Neurobiol Aging. 2021;97:97–105.
52. Perlaki G, Orsi G, Plozer E, Altbacker A, Darnai G, Nagy SA, Horvath R, Toth A, Doczi T, Kovacs N, Bogner P, Schwarcz A, Janszky J. Are there any gender differences in the hippocampus volume after head-size correction? A volumetric and voxel-based morphometric study. Neurosci Lett. 2014;570:119–23. https://doi.org/10.1016/j.neulet.2014.04.013. Epub 2014 Apr 16. PMID: 24746928.
53. Maguire EA, Woollett K, Spiers HJ. London taxi drivers and bus drivers: a structural MRI and neuropsychological analysis. Hippocampus. 2006;16:1091–101.
54. Harkey T, Baker D, Hagen J, Scott H, Palys V. Practical methods for segmentation and calculation of brain volume and intracranial volume: a guide and comparison. Quant Imaging Med Surg. 2022;12(7):3748–61. https://doi.org/10.21037/qims-21-958.
55. Tae W-S, Kim S, Lee K, Nam E-C, Choi J, Park J-I. Hippocampal shape deformation in female patients with unremitting major depressive disorder. AJNR Am J Neuroradiol. 2011;32:671–6. https://doi.org/10.3174/ajnr.A2367.

11 Gene Expression in the Hippocampus

Serpil Taheri and Mai R. S. Abusalim

11.1 Introduction

The hippocampus, a part of the limbic system in the brain, takes control of cognitive and behavioral responses, memory functionalities, and emotions. It is housed in the temporal lobe, which belongs to the cerebral cortex, and is interconnected with different brain regions such as the amygdala and hypothalamus [1]. This part of the brain, besides being responsible for memory consolidation and emotion regulation, also plays a role in the processing of auditory signals, language comprehension, and visual perception [2–4].

The cumulative body of research delineated the many functions in which the hippocampus is involved such as memory encoding and retrieval, spatial navigation, learning, and emotional regulation. Of course, these functions depend on highly regulated processes including synaptic plasticity, neuronal connectivity, and the overall function of neurotransmitter signaling to provide organization in information transmission between neurons [5]. Thus, gene expression in the hippocampus is kept in balance by the control of the dynamic and finely regulated molecular events that shape these structural and functional features.

The Allen Brain Atlas project (Available at; https://www.brain-map.org/) has been paramount in revealing the gene expression lists of several regions of the brain including the hippocampus. Even though ~80% of genes were expressed at the cellular level within the brain, the expression of 70.5% of these genes occurred in <20% of total cells, a property that may account for the functional specificity of genes in the hippocampus. In addition to common housekeeping genes, such as *ACTB*, *GAPDH*, *PGK1*, *RPL13a*, *YWHAZ*, *SDHA*, and others, the hippocampus contains specific genes involved in its development, synaptic plasticity, neurotransmission, and neuroprotection [6].

The hippocampus has diverse gene expression patterns across spatial resolution from region to single cell scale. This heterogeneity corresponds with its various functions, such as the dorsal (posterior in primates) hippocampus contributing most heavily to the formation of spatial memory and ability to navigate, while the ventral (anterior in primates) hippocampus plays a key role in emotional regulation and stress responses. This differential gene expression emphasizes how different hippocampal regions, as well as cell types have specialized roles in the functions behind cognitive and emotional processing [7].

S. Taheri (✉)
Department of Medical Biology, Erciyes University Faculty of Medicine, Kayseri, Türkiye

Betül Ziya Eren Genome and Stem Cell (GENKÖK) Center, Erciyes University, Kayseri, Türkiye
e-mail: serpiltaheri@hotmail.com

M. R. S. Abusalim
Department of Molecular Biology and Genetic, Erciyes University Gevher Nesibe Genome and Stem Cell Institute, Kayseri, Türkiye

Betül Ziya Eren Genome and Stem Cell (GENKÖK) Center, Erciyes University, Kayseri, Türkiye

M. Turgut et al. (eds.), *The Human Hippocampus*, https://doi.org/10.1007/978-3-031-88907-3_11

In particular, the ventral hippocampus is enriched with genes responsible for neuropeptide signaling pathways and behavioral fear responses, which may reflect a functional relevance of this region in emotion, affect stress, and anxiety regulation. For instance, genes including gastrin releasing peptide (*Grp*), galanin receptor1 (*Galr1*), neuropeptide Y receptor Y2 (*Npy2r*), and proenkephalin (*Penk*) are more evident in postnatal rat ventral hippocampus, and they are predicted to affect the neurotransmitter release and strength of synaptic transmission, consequently modulating emotional states and fear-related behaviors. In contrast, genes enriched in the dorsal hippocampus are involved in memory formation, synaptic plasticity, and cognitive functions, indicating its role in encoding, storing, and retrieving, among others, declarative memories. Adenylate cyclase 1 (*Adcy1*), Calcium/calmodulin-dependent protein kinase IV (*Camk4*), and regulator of G protein signaling 14 (*Rgs14*) are among the genes involved in the molecular pathways of long-term memory consolidation and synaptic potentiation [8, 9].

In addition, enrichment with the ephrin receptor (EPH) signaling pathway and circadian rhythms (neurotrophic receptor tyrosine kinase 11 (Ntrk1), EPH receptor A52/73 (Epha5/7), potassium voltage-gated channel subfamily H member 74 (Kcnh7), and tryptophan hydroxylase 25 (Tph2)) in the dorsal hippocampus was also observed in many studies conducted on animals, again reinforcing its role in cognition and retiming memory. Current results highlight the anatomical segregation of hippocampal functioning across the dorsoventral axis reflecting the rich complexity of its modulation of cognitive and emotional processes [8].

Finally, regional gene expression patterns also drive the complexity of the hippocampus, and many genes are heterogeneously expressed within and across hippocampal subregions, highlighting its functional specialization. This heterogeneity is most noticeable in the pyramidal cell layer of cornu ammonis (CA) subfields, and the dentate gyrus' (DG) granule cells (GCs). For example, the crystallin Mu (*CRYM*) gene is involved in thyroid hormone binding and regulation of oxidative stress and exhibits a distal–proximal gradient in the CA3 region. Desmoplakin (*DSP*) expression is more prominent in the dorsal granular cells than in the ventral cells. The *DSP* gene encodes for a protein that acts as an adhesion molecule and serves in different processes such as regulating gene expression, cell differentiation, and migration. A knockdown or suppression of *Dsp* in mice can result in reduced neuronal activity, impaired adult neurogenesis, decreased serotonin responsiveness, and increased anxiety-like behaviors such as reduced exploratory behavior, highlighting its critical role in maintaining the functional integrity of the DG [10]. Meanwhile, Wolfram syndrome 1 (*WFS1*) gene, associated with the endoplasmic reticulum stress response and neuronal health, is specifically expressed in CA1 [11]. These distinct gene expression patterns underscore the specialized functions of each hippocampal subregion; the involvement of CA1 in synaptic plasticity and memory formation, the indicated functions of CA2 in unique stress response and social memory, the essentiality of CA3 for pattern completion and excitatory circuits, and finally the DG's roles in structural organization and connectivity (Fig. 11.1).

Overall, the gene expression of the hippocampus reflects its complex structure and functional specializations, with distinct gene expression profiles driving its different cellular populations and elaborate neural circuits. The identification of these expression profiles is necessary in order to reveal the molecular mechanisms highlighting the implications of the hippocampal involvement in several physiological functions and in the pathology of diverse conditions.

11.1.1 The Genetic Architecture of the Hippocampal Anatomy

The hippocampus is firmly embedded in the brain, nevertheless, a sagittal slice can reveal the anatomy of the hippocampus formation, which consists of the hippocampal proper, as well as the gyrus fasciolaris, indusium griseum, and longitudinal striae [12–14]. The hippocampal

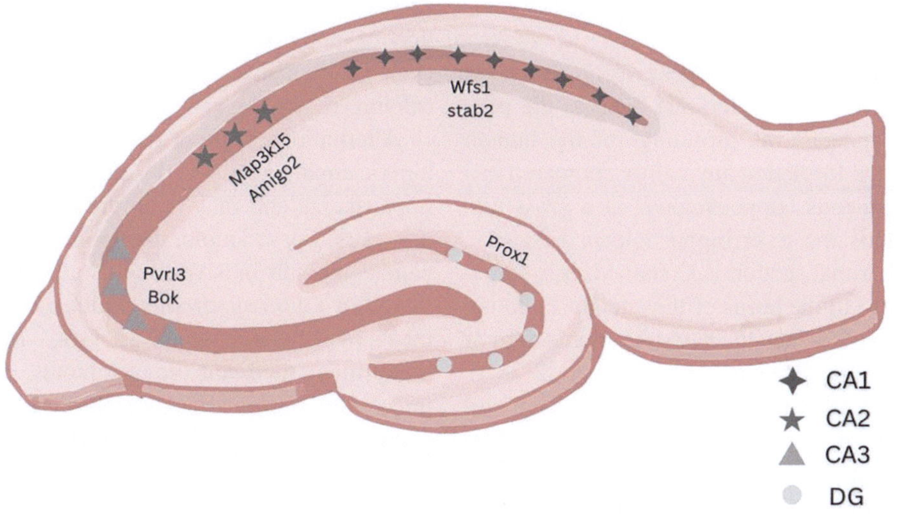

Fig. 11.1 Marker genes in the rodent hippocampus. The rodent CA1 region is defined by the expression of Wfs1 while Map3K15 and Pvrl3 delineate CA2 and CA3 respectively. Satb2 in CA1, Amigo2 in CA2, Bok in CA3, and Prox1 in the DG have been underlined as marker genes for these regions

sulcus divides the DG, a band of grey matter in the hippocampal formation, from the main body of the hippocampus. Adjacent to the hippocampus, within the parahippocampal gyrus lies both the entorhinal cortex (EC) and subiculum linking input/output signaling through the hippocampus [12].

The grey matter of the DG is made up of three layers, the molecular layer, intermediate granular layer, and polymorphic layer [12, 15]. The molecular layer is a key region for information flow where dendrites from the granular cells and interneurons meat axons form the EC inputs, thus, it serves as a site for initial integration of incoming information from the EC and modulatory input from local interneurons [16]. The granular cells, on the other hand, form the principal neurons of the DG and they are responsible for receiving excitatory input from the EC and are believed to have a key role in pattern separation, which is the process of separating similar inputs to create unique representations in memory. Lastly, the polymorphic layer links the hippocampus to the DG. It consists of axons from GCs giving rise to the mossy fibers and several types of interneurons, including basket cells and hilar mossy cells [17]. Through local inhibitory circuits and feedback processes, it likely contributes to the modulation of granule cell activity and information flow within the DG.

In a rat hippocampal transcriptome study the DG was found characterized by high expression levels of ribosome-related genes, RNA degradation-related genes, spliceosome-related genes, neurogenesis markers, such as *forkhead box* (*Fox*) *O31*, *calbindin 12* (*Calb1*), and *glial fibrillary acidic protein3* (*Gfap*), *neurotrophins like brain-derived neurotrophic factor1* (*Bdnf*) and *neurotrophin-32* (*Nt-3*), and alternative splicing factors like *heterogeneous nuclear ribonucleoprotein A1* (*hnRnpa1*). Moreover, studies have revealed a set of genes that are expressed in the DG but not the CA regions including *prospero homeobox 1* (*Prox1*), a gene involved in cell fate determination, gene transcriptional regulation, and progenitor cell regulation across multiple organs, as an established marker gene [18, 19]. Alternations in any of these genes may disrupt the structural and functional properties of the DG. For instance, the knockout or suppression of *Bdnf* or the inhibition of its receptor *tropomyosin receptor kinase B* (*TrKB*) in mice has been shown to lead to a reduction in neurogenesis, and deficits in cognitive functions, and have been associated with stress-related psychiatric disorders [19, 20].

Furthermore, while it appears as an elongated and curved structure lying just beneath the neocortex in rodents, the infoldings of the human hippocampal formation are mainly compared to a seahorse "Genus Hippocampus" in a gross dissection, thus the term hippocampus. Likewise, since its coronal section is C-shaped, it is sometimes linked to the horns of the ancient Egyptian god "Ammon," hence the name Ammon's horn or cornu ammonis [21].

The hippocampus's head, body, and tail regions are histologically subdivided into four sections, respectively, CA1, CA2, CA3, and CA4 [21]. Each of these regions in rodents has distinct gene expression profiles that contribute to their specific functions. For example, it was highlighted in several studies that the expression of *Wfs1* defines the borders of the CA1 region alongside other genes that show special expression in the same region as the established marker gene special AT-rich sequence-binding (*Satb2*) and the potential marker peroxisomal biogenesis factor 5-like (*Pex5l*). The former gene is crucial for neurodevelopment through chromatin remodeling and regulation of gene expression, and the latter gene is essential for establishing a dendritic gradient of the hyperpolarization-activated cyclic nucleotide-gated channel (Hcn1) and for the hyperpolarization in CA1 neurons [19, 22].

Furthermore, mitogen-activated protein kinase 15 (MAP3k15) a protein essential for the cell's response to different stimuli and stressors, delineates the murine CA2 region, with genes like adhesion molecule with Ig like domain 2 (*Amigo2*) showing especially high expression in the same area. This gene encodes for an antiapoptotic protein that promotes neuronal survival and is a downstream target of calcium-dependent survival signals suggesting its neuroprotective role [11, 23, 24]. Meanwhile, the CA3 region is defined by poliovirus receptor-related 3 (*Pvrl3*) a gene significant for neuronal development, synaptic formation, and adhesion, thus is important for cognitive functions such as memory and learning. In addition, gene expression in the regions indicates involvement in biological processes, such as ATP metabolic processes, generation of precursor metabolites, synaptic vesicle cycle genes, such as *SYP*, *VAMP*, *and SYT11*, and calcium-regulating protein [11, 25].

Alternations in these genes can disrupt these hippocampal regions leading to functional impairments and may form the root for various disorders. For example, disruption in the *SATB2* gene can result in SATB2-associated syndromes (SAS) or Glass syndrome, which are characterized by developmental delays, intellectual impairment, and autistic symptoms. Likewise, mice with *Satb2* knockout showed deficits in cognitive functions like spatial learning and memory in early developmental stages [22].

Further, the hippocampus is characterized by its layered structure and diverse cell types, which include pyramidal neurons, interneurons, and glial cells, each contributing to its intricate network and functionality. Pyramidal neurons are the principal excitatory cells in the hippocampus, located in the CA regions. These neurons have a distinctive triangular shape with an apical dendrite that projects toward the stratum radiatum [26, 27]. Interneurons, a heterogeneous group of inhibitory cells, are distributed throughout the hippocampus and play a crucial role in modulating synaptic plasticity and network activity. Subtypes of interneurons, such as chandelier cells, basket cells, and oriens-lacunose-molecular (O-LM) cells, are essential for shaping the hippocampal structure and regulating excitatory input [27, 28].

With equal significance, the hippocampus also houses glial cells, including astrocytes, oligodendrocytes, microglia, and immature oligodendrocyte precursor cells (OPCs). These glial cells play several roles necessary for brain homeostasis including influencing synaptic formation and transmission as well as responding to injury or inflammation [29–32].

A human hippocampal transcriptome atlas revealed the transcriptomic heterogeneity of these glial subtypes, identified enriched cell processes, neurological relevance pathways, and disease specificity, and suggested the possible functional role and gene expression profiles characteristic of each glial subtype [33]. Astrocytes, for instance, are the predominant glial cells that govern the synaptic operation, maintain the blood–brain barrier, and sustain neuronal metabolism. *GFAP* and aquaporin 4 (*AQP4*), a gene that encodes for a water channel protein important for homeostasis, are among the predominantly expressed genes in astrocytes [32]. Such genes exhibit an upregulated expression in AD as a compensatory mechanism to provide structural support and repair the impaired water balance and altered fluid dynamics associated with AD pathology [34, 35].

Oligodendrocytes, on the other hand, myelinate axons, thereby ensuring fast signal speed. For this, they show higher expression of proteins such as myelin basic protein (MBP) and Proteolipid Protein 1 (PLP1) [36, 37].

Microglia, the brain's resident immune cells, constantly monitor and therefore constantly react to local damage or infection. Genes such as triggering receptors expressed on myeloid cells 2 (*Trem2*) and C-X3-C motif chemokine receptor 1 (*Cx3cr1*) were among other genes identified in this subpopulation in mouse model studies [38, 39]. Finally, OPCs serve as progenitor cells that can differentiate into oligodendrocytes which may play a role in hippocampal plasticity and repair and have been implicated in several neurodegenerative diseases such as AD. Together, these diverse cell types form the intricate neural network of the hippocampus, enabling its pivotal role in memory processing besides other functions [40].

11.2 Genes Involved in Hippocampal Functions was intended to serve as the main heading for the following subheadings:

Embryonic Development The whole gamut of hippocampus development is a careful interaction between important patterning events and the selective expression of associated genes. This begins early during embryogenesis with the development of the neural tube, a precursor structure to the central nervous system. A zone of the epithalamus called the cortical hem in the dorsal medial part of the developing brain progresses inside the telencephalon and acts as an essential coordinator of hippocampal development [41]. It is important because this area serves as a landmark for the hippocampal boundaries and gives birth to further development.

Various critical genes that control these processes become active at this time, and each gene is responsible for a specific aspect of shaping up the hippocampal structure. The transcription factors *paired box 6* (*PAX6*) and *empty spiracles homeobox2* (*EMX2*) are two of the most important genes that play a role in determining the borders of the hippocampal CA anlage [42]. While *PAX6* ensures the development of the hippocampus in a medial position, *EMX2* controls the timely growth and proper development of it. In addition to boundaries, Pax6 is also differentially expressed between species and more diffusely distributed in primates, including humans, than in rodents [43]. Simultaneously, Wnt3a cascade activates signaling pathways that regulate the proliferation and differentiation of progenitors throughout the development of the hippocampus. This signaling pathway is critical to the control of self-renewal and differentiation of neural progenitor cells [44, 45].

Further progression in hippocampal development involves a series of critical patterning events, including the division of the pallium into distinct regions such as the neopallium and archipallium. Transcription factors, such as *LIM homeobox 2* (*LHX2*), *FOXG1*, *EMX1*, and *EMX2* subdivide the archipallium into further compartments by promoting cell proliferation and cell differentiation. While the hippocampus is comprised of multiple genes responsible for the development, mutations or alterations in these genes disrupt the hippocampal development leading to structural abnormalities and functional deficits. For example, rodent studies have shown the significance of the *Lhx2* gene for the formation of the hippocampus from cortical stem cells by mediating the balance between neurogenesis and astrogliogenesis [46]. Mutations in this gene can cause an expansion of the cortical hem and ventral pallium at the expense of other pallial structures (e.g. hippocampal formation) [47].

Among the critical players in the hippocampal development is *PROX1*. This gene expression is maintained during the embryonic development of the hippocampus and through adult neurogenesis. It is crucial for the hippocampal cellular composition by regulating cell proliferation and differentiation, through controlling cell cycle and astrogliogenesis [48, 49].

Fetal Development As development advances, neural progenitors within the hippocampus neuroepithelium divide asymmetrically, giving birth to pyramidal neurons and dentate GCs. These neurons travel radially to produce the various layers of the hippocampus, which include the DG, CA1, CA2, and CA3 regions. At this point, genes such as *neuronal differentiation 1* (*NEUROD1*), a gene that encodes for a transcription factor essential for the differentiation of the nervous system, along with *PROX1* perform critical roles indicating the differentiation of newborn neurons, and the extracellular matrix glycoprotein reelin, cyclin-dependent kinase5 (CDK5), and C-X-C motif chemokine ligand 12 (CXCL12) outline their migration, ensuring their integration into the existing circuitry.

Furthermore, the production of synaptic proteins like synaptic cell adhesion molecule 1 (SynCAM1) and Neuroligin promotes the development of synaptic formation and connections, which are required for circuit construction and function [47, 50–52]. The contributions of neuroligin extend to the formation of synapses, regulation of synaptic transmission and plasticity, and regulation of excitatory and inhibitory signals in the hippocampal circuity. This broad range of functions marks neuroligin as a key factor for the hippocampus' functions. Alternations in this gene have been associated with neurodevelopmental disorders such as autism and schizophrenia (SCZ), underlining its significance in proper synaptic connectivity within the hippocampus [53]. *Olduvai Domain* (*DUF1220*) is another major genetic contributor to hippocampal development. A new study shows that the copy number of the DUF1220 protein domain in the human genome, besides serving as an indicator of the evolutionary distance to humans among different species, also serves as a unique but very significant association with brain size, cognitive capacity, and forms of autism [54].

Neonatal Development Several studies have demonstrated that cell genesis in certain species, such as rats, could exhibit a sexually dimorphic regulation. Epigenetic regulators including growth arrest and DNA-damage-inducible alpha (Gadd45α), histone deacetylase (Hdac), and histone acetyltransferase (Hat), among others, were found to be responsible for the higher rate of neurogenesis in male rats compared to that in females [55]. Further, during the neonatal period, the hippocampal maturation process is sustained by the creation of synaptic connections and the evolution of brain circuits. There is a strong expression of neurotransmission and synaptic plasticity genes at this period, including, BDNF, adhesion molecules like SynCAM1, nerve growth factor Beta (*Nfg*-β), Jun proto-oncogene AP-1 transcription factor subunit1 (*C-Jun*), and UNC-5 homolog1/2 (*Unc5h1/2*) and others. These genes are required for the development of spines, synaptic pruning, and dendritic arborization, all of which are crucial for synaptic maturation and plasticity [56–59].

Postnatal Development The hippocampal structure and function changes during postnatal development at many levels. The connection and function of neurons, in turn, are dependent on factors from glial cells like oligodendrocytes (to selectively myelinate axons) or astrocytes. In murine these include Gfap, Nestin of neural progenitors, and Fatty Acid Binding Protein (Flbp) from glial precursors that form the hippocampus glial network. Furthermore, markers of neuronal maturation and differentiation including the neuronal precursor cell marker doublecortin (Dcx) marking early stages of neurogenesis, polysialylated-neuronal cell adhesion molecule (Psa-Ncam), significant for plasticity and neuronal changes, and indicates the ongoing development, Calretinin as a marker for newborn neurons suggests the transitional phase in neuronal differentiation, The *neuronal nuclei* (*NeuN*) as a marker for complete maturation, and lastly, *Calbindin* which is expressed in mature GCs, signifying the fully developed neurons, all increase at this stage demonstrating the continued hippocampal establishment [60]. Notably, these molecules are expressed in the hippocampus of several species including humans with different timing and intensity.

FOXG1 also plays an essential role in the postnatal development of the DG since it regulates the expansion of the precursor cell population and promotes the differentiation of these cells into mature neurons, thus, it is necessary for the maintenance of the DG progenitor pool and may be required for the survival and maturation of postmitotic neurons, its role is preserved event in adult hippocampus. Disruption of FOXG1 results in the loss of the sub-granular zone and a severely disrupted secondary radial glial scaffold, leading to impaired migration of GCs of the DG. This impairment may lead to early developmental issues such as FOXG1 syndrome with symptoms of intellectual disabilities, impaired speech, visual impairments, gastrointestinal issues, and sleep disturbance [61, 62].

Adult Hippocampus The hippocampus continues to develop throughout childhood and reaches maturity by adulthood when the network that supports learning and memory has been laid down. In addition, hippocampus markers, such as *LHX*, *SYNCAM1*, and the *nuclear receptor subfamily 2 group F member 1* (*NR2F1*) are still expressed in adult hippocampus neurons that carry out their functions [9, 56, 63]. Yet, a hallmark of the adult hippocampus is its lifelong ability to feature this flexibility and its capacity for structural remodeling following learning and mature experience.

Furthermore, it has been demonstrated that the DG retains the ability to generate neurons throughout adulthood in its sub-granular zone [64]. This complicated process of neurogenesis is regulated by several genes, including *DCX*, *FOXG1*, *WNT*, *NEUROD1*, *SOX*, *GFAP*, *and PROX1*, all of which play critical roles in neurogenesis at various stages. *DCX*, for example, is expressed by migratory neuroblasts and immature neurons and is one of the most significant neurogenesis markers, whereas Wnt signaling pathways control neural stem cell proliferation and development [54]. *SOX* genes are engaged in several phases of brain development, including cell fate determination similar to *PROX1* which regulates neural progenitor cell fate and neuronal subtype specification [11, 64].

Notably, there are significant differences in the rate of adult hippocampus neurogenesis between species; for instance, rodents have a quicker rate than humans and nonhuman primates. It is still up for question, though, whether adult hippocampal neurogenesis occurs in humans, and if it does, it most certainly happens significantly more slowly than in other species [65, 66]. Adult hippocampal neurogenesis is a significant process that affects behavior and cognitive ability across a range of animals, notwithstanding these variations. Furthermore, a gene recently discovered to be expressed by dentate precursors is the *homeodomain-only protein homeobox* (*HOPX*), a major determinant of murine embryonic and postnatal neurogenesis. Work has shown an array of basic resemblances including perinatal, postnatal, and adult neurogenesis in the mouse DG. Lineage-tracing experiments have shown that clonal HOPX+ similarity indicates the division of the hippocampal lineage into targeted cells residing in either DG or CA regions, indicating its crucial function in determining cell destiny during hippocampal development. The distribution of HOPX+ progenitors at GW11 displays a dorsoventral gradient, indicating that their differentiation is directed by regional determinants as previously observed for CA and DG granule cell neurons during hippocampal development [54].

Several critical patterning events as well as the precise expression of specific genes of importance are ultimately required for the development of the hippocampus and this, in a nutshell, highlights the complexity and high level of regulation involved in hippocampal development. Multiple stages of hippocampal structure are developed and matured through the involvement of a series of transcription factors and signaling pathways from early embryogenesis, through neonatal and postnatal periods to adulthood. The boundaries among them are also delimited by key genes, including *Pax6*, *Emx2*, *Lhx2*, and *Foxg1* among others, which in general provide proliferative and differentiative signals and regional code for their anterior restriction of migration and synaptogenesis. Alternations in these genes can have serious consequences on both the structure and the function of the hippocampus, underlying their significance for hippocampal development. As the hippocampus matures, it preserves its capacity of generating new neurons in the adult DG contributing to its remarkable plasticity and adaptability [54].

11.2.1 A Genetic Blueprint: The Role of the Hippocampus in Memory and Learning

The hippocampus is a crucial region for information processing, cognition, and memory consolidation [67]. It aids the formation of memories and changes short-term sensory information into long-term memories. It is also responsible for the animal's sensory information processing related to its environment and it plays a role in conditioning and learning [68]. The hippocampus is functionally structured along a gradient, with the dorsal area involved in memory and spatial navigation and the ventral region in anxiety-related activities. However, gene expression studies reveal that there are several functional areas along the hippocampus' long axis, calling into question this dichotomy [69].

11.2.2 Genetic Regulation of the Hippocampal Involvement in Cognitive Processes

Studies linking the hippocampus and memory date back to the early 1950s, when the case of Hennry Molaison (HM) sparked interest in this relation. HM, suffering from epileptic seizures underwent surgical removal of his hippocampus, alongside other medial–temporal structures, by the surgeon Beecher Scoville in an attempt to cure his unresponsive seizures, ending up with anterograde amnesia: the loss of the ability to form new memories [70]. This case was one of the first indications of the role of the hippocampus in forming new memories. Subsequent research demonstrated the impact of hippocampal lesions and alternation in hippocampal volume on various forms of memory and learning in humans and animals such as monkeys and rodents. Numerous researchers have proven that hippocampal lesions, especially in the dorsal end, impair spatial memory, including Broadbent's study where he and his colleagues demonstrated that even 30% of lesions to the dorsal hippocampus can cause severe damage to spatial memory. Moreover, they were able to demonstrate that a higher percent of hippocampal damage (75–100%) can impair nonspatial memory too [71]. This and many other studies support the role of the hippocampus in several types of learning and memory including spatial learning and memory, contextual fear conditioning, trace eyeblink conditioning, novel (spatial and contextual) object recognition (NOR), and social recognition [72–76].

Much like other regions in the brain, the hippocampus is a hub of intricate gene expression patterns that give rise to its pivotal role in all memory processes: acquisition, consolidation, and retrieval. Memories are not formed after an event happens; rather, it is an in-vivo process, the brain collects experiences during the event itself. The hippocampus plays a crucial role in memory formation by acting as a collection pot for the information received from various parts of the brain, such as sensory, olfactory, or gustatory information, depending on the type of event experienced. These sensory, emotional, and cognitive components are then encoded into one cohesive memory. The hippocampus further wires these memories back to the cortex for storage. Merging these signals into one integrates several crucial processes such as the regulation of synaptic plasticity, neurogenesis, neuron projection morphogenesis, and cell signaling [77].

The billions of neurons composing the brain can communicate via their synapses by sending and receiving neurotransmitters. A typical neuron can pose millions of synapses with other neurons forming complex networks that correspond to all brain functions, including memory. The memory formation process is mediated by a dynamic network of immediate early genes (IEGs) such as *early growth response 1* (*EGR-1/ZIF268*), *Fos proto-oncogene*, *AP-1 transcription factor subunit* (*C-FOS*), and *activity regulated cytoskeleton associated protein* (*ARC/ARG3.1*) [78]. These genes are rapidly activated by neural activity and serve as molecular beacons that highlight the neurons engaged in memory processing, such as the hippocampal neurons. Their protein products act as transcription factors, modulating the expression of downstream genes that solidify the synaptic changes underlying long-term memory [79]. Behavioral tasks also induce IEG expression in neurons; such IEG-expressing neurons are distributed across a wide variety of brain regions [79].

When neurons in the hippocampus are stimulated by incoming signals, IEGs like *C-FOS* and *ARC* are quickly upregulated. C-FOS, a regulatory transcription factor that influences the activity of other "downstream" genes, acts as a marker for activated neurons, suggesting the initiation of memory processing, whereas ARG3.1 regulates glutamate receptors trafficking, specifically α-amino-3-hydroxy-5-methyl-4-isoxazolepropionic acid receptors (AMPAR), which are critical for synaptic plasticity [78, 80]. This process enhances connections in the synapse, making it easier for neurons to transmit signals efficiently [81].

The regulation of IEG expression is far more complex and depends on the types, duration, and repetitiveness of stimuli, specific experimental conditions, as well as time course of gene induction. This follows the pattern of the hippocampus' heterogeneous expression in general as early works suggested that maze learning was disrupted by dorsal hippocampal lesions, but not by ventral lesions [82]. Additionally, multiple studies have long confirmed that dorsal hippocampal lesions severely impair memory formation in water maze, while equally large ventral lesions do not necessarily produce the same deficits [83, 84]. Together, these findings indicate that the septal and temporal hippocampus encode for distinct functions and may have distinct roles in memory processing, with the former potentially playing a role in modulation and influence, and the latter being more involved in specific memory tasks, such as spatial memory and verbal encoding. Nonetheless, studies have demonstrated that IEG expression can be induced by tasks such as social defeat in both the dorsal and ventral hippocampus. This indicates that this stressor activates neural circuits involved in behavior's cognitive and emotional aspects [85]. Yet some tasks may induce a differential expression of these genes, for example, research that examined ARC mRNA expression rates during sessions of spatial exploration treatments done on mice indicated a higher level of *Arc* mRNA expression due to their spatial behavior within the dorsal hippocampal CA1 and CA3 areas but not so much in the ventral counterpart. Additionally, more GCs expressed *Arc* mRNA in the dorsal compared to the ventral hippocampus in response to spatial behavior. The study highlighted a two- to threefold reduction in the proportion of cells expressing *Arc* in response to exploratory behavior in the more ventral regions of the hippocampus [86].

From Memory to Knowledge When it comes to learning, it is all about observing, imitating, connecting information to form a link, and saving it for repetition or later use. The brain stores newly obtained knowledge by encoding it into memory. Memory, on the other hand, allows the recall and use of this knowledge when needed. By establishing links between newly acquired knowledge and previously acquired information, one may expand their existing knowledge base and establish connections between various concepts or ideas. The hippocampus is involved in several types of memory such as episodic, spatial, and relational by encoding, storing, and retrieval. It is also crucial for concept formation, memory integration, attentional biasing, and memory-based prediction, all essential for learning [87, 88]. Thereby, the hippocampus helps shape and form our learning experiences, allowing us to acquire new skills, knowledge, and understanding.

After forming these memories, the hippocampus utilizes many mechanisms to manage them, the most important of which may be long-term potentiation (LTP). Through this, the hippocampus strengthens the synaptic connections between neurons to stabilize the acquired information [89]. This modifies the wiring or network structure of the brain, rearranging the cortical map. Neurons can create connections throughout life because of growth and rearrangement. On the plus side, LTD (long-term depression) causes synaptic connections to decrease, maintaining brain efficacy [90, 91].

The most studied site where the mechanism of LTP occurs is within the synapses between the presynaptic CA3 Schaffer collateral neurons and the postsynaptic CA1 region [92]. LTP can occur at most excitatory synapses in the brain. Studies have demonstrated how dopaminergic signaling, particularly D1-like receptors, facilitates the potentiation of intrinsic excitability and synaptic LTD in temporoammonic-CA1 neurons. Serotonin (5-HT1A and 5-HT4) especially is abundant in the hippocampus and several studies have shown that the inhibition of 5-HTR results in serious impairment in memory and learning. Likewise, norepinephrine plays a major role in memory and learning by inducing LTP by increasing the transcription of key player mRNAs in this process like Glutamate Ionotropic Receptor AMPA type 1/2 (GLUA1/2) in the CA1 region. Endocannabinoids also can trigger LTP in the hippocampus through the stimulation of cannabinoid receptor 1 (CB1) in astrocytes, contributing to synaptic plasticity [93–96]. Nonetheless, LTP is best studied at the glutamate synapse in the hippocampus. Once a sensory signal is retrieved in a pre-Schaffer collateral neuron through learning, it stimulates and depolarizes the neuron to release neurotransmitters, like glutamate. Glutamate at the synaptic cleft then binds to glutamatergic receptors on the postsynaptic neurons of the CA1 pyramidal cells, mainly AMPA and N-methyl-D-aspartate (NMDA), hence activating them. When the presynaptic neuron is stimulated with a weak signal, only a small amount of the glutamate is released. Although both receptors are bound by glutamate, only AMPA is activated by weak stimulation. This activation allows the sodium to influx through these channels resulting in a slight depolarization of the postsynaptic membrane. NMDA channels remain closed due to a magnesium molecule blogging its pores, however, a strong stimulation or repeated weak signals result in the release of a large amount of glutamate. In this case, AMPA remains open for a longer time, admitting more sodium into the cell, thus, resulting in a greater depolarization. An increased influx of positive charges expels the magnesium from the NMDA channel leading to their activation, allowing not only sodium but also calcium into the neuron. The entry of calcium marks the early phase of LTP. Acting like a secondary messenger, the calcium initiates intracellular signaling cascades. In these pathways, several crucial protein kinases responsible for transcriptional regulation such as Ca^{2+}/calmodulin-dependent protein kinase II and IV (CAMKII/IV), mitogen-activated protein kinase (MAPK), and protein kinase C (PKC) are activated [97–99]. These kinases help in the learning process by enhancing synaptic communication in two ways: (1) they phosphorylate the existing AMPA receptors thereby increasing AMPA's conductance to sodium; (2) they help bring more AMPA receptors from intracellular stores to the postsynaptic membrane. This phase is suggested to form the basis of short-term memory, which lasts for several hours. Following that, in the late phase, new proteins, like CREB are made and gene expression is activated to further enhance the connection between neurons [100]. These proteins induce newly synthesized receptors and expression of other genes and proteins involved in transcription, cell signaling and transportation, neurotransmitters and peptides, cell metabolism, cell structure, and growth factors that induce the growth of new dendritic synapses and synaptic connections [101]. These strong neuronal connections correlate with the formation of long-term memory. This is how a cellular foundation for learning and memory is provided by LTP, which improves information transfer between neurons and the creation of connections between various stimuli or experiences [102]. Importantly, the processes of LTP, and LTD, differ according to the circuits and synapses in which they function, thus they are not unitary [103].

In addition to IEGs, BDNF is an important factor in the process of memory and learning through synaptic plasticity and neurogenesis [104]. In addition to its role in dendritic growth, *BDNF* is highly expressed in the hippocampus and is implicated in the regulation of both excitatory and inhibitory synaptic transmission and in the promotion of activity-dependent plasticity. Additionally, BDNF has been associated with LTP. It enhances memory storage via mechanisms that may include (1) a mechanism that favors changes in spine morphology and thereby stabilizes LTP and (2) facilitating an increase in the density, size, and complexity of dendritic spines through unregulated actin polymerization [105, 106]. In addition to this, BDNF is a key factor for cell survival and proliferation and thus plays a vital role in brain structural and functional plasticity leading to increased neurogenesis [104]. The *BDNF* Val66Met polymorphism, a single nucleotide polymorphism (SNP) in the human *BDNF* gene, is one of the widely studied variants for its influence on memory performance and synaptic plasticity. This SNP results in altered synaptic targeting and activity-dependent release of vesicles containing BDNF. Activity-dependent synaptic plasticity is impaired in individuals carrying the Met allele, especially in Met66 homozygous knock-in mice. In addition, the same polymorphism has been linked to differences in the structure and function of the brain, including a reduction in volumes of various regions including the hippocampus. It is also associated with alterations in verbal and spatial memory tasks, evidencing the relevance of *BDNF* in cognitive functions. Hence, the *BDNF* Val66Met SNP is a prototype of how genetic variations affect memory and learning through the family and synaptic plasticity homeostasis [104].

Other key players in these processes are CREB and CAMKII. Reducing the expression of *CAMKII* for example, through mutation or knockdown, can block LTP and lead to amnesia [107, 108]. Similarly, it was shown that blocking *Creb* reduced AmpaR subunit *GluA1* levels that attenuated LTP but enhanced LTD, and this in turn, impaired contextual fear learning memory in mice [109]. In addition, in a study conducted on a fly strain, a decrease in memory was observed after the inactivation of *Orb2*, a member of the cytoplasmic polyadenylation element-binding (CPEB) family, a day following memory formation. This indicates the key role played by *Orb2* in maintaining memory after consolidation [110].

Further, a major side of *CREB's* role in regulating LTP is due to its ability to induce the translation of the Poli-A tail of dormant mRNAs necessary for memory and learning. Abundant research investigations have emphasized the significance of certain miRNAs in the hippocampus's creation and maintenance of memory. miR-134, a brain-specific miRNA, regulates dendritic spine development via inhibiting Protein kinase LimK1, a serine/threonine kinase that plays a crucial role in regulating actin cytoskeleton dynamics and is involved in neuronal morphology, particularly dendritic spine development, and helps stabilize LTP following induction by targeting Sequence-specific RNA-binding protein Pumilio RNA Binding Family Member (Pum2) a posttranscriptional repressor that regulates RNA during embryonic development and cell differentiation, and thus, it can be considered a regulatory molecule for learning and memory [111, 112].

In this regard, [113] discovered that miR-138-5p controls hippocampal inhibitory synaptic transmission and short term memory, which may have consequences for conditions like SCZ. Likewise, [114] showed that miR-324-5p and miR-24 are involved in the establishment of spatial memory in the ventral striatum and the hippocampal regions, respectively. Further research has revealed that miR-181a plays a significant role in the establishment of hippocampal-dependent memory and that it interacts with Protein Kinase AMP-Activated Catalytic Subunit Alpha 1 (*Prkaa1*), a particular target gene [115].

The normal levels of miR-132 were found to improve cognitive ability in a spatial memory task; conversely, supra-high levels had inhibited learning in mice in a negative feedback fashion.

To further investigate the role of this molecule in hippocampus-dependent learning and memory, researchers suppressed the transgene expression of miR-132, which resulted in the reversion of the impermeant of cognition and elevated levels of dendritic spines caused by high levels of miR-132. This demonstrates the regulatory role of this miRNA in memory and learning [116]. Similarly, an increase in the cluster of miR-183/96/182 levels in the hippocampus was found after a NOR task, which includes hippocampal-dependent mechanisms and is linked to rats' innate attraction to novelty, is typically used to assess short- and long-term memory [117]. Importantly, researchers of the same study have shown that during NOR, miR-183/96/182 overexpressing mice had significantly better memory than control mice. Additionally, in another study, defective synthesis of miR-183/96/182 cluster, was linked to age-related memory impairment, due to elevated protein phosphatase 1 (PP1) (a memory suppressor) signaling. Researchers were able to restore memory in old mice by correcting the impairment by overexpressing miR-183/96/182 in the hippocampus or by environmental enrichment that normalizes *PP1* activity [118]. In the same previous study, was able to show another regulatory role of these molecules as regulatory for plasticity-related genes. The downregulation of numerous genes including ion channels, receptors, one kinase, a phosphatase, and the HDAC9 was linked to the overexpression of miR-183/96/182. The combined findings of these investigations highlight the role miRNAs play in regulating hippocampal memory-related functions [117].

The hippocampus is critical for coordinating the acquisition, consolidation, and recall of memories, as well as learning by supporting synaptic plasticity, neurogenesis, gene expression, and mechanisms that involve IEGs, several transcription factors such as BDNF, and various miRNAs. Investigation of such molecular mechanisms and genetic factors leads not only to knowledge of how memory and learning are controlled at the very basic processes but also opens novel approaches to the development of the treatment of memory-related diseases and the therapy of cognitive enhancement.

11.2.3 Genetic Regulation of Spatial Navigation and Cognitive Mapping

Cognitive processes are mental activities that enable individuals to acquire, process, store, and utilize information. These processes are fundamental to interaction with the world, understanding experiences, and performing daily tasks, encompassing perception, memory, attention, decision-making, problem-solving, and language comprehension [119]. The hippocampus plays a pivotal role in these cognitive functions through its unique capabilities. The extensive neuronal circuitry of the hippocampus connects it to various brain regions, enabling the receipt of inputs from various brain regions, particularly through the EC which acts as a major hub, channeling sensory and cognitive information to the hippocampus. The received inputs undergo various processing via the hippocampal subregions; starting from the DG which helps distinguish between similar input patterns, and following the CA3 neurons that are responsible for pattern completion to reconstruct a whole memory from partial cues [5].

These hippocampal circuities highly depend on several genes. For instance, *Prox1*, as previously described, is essential for the development of GCs in the DG which forms the main input site in the hippocampal circuit. Disruption in this gene will affect DC structure and function, hence impairing the hippocampal circuitry. Furthermore, by strengthening or weakening synaptic plasticity according to the priority of the incoming inputs, LTP and LTD mechanisms in the hippocampus filter stimuli to only keep relevant information [120]. Additionally, neuronal oscillations such as theta and gamma rhythms synchronize neuronal activity across the hippocampus and other brain regions, enhancing selective attention by enabling the brain to focus on relevant stimuli while filtering out irrelevant information and improving memory by efficiently integrating sensory information, combining inputs from various sensory modalities into a coherent perceptual experience [121]. Altogether, it allows the hippocampus to integrate, process, and filter diverse types of stimuli, supporting complex cognitive tasks.

For example, decision-making underpins planning, option appraisal, and goal-directed action. This process of developing and changing representations to envision potential futures and alternatives is a key role that the hippocampus plays. Consequently, with respect to value-based decisions, the hippocampus can contribute to accumulating and fine-tuning choice-outcome associations by evaluating alternative choices whose information might be supported by interactions with both the prefrontal cortex (PFC) and striatum that together enable deliberative processes [122–124]. Individuals suffering from hippocampal loss exhibit a lack of preference for favorable or unfavorable alternatives, instead opting for a straightforward "lose-shift" approach. This suggests that the hippocampus plays a crucial role in the establishment, upkeep, and modification of choice-outcome associations [125]. Furthermore, when making value-based judgments, hippocampal amnesic individuals show longer response times and more stochastic choices [122]. Additionally, individuals with hippocampal lesions exhibit greater variability in their value-based decision-making, and this variability is correlated with the extent of impaired hippocampus tissue [126]. These results highlight the role of hippocampal formation in preserving consistent decision-making processes.

Furthermore, the hippocampus is important for language comprehension and supports the flexible use of multiple information sources, processing, rapid generation as well as interpretability of language [127, 128]. Recent fMRI studies have detected hippocampal activation during linguistic tasks, further supporting its involvement in language comprehension and processing [129]. Moreover, the hippocampus is linked to declarative memory, storing the mental lexicon, including words and meanings, while procedural memory, associated with the basal ganglia and frontal lobe, applies rule-governed sequences for grammar [130]. This highlights its role in relational binding and representational flexibility necessary for language processing.

Much is known about language comprehension and use in patients with hippocampal lesions. These patients typically show right-lateralized or bilateral language region involvement, delayed speech onset, and online communication breakdowns which all underline the necessity of the hippocampus for flexible access to stored information [131, 132]. There is also evidence from discrete language impairments that grammatical impairments can be tightly coupled with procedural memory dysfunction [133]. Previous studies found that patients with hippocampal amnesia displayed reduced verbal play relative to control groups, emphasizing the importance of the hippocampus for online processing and productive use of language [134, 135]. In essence, these findings confirm the critical role of the hippocampus in comprehending language, integrating linguistic information, and documenting new words through a system that utilizes declarative (explicit) and procedural (implicit memory control) during language mastering.

In addition to its involvement in language and decision-making, the hippocampal region is important for several other aspects of cognitive processes. It enables individuals to interpret recent occurrences by combining sensory data with previously stored knowledge. Furthermore, it facilitates learning and memory consolidation by fostering the acquisition of new skills and knowledge through experience and practice. It also helps in adaptive behavior since it supports the brain to adapt to unfamiliar situations and challenges. Additionally, it helps with more complex work and behavior management tasks including scheduling, planning, and preventing inappropriate reactions [136, 137]. This makes the hippocampus fundamental to all cognitive functions, including learning, executive processes, information synthesis, and flexibility among others.

As in all biological processes, cognitive functions also require a complex process of regulation involving several genes and molecules. Genes, such as *BDNF*, *ARC*, *CREB*, and *EGR-1*, are responsible for regulating synaptic plasticity and are known to be upregulated in the hippocampus during functions such as learning and memory [136, 137].

A vast body of research has highlighted the pivotal role of BDNF in cognitive functions, alternations in its expression were associated with various diseases including Alzheimer's disease (AD) and SCZ, and its activation was shown to recover memory deficits and synaptic proteins in animal models, proving its critical role in cognitive function [138, 139]. The demonstrated downregulation of some genes such as cholecystokinin (*CCK*), Tachykinin precursor 1 (*TAC1*), and Calbindin 1 (*CALB1*) in the hippocampus of AD patients also supports their involvement in cognitive dysfunction, nonetheless, it does not necessarily mean that their upregulation is demanded in all cognitive functions [140].

Furthermore, the hypermethylation of the *Transforming Growth Factor Beta 1* (*Tgf-β1*) gene promoter was found to increase the expression of *Potassium Chloride Cotransporter 2* (*Kcc2*), thus enhancing GABAergic synaptic inhibition and resulting in hippocampus-dependent memory deficits in adult mice [141]. The *G Protein Coupled Receptor 17* (*GPR17*) gene was also demonstrated to affect cognition; its downregulation was shown to lessen the generation of *amyloid-β*, lower the expression of the subunit p65 of *the Nuclear Factor-kappa B* (*NF-κB p65*), phosphorylate CREB, raise the expression of *BDNF*, suppress the accumulation of pro-inflammatory cytokines, inhibit glial cells activation, and increase *BCL-2*, *Postsynaptic Density Protein-95* (*PSD-95*), and *Synapsin* (*SYN*), all of which help to mitigate cognitive deficits [142].

More importantly, one of the most significant genes for cognitive processes is *FOXP2*. The protein of this gene regulates other genes necessary for the development of neuronal circuits involved in speech and language processing. Mutations in this gene have been linked to language disorders such as developmental verbal dyspraxia accompanied by a reduction in hippocampal volume [143–145]. Similarly, *Neuregulin 1* (*NRG1*) is a crucial factor in neuronal development, synaptic plasticity, and cognitive functions such as language and attention, whose alterations are linked to deficits in some aspects of cognition found in patients with attention deficit hyperactivity disorder (ADHD) or SCZ [146].

It is important to remember that the regulation of these genes is complex and subject to variation based on a range of circumstances, including the nature of the cognitive task, the person's age and gender, and the presence of any underlying neurological or psychiatric conditions. In this regard, the expression of *C-FOS* shows different patterns in response to different types of learning tasks. Research results showed that *c-Fos's* expression was higher in the CA1 region during spatial learning, while it was higher in the DG during nonspatial learning in a Morris water maze task [147]. Additionally, research indicates that there can be gender differences in the expression of several genes related to synaptic plasticity, such as NMDA receptors. The study revealed that while males rely on m-NMDAR functions for LTP and episodic memory, females tend to use estrogen signaling to substitute for these functions. In addition, male mice in this study showed better performance in spatial tasks, and female animals were better at "what" and "when" elements besides being able to manage better in longer and more complex sequences of information than men, which corresponds to human studies [148].

11.2.4 The Neural Genetics of Hippocampal Function in Mood and Social Interactions

Spatial navigation and cognitive mapping are basic behaviors that enable an organism to effectively interact with its environment [149]. It is the physical act of moving through and comprehending the physical characteristics of a given environment (spatial navigation) as well as the hierarchical, schematic, or cognitive mapping and the corresponding mental representation for negotiating one's environment (cognitive mapping) [150]. These processes are crucial in foraging, predator avoidance, and navigation to specific places.

The hippocampus is involved in spatial navigation and cognitive mapping, particularly, the dorsal (posterior) lobe. Spatial learning and memory deficits have been particularly studied in rats, where hippocampal lesions prevent them from being able to navigate mazes correctly, suggesting the important role of the hippocampus in spatial awareness [84]. In a similar study, using a rat model, McGregor (2004) showed that lesions to the hippocampus abolish response-based navigation in geometrically shaped areas [151]. Whereas in humans, damage to the hippocampus could result in impairment of spatial abilities, perhaps problems such as not being able to find one way through familiar surroundings. One very illustrative example of this is that people with disorders like AD which sustain damage to the hippocampus usually show pronounced spatial disorientation [152].

The hippocampus's involvement in these processes is supported by the intricacy of its physical structure and its extensive connections with other brain areas [5]. The EC, along with other cortical regions receives sensory signals and relays them to the hippocampus through the DG [21, 153]. This information includes external cues like visual, auditory, olfactory, and tactile information and internal physiological cues that involve proprioceptive information about body position in space and the vestibular feedback mechanism regarding balance during gait [154].

The transmission in the hippocampus for forming cognitive maps is through Cellular composition that allows it to combine, process, and utilize space, and time elements [77]. Place cells in the CA3 and CA1 regions fire specifi cally when an organism is in a particular location, encoding spatial information and building cognitive maps. Grid cells in the EC create a sense of distance and direction, while head direction cells provide a stable reference frame for orientation. Other cells, such as boundary cells and object-vector cells, contribute to precise spatial mapping by responding to environmental boundaries and specific objects, respectively [155, 156].

Moreover, hippocampal oscillations, including theta and gamma rhythms, synchronize neural activities during movement and memory processes, integrating sensory inputs and motor feedback into a coherent sense of direction and position. These oscillations are crucial for spatial awareness and memory [157].

Gene regulation in the hippocampus is necessary for such complicated processes. The neural circuits necessary for these cognitive processes are regulated by the expression of different genes including the Nuclear Receptor 4 (*NR4A*) subfamily, which consists of *NR4A1*, *NR4A2*, and *NR4A3*. After spatial learning, the CA1 pyramidal layer of the hippocampus exhibits a strong upregulation of these genes [158]. Members of the NR4A subfamily are transcription factors that control the expression of downstream genes essential for the consolidation of memory [159]. Memory problems and cognitive disorders, such as AD, have been linked to dysregulation of *NR4A* genes [160]. Moreover, it has been demonstrated that increased *Nr4a* expression can ameliorate cognitive deficiencies in mice models of neurodegenerative illnesses, underscoring the potential therapeutic value of this protein [161].

Immediate early genes such as *Homer protein homolog 1* (*HOMER1*), *ARC*, *EGR1*, and *EGR3* are important for synaptic plasticity, which is a basic process in memory and learning. Neuronal activity linked to spatial learning triggers a fast induction of these genes [158]. For example, *EGR1* is required for the transcriptional control of genes involved in LTP and is increased in all hippocampus subregions. In a similar vein, *ARC* plays a crucial role in the synaptic changes that support memory consolidation. Deficits in synaptic plasticity and cognitive function are reported in several neurological illnesses, including autism spectrum disorders and SCZ, where the dysregulation of these IEGs is linked [79].

Moreover, during spatial navigation, the hippocampal expression of *Period Circadian Regulator 1* (*PER1*) and *Dual Specificity Phosphatase 5* (*DUSP5*) is elevated. *PER1*, a fundamental part of the circadian clock, controls when genes are expressed, which affects cognitive functions [162]. Decreases in cognitive function, including orientation, and mood problems, as well as diseases like cerebral small vessel diseases, have been associated with the hypermethylation of *PER1* [163]. The MAPK signaling pathway, which is essential for memory and synaptic plasticity, is also regulated by *DUSP5* [164].

In addition, in response to stressful situations like hypothermia, the cold-inducible RNA-binding protein (CIRBP) and RNA-binding motif Protein 3 (RBM3) are important for neuroprotection and cognitive function. These proteins support the integrity and plasticity of synapses and are increased in the hippocampus [165]. Particularly, *RBM3* may be a target for therapeutic approaches in neurodegenerative illnesses like AD since it has been demonstrated to support synaptic regeneration and guard against neurodegeneration [166]. Both *Cirbp* and *Rbm3* are downregulated in mice after spatial learning and that could be part of a broader gene regulatory network modulating memory consolidation processes [158].

Finally, *Heat Shock Protein Family A Member 5 (HSPA5)* and *Serum/glucocorticoid-regulated kinase 1 (SGK1)* also show an increase in response to spatial learning, due to their roles in protein homeostasis and synaptic plasticity [158]. Overall, the control of these genes in the hippocampus is essential for cognitive mapping and spatial navigation. Since their dysregulation can result in a variety of cognitive problems, it is critical to comprehend the dynamics of gene expression to design effective treatment strategies [159, 166, 167].

11.2.5 The Neural Genetics of Hippocampal Function in Mood and Social Interactions

Due to its widespread connections with other brain regions like the amygdala and the PFC, as well as other limbic areas, the hippocampus plays important roles in emotional regulation and social behaviors and can modulate numerous mental processes [168].

Indeed, studies have shown that hippocampal lesions result in both emotion processing and social behavior deficits in humans and animals. Patients with hippocampus damage often present mood regulation deficits, expressed in higher levels of anxiety, depression, and tendencies to poor stress coping. Consistently, many animal studies of hippocampal lesions exhibit altered aggression levels, social recognition impairment, or reduced affiliative interactions. These discoveries highlight the central importance of the hippocampus in modulating behaviors of the affective system and in shaping social interaction [168].

Emotional regulation and neurogenesis in the hippocampus are in a reciprocal relationship. Impaired neurogenesis of the DG, particularly within the ventral hippocampus, dysregulates its function and plays a role in affective disorders such as depression, anxiety, and post-traumatic stress disorder (PTSD) [169]. While stress or anxiogenic situations activate mature GCs in the ventral DG, adult-born neurons suppress these stress-responsive cells, and this contributes to the protection of chronic social defeat-induced anxiety [170]. Chronic stress blocks hippocampal neurogenesis, and stress hormones such as cortisol decrease the proliferation of neural progenitor cells. This decrease in neurogenesis is linked to stress vulnerability and mood disorder [171, 172]. In contrast, interventions that stimulate neurogenesis, such as pharmacologic agents (e.g. antidepressants, BDNF mimetics), lifestyle changes (e.g. exercise, cognitive stimulation), or dietary factors (e.g. omega-3 FAs, flavonoids) were shown to correlate with decreased anxiety and depressive-like behavior in animal models and improved stress resilience [172]. Together, these mechanisms underpin the ability of the hippocampus to process novel information and to adapt to and learn from changes in the environment, including changes in the social environment caused by social isolation. Animal studies have shown that more neurogenesis is associated with less stress-reactivity and fewer stress-related behaviors. For instance, in rodent models, improved performance in models of resilience to stress (water maze procedure and tail suspension test) is a consequence of genome-wide knockout of the pro-apoptotic gene *Bax*, which increases neurogenesis [173].

Studies exploring the impact of unpredictable chronic mild stress (UCMS) induced gene expression alterations in the hippocampus have shown differential expression of genes that have relevance in adult neurogenesis: *Interleukin 1 beta* (*IL1B*) and *CREB1* are examples of this. On the other hand, the activation of *CREB1* increases neurogenesis; and the activation of *IL1B*, by activating the NF-κB pathway, has a negative effect on it. After UCMS exposure, both genes were upregulated, further illustrating the impact of stress on neurogenesis, synaptic plasticity, and hippocampal dysfunction, ultimately leading to the etiology of mood disorders [174].

Furthermore, reductions in hippocampal telomere length have been reported in a limited number of cohort subjects with Major Depressive Disorder (MDD) compared with controls, and these results have been interpreted as evidence of increased cellular aging and impaired neurogenesis due to stress. Finally, several stress-related and neuroprotection genes, such as *FKBP5*, *CRH*, and *HSPA2* were found to be significantly differentially expressed between MDD subjects and controls, emphasizing the overall impact of stress exposure on cellular aging and brain function in depression [175].

BDNF is also important, promoting neuronal survival, synaptic plasticity, and neurogenesis. Depression, generally, and more specifically in MDD patients, has been strongly related to a deficit of serum BDNF. Furthermore, BDNF levels induced by physical activity can further boost the therapeutic effects of antidepressants and recovery when those drugs depend on BDNF and its receptor TrkB in the hippocampus [176–178].

Depression and other mood disorders are associated with disruptions in hippocampal synaptic plasticity as a result of changes in levels of related genes and proteins, separate from the neurogenic process. The findings from patients with MDD and chronic stress models show increased expression of neuronal nitric oxide synthase (nNOS) but reduced expression and activity of protein kinase A (PKA) subunits in the hippocampus. Both are crucial for synaptic plasticity and neurotransmitter regulation with the addition that the excess of NOS is detrimental to neurons.

Moreover, in the hippocampus, chronic stress downregulates, and chronic antidepressant upregulates *CREB* activity. Other findings include decreased gene expression of synaptic proteins and AMPAR subunits in MDD, and reduced gene expression may also be normalized by chronic treatments with antidepressants. These deregulations may lead to cognitive impairments, emotional dysregulation, and poor stress adaptation that are associated with depression [179].

Immediate early genes such as *EGR1* are required for the consolidation of lasting effects of stress in the hippocampus. Higher expression of *EGR1*gene is associated with the storage of fear memory, whereas induced loss of *EGR1* has been shown to disrupt fear memory. Indeed, in rodent models, chronic stress induced by social isolation decreases *Egr1* gene expression in the hippocampus. Significantly, hippocampal synaptic transmission and plasticity are improved by environmental enrichment, leading to potential protection against behavioral depressive phenotypes [78].

The balance between glutamate and GABA neurotransmission in the hippocampus is critical. In one of the main drivers of emotional responses, glutamate is associated with the activation and potentiation of neural circuits. It acts on several receptors, such as NMDA, AMPA, and kainate receptors, inducing fast excitatory synaptic transmission [180, 181]. Conversely, GABA is necessary for reducing over-the-top responses to emotions such as fear and anxiety [182]. These systems support the control of stress responses and fear extinction by securing the stability of neural circuits. Long-range projections from the ventral hippocampus to the PFC and basolateral amygdala are involved in fear extinction and stress responsiveness [183]. However, if this balance of neuronal excitation and inhibition is lost, this can result in the development of neuropsychiatric disorders such as anxiety, depression, arousal, and post-traumatic stress disorders [184]. The overactivity of glutamatergic neurons can cause excitotoxicity which is implicated in mood disorders [185]. In a similar vein, dysfunction in GABAergic neurotransmission is typically found in individuals with severe anxiety, stress, and emotional dysregulation [186]. Important genes and proteins of the glutamate and GABA systems contribute to increased levels of stress in response to stressors, and problems with the NMDA receptor are implicated in mood and cognitive problems. For example, it has been reported that mouse models of chronic stress or PTSD display an alteration in the ratio of glutamate to Gaba in the hippocampus associated with increased anxiety and social deficits [187].

Functionally, the connectivity of the hippocampus is regulated by various neurotransmitters, including serotonin (5-HT) and dopamine. Mood, anxiety, and social behavior are regulated by serotonin acting on multiple receptors including 5-HT1A and 5-HT2A [188, 189]. On the one hand, chronic treatment with selective serotonin reuptake inhibitors (SSRIs) was reported to enhance neurogenesis, especially in the hippocampus, and was also associated with improved mood, whereas, on the other hand, genetic variations in the *SLC6A4* gene that codes for the *Serotonin Transporter* (*SERT*) have been found to influence serotonin levels, with certain polymorphisms being linked with an increased risk for depression upon exposure to stressful life events [189, 190]. Furthermore, miR-16, a microRNA that affects serotonin transporter expression, influences stress resistance by altering serotonin levels in the hippocampus [191]. Dopamine, on the other hand, deals with motivation, reward-related behaviors, and emotional regulation and acts through numerous receptors, whereby D1 and D2 receptors in the hippocampus are key to these behaviors [192]. Dopamine receptor expression in the brain life changes after Chronic stress and is connected with anxiety and depressive behaviors. Thus, the interaction of serotonin and dopamine systems in the hippocampus is critical to maintaining emotional stability and responding to stress. Dysregulation in either system can result in mood disorders, such as anxiety, sadness, and PTSD, highlighting their role in hippocampus function and general mental health [193].

One main part of the body's stress response system Hypothalamic-Pituitary-Adrenal (HPA) axis is intimately involved in emotional regulation and social behavior [194, 195]. The hypothalamus, pituitary gland, and adrenal cortex interact in a complicated way that leads to cortisol release, a crucial stress hormone [195]. Cortisol levels are carefully controlled, and dysregulation is related to a variety of mental disorders [196]. Further, glucocorticoid receptors (GRs), found abundantly in the DG, mediate the effects of cortisol in the hippocampus. Studies have demonstrated hippocampal hyperexcitability in the DG and circuit disruptions important for fear responses associated with *Gr* knockdown or depletion in newborn GCs [197]. Results of chronic stress on the HPA axis include persistent high levels of cortisol, which can provide impaired comments disinhibition, impacting *Gr* expression and activity, and shifting the relationship between neurogenesis and synaptic plasticity [198].

The interaction between the HPA axis and the hippocampus also involves corticotropin-releasing hormone (CRH), which is a principal regulator of the HPA axis. Increased levels of CRH can induce anxiety and attenuate social behavior [197]. The HPA axis affects the connectivity of the hippocampus to other regions of the brain, such as the anterior cingulate gyrus, amygdala, and PFC. Ventral hippocampal projections to PFC mediate adrenergic potentiation of plasticity of glutamatergic transmission that is necessary for fear extinction and stress responsiveness [199]. Damage to these pathways can result in emotional dysregulation and reduction of social behaviors as seen in current studies. Research indicates that children who have undergone early life stress have altered brain pathways, which can lead to long-term vulnerabilities in emotional and social functioning. Early childhood stress can cause hyperarousal of the sympathetic nervous system and prolonged activation of the HPA axis, resulting in structural and functional alterations in the hippocampus, amygdala, and medial PFC. These alterations emerge as greater stress sensitivity, reduced fear extinction, and disrupted emotional regulation, all of which indicate a higher risk of adult mental illnesses such as depression, anxiety, and PTSD. This supports the concept that the HPA axis not only regulates the stress response phenotypically but also shapes social behavior by controlling the degree and/or pattern of hippocampal connectivity with other regions of the brain [200].

Inflammation also regulates emotions and social behavior through interactions with the hippocampus. Hippocampal function is significantly influenced by appropriately controlled inflammatory processes; its dysregulation can result in mood disorders such as depression. Pro-inflammatory cytokines like IL-1β, interleukin-6 (IL-6), and TNF-α have been identified as potential mediators that link inflammation with diminution of hippocampal function [201]. The excessive production of these cytokines prevents neurogenesis in the hippocampus inhibits synaptic plasticity and alters the functions of larger neurotransmitter systems, which collectively affect the ability to regulate emotions and perform meaningful social interactions [202, 203].

In addition, the activity of inflammasomes, especially NOD-like receptor pyrin domain-containing-3 (NLRP3) inflammasome, in the hippocampus, enhances neuroinflammation and depressive-like behaviors [204, 205]. Amplification of inflammatory responses and increasing damage to hippocampal integrity is mediated, in part, by a molecular complex essential for the maturation and release of IL-1β in response to pro-inflammatory stimuli [206]. Furthermore, the key mediators that link inflammation with hippocampal dysfunction are the enzymes indoleamine 2,3-dioxygenase (IDO) and cyclooxygenase-2 (COX-2). IDO-driven metabolism of tryptophan, on the other hand, promotes the kynurenine pathway (KP) reducing tryptophan available for serotonin production and influencing mood regulation [201, 207]. COX-2, however, leads to inflammation and oxidative stress, both of which can further impair hippocampal function and emotional regulation [208].

Results of studies in patients and animal models convincingly demonstrate that inflammation induces hippocampal dysfunction which in turn may influence mood and social behavior. In a clinical study, for example, higher levels of pro-inflammatory cytokines were linked to decreased hippocampal volume and higher difficulties with emotional processing in depressive patients [202, 209, 210]. Studies using animal models of depression also support the idea that correcting inflammation has beneficial effects on hippocampal function and behavior changes [211, 212]. Together these results reveal a complex interplay between inflammation and hippocampal processing in mediating changes in affect and social behavior, in addition to identifying novel therapeutic targets in mood disorders [201].

11.3 Conclusion

The hippocampus governs a broad range of functions including memory formation and consolidation, cognitive processes, spatial navigation, and mood regulation. This involvement is mainly dependent on the hippocampus's neuronal characteristics and vast connections with other brain regions such as the EC, PFC, and amygdala. Naturally, to maintain the structure of information transmission between neurons, all of these tasks depend on strictly controlled processes such as synaptic plasticity, neural connection, and neurotransmitter signaling. The hippocampus's dynamic and tightly regulated process of gene expression is intended to maintain the appropriate functioning and development of these anatomical and functional characteristics. Alternations in the involved genes can disrupt the functioning of the hippocampus leading to serious health issues such as neurodegenerative diseases and psychiatric disorders [5].

References

1. Toyoda H, Li X-Y, Wu L-J, Zhao M-G, Descalzi G, Chen T, Koga K, Zhuo M. Interplay of amygdala and cingulate plasticity in emotional fear. Neural Plast. 2011;2011:813749.
2. Herlin B, Navarro V, Dupont S. The temporal pole: from anatomy to function—a literature appraisal. J Chem Neuroanat. 2021;113:101925.
3. Mega MS, Cummings JL, Salloway S, Malloy P. The limbic system: an anatomic, phylogenetic, and clinical perspective. J Neuropsychiatry Clin Neurosci. 1997;9(3):315–30.
4. Wang C, Grohme MA, Mali B, Schill RO, Frohme M. Towards decrypting cryptobiosis—analyzing anhydrobiosis in the tardigrade Milnesium tardigradum using transcriptome sequencing. PLoS One. 2014;9(3):e92663.

5. Andersen P, Morris R, Amaral D, Bliss T, O'Keefe J. The hippocampus book. Oxford University Press; 2006. https://academic.oup.com/book/25965
6. Schwarz AP, Nikitina VA, Krytskaya DU, Shcherbakova KP, Trofimov AN. Reference gene expression stability within the rat brain under mild intermittent ketosis induced by supplementation with medium-chain triglycerides. PLoS One. 2023;18(2):e0273224.
7. Moser MB, Moser EI. Distributed encoding and retrieval of spatial memory in the hippocampus. J Neurosci. 1998;18(18):7535–42.
8. Lee A-R, Kim J-H, Cho E, Kim M, Park M. Dorsal and ventral hippocampus differentiate in functional pathways and differentially associate with neurological disease-related genes during postnatal development. Front Mol Neurosci. 2017;10:331.
9. Yang X, Wan R, Liu Z, Feng S, Yang J, Jing N, Tang K. The differentiation and integration of the hippocampal dorsoventral axis are controlled by two nuclear receptor genes. eLife. 2023;12:RP86940.
10. Otsubo K, Sakashita N, Nishimoto Y, Sato Y, Kobayashi K, Suzuki K, Segi-Nishida E. The role of desmoplakin for neuronal function in the dentate gyrus and anxiety-related behavior. bioRxiv. 2023;2023:2023.11.17.567646.
11. Lein ES, Hawrylycz MJ, Ao N, Ayres M, Bensinger A, Bernard A, Boe AF, Boguski MS, Brockway KS, Byrnes EJ, Chen L, Chen L, Chen T-M, Chi Chin M, Chong J, Crook BE, Czaplinska A, Dang CN, Datta S, et al. Genome-wide atlas of gene expression in the adult mouse brain. Nature. 2007;445(7124):168–76.
12. Chauhan P, Jethwa K, Rathawa A, Chauhan G, Mehra S. The anatomy of the hippocampus. In: Laboratory of Ischemic and Neurodegenerative Brain Research, Mossakowski Medical Research Institute, Polish Academy of Sciences, Warsaw, Poland, Ryszard P, editors. Cerebral Ischemia. Exon Publications; 2021. p. 17–30.
13. Tatu L, Vuillier F. Structure and vascularization of the human hippocampus. Front Neurol Neurosci. 2014;34:18–25.
14. Van Hoeymissen E, Philippaert K, Vennekens R, Vriens J, Held K. Horizontal hippocampal slices of the mouse brain. J Vis Exp. 2020;163 https://doi.org/10.3791/61753.
15. Laplante F, Mnie-Filali O, Sullivan RM. A neuroanatomical and neurochemical study of the indusium griseum and anterior hippocampal continuation: comparison with dentate gyrus. J Chem Neuroanat. 2013;50–51:39–47.
16. Yu J, Swietek B, Proddutur A, Santhakumar V. Dentate total molecular layer interneurons mediate cannabinoid-sensitive inhibition. Hippocampus. 2015;25(8):884–9.
17. Seri B, García-Verdugo JM, Collado-Morente L, McEwen BS, Alvarez-Buylla A. Cell types, lineage, and architecture of the germinal zone in the adult dentate gyrus. J Comp Neurol. 2004;478(4):359–78.
18. Hamilton DJ, White CM, Rees CL, Wheeler DW, Ascoli GA. Molecular fingerprinting of principal neurons in the rodent hippocampus: a neuroinformatics approach. J Pharm Biomed Anal. 2017;144:269–78.
19. Machado JPD, Athie MCP, Matos AHB, Lopes-Cendes I, Vieira A, S. The transcriptome of rat hippocampal subfields. IBRO Neurosci Rep. 2022;13:322–9.
20. Sun Y-X, Su Y-A, Wang Q, Zheng J-Y, Zhang C-C, Wang T, Liu X, Ma Y-N, Li X-X, Zhang X-Q, Xie X-M, Wang X-D, Li J-T, Si T-M. The causal involvement of the BDNF-TrkB pathway in dentate gyrus in early-life stress-induced cognitive deficits in male mice. Transl Psychiatry. 2023;13(1):173.
21. Knierim JJ. The hippocampus. Curr Biol. 2015;25(23):R1116–21.
22. Guo Q, Wang Y, Wang Q, Qian Y, Jiang Y, Dong X, Chen H, Chen X, Liu X, Yu S, Zhu J, Shan S, Wu B, Zhou W, Wang H. In the developing cerebral cortex: axonogenesis, synapse formation, and synaptic plasticity are regulated by SATB2 target genes. Pediatr Res. 2023;93(6):1519–27.
23. Laeremans A, Nys J, Luyten W, D'Hooge R, Paulussen M, Arckens L. AMIGO2 mRNA expression in hippocampal CA2 and CA3a. Brain Struct Funct. 2013;218(1):123–30.
24. Peng S, Gu W, Zhu W, Zhuang Y, Yang X, Lv Y, Meng S, Xie W, Li M. A new AAV tool for highly preferentially targeting hippocampal CA2. Mol Brain. 2023;16(1):50.
25. Wang X-X, Li J-T, Xie X-M, Gu Y, Si T-M, Schmidt MV, Wang X-D. Nectin-3 modulates the structural plasticity of dentate granule cells and long-term memory. Transl Psychiatry. 2017;7(9):e1228.
26. Böhm C, Peng Y, Geiger JRP, Schmitz D. Routes to, from and within the subiculum. Cell Tissue Res. 2018;373(3):557–63.
27. Stark E, Roux L, Eichler R, Senzai Y, Royer S, Buzsáki G. Pyramidal cell-interneuron interactions underlie hippocampal ripple oscillations. Neuron. 2014;83(2):467–80.
28. Honoré E, Khlaifia A, Bosson A, Lacaille J-C. Hippocampal somatostatin interneurons, long-term synaptic plasticity and memory. Front Neural Circuits. 2021;15:687558.
29. Cammarota M, Boscia F. Contribution of oligodendrocytes, microglia, and astrocytes to myelin debris uptake in an explant model of inflammatory demyelination in rats. Cells. 2023;12(17):2203.
30. de Waard DM, Bugiani M. Astrocyte-oligodendrocyte-microglia crosstalk in astrocytopathies. Front Cell Neurosci. 2020;14:608073.
31. Liu Y, Shen X, Zhang Y, Zheng X, Cepeda C, Wang Y, Duan S, Tong X. Interactions of glial cells with neuronal synapses, from astrocytes to microglia and oligodendrocyte lineage cells. Glia. 2023;71(6):1383–401.
32. Mederos S, González-Arias C, Perea G. Astrocyte-neuron networks: a multilane highway of signal-

ing for homeostatic brain function. Front Synaptic Neurosci. 2018;10:45.
33. Su Y, Zhou Y, Bennett ML, Li S, Carceles-Cordon M, Lu L, Huh S, Jimenez-Cyrus D, Kennedy BC, Kessler SK, Viaene AN, Helbig I, Gu X, Kleinman JE, Hyde TM, Weinberger DR, Nauen DW, Song H, Ming G-L. A single-cell transcriptome atlas of glial diversity in the human hippocampus across the postnatal lifespan. Cell Stem Cell. 2022;29(11):1594–1610.e8.
34. Yang Z, Sreenivasan K, Toledano Strom EN, Osse AML, Pasia LG, Cosme CG, Mugosa MRN, Chevalier EL, Ritter A, Miller JB, Cordes D, Cummings JL, Kinney JW. Clinical and biological relevance of glial fibrillary acidic protein in Alzheimer's disease. Alzheimers Res Ther. 2023;15(1):190.
35. Zeppenfeld DM, Simon M, Haswell JD, D'Abreo D, Murchison C, Quinn JF, Grafe MR, Woltjer RL, Kaye J, Iliff JJ. Association of perivascular localization of aquaporin-4 with cognition and Alzheimer disease in aging brains. JAMA Neurol. 2017;74(1):91–9.
36. Fan C, Kim D, An H, Park Y. Identifying an oligodendrocyte enhancer that regulates Olig2 expression. Hum Mol Genet. 2023;32(5):835–46.
37. Galiano MR, Andrieux A, Deloulme JC, Bosc C, Schweitzer A, Job D, Hallak ME. Myelin basic protein functions as a microtubule stabilizing protein in differentiated oligodendrocytes. J Neurosci Res. 2006;84(3):534–41.
38. Cho S-H, Sun B, Zhou Y, Kauppinen TM, Halabisky B, Wes P, Ransohoff RM, Gan L. CX3CR1 protein signaling modulates microglial activation and protects against plaque-independent cognitive deficits in a mouse model of Alzheimer disease. J Biol Chem. 2011;286(37):32713–22.
39. Rachmian N, Medina S, Cherqui U, Akiva H, Deitch D, Edilbi D, Croese T, Salame TM, Ramos JMP, Cahalon L, Krizhanovsky V, Schwartz M. Identification of senescent, TREM2-expressing microglia in aging and Alzheimer's disease model mouse brain. Nat Neurosci. 2024;27(6):1116–24.
40. Zou P, Wu C, Liu TC-Y, Duan R, Yang L. Oligodendrocyte progenitor cells in Alzheimer's disease: from physiology to pathology. Transl Neurodegener. 2023;12(1):52.
41. Kier EL, Kim JH, Fulbright RK, Bronen RA. Embryology of the human fetal hippocampus: MR imaging, anatomy, and histology. AJNR Am J Neuroradiol. 1997;18(3):525–32.
42. Kempermann G, Jessberger S, Steiner B, Kronenberg G. Milestones of neuronal development in the adult hippocampus. Trends Neurosci. 2004;27(8):447–52.
43. Duan D, Fu Y, Paxinos G, Watson C. Spatiotemporal expression patterns of Pax6 in the brain of embryonic, newborn, and adult mice. Brain Struct Funct. 2013;218(2):353–72.
44. Grove EA, Tole S, Limon J, Yip L, Ragsdale CW. The hem of the embryonic cerebral cortex is defined by the expression of multiple Wnt genes and is compromised in Gli3-deficient mice. Development. 1998;125(12):2315–25.
45. Grove EA, Tole S. Patterning events and specification signals in the developing hippocampus. Cerebr Cortex. 1999;9(6):551–61.
46. Subramanian L, Sarkar A, Shetty AS, Muralidharan B, Padmanabhan H, Piper M, Monuki ES, Bach I, Gronostajski RM, Richards LJ, Tole S. Transcription factor Lhx2 is necessary and sufficient to suppress astrogliogenesis and promote neurogenesis in the developing hippocampus. Proc Natl Acad Sci USA. 2011;108(27):E265–74.
47. Li G, Pleasure SJ. The development of hippocampal cellular assemblies. Wiley Interdiscip Rev Dev Biol. 2014;3(2):165–77.
48. Lavado A, Lagutin OV, Chow LML, Baker SJ, Oliver G. Prox1 is required for granule cell maturation and intermediate progenitor maintenance during brain neurogenesis. PLoS Biol. 2010;8(8):e1000460.
49. Stergiopoulos A, Elkouris M, Politis PK. Prospero-related homeobox 1 (Prox1) at the crossroads of diverse pathways during adult neural fate specification. Front Cell Neurosci. 2014;8:454.
50. Burton SD, Johnson JW, Zeringue HC, Meriney SD. Distinct roles of neuroligin-1 and SynCAM1 in synapse formation and function in primary hippocampal neuronal cultures. Neuroscience. 2012;215:1–16.
51. Ishii K, Kubo K, Nakajima K. Reelin and neuropsychiatric disorders. Front Cell Neurosci. 2016;10:229.
52. Urbán N, Guillemot F. Neurogenesis in the embryonic and adult brain: same regulators, different roles. Front Cell Neurosci. 2014;8:396.
53. Liu X, Hua F, Yang D, Lin Y, Zhang L, Ying J, Sheng H, Wang X. Roles of neuroligins in central nervous system development: focus on glial neuroligins and neuron neuroligins. J Transl Med. 2022;20(1):418.
54. Zhong S, Ding W, Sun L, Lu Y, Dong H, Fan X, Liu Z, Chen R, Zhang S, Ma Q, Tang F, Wu Q, Wang X. Decoding the development of the human hippocampus. Nature. 2020;577(7791):531–6.
55. Stockman SL, Kight KE, Bowers JM, McCarthy MM. Neurogenesis in the neonatal rat hippocampus is regulated by sexually dimorphic epigenetic modifiers. Biol Sex Differ. 2022;13(1):9.
56. Biederer T, Sara Y, Mozhayeva M, Atasoy D, Liu X, Kavalali ET, Südhof TC. SynCAM, a synaptic adhesion molecule that drives synapse assembly. Science (New York, N.Y.). 2002;297(5586):1525–31.
57. Oliva AA, Atkins CM, Copenagle L, Banker GA. Activated c-Jun N-terminal kinase is required for axon formation. J Neurosci. 2006;26(37):9462–70.
58. Roback JD, Large TH, Otten U, Wainer BH. Nerve growth factor expression in the developing hippocampus isolated in vitro. Dev Biol. 1990;137(2):451–5.
59. Williams ME, Wu SC-Y, McKenna WL, Hinck L. Surface expression of the netrin receptor UNC5H1 is regulated through a protein kinase C-interacting protein/protein kinase-dependent mechanism. J Neurosci. 2003;23(36):11279–88.

60. Knoth R, Singec I, Ditter M, Pantazis G, Capetian P, Meyer RP, Horvat V, Volk B, Kempermann G. Murine features of neurogenesis in the human hippocampus across the lifespan from 0 to 100 years. PLoS One. 2010;5(1):e8809.
61. Wang J, Zhai H-R, Ma S-F, Shi H-Z, Zhang W-J, Yun Q, Liu W-J, Liu Z-Z, Zhang W-N. FOXG1 contributes adult hippocampal neurogenesis in mice. Int J Mol Sci. 2022;23(23):14979.
62. Wong L-C, Singh S, Wang H-P, Hsu C-J, Hu S-C, Lee W-T. FOXG1-related syndrome: from clinical to molecular genetics and pathogenic mechanisms. Int J Mol Sci. 2019;20(17):4176.
63. Hevner RF. Evolution of the mammalian dentate gyrus. J Comp Neurol. 2016;524(3):578–94.
64. Abbott LC, Nigussie F. Adult neurogenesis in the mammalian dentate gyrus. Anat Histol Embryol. 2020;49(1):3–16.
65. Charvet CJ, Finlay BL. Comparing adult hippocampal neurogenesis across species: translating time to predict the tempo in humans. Front Neurosci. 2018;12:706.
66. Spalding KL, Bergmann O, Alkass K, Bernard S, Salehpour M, Huttner HB, Boström E, Westerlund I, Vial C, Buchholz BA, Possnert G, Mash DC, Druid H, Frisén J. Dynamics of hippocampal neurogenesis in adult humans. Cell. 2013;153(6):1219–27.
67. Sweatt JD. Chapter 3—The hippocampus serves a role in multimodal information processing, and memory consolidation. In: Sweatt JD, editor. Mechanisms of memory. Academic; 2003. p. 61–90.
68. Wu Y, Sutherland RJ. Hippocampal evoked potentials in novel environments: a behavioral clamping method. Behav Brain Res. 2006;172(1):63–71.
69. Strange BA, Witter MP, Lein ES, Moser EI. Functional organization of the hippocampal longitudinal axis. Nat Rev Neurosci. 2014;15(10):655–69.
70. Scoville WB, Milner B. Loss of recent memory after bilateral hippocampal lesions. J Neurol Neurosurg Psychiatry. 1957;20(1):11–21.
71. Broadbent NJ, Squire LR, Clark RE. Spatial memory, recognition memory, and the hippocampus. Proc Natl Acad Sci USA. 2004;101(40):14515–20.
72. Aggleton JP, Hunt PR, Rawlins JN. The effects of hippocampal lesions upon spatial and nonspatial tests of working memory. Behav Brain Res. 1986;19(2):133–46.
73. Anagnostaras SG, Maren S, Fanselow MS. Temporally graded retrograde amnesia of contextual fear after hippocampal damage in rats: within-subjects examination. J Neurosci. 1999;19(3):1106–14.
74. Kogan JH, Frankland PW, Silva AJ. Long-term memory underlying hippocampus-dependent social recognition in mice. Hippocampus. 2000;10(1):47–56.
75. Mumby DG, Gaskin S, Glenn MJ, Schramek TE, Lehmann H. Hippocampal damage and exploratory preferences in rats: memory for objects, places, and contexts. Learn Mem (Cold Spring Harbor, N.Y.). 2002;9(2):49–57.
76. Tseng W, Guan R, Disterhoft JF, Weiss C. Trace eyeblink conditioning is hippocampally dependent in mice. Hippocampus. 2004;14(1):58–65.
77. Olsen RK, Moses SN, Riggs L, Ryan JD. The hippocampus supports multiple cognitive processes through relational binding and comparison. Front Hum Neurosci. 2012;6:146.
78. Gallo FT, Katche C, Morici JF, Medina JH, Weisstaub NV. Immediate early genes, memory and psychiatric disorders: focus on c-Fos, Egr1 and Arc. Front Behav Neurosci. 2018;12:79.
79. Minatohara K, Akiyoshi M, Okuno H. Role of immediate-early genes in synaptic plasticity and neuronal ensembles underlying the memory trace. Front Mol Neurosci. 2015;8:78.
80. Matsuzaki M, Honkura N, Ellis-Davies GCR, Kasai H. Structural basis of long-term potentiation in single dendritic spines. Nature. 2004;429(6993):761–6.
81. Newpher TM, Harris S, Pringle J, Hamilton C, Soderling S. Regulation of spine structural plasticity by Arc/Arg3.1. Semin Cell Dev Biol. 2018;77:25–32.
82. Hughes KR. Dorsal and ventral hippocampus lesions and maze learning: influence of preoperative environment. Can J Psychol. 1965;19(4):325–32.
83. Moser E, Moser MB, Andersen P. Spatial learning impairment parallels the magnitude of dorsal hippocampal lesions, but is hardly present following ventral lesions. J Neurosci. 1993;13(9):3916–25.
84. Moser MB, Moser EI, Forrest E, Andersen P, Morris RG. Spatial learning with a minislab in the dorsal hippocampus. Proc Natl Acad Sci USA. 1995;92(21):9697–701.
85. Kubik S, Miyashita T, Guzowski JF. Using immediate-early genes to map hippocampal subregional functions. Learn Mem (Cold Spring Harbor, N.Y.). 2007;14(11):758–70.
86. Chawla MK, Sutherland VL, Olson K, McNaughton BL, Barnes CA. Behavior-driven arc expression is reduced in all ventral hippocampal subfields compared to CA1, CA3, and dentate gyrus in rat dorsal hippocampus. Hippocampus. 2018;28(2):178–85.
87. Brem A-K, Ran K, Pascual-Leone A. Learning and memory. Handb Clin Neurol. 2013;116:693–737.
88. Mack ML, Love BC, Preston AR. Building concepts one episode at a time: the hippocampus and concept formation. Neurosci Lett. 2018;680:31–8.
89. Pastalkova E, Serrano P, Pinkhasova D, Wallace E, Fenton AA, Sacktor TC. Storage of spatial information by the maintenance mechanism of LTP. Science (New York, N.Y.). 2006;313(5790):1141–4.
90. Bliss TVP, Collingridge GL. A synaptic model of memory: long-term potentiation in the hippocampus. Nature. 1993;361(6407):31–9.
91. Lüscher C, Malenka RC. NMDA receptor-dependent long-term potentiation and long-term depression (LTP/LTD). Cold Spring Harb Perspect Biol. 2012;4(6):a005710.

92. Kumar A. Long-term potentiation at CA3-CA1 hippocampal synapses with special emphasis on aging, disease, and stress. Front Aging Neurosci. 2011;3:7.
93. Glikmann-Johnston Y, Saling MM, Reutens DC, Stout JC. Hippocampal 5-HT1A receptor and spatial learning and memory. Front Pharmacol. 2015;6:289.
94. Kim H-H, Lee S-H, Ho W-K, Eom K. Dopamine receptor supports the potentiation of intrinsic excitability and synaptic LTD in temporoammonic-CA1 synapse. Exp Neurobiol. 2022;31(6):361–75.
95. Maity S, Rah S, Sonenberg N, Gkogkas CG, Nguyen PV. Norepinephrine triggers metaplasticity of LTP by increasing translation of specific mRNAs. Learn Mem (Cold Spring Harbor, N.Y.). 2015;22(10):499–508.
96. Piette C, Cui Y, Gervasi N, Venance L. Lights on endocannabinoid-mediated synaptic potentiation. Front Mol Neurosci. 2020;13:132.
97. Baltaci SB, Mogulkoc R, Baltaci AK. Molecular mechanisms of early and late LTP. Neurochem Res. 2019;44(2):281–96.
98. Bashir ZI, Bortolotto ZA, Davies CH, Berretta N, Irving AJ, Seal AJ, Henley JM, Jane DE, Watkins JC, Collingridge GL. Induction of LTP in the hippocampus needs synaptic activation of glutamate metabotropic receptors. Nature. 1993;363(6427):347–50.
99. Miyamoto E. Molecular mechanism of neuronal plasticity: induction and maintenance of long-term potentiation in the hippocampus. J Pharmacol Sci. 2006;100(5):433–42.
100. Mayr B, Montminy M. Transcriptional regulation by the phosphorylation-dependent factor CREB. Nat Rev Mol Cell Biol. 2001;2(8):599–609.
101. Alberini CM. Transcription factors in long-term memory and synaptic plasticity. Physiol Rev. 2009;89(1):121–45.
102. Hayashi Y. Molecular mechanism of hippocampal long-term potentiation—towards multiscale understanding of learning and memory. Neurosci Res. 2022;175:3–15.
103. Malenka RC, Bear MF. LTP and LTD: an embarrassment of riches. Neuron. 2004;44(1):5–21.
104. Miranda M, Morici JF, Zanoni MB, Bekinschtein P. Brain-derived neurotrophic factor: a key molecule for memory in the healthy and the pathological brain. Front Cell Neurosci. 2019;13:363.
105. Azman KF, Zakaria R. Recent advances on the role of brain-derived neurotrophic factor (BDNF) in neurodegenerative diseases. Int J Mol Sci. 2022;23(12):6827.
106. Cunha C, Brambilla R, Thomas KL. A simple role for BDNF in learning and memory? Front Mol Neurosci. 2010;3:1.
107. Cook SG, Rumian NL, Bayer KU. CaMKII T286 phosphorylation has distinct essential functions in three forms of long-term plasticity. J Biol Chem. 2022;298(9):102299.
108. Silva AJ, Stevens CF, Tonegawa S, Wang Y. Deficient hippocampal long-term potentiation in alpha-calcium-calmodulin kinase II mutant mice. Science (New York, N.Y.). 1992;257(5067):201–6.
109. Middei S, Houeland G, Cavallucci V, Ammassari-Teule M, D'Amelio M, Marie H. CREB is necessary for synaptic maintenance and learning-induced changes of the AMPA receptor GluA1 subunit. Hippocampus. 2013;23(6):488–99.
110. Li L, Sanchez CP, Slaughter BD, Zhao Y, Khan MR, Unruh JR, Rubinstein B, Si K. A putative biochemical engram of long-term memory. Curr Biol. 2016;26(23):3143–56.
111. Earls LR, Westmoreland JJ, Zakharenko SS. Non-coding RNA regulation of synaptic plasticity and memory: implications for aging. Ageing Res Rev. 2014;17:34–42.
112. Ryan B, Joilin G, Williams JM. Plasticity-related microRNA and their potential contribution to the maintenance of long-term potentiation. Front Mol Neurosci. 2015;8:4.
113. Daswani R, Gilardi C, Soutschek M, Nanda P, Weiss K, Bicker S, Fiore R, Dieterich C, Germain P-L, Winterer J, Schratt G. MicroRNA-138 controls hippocampal interneuron function and short-term memory in mice. eLife. 2022;11:e74056.
114. Capitano F, Camon J, Ferretti V, Licursi V, De Vito F, Rinaldi A, Vincenti S, Mannironi C, Fragapane P, Bozzoni I, Oliverio A, Negri R, Presutti C, Mele A. microRNAs modulate spatial memory in the hippocampus and in the ventral striatum in a region-specific manner. Mol Neurobiol. 2016;53(7):4618–30.
115. Zhang S, Chen J, Zhang J, Xu J. miR-181a involves in the hippocampus-dependent memory formation via targeting PRKAA1. Sci Rep. 2017;7(1):8480.
116. Hansen KF, Karelina K, Sakamoto K, Wayman GA, Impey S, Obrietan K. miRNA-132: a dynamic regulator of cognitive capacity. Brain Struct Funct. 2013;218(3):817–31.
117. Woldemichael BT, Jawaid A, Kremer EA, Gaur N, Krol J, Marchais A, Mansuy IM. The microRNA cluster miR-183/96/182 contributes to long-term memory in a protein phosphatase 1-dependent manner. Nat Commun. 2016;7:12594.
118. Jawaid A, Woldemichael BT, Kremer EA, Laferriere F, Gaur N, Afroz T, Polymenidou M, Mansuy IM. Memory decline and its reversal in aging and neurodegeneration involve miR-183/96/182 biogenesis. Mol Neurobiol. 2019;56(5):3451–62.
119. Newen A. What are cognitive processes? An example-based approach. Synthese. 2017;194(11):4251–68.
120. Bliss TVP, Collingridge GL, Morris RGM, Reymann KG. Long-term potentiation in the hippocampus: discovery, mechanisms and function. Neuroforum. 2018;24(3):A103–20.
121. Lisman JE, Jensen O. The theta-gamma neural code. Neuron. 2013;77(6):1002–16.
122. Bakkour A, Palombo DJ, Zylberberg A, Kang YH, Reid A, Verfaellie M, Shadlen MN, Shohamy D. The

hippocampus supports deliberation during value-based decisions. eLife. 2019;8:e46080.
123. Johnson A, van der Meer MAA, Redish AD. Integrating hippocampus and striatum in decision-making. Curr Opin Neurobiol. 2007;17(6):692–7.
124. Yu JY, Frank LM. Hippocampal-cortical interaction in decision making. Neurobiol Learn Mem. 2015;117:34–41.
125. Palombo DJ, Keane MM, Verfaellie M. How does the hippocampus shape decisions? Neurobiol Learn Mem. 2015;125:93–7.
126. Enkavi AZ, Weber B, Zweyer I, Wagner J, Elger CE, Weber EU, Johnson EJ. Evidence for hippocampal dependence of value-based decisions. Sci Rep. 2017;7:17738.
127. Alamri S. The contribution of the hippocampus to language processing. Universitat de Barcelona; n.d.
128. Duff MC, Brown-Schmidt S. The hippocampus and the flexible use and processing of language. Front Hum Neurosci. 2012;6:69.
129. Covington NV, Duff MC. Expanding the language network: direct contributions from the hippocampus. Trends Cogn Sci. 2016;20(12):869–70.
130. Ullman MT. A neurocognitive perspective on language: the declarative/procedural model. Nat Rev Neurosci. 2001;2(10):717–26.
131. Knecht S. Does language lateralization depend on the hippocampus? Brain. 2004;127(Pt 6):1217–8.
132. Liégeois F, Connelly A, Cross JH, Boyd SG, Gadian DG, Vargha-Khadem F, Baldeweg T. Language reorganization in children with early-onset lesions of the left hemisphere: an fMRI study. Brain. 2004;127(Pt 6):1229–36.
133. Newman AJ, Pancheva R, Ozawa K, Neville HJ, Ullman MT. An event-related fMRI study of syntactic and semantic violations. J Psycholinguist Res. 2001;30(3):339–64.
134. Duff MC, Hengst JA, Tengshe C, Krema A, Tranel D, Cohen NJ. Hippocampal amnesia disrupts the flexible use of procedural discourse in social interaction. Aphasiology. 2008;22(7–8):866–80.
135. Rubin RD, Brown-Schmidt S, Duff MC, Tranel D, Cohen NJ. How do I remember that I know you know that I know? Psychol Sci. 2011;22(12):1574–82.
136. Rubin RD, Watson PD, Duff MC, Cohen NJ. The role of the hippocampus in flexible cognition and social behavior. Front Hum Neurosci. 2014;8:742.
137. Varela C, Weiss S, Meyer R, Halassa M, Biedenkapp J, Wilson MA, Goosens KA, Bendor D. Tracking the time-dependent role of the hippocampus in memory recall using DREADDs. PLoS One. 2016;11(5):e0154374.
138. Carlino D, De Vanna M, Tongiorgi E. Is altered BDNF biosynthesis a general feature in patients with cognitive dysfunctions? Neuroscientist. 2013;19(4):345–53.
139. de Pins B, Cifuentes-Díaz C, Farah AT, López-Molina L, Montalban E, Sancho-Balsells A, López A, Ginés S, Delgado-García JM, Alberch J, Gruart A, Girault J-A, Giralt A. Conditional BDNF delivery from astrocytes rescues memory deficits, spine density, and synaptic properties in the 5xFAD mouse model of Alzheimer disease. J Neurosci. 2019;39(13):2441–58.
140. Liu Y-J, Liu T-T, Jiang L-H, Liu Q, Ma Z-L, Xia T-J, Gu X-P. Identification of hub genes associated with cognition in the hippocampus of Alzheimer's Disease. Bioengineered. 2021;12(2):9598–609.
141. Rong J, Yang Y, Liang M, Zhong H, Li Y, Zhu Y, Sha S, Chen L, Zhou R. Neonatal inflammation increases hippocampal KCC2 expression through methylation-mediated TGF-β1 downregulation leading to impaired hippocampal cognitive function and synaptic plasticity in adult mice. J Neuroinflammation. 2023;20(1):15.
142. Liang Y, Kang X, Zhang H, Xu H, Wu X. Knockdown and inhibition of hippocampal GPR17 attenuates lipopolysaccharide-induced cognitive impairment in mice. J Neuroinflammation. 2023;20:271.
143. den Hoed J, Devaraju K, Fisher SE. Molecular networks of the FOXP2 transcription factor in the brain. EMBO Rep. 2021;22(8):e52803.
144. Lai CS, Fisher SE, Hurst JA, Vargha-Khadem F, Monaco AP. A forkhead-domain gene is mutated in a severe speech and language disorder. Nature. 2001;413(6855):519–23.
145. Liégeois FJ, Hildebrand MS, Bonthrone A, Turner SJ, Scheffer IE, Bahlo M, Connelly A, Morgan AT. Early neuroimaging markers of FOXP2 intragenic deletion. Sci Rep. 2016;6(1):35192.
146. Shi L, Bergson CM. Neuregulin 1: an intriguing therapeutic target for neurodevelopmental disorders. Transl Psychiatry. 2020;10(1):190.
147. Mahringer D, Petersen AV, Fiser A, Okuno H, Bito H, Perrier J-F, Keller GB. Expression of c-Fos and Arc in hippocampal region CA1 marks neurons that exhibit learning-related activity changes. bioRxiv. 2019;2019:644526.
148. Le AA, Lauterborn JC, Jia Y, Cox CD, Lynch G, Gall CM. Metabotropic NMDA receptor signaling contributes to sex differences in synaptic plasticity and episodic memory. bioRxiv. 2024;2024:2024.01.26.577478.
149. Foster DJ, Knierim JJ. Sequence learning and the role of the hippocampus in rodent navigation. Curr Opin Neurobiol. 2012;22(2):294–300.
150. O'Keefe J, Nadel L. The hippocampus as a cognitive map. Clarendon Press: Oxford University Press; 1978.
151. McGregor A, Hayward AJ, Pearce JM, Good MA. Hippocampal lesions disrupt navigation based on the shape of the environment. Behav Neurosci. 2004;118(5):1011–21.
152. Rao YL, Ganaraja B, Murlimanju BV, Joy T, Krishnamurthy A, Agrawal A. Hippocampus and

its involvement in Alzheimer's disease: a review. 3 Biotech. 2022;12(2):55.
153. Petersen RP, Moradpour F, Eadie BD, Shin JD, Kannangara TS, Delaney KR, Christie BR. Electrophysiological identification of medial and lateral perforant path inputs to the dentate gyrus. Neuroscience. 2013;252:154–68.
154. Mao D. Neural correlates of spatial navigation in primate hippocampus. Neurosci Bull. 2023;39(2):315–27.
155. Bicanski A, Burgess N. A neural-level model of spatial memory and imagery. eLife. 2018;7:e33752.
156. Moser M-B, Rowland DC, Moser EI. Place cells, grid cells, and memory. Cold Spring Harb Perspect Biol. 2015;7(2):a021808.
157. Jutras MJ, Fries P, Buffalo EA. Oscillatory activity in the monkey hippocampus during visual exploration and memory formation. Proc Natl Acad Sci USA. 2013;110(32):13144–9.
158. Vanrobeys Y, Mukherjee U, Langmack L, Bahl E, Lin L-C, Michaelson JJ, Abel T, Chatterjee S. Mapping the spatial transcriptomic signature of the hippocampus during memory consolidation. bioRxiv. 2023;2023:2023.01.18.524576.
159. Català-Solsona J, Miñano-Molina AJ, Rodríguez-Álvarez J. Nr4a2 transcription factor in hippocampal synaptic plasticity, memory and cognitive dysfunction: a perspective review. Front Mol Neurosci. 2021;14:786226.
160. Chatterjee S, Bahl E, Mukherjee U, Walsh EN, Shetty MS, Yan AL, Vanrobaeys Y, Lederman JD, Giese KP, Michaelson J, Abel T. Endoplasmic reticulum chaperone genes encode effectors of long-term memory. Sci Adv. 2022;8(12):eabm6063.
161. Chatterjee S, Walsh E, Yan A, Giese KP, Safe S, Abel T. Pharmacological activation of Nr4a rescues age-associated memory decline. Neurobiol Aging. 2020;85:140–4.
162. Bellfy L, Smies CW, Bernhardt AR, Bodinayake KK, Sebastian A, Stuart EM, Wright DS, Lo C-Y, Murakami S, Boyd HM, von Abo MJ, Albert I, Kwapis JL. The clock gene Per1 may exert diurnal control over hippocampal memory consolidation. Neuropsychopharmacology. 2023;48(12):1789–97.
163. Xu Y, Wang Y, Jiang Y, Liu M, Zhong W, Ge Z, Sun Z, Shen X. Relationship between cognitive dysfunction and the promoter methylation of PER1 and CRY1 in patients with cerebral small vessel disease. Front Aging Neurosci. 2023;15:1174541.
164. Albert-Gascó H, Ros-Bernal F, Castillo-Gómez E, Olucha-Bordonau FE. MAP/ERK signaling in developing cognitive and emotional function and its effect on pathological and neurodegenerative processes. Int J Mol Sci. 2020;21(12):4471.
165. Zhu X, Guan R, Zou Y, Li M, Chen J, Zhang J, Luo W. Cold-inducible RNA binding protein alleviates iron overload-induced neural ferroptosis under perinatal hypoxia insult. Cell Death Differ. 2024;31(4):524–39.
166. Peretti D, Smith HL, Verity N, Humoud I, de Weerd L, Swinden DP, Hayes J, Mallucci GR. TrkB signaling regulates the cold-shock protein RBM3-mediated neuroprotection. Life Sci Alliance. 2021;4(4):e202000884.
167. Chen L, Fan F, Wu L, Zhao Y. The nuclear receptor 4A family members: mediators in human disease and autophagy. Cell Mol Biol Lett. 2020;25(1):48.
168. Ghasemi M, Navidhamidi M, Rezaei F, Azizikia A, Mehranfard N. Anxiety and hippocampal neuronal activity: relationship and potential mechanisms. Cogn Affect Behav Neurosci. 2022;22(3):431–49.
169. Anacker C, Hen R. Adult hippocampal neurogenesis and cognitive flexibility—linking memory and mood. Nat Rev Neurosci. 2017;18(6):335–46.
170. Anacker C, Luna VM, Stevens GS, Millette A, Shores R, Jimenez JC, Chen B, Hen R. Hippocampal neurogenesis confers stress resilience by inhibiting the ventral dentate gyrus. Nature. 2018;559(7712):98–102.
171. Levone BR, Cryan JF, O'Leary OF. Role of adult hippocampal neurogenesis in stress resilience. Neurobiol Stress. 2014;1:147–55.
172. Schoenfeld TJ, Gould E. Stress, stress hormones, and adult neurogenesis. Exp Neurol. 2012;233(1):12–21.
173. Planchez B, Lagunas N, Le Guisquet A-M, Legrand M, Surget A, Hen R, Belzung C. Increasing adult hippocampal neurogenesis promotes resilience in a mouse model of depression. Cells. 2021;10(5):972.
174. Malki K, Mineur YS, Tosto MG, Campbell J, Karia P, Jumabhoy I, Sluyter F, Crusio WE, Schalkwyk LC. Pervasive and opposing effects of Unpredictable Chronic Mild Stress (UCMS) on hippocampal gene expression in BALB/cJ and C57BL/6J mouse strains. BMC Genomics. 2015;16(1):262.
175. Mamdani F, Rollins B, Morgan L, Myers RM, Barchas JD, Schatzberg AF, Watson SJ, Akil H, Potkin SG, Bunney WE, Vawter MP, Sequeira PA. Variable telomere length across post-mortem human brain regions and specific reduction in the hippocampus of major depressive disorder. Transl Psychiatry. 2015;5(9):e636.
176. Matrisciano F, Bonaccorso S, Ricciardi A, Scaccianoce S, Panaccione I, Wang L, Ruberto A, Tatarelli R, Nicoletti F, Girardi P, Shelton RC. Changes in BDNF serum levels in patients with major depression disorder (MDD) after 6 months treatment with sertraline, escitalopram, or venlafaxine. J Psychiatr Res. 2009;43(3):247–54.
177. Polyakova M, Stuke K, Schuemberg K, Mueller K, Schoenknecht P, Schroeter ML. BDNF as a biomarker for successful treatment of mood disorders: a systematic & quantitative meta-analysis. J Affect Disord. 2015;174:432–40.
178. Zhao G, Zhang C, Chen J, Su Y, Zhou R, Wang F, Xia W, Huang J, Wang Z, Hu Y, Cao L, Guo X, Yuan C, Wang Y, Yi Z, Lu W, Wu Y, Wu Z, Hong W, et al. Ratio of mBDNF to proBDNF for differential diagnosis of major depressive disorder and bipolar depression. Mol Neurobiol. 2017;54(7):5573–82.

179. Marsden WN. Synaptic plasticity in depression: molecular, cellular and functional correlates. Prog Neuro-Psychopharmacol Biol Psychiatry. 2013;43:168–84.
180. Bowie D. Ionotropic glutamate receptors & CNS disorders. CNS Neurol Disord Drug Targets. 2008;7(2):129–43.
181. Jin Z, Bhandage AK, Bazov I, Kononenko O, Bakalkin G, Korpi ER, Birnir B. Selective increases of AMPA, NMDA, and kainate receptor subunit mRNAs in the hippocampus and orbitofrontal cortex but not in prefrontal cortex of human alcoholics. Front Cell Neurosci. 2014;8:11.
182. Wager TD, Kang J, Johnson TD, Nichols TE, Satpute AB, Barrett LF. A Bayesian model of category-specific emotional brain responses. PLoS Comput Biol. 2015;11(4):e1004066.
183. Kupferschmidt DA, Cummings KA, Joffe ME, MacAskill A, Malik R, Sánchez-Bellot C, Tejeda HA, Castillo HY. Prefrontal interneurons: populations, pathways, and plasticity supporting typical and disordered cognition in rodent models. J Neurosci. 2022;42(45):8468–76.
184. Sohal VS, Rubenstein JLR. Excitation-inhibition balance as a framework for investigating mechanisms in neuropsychiatric disorders. Mol Psychiatry. 2019;24(9):1248–57.
185. Haroon E, Miller AH, Sanacora G. Inflammation, glutamate, and glia: a trio of trouble in mood disorders. Neuropsychopharmacology. 2017;42(1): 193–215.
186. Jie F, Yin G, Yang W, Yang M, Gao S, Lv J, Li B. Stress in regulation of GABA amygdala system and relevance to neuropsychiatric diseases. Front Neurosci. 2018;12:562.
187. Engin E, Smith KS, Gao Y, Nagy D, Foster RA, Tsvetkov E, Keist R, Crestani F, Fritschy J-M, Bolshakov VY, Hajos M, Heldt SA, Rudolph U. Modulation of anxiety and fear via distinct intrahippocampal circuits. eLife. 2016;5:e14120.
188. Kjaerby C, Athilingam J, Robinson SE, Iafrati J, Sohal VS. Serotonin 1B receptors regulate prefrontal function by gating callosal and hippocampal inputs. Cell Rep. 2016;17(11):2882–90.
189. Yohn CN, Gergues MM, Samuels BA. The role of 5-HT receptors in depression. Mol Brain. 2017;10(1):28.
190. Munafò MR. The serotonin transporter gene and depression. Depress Anxiety. 2012;29(11):915–7.
191. Lopizzo N, Zonca V, Cattane N, Pariante CM, Cattaneo A. miRNAs in depression vulnerability and resilience: novel targets for preventive strategies. J Neural Transm. 2019;126(9):1241–58.
192. Jones-Tabah J, Mohammad H, Paulus EG, Clarke PBS, Hébert TE. The signaling and pharmacology of the dopamine D1 receptor. Front Cell Neurosci. 2021;15:806618.
193. Baik J-H. Stress and the dopaminergic reward system. Exp Mol Med. 2020;52(12):1879–90.
194. Packard AEB, Egan AE, Ulrich-Lai YM. HPA axis—interaction with behavioral systems. Compr Physiol. 2016;6(4):1897–934.
195. Smith SM, Vale WW. The role of the hypothalamic-pituitary-adrenal axis in neuroendocrine responses to stress. Dialogues Clin Neurosci. 2006;8(4):383–95.
196. Dziurkowska E, Wesolowski M. Cortisol as a biomarker of mental disorder severity. J Clin Med. 2021;10(21):5204.
197. Kino T. Stress, glucocorticoid hormones, and hippocampal neural progenitor cells: implications to mood disorders. Front Physiol. 2015;6:230.
198. Hannibal KE, Bishop MD. Chronic stress, cortisol dysfunction, and pain: a psychoneuroendocrine rationale for stress management in pain rehabilitation. Phys Ther. 2014;94(12):1816–25.
199. Bhardwaj SK, Tse YC, Ryan R, Wong TP, Srivastava LK. Impaired adrenergic-mediated plasticity of prefrontal cortical glutamate synapses in rats with developmental disruption of the ventral hippocampus. Neuropsychopharmacology. 2014;39(13):2963–73.
200. Murphy F, Nasa A, Cullinane D, Raajakesary K, Gazzaz A, Sooknarine V, Haines M, Roman E, Kelly L, O'Neill A, Cannon M, Roddy DW. Childhood trauma, the HPA axis and psychiatric illnesses: a targeted literature synthesis. Front Psych. 2022;13:748372.
201. Troubat R, Barone P, Leman S, Desmidt T, Cressant A, Atanasova B, Brizard B, El Hage W, Surget A, Belzung C, Camus V. Neuroinflammation and depression: a review. Eur J Neurosci. 2021;53(1):151–71.
202. Liu F, Yang Y, Fan X-W, Zhang N, Wang S, Shi Y-J, Hu W-J, Wang C-X. Impacts of inflammatory cytokines on depression: a cohort study. BMC Psychiatry. 2024;24(1):195.
203. Zhang W, Xiao D, Mao Q, Xia H. Role of neuroinflammation in neurodegeneration development. Signal Transduct Target Ther. 2023;8:267.
204. Li J-M, Liu L-L, Su W-J, Wang B, Zhang T, Zhang Y, Jiang C-L. Ketamine may exert antidepressant effects via suppressing NLRP3 inflammasome to upregulate AMPA receptors. Neuropharmacology. 2019;146:149–53.
205. Ślusarczyk J, Trojan E, Głombik K, Piotrowska A, Budziszewska B, Kubera M, Popiołek-Barczyk K, Lasoń W, Mika J, Basta-Kaim A. Targeting the NLRP3 inflammasome-related pathways via tianeptine treatment-suppressed microglia polarization to the M1 phenotype in lipopolysaccharide-stimulated cultures. Int J Mol Sci. 2018;19(7):1965.
206. Chen M-Y, Ye X-J, He X-H, Ouyang D-Y. The signaling pathways regulating NLRP3 inflammasome activation. Inflammation. 2021;44(4):1229–45.
207. Badawy AA-B. Kynurenine pathway of tryptophan metabolism: regulatory and functional aspects. Int J Tryptophan Res. 2017;10:1178646917691938.
208. Minghetti L. Cyclooxygenase-2 (COX-2) in inflammatory and degenerative brain diseases. J Neuropathol Exp Neurol. 2004;63(9):901–10.

209. Felger JC, Lotrich FE. Inflammatory cytokines in depression: neurobiological mechanisms and therapeutic implications. Neuroscience. 2013;246:199–229.
210. Mancini A, Gaetani L, Di Gregorio M, Tozzi A, Ghiglieri V, Calabresi P, Di Filippo M. Hippocampal neuroplasticity and inflammation: relevance for multiple sclerosis. Multiple Sclerosis Demyelinating Disord. 2017;2(1):2.
211. Bai S, Guo W, Feng Y, Deng H, Li G, Nie H, Guo G, Yu H, Ma Y, Wang J, Chen S, Jing J, Yang J, Tang Y, Tang Z. Efficacy and safety of anti-inflammatory agents for the treatment of major depressive disorder: a systematic review and meta-analysis of randomised controlled trials. J Neurol Neurosurg Psychiatry. 2020;91(1):21–32.
212. Majidi J, Kosari-Nasab M, Salari A-A. Developmental minocycline treatment reverses the effects of neonatal immune activation on anxiety- and depression-like behaviors, hippocampal inflammation, and HPA axis activity in adult mice. Brain Res Bull. 2016;120:1–13.

Surgical Approaches for the Hippocampal Lesions

12

Sinan Sağıroğlu and Mehmet Turgut

12.1 Introduction

The medial temporal lobe contains the hippocampus, which has long been the focus of study due to its significance in memory and cognitive processes. It is a component of the limbic system and is essential for memory formation and storage, as well as for the recall and retention of experiences and event-based information. Additionally, it plays a significant role in spatial navigation, contextual processing, and the regulation of emotional responses [1]. The hippocampus is richly connected to other brain regions through various pathways. This intricate network allows the hippocampus to play a vital role in various cognitive functions, including episodic memory, spatial navigation, and emotional regulation [2].

Mesial temporal sclerosis is the major etiology of pharmacoresistant focal epilepsy in adults. Hippocampal sclerosis represents a condition characterized by selective neuronal loss and gliosis, often accompanying temporal lobe epilepsies. Alterations in the hippocampus's structure and function have been correlated with various neurological and psychiatric disorders, like post-traumatic stress disorder, depressive disorder, and Alzheimer's disease [2, 3].

Besides hippocampal sclerosis, tumors, hamartomas, tuberous sclerosis, cavernous malformations, and aneurysms can be symptomatic within the hippocampal region; hence, surgery remains the gold standard treatment [4–10].

12.2 Anatomy of the Temporal Lobe and Hippocampus

The temporal lobe, which constitutes 17% of the cerebral cortex, is found only in primates and is the largest in humans. The temporal lobe, located beneath the lateral fissure, is characterized by the superior (T1), middle (T2), and inferior (T3) temporal gyri, with these gyri being defined by the superior and inferior temporal sulci. The primary auditory cortex (Heschl's gyrus) and Wernicke's area are found within the superior temporal gyrus. On the temporal lobe's inferior surface, the inferior gyrus curves and continues as the lateral occipitotemporal gyrus. The occipitotemporal sulcus separates the medial (T4, fusiform)

S. Sağıroğlu
Department of Neurosurgery, Kızıltepe State Hospital, Mardin, Türkiye

M. Turgut (✉)
Department of Neurosurgery, Aydın Adnan Menderes University Faculty of Medicine, Aydın, Türkiye

Department of Histology and Embryology, Aydın Adnan Menderes University Health Sciences Institute, Aydın, Türkiye
e-mail: drmturgut@yahoo.com

M. Turgut et al. (eds.), *The Human Hippocampus*, https://doi.org/10.1007/978-3-031-88907-3_12

and occipitotemporal gyri. Adjacent to it, the collateral sulcus separates the fusiform gyrus from the parahippocampal (T5) gyrus. Typically, the rhinal sulcus represents the anterior extension of the collateral sulcus. The parahippocampal gyrus contains the entorhinal cortex, which serves as a crucial hub for memory and navigation, connecting the hippocampus with other cortical areas. Medial to T1 is the circular (peri-insular) sulcus, which separates the insula from the temporal lobe [11–13].

Anatomically, the hippocampal formation is located on the medial surface of the temporal lobe. This formation encompasses the parahippocampal gyrus, hippocampus, dentate gyrus, subiculum, and the associated white matter tracts. Particularly, the fimbria is a prominent white matter tract arising from the hippocampus and continuing into the fornix. Superoanterior to the hippocampus, the brown-grayish amygdala is localized on the medial aspect of the temporal lobe. An important area in this region that has recently been under the spotlight of researchers is the piriform cortex. This small area, located on the amygdala, forms the medial surface of the temporal stem and serves as a cortico-subcortical hub. The importance of this region has been emphasized in recent epilepsy research [2, 11, 14, 15].

The temporal lobe is interwoven with a complex network of subcortical white matter bundles, including the arcuate fasciculus, which courses above the insula and links Broca's area to Wernicke's area. The inferior longitudinal fasciculus courses inferolaterally to the temporal horn, connecting the temporal lobe to the occipital lobe and is important for visual information processing in the brain. The uncinate fasciculus, coursing through the temporal stem, connects the temporal pole to the frontal lobe. The anterior bundle of the optic tract (Meyer's loop) borders the temporal horn of the lateral ventricle and extends into the temporal lobe [11, 16] (Fig. 12.1).

The earlier chapters of this book provide detailed information on the hippocampus' anatomy and embryology. Briefly, the hippocampus is a bilaminar gray matter structure. It extends into the temporal horn of the lateral ventricle, occupying the medial part of its floor, and two interlocking gray matter structures make up the hippocampus: the cornu ammonis, called the hippocampus proper, and the dentate gyrus [2, 11].

Viewed from the axial plane, the hippocampus resembles a seahorse. It can be divided into three distinct sections in both the axial and sagittal planes: the head (anterior), the body (intermediate), and the tail (posterior). The head of the hippocampus is broad and includes the pes hippocampi, which looks like the paw of an animal, reflecting its convoluted surface. The body narrows as it extends posteriorly, and the tail curves upwards, blending into the parahippocampal gyrus [2, 11] (Fig. 12.2).

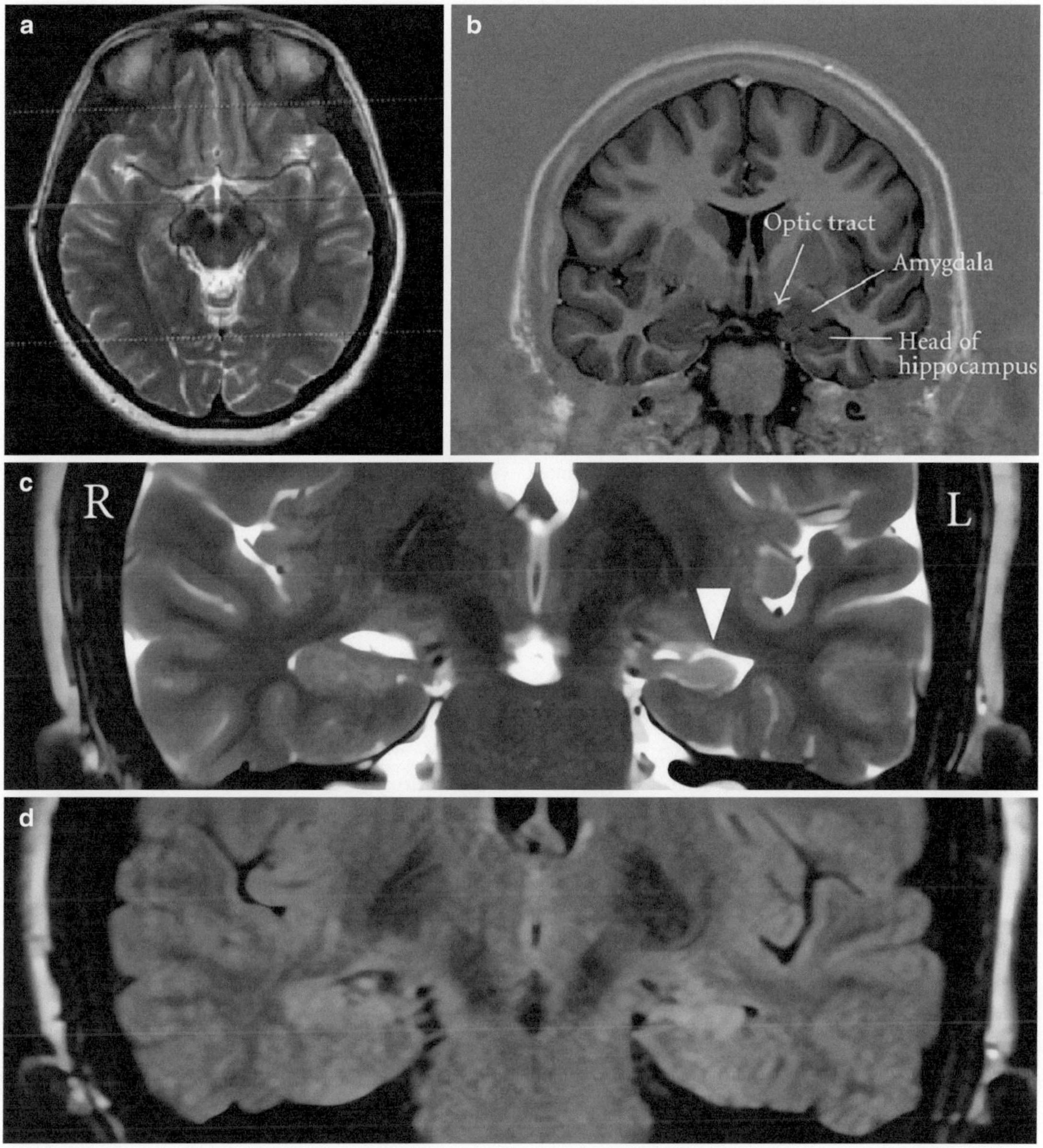

Fig. 12.1 (**a**, **b**) Coronal inversion recovery MRI at the level of the optic tract (blue line), illustrating the anatomical relationship between the amygdala and the optic tract. (**c**, **d**) Coronal T2 and FLAIR magnetic resonance images (MRI) showing left mesial temporal sclerosis. (Modified image from Al-Otaibi et al. [26])

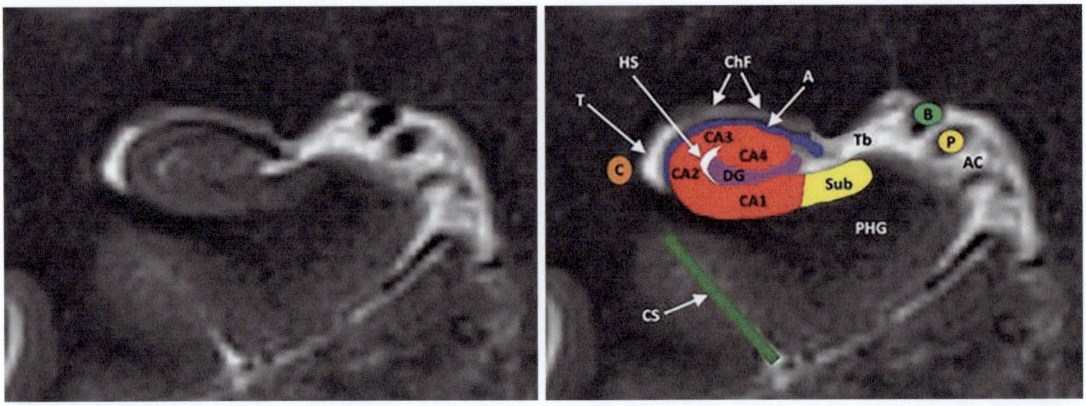

Fig. 12.2 Anatomical features of the hippocampal formation at the level of the hippocampal body on a coronal T2 MRI. The hippocampal formation includes the cornu ammonis (hippocampus proper), which is histologically divided into the four Sommer sectors (CA1–CA4), and the dentate gyrus (DG). A small hippocampal cyst (Hs) indicates the location of the mostly obliterated hippocampal sulcus. *A* alveus, *Ac* ambient cistern, *B* basal vein of Rosenthal, *C* tail of the caudate nucleus, *ChF* choroid fissure, *CS* collateral sulcus, *DG* dentate gyrus, *P* posterior cerebral artery, *PHG* parahippocampal gyrus, *Sub* subiculum, *T* temporal horn of the lateral ventricle, *Tb* transverse fissure of Bichat [2]

12.3 Surgical Management of Hippocampal Lesions

Besides the lesions mentioned above, surgery in this region is mostly indicated for medically refractory epilepsy. While there is no single definition of medically refractory epilepsy, it is commonly described as epilepsy that does not respond to at least two different antiepileptic drug monotherapies and one combination therapy, all administered at therapeutic doses over a period of 1–2 years. However, many patients have undergone more extensive and prolonged medication trials [17, 18].

Selective amygdalohippocampectomy may not be appropriate for some patients with clearly defined mesial temporal onset seizures. This group encompasses individuals with independent bitemporal onset seizures and those at significant risk for severe global memory impairment after the procedure. Bilateral temporal lobectomy results in Kluver-Bucy syndrome and is contraindicated [11]. Patients with dominant temporal lobe foci are particularly prone to experiencing a decline in verbal memory following surgery. This is especially true for those who show strong verbal memory performance before the operation, have a normal hippocampal volume, and had a later onset of seizures during adulthood [17, 19, 20]. Deficits can also occur with similar patterns in the nondominant temporal lobe, but they are generally less severe.

The Wada test, in which amobarbital is administered sequentially to both internal carotid arteries, can be used to determine the dominant side for language [21]. Patients afflicted by severe bilateral hippocampal atrophy who fail to demonstrate memory function support on the Wada test on the proposed surgical side may be prone to significant memory impairment, albeit in rare instances [17].

For conditions such as idiopathic generalized seizures, temporal neocortical foci, extratemporal focal epilepsy, or temporal lobe epilepsy with an unclear source localized to the mesial temporal regions, a radical resection is advised over selective approaches. Additionally, individuals with solely psychogenic nonepileptic seizures are not eligible, and those who have both psychogenic nonepileptic seizures and mesial temporal lobe epilepsy need to be carefully evaluated, with treatment of the psychological component being crucial for a satisfactory result [17, 20].

12.3.1 Anterior Temporal Lobectomy with Amygdalohippocampectomy

A temporal lobectomy is typically performed for the treatment of drug-resistant temporal lobe epilepsy or to address various space-occupying lesions, such as tumors or cavernous malformations. The primary objective of this procedure is to excise the epileptogenic focus or pathological tissue responsible for the patient's neurological symptoms, while minimizing the risk of postoperative deficits and preserving essential brain functions. The preoperative evaluation of candidates for temporal lobectomy entails a comprehensive assessment of the patient's clinical history, neurological examination, video-electroencephalographic (EEG) findings, and neuroimaging studies, including single-photon emission computed tomography (SPECT) and magnetic resonance imaging (MRI). In order to precisely localize the seizure focus and determine the extent of the surgical resection, magnetoencephalography, functional MRI, positron emission tomography (PET), and, in some cases, invasive methods such as stereo-electroencephalography (SEEG) or subdural EEG recordings are utilized [22–24].

The patient positioned in a semi-recumbent position with the head rotated slightly toward the contralateral side. A curvilinear or question-mark incision is made in the scalp based on the surgeon's preference, and a cuff for the temporalis muscle is preserved. A pterional craniotomy is undertaken with the objective of exposing the temporal lobe. Following exposure, the dura is carefully opened, allowing for the precise dissection and mobilization of the temporal lobe.

Resection limits are determined 5–6 cm posteriorly from the temporal tip on the nondominant side and 3–4.5 cm from the dominant side to avoid language deficits and visual defects. The vein of Labbé is another limiting factor. Penfield number two can also be used to measure the extent of the corticotomy. The primary auditory cortex resides in posterior T1, and surgical insult should be avoided in this area. The safe posterior limit for the resection can also be determined by the central artery located in the central sulcus. Coagulation with bipolar cautery starts at the determined posterior limit of the resection of all three gyri on the lateral surface and advances to the temporal stem. Subpial dissection with gentle suction or an ultrasonic aspirator is critical to avoid arterial injury to the branches of the middle cerebral artery. The resection is extended through the fusiform gyrus and by visualizing the collateral sulcus into the temporal horn of the lateral ventricle. Meyer's loop resides at the superolateral border of the temporal horn, and its preservation is crucial [11, 16]. The stem is resected through the inferior circular sulcus to the lateral choroidal fissure. With the stem resected, the temporal lobe can be externalized with further resection of the structures lateral to the choroidal fissure [7, 20, 25, 26].

At this point, the insula, hippocampus through the temporal horn, the superoanterior part of it, the amygdala, and, inferiorly, the parahippocampal gyrus are visible. An incision into the fimbria opens the choroidal fissure, and the superior choroidal point, which is the anterior end of the choroid plexus, can be identified. This is the superior-medial limit of the resection to preserve the optic tract and lateral geniculate body. Along a line connecting the choroidal point to the limen insulae, part of the amygdala is resected. Another safe anatomical point is the line between the middle cerebral artery bifurcation and the inferior choroidal point [27]. Subpial resection is carried out toward the Sylvian fissure, marking the anterior border. Further superior and medial resection at this stage compromises the globus pallidus and should be avoided. An arbitrary line from the proximal middle cerebral artery to the choroidal point marks the superoposterior safe resection margin for the amygdala [8, 25, 26].

The inferoposteromedial part of the amygdala and the head of the hippocampus are dissected from the uncal recess and advanced toward the tail by following the choroidal fissure to the tentorium at the lateral margin. The arachnoid must be preserved medially to protect the posterior cerebral artery, basal vein of Rosenthal, oculomotor nerve, and mesencephalon. The fimbria is dissected from the arachnoid, and the parahippocampal gyrus and entorhinal cortex are resected subpially by dissection. The en bloc resection of the head and anterior third of the hippocampus is considered optimal for histologic studies [28, 29]. In tumors in the amygdala or uncal region, attention should be paid to the resection margins because the border with the basal ganglia is not clear. Using neuronavigation in these areas can be beneficial in preventing complications. However, in diffuse infiltrative tumors that invade the brainstem, the subpial plane might be compromised, requiring subtotal resection. With careful coagulation of hippocampal feeders (Ammon's horn) while avoiding injury to the posterior cerebral artery, the resection is completed [25, 26, 30].

Spencer et al. described a modified version of the classical anterior temporal lobectomy (ATL), in which similar steps are followed, and the sphenoid wing is flattened to expose the temporal tip. For the dominant side, 3 cm of T1 is removed from the temporal tip and 4.5 cm for the nondominant side. The rationale in this method is to prevent speech complications. Resection of the corticotomy and mesial temporal structures is performed in a similar fashion after identification of the temporal horn [31].

This procedure is well established and successful in removing epileptogenic zones while offering good exposure to preserve critical structures. There are numerous differences and nuances regarding surgical techniques reported in the literature, while the general aspects and steps of the surgery are common.

12.3.1.1 Potential Outcomes and Complications

While generally considered safe with a seizure-free rate of 70%, temporal lobectomy has a mortality risk of 1%, with a morbidity risk reaching up to 17%. Surgical complications such as hematoma, hydrocephalus, pseudomeningocele, and cerebrospinal fluid (CSF) leakage are rare, with a cumulative rate of 1–2%. Psychiatric disorders have a risk of 7%, and cognitive impairment has a 5% risk, especially on the dominant side. Hemiparesis and speech difficulties occur in about 4% of patients due to direct insult to the uncinate or arcuate fasciculi or vasospasm [30]. Extensive resections on the left side have shown significantly greater deterioration in verbal and attention-related performance compared to selective resections [19]. Visual field defects, such as homonymous quadrantanopsia, affect at least 4–6% of patients and are linked to damage in Meyer's loop. Preoperative diffusion tensor imaging (DTI) MRI is a valuable tool for identifying Meyer's loop. Electrocorticography-guided resection of the temporal lobe can be beneficial to selectively resect epileptogenic dysplastic tissue to minimize complications [20, 30, 32, 33].

To minimize complications, surgeons have refined their techniques to limit the resection based on the specific pathology, such as selectively removing epileptogenic zones or tumors. Numerous invasive and noninvasive diagnostic techniques have been developed to achieve this goal. We will examine several surgical methods created for this purpose in the following section.

12.3.2 Procedures for Selective Amygdalohippocampectomy

12.3.2.1 Transsylvian Approach

Wieser and Yaşargil [34] described the transsylvian technique for hippocampal lesions. The head is tilted to the contralateral side, and the neck is extended to maintain the zygoma in a superior position. After a curvilinear incision and a pterional craniotomy with the sphenoid ridge flattened, the dura is incised, and the Sylvian sulcus is identified and dissected. The limen insulae is identified with the middle cerebral artery bifurcation, mesial uncus, and temporal pole. To reach the amygdala and the temporal horn, a 1–2 cm incision is made lateral to M1 and anteromedial to M2 at the beginning of the anterior point of the inferior peri-insular sulcus just at the level of the limen insulae. Opening the arachnoid and CSF removal at this step are avoided to prevent the collapse of the temporal horn, which would complicate later stages of the operation. After careful suction, the lateral parts of the amygdala are removed without advancing too medially to protect the optic tract. After resection of at least 70% of the amygdala, the piriform cortex is removed until the arachnoid of the ambient cistern is visible. The temporal horn is visualized and entered anteromedially. The rest of the amygdala is not explored to avoid damaging the optic tract.

The choroid plexus is reflected medially, and after the tela is opened, the hippocampus is reached. The uncus is resected with care to avoid damaging the

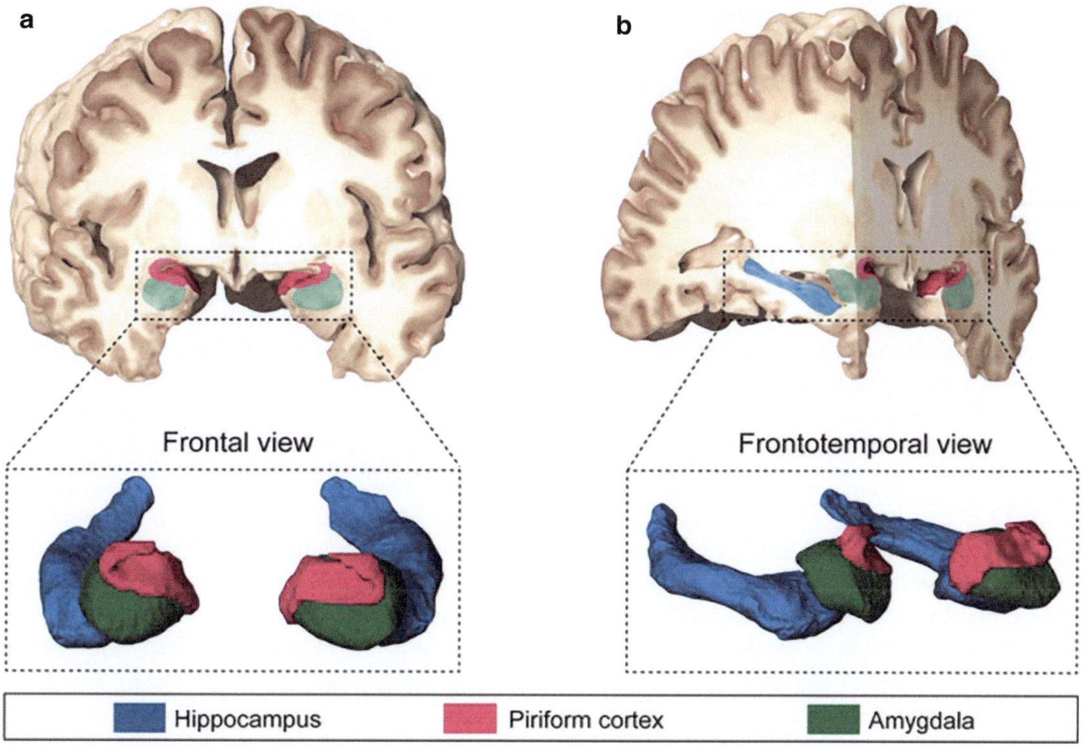

Fig. 12.3 Schematic illustration of mesiotemporal structures including the hippocampus, amygdala, and piriform cortex [15]

posterior cerebral artery and the oculomotor nerve. The hippocampus is dissected from medial to lateral, from head to tail. The anterior choroidal artery must be preserved carefully, while its lateral branches to the uncus and hippocampus are coagulated. The inferior choroidal point, where the anterior choroidal artery enters the ventricle, marks the end of dissection. The fimbria is incised, and the hippocampus with the parahippocampal gyrus is suctioned subpially using a low-setting ultrasonic aspirator. The collateral sulcus is the lateral border of the resection to the base of the temporal fossa. The remaining posterior hippocampus can be resected with a low-setting ultrasonic aspirator [13, 15, 25, 34, 35].

Yeni et al. [36] reported a high rate of 36.6% visual field deficits with the transsylvian approach. The authors recommend a modified pial incision at the most anterior edge of the inferior limiting sulcus to prevent injury to Meyer's loop. However, Yaşargil comments that with piecemeal removal of the amygdala and gentle traction with cotton sponges while sparing the dorsolateral wall of the temporal horn, any possible damage to Meyer's loop can be avoided [13, 35]. Commentators suggest that traction to the tissue and/or insult to the anterior choroidal artery, either by vasospasm or traction, may be the cause of the visual deficit [13, 35]. These discussions highlight the difficulty of the technique and the importance of precise execution.

The authors propose that while this technique is elegant and respectful of tissue, it is a challenge even for experienced surgeons, and the learning curve is steep. Retractors are obsolete in this technique with adequate Sylvian dissection; hence, traction injury is avoided. The highly variable nature of the arterial anatomy in the region, the risk of vasospasm due to vascular insult, and resulting complications must be kept in mind. Recent studies emphasize the role of the piriform cortex, and its resection is correlated with better seizure control [14, 15]. This structure is resected naturally due to the surgical corridor in this approach and may explain the comparable success rate to temporal resection (Fig. 12.3).

12.3.2.2 Transcortical Approaches

With advancing technology and the aim of reducing the potential risks of extensive surgical resections, transcortical approaches have been defined. The general principle in all these approaches involves planning a surgical corridor directed at the lesion identified with MRI imaging while avoiding arteries and deep brain structures. This approach aims to reduce surgical duration and postoperative hospital stay. In almost all approaches, neuronavigation-guided minicraniotomy is a common practice. One significant disadvantage of these techniques is the risk of retraction injury to the surrounding cortex (Fig. 12.4).

Trans T1

Under general anesthesia, the head is tilted, and the neck is extended to ensure the zygomatic arch is the highest point of the operative field. A linear or curvilinear incision is made. Burr holes are drilled on each side of the sphenoid wing, and a small temporal craniotomy is performed to uncover the superior temporal gyrus (STG) and the Sylvian fissure. The sphenoid wing is then smoothed using a rongeur to ensure sufficient visibility of the anterior segment of T1. Hemostasis of the middle meningeal artery is achieved using cautery and bone wax. A craniotomy of about 3 cm in diameter, centered approximately 4 cm posterior to the temporal pole, is made. The dura is incised in a semicircular manner and folded back over the temporalis muscle. A corticectomy, approximately 2 cm in length, is performed along the upper edge of T1, situated just beneath the Sylvian fissure [28, 37]. The typical distance between the tip of the temporal lobe and the anterior edge of Meyer's loop averages 27 ± 3.5 mm, with a range of 22–37 mm [16]. The resection should not extend posteriorly above the temporal horn to avoid damage to Meyer's loop.

An ultrasonic aspirator is used to establish a pathway along the Sylvian fissure, which extends to the temporal horn. The roof of the temporal horn is reached approximately 30 mm under T1 [16]. When the ventricle is opened, retractors are placed into the temporal horn to enhance visibility of the choroid plexus and choroidal fissure, which marks the medial edge for hippocampal dissection. The hippocampus is then exposed, and the mesial structures are resected subpially. The procedure begins by accessing the lateral ventricular sulcus at the point where the hippocampus meets the collateral eminence, followed by a lateral disconnection along the ventricular margin in the anteroposterior direction, toward the collateral sulcus. This approach allows for the lateral retraction of the hippocampus and the dissection of the fimbria from the choroidal fissure to achieve mesial disconnection [28, 37, 38].

The medial region of the parahippocampal gyrus is distinguished from the hippocampal sulcus's arachnoid. A mediolateral transverse cut is performed to separate the hippocampus and parahippocampal gyrus from the hippocampal tail at the posterior end. Subsequently, the arachnoidal covering is detached by rolling the structure from the mesial to the lateral side, which releases the tissue from its vascular pedicle. The hippocampal formation is removed en bloc for optimal pathological examination. The standard posterior resection limit is situated at the midbrain tectum, whereas the anterior boundary extends to the rhinal sulcus, encompassing the uncus. The removal of the uncus and amygdala is performed subpially [28, 37, 38].

The advantage of this technique is that it can be performed with a good understanding of anatomy and anatomical landmarks without the need for neuronavigation. The piriform cortex is resected because it is approached anterosuperiorly to the hippocampus. However, the disadvantage is that the anterior temporal stem is cut similarly to the transsylvian technique, and there is a risk of damaging Meyer's loop if suction is extended posteriorly at the roof of the temporal horn during corticectomy. The anterosuperior approach to the temporal horn in this technique minimizes the risk of Meyer's loop damage compared to lateral approaches.

Trans T2

Originally described by Niemeyer [39], transcortical resection through the middle temporal gyrus in selective approaches has historically been the starting point of selective approaches.

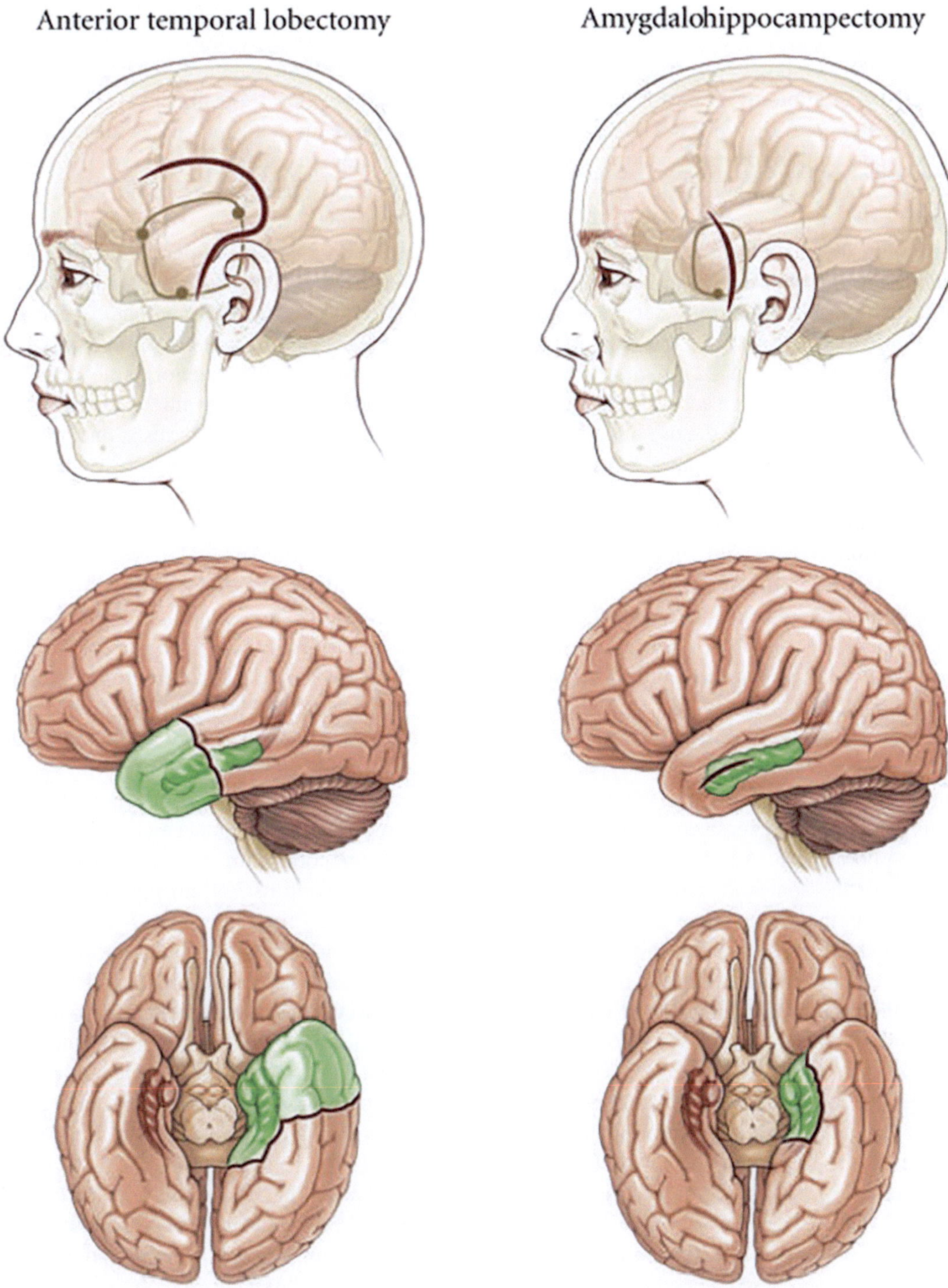

Fig. 12.4 Comparison of anterior temporal lobectomy and selective amygdalohippocampectomy [17]

The head and incision are prepared similarly to the previous approach, but this time, neuronavigation is used to target the middle temporal gyrus, and a craniotomy is performed 2–5 cm behind the temporal tip, with a 2–2.5 cm corticotomy made with bipolar and suction or ultrasonic aspirator [17]. Similarly, trans-superior temporal sulcus approaches have been described for reaching the temporal horn [37]. Reaching the temporal horn and mastering its anatomical location is the most important step in all these surgical approaches, and the middle temporal gyrus, with a distance of 22–26 mm from the cortex, provides the easiest access to the temporal horn [16]. At this stage, the use of neuronavigation is crucial, as the temporal horn can be small and missing it while advancing superiorly can result in damage to Meyer's loop, Sylvian vessels, temporal stem, and even basal ganglia structures [17, 29, 30].

The temporal horn wall is thin, and once identified with low-power suction, a retractor is placed. The parahippocampal gyrus and hippocampus are then removed en bloc, similar to previous techniques. Posteriorly, the tail of the hippocampus and anteriorly, the amygdala and uncus, are resected subpially [17, 26, 30].

This technique offers ease due to its proximity to the temporal horn. Additionally, there is no need to drill the sphenoid wing as in trans-T1 approaches. In this technique, selective access to hippocampal lesions is possible while leaving the amygdala intact. The disadvantage of this technique is the risk of damaging Meyer's loop and cortical injury due to retraction [16, 26].

Trans T3 and Subtemporal Approaches

Hori et al. [40] described the subtemporal approach to the mesial hippocampal structures. Similar to the previous technique, the head is positioned and the incision is planned to reach the basal temporal area. If the subtemporal approach is planned based on pathology, the craniotomy or minicraniotomy is performed, and the lower limit of the craniotomy is flattened to the basal limit of the middle fossa. If the Trans T3 intervention is planned, there is no need to flatten the bone [29]. At this stage, air cells are waxed if opened. To minimize traction injury, CSF drainage is important, and mannitol infusion is recommended before dura opening [41, 42]. The temporal horn is situated at a distance of 10–14 mm from the middle fossa base. The application of these maneuvers permits the use of minimal traction [16].

Once the dura has been opened, the temporal lobe is retracted with a retractor, allowing for the visualization of the fusiform and parahippocampal gyri, uncus, and the ambient cistern and tentorium. In the event that traction is not feasible due to the presence of basal veins, a partial resection of the inferior temporal gyrus is undertaken, employing a transcortical approach as the preferred method [43]. A surgical incision is made to the uncus at a point 10–15 mm posterior to the intersection of the third cranial nerve and the tentorial edge, and the tissues are cleaned with suction to access the temporal horn from below. Alternatively, a technique involving an incision of the fusiform gyrus and corticectomy, aiming for a wider operative field and less retraction, has also been described [8, 42]. To avoid retraction completely, corticectomy through T3 with neuronavigation guidance can also be selected [25]. After identifying the amygdala and hippocampus within the temporal horn, the parahippocampal gyrus is excised up to 3–4 cm posteriorly from the uncus. The hippocampus's head is then removed en bloc subpially. Suction or ultrasonic aspirator is used to extend the hippocampal resection posteriorly, then the anterior hippocampus and amygdala are resected subpially until the middle cerebral artery is visible subpially. Excessive superomedial extension in amygdala resection is avoided for the reasons previously mentioned [41].

This approach has the benefit of reducing the risk of visual field defects by accessing the temporal horn from beneath. The temporal stem remains intact in this approach. Important considerations include the risk of cortical injury due to retraction, damage to the basal temporal veins and the vein of Labbé, and injury to the frontal branch of the facial nerve [26, 42].

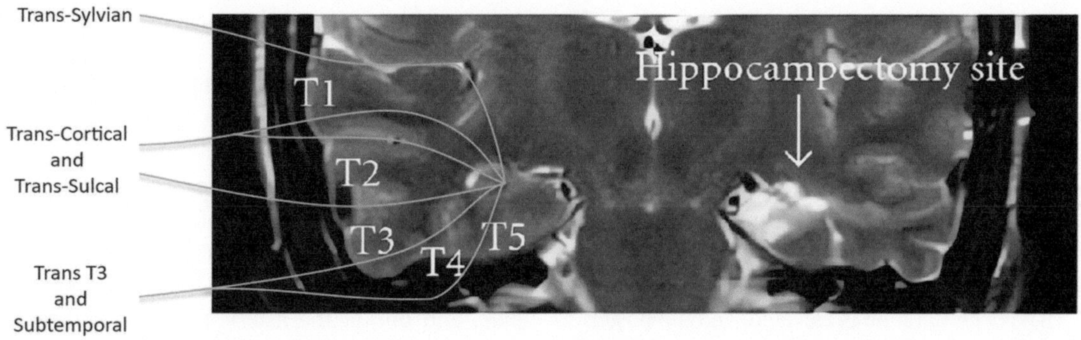

Fig. 12.5 Coronal T2 MRI depicting the resection site after selective amygdalohippocampectomy on the left side. The right side highlights the temporal gyri—superior (T1), middle (T2), inferior (T3), fusiform gyrus (T4), and parahippocampal gyrus (T5)—and surgical corridors for lateral approaches. (Modified image from Al-Otaibi et al. [26])

Summary of Transcortical Approaches

When choosing transcortical approaches, the surgeon's familiarity with the technique, experience, and anatomical knowledge is crucial. Additionally, if temporal dysplasia is detected during preoperative MRI evaluation, it is recommended to use intraoperative electrocorticographic guidance to identify and resect the dysplastic region and select the appropriate surgical technique [30] (Fig. 12.5).

12.3.3 Posterior Approaches for Mesial Temporal Lobe

12.3.3.1 Retrolabyrinthine Presigmoid Transpetrosal Approach

Hori et al. [24] described this posterior approach to mesial structures as an alternative to alleviate and avoid retraction injuries to the temporal cortex. The patient is positioned in the supine position with the head turned 90°. A reverse U-shaped incision is made from the tragus to the asterion above the ear. Burr holes are placed anterior to the petrous bone and on each side of the sphenoid and transverse sinuses. The bone over the sinuses is drilled until the superficial cortex is reached to visualize the sinuses. The superficial cortex of the mastoid tip may be removed to increase exposure. After craniotomy, the dura is incised laterally to the superior petrosal sinus. The retrolabyrinth is drilled while protecting the superior petrosal sinus [24].

With retraction, the temporal lobe is elevated, and the ambient cistern is incised to release CSF. After this step, the surgeon turns anteriorly to identify the fusiform gyrus near the tentorial edge. To prevent excessive traction on the bridging veins and the vein of Labbé, the dura is cut laterally to the superior petrosal sinus, and the posterior temporal lobe is covered [24].

Using neuronavigation, the fusiform gyrus is resected subpially with an ultrasonic aspirator, allowing access to the temporal horn. The medial fusiform gyrus, parahippocampal gyrus, hippocampus, and uncus are resected as previously described. Amygdala resection can then be performed [24].

In this approach, the temporal neocortex is completely preserved, and isolated hippocampectomy with the medial fusiform gyrus is possible. Although the amygdala can be resected with this technique, it is difficult to access. The disadvantages of the approach are the difficulty of the surgical steps and the length of the surgical corridor, as well as the risk of damage due to traction.

12.3.3.2 Supratentorial-Infraoccipital Approach

Smith and Spetzler [44] described this approach as an alternative route to achieve selective lesionectomy while avoiding traction injuries and visual deficits. The patient lies in a prone position with the neck flexed and the table in a slight reverse Trendelenburg position to avoid venous congestion and use gravity retraction on the occipital lobe during the later stages of the operation. An occipital craniotomy, including the transverse sinus, the superior sagittal sinus, and the torcula, is performed, and dural tuck-up sutures are placed. The dura is opened obliquely to flap over the superior sagittal sinus and transverse sinuses. A lumbar drain is used to release CSF at this stage to alleviate traction [44].

The occipital lobe is elevated, and the posterior cerebral artery is followed to the ambient cistern to further release CSF. This maneuver relaxes the occipital lobe to minimize retraction. The superior cerebellar artery branches, trochlear nerve, cerebral peduncle, and quadrigeminal plate are visible medially, and resection of mesial structures is performed with the aid of navigation [44].

This approach has been rarely used in the literature, and comprehensive studies on its outcomes are lacking. Theoretical reasons for its limited use include the risk of sinus injury, air embolism, visual field defects due to occipital lobe traction, and a long, narrow working corridor. Additionally, advancement through the cistern increases the risk of damaging delicate vascular and neurological structures.

12.3.3.3 Paramedian Supracerebellar-Transtentorial Approach

Most of the approaches described to access the mesial temporal lobe either require traction, necessitate sacrificing the non-diseased neocortex, or pose difficulty in reaching the amygdala anteriorly or the tail of the hippocampus posteriorly. Voight and Yaşargil [5] described the supracerebellar-transtentorial approach to reach the area. Yonekawa et al. [10] released a series of cases to reach the posterior hippocampal region and modified the technique as the paramedian supracerebellar approach to emphasize it as the best surgical corridor to reach the posterior hippocampal region. Later, Türe et al. [9] demonstrated through cadaver dissections and case series that entire mesial temporal resection, including the amygdala, is possible through this route with an endoscope.

In the sitting position, with the neck flexed, a vertical, linear paramedian incision is made between the inion and the mastoid process. The incision extends a third of the way above and two-thirds of the way below the superior nuchal line. During this process, the occipital artery is encountered twice and should be coagulated and divided. Muscles are retracted to expose the inion medially and the asterion laterally. Three burr holes are drilled: one above the superior nuchal line, one lateral to the external occipital protuberance, and one at the asterion. The bone flap is turned with a craniotome, extending third of the way above and two-thirds of the way below the transverse sinus. Bone wax is applied to prevent CSF leakage [9, 10].

Using an operating microscope, a transverse dural opening of about 15 mm is made to access the cisterna magna, allowing CSF drainage and relaxation of the posterior fossa. The exact location of the transverse sinus is confirmed with a micro-Doppler probe before making a curvilinear dural opening 15 mm below the sinus [9]. This creates a wide mediolateral corridor for optimal visualization and maneuvering. The cerebellum is dissected from the tentorium, during which bridging veins may be sacrificed. The lateral petrosal vein and tentorial veins around the midline must be preserved as they are the main draining veins of the posterior fossa. The posterior quadrigeminal and ambient cisterns are opened, identifying landmarks such as the superior cerebellar artery, the trochlear nerve, the pineal gland, and the galenic venous system [9, 10].

Once the trochlear nerve has been identified, it can be observed to progress between the posterior cerebral and superior cerebellar arteries in a lateral direction, and the tentorium is coagulated and incised, creating space for surgical access. The incision is extended from the posterior tentorial incisura to the middle of the petrous ridge. It is advised that large venous lakes within the tentorium be secured with titanium clips, thus preventing the potential occurrence of an air embolism [10]. The anterior leaflet of the tentorium is retracted downward, and the posterior leaflet is pulled up, opening more space for visualization and manipulation [9].

After some exploration of the region, the parahippocampal gyrus, collateral sulcus, and fusiform gyrus can be identified. After performing an intraoperative EEG, a depth electrode is inserted into the parahippocampal gyrus and the hippocampal area. The inferior colliculus marks the beginning of the parahippocampal gyrus resection, and it marks the posterior limit of resection in epilepsy patients to protect the visual cortex. In patients with tumors, dissection begins anterior to this point, and the posterior part is dealt with at the end of the surgery. The medial limit of resection is delineated by the midbrain and the choroidal fissure, while the collateral sulcus indicates the lateral limit. A vertical incision is created using bipolar forceps at the inferior colliculus level, extending outward toward the collateral sulcus. With great care, the major inferior temporal arteries from the P3 and P2 segments of the posterior cerebral artery are preserved during subpial dissection. This dissection continues forward to the uncus. The hippocampal and uncal arteries are identified, coagulated, and cut. The dentate gyrus, identified by its grayish color, is exposed by gently dissecting the inferior aspect of the parahippocampal gyrus. This method offers a distinctive perspective of the roof of the temporal horn. Dissection progresses between the fimbria and choroid plexus to the tip of the temporal horn. A high-definition neuroendoscope and a curved suction are used to visualize and remove the inferior and anterior aspects of the parahippocampal gyrus, which are not well seen through the microscope. Resection of the most anterior and inferior portions is possible with improved visibility provided by a cotton ball positioned beneath the gyrus. To prevent damaging the anterior choroidal artery and optic tract, gentle suction is applied [9].

The cranial nerves, the internal carotid artery, the anterior choroidal artery, the posterior communicating artery, and several parts of the superior cerebellar and posterior cerebral arteries are recognized. The high-definition neuroendoscope assists in visualizing and resecting the remaining tissue. The tentorium is reapproximated, the dura is closed watertight, and the bone flap is fixated [9].

This approach is anatomically ideal for selectively reaching mesial structures, especially posterior ones, with minimal retraction. With this method, there is theoretically no risk of visual field deficiencies. Air embolism risk, a challenging learning curve, a lengthy operative corridor, and the need to use several instruments during the procedure are among the drawbacks.

12.4 Minimally Invasive Interventions for the Hippocampus

For patients who are not candidates for traditional surgery or who would prefer less invasive options, a number of minimally invasive techniques have been developed through recent medical advancements. These techniques primarily focus on lesions in the mesial temporal region, especially in drug-resistant epilepsy. In this section, we will briefly mention them.

12.4.1 Multiple Hippocampal Transections

To provide patients with epilepsy a less invasive option than ATL, multiple hippocampal transections (MHT) were described by Shimizu et al. in 2006 [45]. This technique involves making several incisions along the longitudinal fibers of the hippocampus while preserving the circular fibers crucial for memory function. By using this method, the likelihood of hippocampal damage and memory impairments is reduced [23, 45].

Transsylvian, transcortical, or transsulcal methods can be used for MHT. Along the hippocampal region, several transsections are made, with an average of 5–8 cuts. Further incisions or surgical procedures might be required if seizures are attributed to other temporal lobe structures, such as the parahippocampal gyrus or amygdala [45].

Studies show that MHT efficiently interferes with epileptogenic circuits, resulting in a 75% seizure-free rate [45]. Although MHT seems to be a promising treatment for intractable epilepsy with little effect on cognitive function, further investigation is needed to validate its long-term advantages.

12.4.2 Laser Interstitial Thermal Therapy

Another minimally invasive method, called laser interstitial thermal therapy (LITT), uses laser energy to produce thermal energy that causes coagulative necrosis in the targeted tissue. Guided by real-time MRI thermography, it offers precise control and minimal damage to surrounding tissues. Under neuronavigation guidance, a 5 mm incision is made in the occipital scalp, and a burr hole is created. A rigid guide is attached, and the fiber-optic laser probe is inserted through the burr hole along a preplanned path. The patient is then placed in an MRI scanner, either by transporting them or using an intraoperative MRI. After confirming the probe's position, ablation begins, interspersed with 2D T1 MRI images showing brain temperature. After ablation, the probe is removed, the incision is closed, and the patient is monitored for at least 24 h before discharge [46].

Complications from LITT occur in 17% of cases, with 5% being permanent and 10% temporary. Visual field deficits are the most common problem, affecting up to 37.5% of patients [47]. There have been reports of memory declines in the verbal and visual domains in 24.2% and 25.2% of patients, respectively [48]. Research indicates that between 52.5% and 75% of patients experience seizure freedom after at least a year of follow-up [47–49]. Overall, although LITT appears to be a promising short-term seizure control method, more research is needed to determine its long-term effectiveness and compare it to open surgical techniques.

12.4.3 Stereoencephalography-Guided Radiofrequency Ablation

The stereoencephalography-guided radiofrequency ablation (SEEG-RFA) technique begins with extensive preoperative preparation with MRI. Under local anesthesia, electrodes are implanted, and a postoperative CT scan verifies the exact placement. Through multidisciplinary discussion, the seizure onset zones are determined following the recording of three habitual seizures. Using a device set to 7.5 W power, radiofrequency thermocoagulation is administered to the identified zones for a total of 90 s, divided into two 30-s sessions with a 30-s break in between. Since a single coagulation only covers 5–7 mm, multiple coagulations are necessary for targets bigger than 7–10 mm. The patient is closely observed for language, movement, and other functions during the surgery [23, 50].

The operation should be stopped if the patient feels any discomfort, such as heat, electromagnetic noises, or slight pain. If a seizure occurs during RF-TC, the procedure must be stopped immediately. After RF-TC, seizure recording should continue to assess treatment efficacy before removing the electrodes without anesthesia. An MRI within 48 h is used to evaluate the results [23, 50].

SEEG-RFA has a relatively low success rate in large lesions due to the limited ablation radius of the technique. Recent studies highlight that it reduces seizure frequency by up to 50% and delays the need for secondary interventions [23]. This technique shows better results for smaller lesions [51].

12.5 Neuromodulation

In cases where resection is contraindicated, such as in patients with generalized epilepsy syndromes or poorly defined, multifocal, or functionally significant brain areas, palliative surgical options are considered. These include responsive neurostimulation (RNS), deep brain stimulation (DBS), and vagus nerve stimulation (VNS).

All three of these approaches have been approved by the FDA for the treatment of medically resistant epilepsy. For some individuals, these methods are vital, even though they typically result in less control over seizures than resective surgery. In addition, alternative neuromodulation modalities such as targeted ultrasound and transcranial magnetic stimulation are still being investigated [52].

12.5.1 Responsive Neurostimulation

The closed-loop responsive neurostimulation system aims to prevent seizures by continuously monitoring the electrocorticogram at the seizure focus and administering brief bursts of high-frequency electrical stimulation. The RNS system utilizes implanted subdural and depth electrodes to record and analyze regional electrocorticographic signals in real-time, subsequently triggering stimulation to prevent the occurrence of clinical seizures. FDA-approved in 2013 for treating intractable partial epilepsy in adults [7]. In cases of bilateral mesial temporal lobe epilepsy where surgical intervention is contraindicated, bilateral hippocampal electrode placement can be employed as an alternative [53].

Long-term studies have shown its efficacy, with a median seizure reduction rate of 44% to 66% over 1–6 years, and a 9-year follow-up demonstrating a 75% reduction in seizure frequency [54, 55].

12.5.2 Deep Brain Stimulation

In order to regulate aberrant activity, DBS involves inserting electrodes into targeted areas of the brain. Typically, the system operates in an open-loop configuration, wherein the stimulation parameters are predetermined and remain unaltered in response to input received in real time. After it was found that injury to the anterior thalamic nucleus (ANT) in animal models decreased epileptogenic activity, perhaps by raising the threshold for cortical epileptic discharges, DBS became a viable treatment for epilepsy [52]. The ANT is part of the Papez circuit, involved in limbic system connectivity [52]. In the SANTE trial, patients experienced a significant 40% reduction in seizure frequency during the initial 3 months of stimulation, which improved to a 56% reduction over 2 years. Additionally, 54% of patients achieved at least a 50% reduction in seizures during the study period [56].

Apart from the ANT, other brain structures have been subjected to evaluation for potential DBS targeting, including the centromedian thalamus and hippocampus. DBS targeting the centromedian thalamic nucleus has demonstrated potential due to its connections with the medial frontal and cingulate gyri. While this was not statistically significant, some studies have reported a 30% decrease in tonic-clonic seizures. Nevertheless, the results have been inconsistent. In a comparable direction, research on hippocampus DBS have revealed that 87.5% of patients saw a reduction in seizure frequency of more than 50% [33].

The precision with which electrodes are placed has a significant impact on DBS' effectiveness. The accuracy of the procedure has been enhanced through the utilization of high-resolution MRI and neuronavigation, which has resulted in a reduction in electrode placement variability and an improvement in outcomes. Despite these developments, DBS carries some risk factors, such brain paresthesias, pain at the implant site, infections, and, in rare cases, depression. Long-term follow-up studies have demonstrated that although DBS can be quite beneficial, especially for ANT stimulation, further research and technique improvement are required to maximize safety and efficacy [7, 33, 52].

12.5.3 Vagus Nerve Stimulation

VNS was approved by the US FDA in 1997 as an adjunctive treatment for pharmacoresistant epilepsy. It is frequently employed in cases of intractable epilepsy where the patient is unsuitable for, or has not responded to, resective surgery. This includes individuals with persistent seizures after temporal lobe resection or those who refuse surgery [7, 52].

This extracranial procedure involves the implantation of a programmable pulse generator device in the infraclavicular region, subcutaneously. Lead wires are attached to the midcervical vagus nerve via a subcutaneous tunnel through a second neck incision. The device continuously operates but can be controlled by the patient or programmer. To avoid damaging the sinoatrial node, which is innervated by the right vagus nerve, VNS is generally placed on the left side [57]. MR imaging can be performed on patients who have an implanted VNS device, although specific conditions apply for newer models and receive-only coils [52].

In clinical trials, the high-stimulation group experienced a 52% seizure reduction over 16–18 months. Long-term studies revealed that 35% of patients had more than a 50% seizure reduction, and 20% had more than a 75% reduction at a 12-month follow-up [57]. A large analysis by Englot et al. [58] reported that 49% of patients achieved ≥50% seizure reduction within 4 months and 63% between 24 and 48 months. Gradually progressive responses in patients treated with VNS are supported by controlled trials and follow-up studies.

References

1. Zeithamova D, Schlichting ML, Preston AR. The hippocampus and inferential reasoning: building memories to navigate future decisions. Front Hum Neurosci. 2012; https://doi.org/10.3389/fnhum.2012.00070/abstract.
2. Dekeyzer S, De Kock I, Nikoubashman O, Vanden Bossche S, Van Eetvelde R, De Groote J, Acou M, Wiesmann M, Deblaere K, Achten E. "Unforgettable"—a pictorial essay on anatomy and pathology of the hippocampus. Insights Imaging. 2017;8(2):199–212.
3. Anand K, Dhikav V. Hippocampus in health and disease: an overview. Ann Indian Acad Neurol. 2012;15(4):239.
4. Diehl B, Prayson R, Najm I, Ruggieri P. Hamartomas and epilepsy: clinical and imaging characteristics. Seizure. 2003;12(5):307–11.
5. Voigt K, Yaşargil M. Cerebral cavernous haemangiomas or cavernomas. Minim Invasive Neurosurg. 1976;19(02):59–68.
6. Liu T, Ding J, Zhang S, Wang Y, Xu J, Yuan L, Zhai F, Zhang G, Liang S. Independent temporal lobe epilepsy in patients with tuberous sclerosis complex. Pediatric Investig. 2022;6(1):23–8.
7. Chang EF, Englot DJ, Vadera S. Minimally invasive surgical approaches for temporal lobe epilepsy. Epilepsy Behav. 2015;47:24–33.
8. Przybylowski CJ, Whiting AC, Preul MC, Smith KA. Anatomical subpial resection of tumors in the amygdala and hippocampus. World Neurosurg. 2021;151:e652–62.
9. Türe U, Harput MV, Kaya AH, Baimedi P, Firat Z, Türe H, Bingöl CA. The paramedian supracerebellar-transtentorial approach to the entire length of the mediobasal temporal region: an anatomical and clinical study. Laboratory investigation. J Neurosurg. 2012;116(4):773–91.
10. Yonekawa Y, Imhof HG, Taub E, Curcic M, Kaku Y, Roth P, Wieser HG, Groscurth P. Supracerebellar transtentorial approach to posterior temporomedial structures: technical note. J Neurosurg. 2001;94(2):339–45.
11. Kiernan JA. Anatomy of the temporal lobe. Epilepsy Res Treat. 2012;2012:1–12.
12. Destrieux C, Bourry D, Velut S. Surgical anatomy of the hippocampus. Neurochirurgie. 2013;59(4–5):149–58.
13. Yaşargil MG, Krayenbühl N, Roth P, Hsu SPC, Yaşargil DCH. The selective amygdalohippocampectomy for intractable temporal limbic seizures: historical vignette. J Neurosurg. 2010;112(1):168–85.
14. Galovic M, Baudracco I, Wright-Goff E, Pillajo G, Nachev P, Wandschneider B, Woermann F, Thompson P, Baxendale S, McEvoy AW, Nowell M, Mancini M, Vos SB, Winston GP, Sparks R, Prados F, Miserocchi A, De Tisi J, Van Graan LA, Rodionov R, Wu C, Alizadeh M, Kozlowski L, Sharan AD, Kini LG, Davis KA, Litt B, Ourselin S, Moshé SL, Sander JWA, Löscher W, Duncan JS, Koepp MJ. Association of piriform cortex resection with surgical outcomes in patients with temporal lobe epilepsy. JAMA Neurol. 2019;76(6):690.
15. Borger V, Schneider M, Taube J, Potthoff A, Keil VC, Hamed M, Aydin G, Ilic I, Solymosi L, Elger CE, Güresir E, Fimmers R, Schuss P, Helmstaedter C, Surges R, Vatter H. Resection of piriform cortex predicts seizure freedom in temporal lobe epilepsy. Ann Clin Transl Neurol. 2021;8(1):177–89.
16. Ebeling U, Reulen HJ. Neurosurgical topography of the optic radiation in the temporal lobe. Acta Neurochir. 1988;92(1–4):29–36.
17. Spencer D, Burchiel K. Selective amygdalohippocampectomy. Epilepsy Res Treat. 2012;2012:1–8.
18. Ni XJ, Zhong H, Liu YX, Lin HW, Gu ZC. Current trends and hotspots in drug-resistant epilepsy research: insights from a bibliometric analysis. Front Neurol. 2022;13:1023832.
19. Clusmann H, Schramm J, Kral T, Helmstaedter C, Ostertun B, Fimmers R, Haun D, Elger CE. Prognostic factors and outcome after different types of resection for temporal lobe epilepsy. J Neurosurg. 2002;97(5):1131–41.
20. Brotis AG, Giannis T, Kapsalaki E, Dardiotis E, Fountas KN. Complications after anterior temporal lobectomy for medically intractable epilepsy: a systematic review and meta-analysis. Stereotact Funct Neurosurg. 2019;97(2):69–82.
21. Hori T, Yamane F, Ochiai T, Kondo S, Shimizu S, Ishii K, Miyata H. Selective subtemporal amygdalohippocampectomy for refractory temporal lobe epilepsy: operative and neuropsychological outcomes. J Neurosurg. 2007;106(1):134–41.
22. Rosenow F, Bast T, Czech T, Feucht M, Hans VH, Helmstaedter C, Huppertz H, Noachtar S, Oltmanns

F, Polster T, Seeck M, Trinka E, Wagner K, Strzelczyk A. Revised version of quality guidelines for presurgical epilepsy evaluation and surgical epilepsy therapy issued by the Austrian, German, and Swiss working group on presurgical epilepsy diagnosis and operative epilepsy treatment. Epilepsia. 2016;57(8):1215–20.
23. Winter F, Krueger MT, Delev D, Theys T, Van Roost DM, Fountas K, Schijns OEMG, Roessler K. Current state of the art of traditional and minimal invasive epilepsy surgery approaches. Brain Spine. 2024;4:102755.
24. Hori T, Kondo S, Takenobu A, Hirao J, Kohaya N, Takeuchi H, Watanabe T. Retrolabyrinthine presigmoid transpetrosal approach for selective subtemporal amygdalohippocampectomy. Neurol Med Chir (Tokyo). 1999;39(3):214–25.
25. Alonso Vanegas MA, Lew SM, Morino M, Sarmento SA. Microsurgical techniques in temporal lobe epilepsy. Epilepsia. 2017;58(S1):10–8.
26. Al-Otaibi F, Baeesa SS, Parrent AG, Girvin JP, Steven D. Surgical techniques for the treatment of temporal lobe epilepsy. Epilepsy Res Treat. 2012;2012:1–13.
27. Tubbs RS, Miller JH, Cohen-Gadol AA, Spencer DD. Intraoperative anatomic landmarks for resection of the amygdala during medial temporal lobe surgery. Neurosurgery. 2010;66(5):974–7.
28. Mathon B, Clemenceau S. Selective amygdalohippocampectomy via trans-superior temporal gyrus keyhole approach. Acta Neurochir. 2016;158(4):785–9.
29. Yang PF, Zhang HJ, Pei JS, Lin Q, Mei Z, Chen ZQ, Jia YZ, Zhong ZH, Zheng ZY. Keyhole epilepsy surgery: corticoamygdalohippocampectomy for mesial temporal sclerosis. Neurosurg Rev. 2016;39(1):99–108.
30. Muzumdar D, Patil M, Goel A, Ravat S, Sawant N, Shah U. Mesial temporal lobe epilepsy—an overview of surgical techniques. Int J Surg. 2016;36:411–9.
31. Spencer DD, Spencer SS, Mattson RH, Williamson PD, Novelly RA. Access to the posterior medial temporal lobe structures in the surgical treatment of temporal lobe epilepsy. Neurosurgery. 1984;15(5):667–71.
32. Marathe K, Alim-Marvasti A, Dahele K, Xiao F, Buck S, O'Keeffe AG, Duncan JS, Vakharia VN. Resective, ablative and radiosurgical interventions for drug resistant mesial temporal lobe epilepsy: a systematic review and meta-analysis of outcomes. Front Neurol. 2021;12:777845.
33. Shamim D, Nwabueze O, Uysal U. Beyond resection: neuromodulation and minimally invasive epilepsy surgery. Noro Psikiyatr Ars. 2022;59(Suppl 1):S81–90.
34. Wieser HG, Yaşargil MG. Selective amygdalohippocampectomy as a surgical treatment of mesiobasal limbic epilepsy. Surg Neurol. 1982;17(6):445–57.
35. Yasargil MG, Teddy PJ, Roth P. Selective amygdalohippocampectomy operative anatomy and surgical technique. In: Symon L, Brihaye J, Guidetti B, Loew F, Miller JD, Nornes H, Pásztor E, Pertuiset B, Yaşargil MG, editors. Advances and technical standards in neurosurgery, vol. 12. Vienna: Springer Vienna; 1985. p. 93–123. https://doi.org/10.1007/978-3-7091-7008-3_2.
36. Yeni SN, Tanriover N, Uyanik Ö, Ulu MO, Özkara Ç, Karaağaç N, Ozyurt E, Uzan M. Visual field defects in selective amygdalohippocampectomy for hippocampal sclerosis: the fate of Meyer's loop during the transsylvian approach to the temporal horn. Neurosurgery. 2008;63(3):507–15.
37. Liu Y, Ren Y, Meng Q, Wu H, Dong S, Li H, Liu X, Li K, Du C, Zhang H. Functional anterior temporal lobectomy for temporal lobe epilepsy: a novel, alternative, and less invasive therapy. Oper Neurosurg. 2023;24(1):111–8.
38. Usui N, Kondo A, Nitta N, Tottori T, Inoue Y. Surgical resection of amygdala and uncus. Neurol Med Chir (Tokyo). 2018;58(9):377–83.
39. Niemeyer P. The transventricular amygdalohippocampectomy in temporal lobe epilepsy. In: Baldwin M, Bailey P, editors. Temporal lobe epilepsy. Illinois: CC Thomas Springfield; 1958. p. 461–82.
40. Hori T, Tabuchi S, Kurosaki M, Kondo S, Takenobu A, Watanabe T. Subtemporal amygdalohippocampectomy for treating medically intractable temporal lobe epilepsy. Neurosurgery. 1993;33(1):50–6; discussion 56–57.
41. Park TS, Bourgeois BFD, Silbergeld DL, Dodson WE. Subtemporal transparahippocampal amygdalohippocampectomy for surgical treatment of mesial temporal lobe epilepsy: technical note. J Neurosurg. 1996;85(6):1172–6.
42. Little AS, Smith KA, Kirlin K, Baxter LC, Chung S, Maganti R, Treiman DM. Modifications to the subtemporal selective amygdalohippocampectomy using a minimal-access technique: seizure and neuropsychological outcomes: technical note. J Neurosurg. 2009;111(6):1263–74.
43. Delev D, Schramm J, Clusmann H. How I do it—selective amygdalohippocampectomy via a navigated temporobasal approach, when veins forbid elevation of the temporal lobe. Acta Neurochir. 2018;160(3):597–601.
44. Smith KA, Spetzler RF. Supratentorial—infraoccipital approach for posteromedial temporal lobe lesions. J Neurosurg. 1995;82(6):940–4.
45. Abramov I, Jubran JH, Houlihan LM, Park MT, Howshar JT, Farhadi DS, Loymak T, Cole TS, Pitskhelauri D, Preul MC. Multiple hippocampal transection for mesial temporal lobe epilepsy: a systematic review. Seizure. 2022;101:162–76.
46. Patel B, Kim AH. Laser interstitial thermal therapy. Mo Med. 2020;117(1):50–5.
47. Kerezoudis P, Parisi V, Marsh WR, Kaufman TJ, Lehman VT, Worrell GA, Miller KJ, Van Gompel JJ. Surgical outcomes of laser interstitial thermal therapy for temporal lobe epilepsy: systematic review and meta-analysis. World Neurosurg. 2020;143:527–536.e3.
48. Alomar SA, Moshref RH, Moshref LH, Sabbagh AJ. Outcomes after laser interstitial thermal ablation for temporal lobe epilepsy: a systematic review and meta-analysis. Neurosurg Rev. 2023;46(1):261.
49. Youngerman BE, Banu MA, Khan F, McKhann GM, Schevon CA, Jagid JR, Cajigas I, Theodotou CB, Ko

A, Buckley R, Ojemann JG, Miller JW, Laxton AW, Couture DE, Popli GS, Buch VP, Halpern CH, Le S, Sharan AD, Sperling MR, Mehta AD, Englot DJ, Neimat JS, Konrad PE, Sheth SA, Neal EG, Vale FL, Holloway KL, Air EL, Schwalb JM, D'Haese PF, Wu C. Long-term outcomes of mesial temporal laser interstitial thermal therapy for drug-resistant epilepsy and subsequent surgery for seizure recurrence: a multicentre cohort study. J Neurol Neurosurg Psychiatry. 2023;94(11):879–86.

50. Wang L, Jin W, Zhang Y, Wang S, Li Q, Qin J, Li Z, Cheng Y, Feng K, Yin S. Stereoelectroencephalography-guided radiofrequency thermocoagulation in drug-resistant focal epilepsy. Ann Transl Med. 2022;10(4):192.
51. Castellano JF, Singla S, Barot N, Aronson JP. Stereoelectroencephalography-guided radiofrequency thermocoagulation: diagnostic and therapeutic implications. Brain Sci. 2024;14(2):110.
52. Adin ME, Spencer DD, Damisah E, Herlopian A, Gerrard JL, Bronen RA. Imaging of neuromodulation and surgical interventions for epilepsy. AJNR Am J Neuroradiol. 2021;2021:ajnr;ajnr.A7222v1.
53. Spencer D, Gwinn R, Salinsky M, O'Malley JP. Laterality and temporal distribution of seizures in patients with bitemporal independent seizures during a trial of responsive neurostimulation. Epilepsy Res. 2011;93(2–3):221–5.
54. Nair DR, Laxer KD, Weber PB, Murro AM, Park YD, Barkley GL, Smith BJ, Gwinn RP, Doherty MJ, Noe KH, Zimmerman RS, Bergey GK, Anderson WS, Heck C, Liu CY, Lee RW, Sadler T, Duckrow RB, Hirsch LJ, Wharen RE, Tatum W, Srinivasan S, McKhann GM, Agostini MA, Alexopoulos AV, Jobst BC, Roberts DW, Salanova V, Witt TC, Cash SS, Cole AJ, Worrell GA, Lundstrom BN, Edwards JC, Halford JJ, Spencer DC, Ernst L, Skidmore CT, Sperling MR, Miller I, Geller EB, Berg MJ, Fessler AJ, Rutecki P, Goldman AM, Mizrahi EM, Gross RE, Shields DC, Schwartz TH, Labar DR, Fountain NB, Elias WJ, Olejniczak PW, Villemarette-Pittman NR, Eisenschenk S, Roper SN, Boggs JG, Courtney TA, Sun FT, Seale CG, Miller KL, Skarpaas TL, Morrell MJ, on behalf of the RNS System LTT Study. Nine-year prospective efficacy and safety of brain-responsive neurostimulation for focal epilepsy. Neurology. 2020;95(9):e1244–56. https://doi.org/10.1212/WNL.0000000000010154.
55. Bergey GK, Morrell MJ, Mizrahi EM, Goldman A, King-Stephens D, Nair D, Srinivasan S, Jobst B, Gross RE, Shields DC, Barkley G, Salanova V, Olejniczak P, Cole A, Cash SS, Noe K, Wharen R, Worrell G, Murro AM, Edwards J, Duchowny M, Spencer D, Smith M, Geller E, Gwinn R, Skidmore C, Eisenschenk S, Berg M, Heck C, Van Ness P, Fountain N, Rutecki P, Massey A, O'Donovan C, Labar D, Duckrow RB, Hirsch LJ, Courtney T, Sun FT, Seale CG. Long-term treatment with responsive brain stimulation in adults with refractory partial seizures. Neurology. 2015;84(8):810–7.
56. Fisher R, Salanova V, Witt T, Worth R, Henry T, Gross R, Oommen K, Osorio I, Nazzaro J, Labar D, Kaplitt M, Sperling M, Sandok E, Neal J, Handforth A, Stern J, DeSalles A, Chung S, Shetter A, Bergen D, Bakay R, Henderson J, French J, Baltuch G, Rosenfeld W, Youkilis A, Marks W, Garcia P, Barbaro N, Fountain N, Bazil C, Goodman R, McKhann G, Babu Krishnamurthy K, Papavassiliou S, Epstein C, Pollard J, Tonder L, Grebin J, Coffey R, Graves N, the SANTE Study Group. Electrical stimulation of the anterior nucleus of thalamus for treatment of refractory epilepsy. Epilepsia. 2010;51(5):899–908.
57. Abdennadher M, Rohatgi P, Saxena A. Vagus nerve stimulation therapy in epilepsy: an overview of technical and surgical method, patient selection, and treatment outcomes. Brain Sci. 2024;14(7):675.
58. Englot DJ, Rolston JD, Wright CW, Hassnain KH, Chang EF. Rates and predictors of seizure freedom with vagus nerve stimulation for intractable epilepsy. Neurosurgery. 2016;79(3):345–53.

Hippocampal Changes in Hydrocephalus

13

Iddi Kassim Mkopi

13.1 Introduction

The medial temporal lobe contains the hippocampus, a region of the brain. It was named after a seahorse because of its similar appearance when grossly dissected. The brain uses the hippocampus to create a spatiotemporal framework that houses different sensory, emotional, and cognitive elements, which is crucial for memory formation. The dentate gyrus, which is important for pattern separation; the cornu ammonis, which is central to pattern completion; and the subiculum, which is important for memory retrieval, are the three regions that make up the hippocampus [1]. There has also been evidence linking the hippocampus to cognition [2, 3].

The abnormal dilatation of the cerebral ventricles caused by a blockage in the flow or malabsorption of cerebrospinal fluid (CSF) is referred to as hydrocephalus. The etiology of hydrocephalus can be genetic, congenital, or due to some central nervous system infections. In hydrocephalus, ventricles dilate, and pressure inside the cranium increases, leading to symptoms including headache, vomiting, nausea, loss of consciousness, or even coma. As the hydrocephalic clinic continues, CSF dynamics normalize with the increase in the size of the cerebral ventricles. This condition has to be intervened as soon as possible according to the clinical symptoms. If it remains untreated, it can result in cognitive problems, attention deficits, information processing issues, and executive function impairments similar to subcortical dementia. Hydrocephalus can also cause swelling of dendrites and axons, a decrease in dendritic spines and branching, or even a reduction of neuronal dendrites in the cortex and hippocampus. Clinical signs and symptoms of hydrocephalus differ with chronicity and usually present as cognitive impairment.

Dendritic spine densities on cortical and hippocampal pyramidal neurons have been demonstrated to be reduced in certain hydrocephalic patient investigations. These findings are consistent with decreased expressions of synaptophysin and postsynaptic density protein 95. Reduced excitatory connections were seen in the cortical and hippocampal regions, which may be the reason for learning and memory impairments in hydrocephalus patients.

In some studies of hydrocephalic patients, it has been shown that dendritic spine densities on both cortical and hippocampal pyramidal neurons were decreased, corresponding to the findings that the expressions of both synaptophysin and postsynaptic density protein 95 were reduced. The findings of cortical and hippocampal changes indicated reductions in excitatory connectivity,

I. K. Mkopi (✉)
Department of Neurosurgery, Nizamiye Hospital, Abuja, Nigeria
e-mail: iddik20@yahoo.com

M. Turgut et al. (eds.), *The Human Hippocampus*, https://doi.org/10.1007/978-3-031-88907-3_13

which could be the cause of learning and memory deficits in hydrocephalus.

It is well known that hydrocephalus can induce axonal degeneration and demyelination of periventricular white matter [4–6] and change cerebral neurotransmitter levels [7]. Nevertheless, whether hydrocephalus affects central synaptic density and neuronal structure remains largely unknown.

It is also well known that hydrocephalus affects the amounts of neurotransmitters in the brain [7], causes periventricular white matter demyelination, and leads to axonal degeneration. However, it is still mostly unclear if hydrocephalus has an impact on neuronal structure and central synaptic density.

13.2 Discussion

The neurological disorder known as hydrocephalus is characterized by abnormalities in CSF flow, which can result in an accumulation of CSF inside the skull. Changes in the cortex, subcortex, and the volume and structure of the hippocampal regions can occur in hydrocephalus. A number of factors, such as the genesis of hydrocephalus, the size of the ventricles, and the onset of the changes, affect the neuropathological changes associated with hydrocephalus. These changes result from both physical and metabolic abnormalities. The cranium's brain tissue is compressed as the ventricles get bigger, raising intracranial pressure and causing brain injury. Surgery, including ventricular shunting and/or endoscopic third ventriculostomy, is one method of treating hydrocephalus. Endoscopic third ventriculostomy (ETV), but these kinds of surgeries are associated with both similar and different complications. Nonsurgical pharmacological approaches for treatment are being widely investigated through preclinical animal models of hydrocephalus. In this chapter, we are going to discuss some changes that occur in the hippocampus due to hydrocephalus. Hippocampal volume segmentation can be used to study the relationship between hippocampal volume and various neurological and psychiatric disorders. For example, studies have shown that hippocampal volume is reduced in people with Alzheimer's disease, schizophrenia, and depression. Damage to the hippocampus can lead to profound memory deficits, such as those seen in patients with Alzheimer's disease or other forms of amnesia.

Here are some notable changes that can occur in the hippocampus due to hydrocephalus:

1. **Compression and Atrophy**: The increased CSF pressure can compress the hippocampus, leading to tissue damage and atrophy over time. This compression can disrupt the normal architecture of the hippocampus, affecting its function.
2. **Neuronal Damage**: The pressure from accumulated CSF can cause neuronal damage within the hippocampus. Neurons may undergo degeneration or dysfunction, impairing the hippocampus's ability to process and store memories.
3. **Functional Impairments**: Damage to the hippocampus can result in significant cognitive impairments, particularly in memory formation and spatial navigation. Patients with hydrocephalus may have trouble learning new information or recalling past events.
4. **Vascular Changes**: The increased pressure can also affect the blood supply to the hippocampus, leading to reduced oxygen and nutrient delivery. This ischemic condition can further contribute to hippocampal damage and functional deficits.
5. **Gliosis**: Chronic hydrocephalus can trigger reactive gliosis, where glial cells (support cells in the brain) become activated and proliferate. This response is part of the brain's attempt to repair damage but can also contribute to tissue scarring and further disruption of hippocampal function.
6. **Impact on Neurogenesis**: The hippocampus is among the few parts of the adult brain where neurogenesis (formation of new neurons) occurs. Hydrocephalus-related changes can disrupt this process, further compromising the hippocampus's ability to regenerate and adapt.
7. **Volume Changes**: Hydrocephalus can lead to changes in the size and shape of the hippo-

campus. Studies have shown that in some cases of hydrocephalus, the hippocampus may be compressed or displaced due to the accumulation of CSF.

These changes in the hippocampus can vary depending on the severity, duration, and type of hydrocephalus (such as congenital or acquired). Early diagnosis and treatment of hydrocephalus are crucial to minimizing brain damage and preserving cognitive function, including interventions like shunt placement or other surgical procedures aimed at relieving CSF pressure.

Research continues to explore the mechanisms behind hippocampal changes in hydrocephalus and potential therapeutic strategies to mitigate these effects and improve patient outcomes.

Severe hydrocephalus can finally cause swelling of the dendrites and axons, a decrease in dendritic spines and branching, or even a reduction of neuronal dendrites in the cortex and hippocampus [2], correlating with an accumulation or loss of synaptic vesicle proteins (such as synaptophysin) and a loss of synapses [7].

For the time being, in the hydrocephalic condition, neurochemical changes in the septohippocampal system of hydrocephalic patients [8, 9] and animals are associated with memory and learning problems [10–13]. Nevertheless, one research study found that memory dysfunction in hydrocephalic adults was associated with septohippocampal changes when it was due to aqueductal stenosis, while memory deficits in normal-pressure hydrocephalus seemed related to prefrontal structural damage [14].

In another study that investigated the abnormalities of hippocampal shape in idiopathic normal-pressure hydrocephalus (INPH) patients, interesting results were found regarding hippocampal changes.

A vertex-wise shape analysis study showed significant bilateral deformity in sufferers with INPH, distinguished from healthy controls after FDR correction. Both bilateral areas of the hippocampus were more elongated and narrowed in INPH patients than in the healthy controls. In both hemispheres, INPH sufferers demonstrated an upward bending of the middle and posterior hippocampus, indicating a bilateral upright dislocation from body to tail. Their results suggested that subjects with INPH exhibit various patterns of hippocampal shape changes [15].

In another study in a porcine model of acquired hydrocephalus, it has been shown that the width of the hippocampus and the thickness of the perirhinal cortex decreased in hydrocephalic pigs after some days following induction. Also, part of the dorsal hippocampus showed a reduction of 23.4% in the hydrocephalic pigs at sacrifice. These results showed that acquired hydrocephalus caused morphological changes, reduced neurogenesis, and increased reactive astrocytosis in the hippocampus and perirhinal cortex [16].

In another research study that examined the effects of vanadium on memory and spatial learning in hydrocephalic mice, the results showed that this chemical element has a dose-dependent protective effect on the pyramidal cells of the hippocampus and on memory and spatial learning functions in juvenile hydrocephalic mice [17].

Furthermore, hydrocephalus has been linked to tissue compression, periventricular white matter ischemia, and reduced cerebral blood flow. Studies have also shown that periventricular white matter undergoes oxidative, hypoxic, and nitrosylative changes [18].

Some animal studies show diminished cerebral blood flow and ischemic changes in hydrocephalus, in addition to changes in oxidative metabolism in the cortex and subcortical domains, including the hippocampus [19–21], which are more prominent during the early stages while ventricular enlargement is actively progressing [22, 23].

Hydrocephalus affects the cerebral cortex and subcortical structures, especially as ventriculomegaly increases further. Stretching of the septum pellucidum results in cortical thinning and distension [24].

Severe hydrocephalus can cause a decrease in dendrites and axons, dendritic spines, and branching numbers, and even lead to the disappearance of neuronal dendrites in the cortex and hippocampus, resulting in atrophy of the cerebral cortex and basal ganglia [3]. Chronic severe hydrocephalus is also associated with reactive astrogliosis in the cortex and neurofibrillary tangles in cortical, hippocampal, and brainstem neurons.

Experimental animal studies also show that there are changes in oxidative metabolism in the hippocampus as well as ischemic changes caused by decreased cerebral blood flow in hydrocephalus [24].

13.3 Conclusion

Hydrocephalus is an affecting animate neurological condition resulting from changes in CSF flow, leading to a build-up of CSF inside the skull. The neuropathological changes associated with hydrocephalus arise from both physical forces and metabolic changes, and they vary due to different factors, including the onset of symptoms, ventricular enlargement, and etiology.

Various studies have demonstrated that in hydrocephalic conditions, many changes occur in the size and shape of the hippocampus. Hydrocephalus causes morphological changes, lowers neurogenesis, and increases reactive astrocytosis in the hippocampus and perirhinal area of the brain.

Hydrocephalus has been shown to be one of the causes of memory disabilities and learning problems due to its effects on the neurons of the hippocampus, particularly pyramidal neurons.

Further research is needed to explore changes in the size, shape and function of the hippocampus due to various forms of hydrocephalus, in order to identify appropriate interventions for the condition.

References

1. Kahhale I, Buser NJ, Madan CR, et al. Quantifying numerical and spatial reliability of hippocampal and amygdala subdivisions in FreeSurfer. Brain Inf. 2023;10:9.
2. Borit A, Sidman RL. New mutant mouse with communicating hydrocephalus and secondary aqueductal stenosis. Acta Neuropathol. 1972;21:316–31.
3. McAllister JP II, Maugans TA, Shah MV, Truex RC Jr. Neuronal effects of experimentally induced hydrocephalus in newborn rats. J Neurosurg. 1985;63:776–83.
4. Del Bigio MR. Hydrocephalus and periventricular white matter water content. J Neurosurg. 1987;67:955–6.
5. Del Bigio MR. Neuropathology and structural changes in hydrocephalus. Dev Disabil Res Rev. 2010;16:16–22.
6. Del Bigio MR, da Silva MC, Drake JM, Tuor UI. Acute and chronic cerebral white matter damage in neonatal hydrocephalus. Can J Neurol Sci. 1994;21:299–305.
7. Del Bigio MR, Zhang YW. Cell death, axonal damage, and cell birth in the immature rat brain following induction of hydrocephalus. Exp Neurol. 1998;154:157–69.
8. Barf HA, Verhoef M, Jennekens-Schinkel A, Post MW, Gooskens RH, Prevo AJ. Cognitive status of young adults with spina bifida. Dev Med Child Neurol. 2003;45:813–20.
9. Vachha B, Adams RC, Rollins NK. Limbic tract anomalies in pediatric myelomeningocele and Chiari II malformation: anatomic correlations with memory and learning—initial investigation. Radiology. 2006;240:194–202.
10. Hawkins D, Bowers TM, Bannister CM, Miyan JA. The functional outcome of shunting H-Tx rat pups at different ages. Eur J Pediatr Surg. 1997;7:31–4.
11. Jones HC, Rivera KM, Harris NG. Learning deficits in congenitally hydrocephalic rats and prevention by early shunt treatment. Childs Nerv Syst. 1995;11:655–60.
12. Miyazawa T, Sato K. Learning disability and impairment of synaptogenesis in HTX-rats with arrested shunt-dependent hydrocephalus. Childs Nerv Syst. 1991;7:121–8.
13. Shim I, Ha Y, Chung JY, Lee HJ, Yang KH, Chang JW. Association of learning and memory impairments with changes in the septohippocampal cholinergic system in rats with kaolin-induced hydrocephalus. Neurosurgery. 2003;53:416–25.
14. Donnet A, Schmitt A, Dufour H, Giorgi R, Grisoli F. Differential patterns of cognitive impairment in patients with aqueductal stenosis and normal pressure hydrocephalus. Acta Neurochir. 2004;146:1301.
15. Kang K, Kim J, Yoon U, Park K-S, Jeong SY, Lee S-W. Abnormalities of hippocampal shape in idiopathic normal-pressure hydrocephalus patients. Alzheimers Dementia. 2023;19:e075480.
16. Garcia-Bonilla M, Nair A, Moore J, Castaneyra-Ruiz L, Zwick SH, Dilger RN, Fleming SA, Golden RK, Talcott MR, Isaacs AM, Limbrick DD Jr, McAllister JP II. Impaired neurogenesis with reactive astrocytosis in the hippocampus in a porcine model of acquired hydrocephalus. Exp Neurol. 2023;363:114354.

17. Femi-Akinlosotu OM, Olopade FE, Obiako J, Olopade JO, Shokunbi MT. Vanadium improves memory and spatial learning and protects the pyramidal cells of the hippocampus in juvenile hydrocephalic mice. Front Neurol. 2023;14:1116727.
18. Del Bigio MR, Khan OH, da Silva Lopes L, Packiasamy ARJ. Cerebral white matter oxidation and nitrosylation in young rodents with kaolin-induced hydrocephalus. J Neuropathol Exp Neurol. 2012;71:274–88.
19. Chumas PD, Drake JM, Del Bigio MR, Da Silva M, Tuor UI. Anaerobic glycolysis preceding white-matter destruction in experimental neonatal hydrocephalus. J Neurosurg. 1994;80:491–501.
20. Higashi K, Asahisa H, Ueda N, Kobayashi K, Hara K, Noda Y. Cerebral blood flow and metabolism in experimental hydrocephalus. Neurol Res. 1986;8:169–76.
21. Socci D, Bjugstad KB, Jones HC, Pattisapu JV, Arendash GW. Evidence that oxidative stress is associated with the pathophysiology of inherited hydrocephalus in the H-Tx rat model. Exp Neurol. 1999;155:109–17.
22. Hochwald GM. Animal models of hydrocephalus: recent developments. Proc Soc Exp Biol Med. 1985;178:1–11.
23. Hochwald GM, Boal RD, Marlin AE, Kumar AJ. Changes in regional blood-flow and water content of brain and spinal cord in acute and chronic experimental hydrocephalus. Dev Med Child Neurol. 1975;35:42–50.
24. Di Curzio D. Neuropathological changes in hydrocephalus—a comprehensive review. Open J Modern Neurosurg. 2018;8:1–29. https://doi.org/10.4236/ojmn.2018.81001.

14 Importance of the Hippocampus in Anxiety-Related Behaviors, Aging, and Alzheimer Disease

Gonca Aşut and Yunus Hacımusalar

14.1 Introduction

14.1.1 Overview of Anxiety: Understanding Physiological and Psychological Aspects

Anxiety is an emotion marked by heightened awareness and sensitivity to threats, prompting defensive behaviors to prevent harm [1]. It includes both trait and state components: trait anxiety is a long-term predisposition, while state anxiety is a short-lived reaction to immediate dangers. Although state anxiety aids in avoiding threats, persistent trait anxiety can become pathological [2, 3].

Anxiety relates to future dangers and is accompanied by worry and tension. Fear, on the other hand, is an emotional response to an imminent threat that triggers a "fight-or-flight" reaction characterized by increased arousal and physiological changes [4, 5]. According to Davis et al., fear is a short-term (phasic) response to specific, often conditioned, aversive cues, while anxiety is a longer-term (tonic) response to vague, often unconditioned, aversive cues or situations [6]. McNaughton and Gray propose that fear triggers active avoidance or escape behaviors, whereas anxiety arises from diffuse aversive stimuli involving conflict and uncertainty. They suggest that fear can lead to anxiety, with anxiety systems inhibiting fear systems [7]. From an evolutionary perspective, both fear and anxiety are adaptive responses. Adaptive anxiety helps navigate environments safely, but excessive anxiety leads to maladaptive avoidance behaviors, which are common in anxiety disorders [8, 9].

In childhood and adolescence, anxiety often appears as part of normal development, making it challenging to identify and diagnose pathological anxiety. During this sensitive phase, temporary neurobiological, cognitive, affective, and behavioral dysfunctions can escalate, leading to psychopathologic anxiety with distress, impairment, and negative developmental outcomes [10, 11].

Global cross-sectional surveys indicate that anxiety disorders, such as generalized anxiety disorder (GAD), panic disorder (PD), specific phobias, and separation anxiety disorder (SAD), have the highest prevalence among mental disorders, with up to 30% of people experiencing them in their lifetime and 14% in the past year [10, 12]. Epidemiological studies show higher rates of anxiety in women compared to men [12, 13], though the neurobiological reasons for this are not well understood. Anxiety symptoms also generally increase with age, with GAD being

G. Aşut (✉)
Department of Psychiatry, Başkent University Faculty of Medicine, Ankara, Türkiye
e-mail: gasut@baskent.edu.tr

Y. Hacımusalar
Department of Psychiatry, Private Practice, Kayseri, Türkiye

M. Turgut et al. (eds.), *The Human Hippocampus*, https://doi.org/10.1007/978-3-031-88907-3_14

more common in older adults than in younger ones [6, 12]. These disorders arise from a mix of genetic susceptibility and psychosocial stressors like early-life adversity. Both DSM-5 and ICD-11 have reclassified obsessive-compulsive and trauma-related disorders separately from anxiety disorders while still emphasizing their symptom overlap [11].

Clinically, anxiety disorders often involve overgeneralized fear, and it is crucial to differentiate normal functional anxiety from pathological anxiety for the management of these disorders. Studies suggest a link between impaired pattern separation and anxiety disorders, possibly due to the overgeneralization of fear [6, 8]. We will look at these studies in more detail later in this chapter.

14.1.2 Significance of Studying the Hippocampus: Why Is the Hippocampus a Key Area of Interest in Anxiety Research?

The hippocampus, a vital component of the limbic system, has long been recognized for its role in memory formation and spatial navigation. However, recent studies of hippocampal neuroplasticity, neurotransmitter systems, and its interactions with other brain regions have shown that its importance extends beyond these functions, making it a key area of interest in anxiety research.

The hippocampus is an important site of neuroplasticity. Hippocampal neuroplasticity is influenced by various factors, including stress, environmental stimuli, and pharmacological agents [14]. Chronic stress, a well-known trigger of anxiety disorders, adversely affects hippocampal plasticity and neurogenesis, particularly in the dentate gyrus [15, 16]. This results in structural and functional alterations that are associated with anxiety-like behaviors. Understanding the mechanisms underlying hippocampal neurogenesis and its disruption in anxiety disorders can provide insights into developing strategies to enhance resilience and recovery through targeted treatments.

The hippocampus is also rich in various neurotransmitter systems, including serotonergic, glutamatergic, GABAergic, and noradrenergic systems, each contributing to anxiety regulation [17]. Abnormalities in these neurotransmitter systems within the hippocampus have been implicated in the pathophysiology of anxiety disorders. By studying these systems, researchers can better understand the neurochemical underpinnings of anxiety and develop pharmacological treatments that specifically target these pathways.

In summary, the hippocampus is a crucial area of interest in anxiety research due to its role in neuroplasticity, extensive neural interactions, and rich neurotransmitter milieu. Investigating the hippocampus's functions and its dysregulation in anxiety disorders offers promising avenues for understanding the etiology of these conditions and developing effective therapeutic interventions.

14.2 Anatomy and Function of the Hippocampus

14.2.1 General Anatomy of the Hippocampus

The hippocampus is part of the limbic system and is located in the medial temporal lobe. The hippocampal region comprises the hippocampal formation and the parahippocampal region. It appears as two interlocking Cs resembling a sea horse on coronal plane images. This distinctive shape inspired the name "hippocampus." In adult humans, the hippocampus has a volume of about 3.0–3.5 cm^3 [18, 19].

In rodents, the hippocampal formation is divided into the dentate gyrus (DG), hippocampus proper (CA3, CA2, CA1), and subiculum (SUB). In humans, the hippocampus consists of DG, CA1, CA2, CA3, CA4, and SUB regions. On coronal anatomic and histologic sections, the subdivisions CA1, CA2, CA3, and CA4 are identifiable [20–22].

The parahippocampal gyrus (PG), which is positioned beneath the hippocampus, is a six-layered cortical area that consists of the presubiculum, parasubiculum, entorhinal cortex (EC),

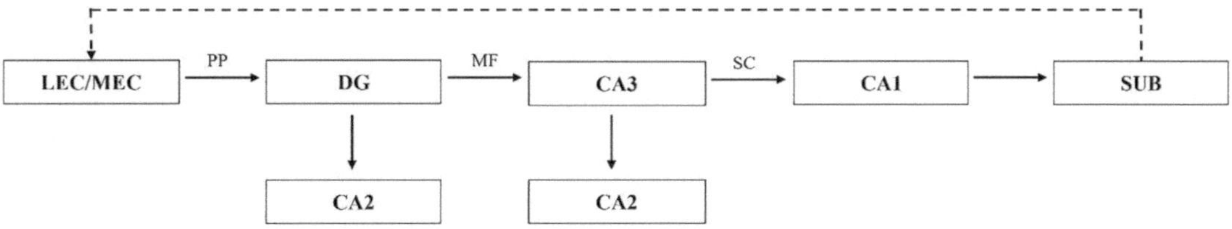

Fig. 14.1 *LEC* lateral entorhinal cortex, *MEC* medial entorhinal cortex, *DG* dentate gyrus, *CA* cornu ammonis, *SUB* subiculum, *PP* perforant pathway, *MF* mossy fibers, *SC* Schaffer collateral pathway

perirhinal cortex, and, in non-primate mammals, the postrhinal cortex (known as the parahippocampal cortex in primates) [20]. The anterior part of the PG bends over and rests on itself and forms the uncus [23].

Research in rodents has shown that the hippocampus is divided into the dorsal (posterior in primates) and ventral (anterior in primates) regions with distinct functions. The dorsal hippocampus (dHPC), starting at the septal pole, is crucial for spatial learning and memory, while the ventral hippocampus (vHPC), beginning at the temporal pole, significantly influences anxiety-related behaviors [6, 9]. This function differs from the amygdala, which is more associated with fear [24, 25]. To understand the arrangement of the subfields in the anterior hippocampus in primates, we need to be aware of its three-dimensional structure. The main body of the hippocampus curves medially in its anterior portion. This forms the extraventricular part of the anterior hippocampus, which is located within the uncus. Therefore, CA1 and SUB must bend around a wider radius than the other subfields. As a result, the anterior part of the hippocampus consists mainly of CA1 and SUB, while the posterior part contains only the DG, CA2, and CA3 [23].

14.2.2 Hippocampal Connectivity

The hippocampus's connectivity forms a "trisynaptic loop" [26]. The DG acts as the main entry point to the hippocampus, receiving inputs from various brain regions. Its principal neurons, the granule cells (GCs) in the stratum granulosum, receive excitatory signals from the lateral and medial entorhinal cortex (LEC/MEC) layer II via perforant path axons and send projections to the CA3 region through mossy fibers (MF). CA3 cells then project to CA1 via the Schaffer collateral (SC) pathway, which, in turn, projects to SUB. Integrating inputs from CA1 and the EC layer III, the SUB sends information along the hippocampal output pathways, such as the prefrontal cortex (PFC) and hypothalamus [19]. DG and CA3 neurons also project to CA2 [26] (Fig. 14.1).

The vHPC (especially vSUB and vCA1) has distinct connections, projecting to the PFC and closely linking with the bed nucleus of the stria terminalis (BNST) and the amygdala, which are associated with the hypothalamic-pituitary-adrenal (HPA) axis [21, 24, 27]. These connections suggest a role for the vHPC in anxiety and fear regulation.

14.2.3 Neurotransmitter Systems in the Hippocampus

Hippocampal neurons consist of glutamatergic excitatory cells (GCs in the DG and pyramidal cells in the hippocampus proper and SUB) and GABAergic inhibitory interneurons [19]. Place cells contribute to spatial representation and are more abundant, stable, and finely tuned in the dHPC than in the vHPC [6]. In the parahippocampal region, excitatory neurons are mainly glutamatergic pyramidal cells, while inhibitory GABAergic neurons differ in morphology, neurochemistry, and connections. Pyramidal cells are primarily located in layers II, III, V, and VI, while GABAergic interneurons are distributed throughout all six layers [20].

GABAergic interneurons contribute to the sparse neural activity of GCs, which is essential for pattern separation and anxiety responses [28]. GABA also plays an important role in controlling hippocampal neurogenesis. GABAergic interneurons establish initial connections with adult-born GCs before the onset of glutamatergic synaptogenesis, which is crucial for the maturation and integration of newborn GCs [29, 30]. An imbalance between GABA and glutamate neurotransmission may cause deficits in the proper integration of newborn GCs, contributing to psychiatric disorders, including anxiety [6].

Hilar glutamatergic mossy cells directly project to mature GCs and indirectly affect them via local GABAergic interneurons [6]. Mossy cells excite both GCs and inhibitory GABAergic neurons, but the net effect on GCs is inhibitory [31]. This supports the role of hilar glutamatergic neurons in the DG's sparse activity and pattern separation, contributing to the anxiolytic effects of mossy cells [32].

The hippocampus receives dense serotonergic innervation from the median raphe nucleus (MRN), which primarily targets the DG and CA1/CA3 fields, while those from the dorsal raphe nucleus (DRN) preferentially target GABAergic interneurons in the hippocampus proper and ventral SUB [20]. Several 5-HT receptors, including 5-HT1A, 5-HT1B, 5-HT2C, 5-HT3, and 5-HT4, and the 5-HT transporter (SERT), regulate 5-HT neurotransmission in the hippocampus [33].

Techniques such as immunohistochemistry, autoradiography, and in situ hybridization have been used to explore 5-HT receptor distribution [20]. Serotonin regulates CA1 neuronal excitability and reduces anxiety by suppressing vHPC inputs to the mPFC via presynaptic 5-HT1B receptors [34]. Additionally, 5-HT hyperpolarizes CA1 pyramidal neurons through 5-HT1A receptors, but chronic high corticosterone exposure weakens this response due to a decreased mineralocorticoid receptor/glucocorticoid receptor (MR/GR) mRNA ratio, impairing 5-HT and stress hormone interactions [6]. The 5-HT2C receptor in the hippocampus influences mood, motor behavior, memory, and conditions such as anxiety and depression. Its mRNA follows a diurnal rhythm, peaking in the morning, which is suppressed by chronic stress [20]. This may suggest that increasing MR expression could effectively treat anxiety disorders.

Additionally, 5-HT modulates acetylcholine (ACh) and noradrenaline (NE) release in the hippocampus. The 5-HT1A receptor excites ACh release, while 5-HT1B receptors inhibit it, likely through direct presynaptic influence on cholinergic terminals [20]. A selective serotonin reuptake inhibitor (SSRI) fluoxetine reduces K^+-evoked NE release, and this effect is abolished by the 5-HT3 antagonist ondansetron [35]. Additionally, 5-HT4 receptors facilitate neurotransmission and neurogenesis in the DG [36].

14.3 Hippocampus and Anxiety: Mechanisms and Pathways

The hippocampus plays a critical role in anxiety, first functioning as part of the Behavioral Inhibition System (BIS), which governs responses to threats and punishment, influencing individual anxiety levels. Gray's theory suggested that the BIS controls avoidance behaviors, with excessive activity leading to higher anxiety-related behaviors [37].

The hippocampus and amygdala mediate different aspects of anxiety [7] and promote fight/flight/freeze behaviors [38]. Anxiety disorders, characterized by excessive or persistent fear, often involve impaired fear extinction and deficits in contextual processing [6]. The hippocampus is crucial for processing contextual fear. The DG, CA3, and CA1 regions are involved in acquiring contextual fear; the DG and CA1 also play roles in retrieving and extinguishing contextual fear memory; and the DG is critical for fear generalization and discrimination [28, 39].

Fear response generalization, while adaptive, can become maladaptive in anxiety disorders. Proper discrimination among similar experiences is essential to prevent inappropriate responses, a process involving hippocampal pattern separation [40]. This separation helps encode details of environments, objects, or events to distinguish between similar memories, primarily supported by the DG-CA3 circuit [8]. Impaired pattern separation may contribute to the development or persistence of anxiety disorders [9].

Rodent studies reveal that both innate and learned fear and anxiety responses are linked to hippocampal activity. Suppressing fear reduces hippocampal excitability, whereas anxiogenic environments increase excitability in DG GCs [6]. Specific DG granule cell populations, or "fear engram cells," are activated during fear acquisition and extinction [41]. The competition between these neurons determines fear expression post-extinction, indicating potential targets for anxiety disorder treatments.

14.3.1 Experimental Evidence Linking the Hippocampus to Anxiety

14.3.1.1 Animal Models of Anxiety

Animal models of anxiety often use conflict-based avoidance tasks and open-field tests, especially in rodents. The elevated plus maze (EPM) is a well-established test for assessing innate anxiety, providing insights into how different brain regions and processes contribute to anxiety-related behaviors and anxiety modulation.

The EPM consists of four arms arranged in a plus shape: two enclosed arms with walls and two open arms exposed to the environment [42]. Wild-type mice typically show a preference for enclosed arms, reflecting an aversion to open spaces. This behavior is considered a manifestation of anxiety [43]. Numerous selective lesion studies have demonstrated that the mPFC [44] and vHPC [24, 45] are critical for normal anxiety-related behaviors observed in the EPM. Interestingly, research suggests a functional distinction between the vHPC and dHPC in anxiety processing. Studies have shown that damage to the vHPC, but not the dHPC, significantly disrupts anxiety-like behaviors in the EPM [24, 46]. Furthermore, studies have identified specific "anxiety cells" within the vCA1 subregion of the vHPC that exhibit increased activity during exploration of the open arms, suggesting their role in representing anxiety-related information [47].

Another rodent anxiety model is the cold water swim stress test. In this test, stress increases DG GC activity and anxiety-like behavior. Schoenfeld et al. showed that despite enhanced neurogenesis from exercise, new born neurons did not alleviate anxiety-like behavior or reduce DG activity. Instead, exercise increased local inhibitory mechanisms in the ventral DG, reducing GC excitability, suggesting that local inhibition plays a key role in anxiety regulation independent of neurogenesis [48, 49].

Optogenetic studies also have shown that stimulating the projections from vCA1 to subcortical structures such as the basolateral amygdala (BLA) and lateral hypothalamus (LHA) can directly influence anxiety-like behaviors in real time [50]. Additionally, rats with selective vHPC lesions show increased social interaction, quicker transitions in anxiety-provoking environments, and reduced hyponeophagia [51].

Anxiety is a complex human experience influenced by numerous factors beyond biology. It is important to note that animal models, while valuable, have limitations. Future research can explore how these findings in rodents translate to a better understanding and treatment of anxiety disorders in humans.

14.3.1.2 Genetic and Developmental Studies

The heritability of gray matter volume in the hippocampus ranges from 40% to 69%, indicating a genetic influence on hippocampal structure [52]. Increased gray matter volume in the right hippocampus, suggesting heightened activity in this region, is linked to higher levels of BIS-based anxiety [38]. Studies have found a positive association between high *sensitivity to punishment scale* scores and hippocampal volume [53], supporting the idea that structural variations in hippocampal volume, particularly the relative sizes of the left and right hippocampi, may predispose individuals to BIS-related anxiety [38, 53, 54].

The serotonin1A receptor (5-HT1AR) plays an important role in modulating anxiety-like behaviors. The 5-HT1AR is expressed as an autoreceptor on *serotonin-containing neurons* in the raphe nuclei and as a heteroreceptor on *non-serotonin-containing neurons* in the forebrain, including the hippocampus, septum, and cortex [55]. Activation of 5-HT1AR in both neuronal populations leads to membrane hyperpolarization and reduced neuronal excitability through coupling to the GIRK2 potassium channel for hippocampal heteroreceptors [17]. The function of 5-HT1AR during early development is crucial for establishing normal anxiety-like behavior in adulthood. Interestingly, knockout phenotypes are not mimicked by removing or blocking 5-HT1AR function in adult animals, implying that these receptors function during early development to establish normal anxiety-like behavior in adulthood [55].

5-HT1AR agonists demonstrate anxiolytic properties in both human and animal models [55]. Intracellular recordings in hippocampal slices have shown that 5-carboxytryptamine (5-CT), a 5-HT1AR full agonist, affects CA1 pyramidal neurons, highlighting the receptor's role in modulating neuronal activity and anxiety-like behavior. 5-HT1AR knockout mice exhibit increased anxiety-like behaviors across various tests [55, 56].

Reduced brain-derived neurotrophic factor (BDNF) levels decrease dendritic complexity in DG neurons, impair fear extinction, and increase anxiety-like behavior [6]. Deletion of TrkB in adult progenitors has similarly been shown to impact dendritic growth and synaptic plasticity, leading to heightened anxiety [57]. Studies have shown that parental morphine exposure causes dendritic retraction in the offspring's DG, which is associated with increased anxiety-like behavior in rats [58]. Since hippocampal-dependent memory requires both synaptic and dendritic plasticity, alterations in synaptic and dendritic plasticity can affect anxiety behavior by impairing fear extinction [6].

These findings highlight the complex interaction between genes, early development, and hippocampal function in shaping vulnerability to anxiety. Future research can explore how these factors interact and identify potential targets for intervention to prevent or reduce anxiety in vulnerable individuals.

14.3.2 Role of the Ventral Hippocampus in Anxiety

Recent studies highlight the vHPC's significant role in anxiety-related behaviors [6] through its influence on downstream targets like the mPFC and the amygdala [46]. vHPC—mPFC connection is essential for anxiety regulation, as the mPFC receives direct projections from the vHPC, amygdala, and other limbic structures [59] but only indirect inputs from the dHPC [46]. The vHPC contains "place cells" with larger fields than those in the dHPC, potentially guiding emotional behavior through less spatially discrete stimuli. It provides contextual information to the mPFC, which modulates anxiety and decision-making related to defensive and exploratory behaviors [60].

Lesion studies have provided further insights into the vHPC's role in anxiety and fear. Damaging the vHPC disrupts contextual fear conditioning, similar to lesions of the entire hippocampus [24]. Interestingly, lesions in the dHPC, which are more involved in spatial learning, do not affect fear conditioning [25]. This suggests that the vHPC plays a specific role in emotional learning and fear processing.

Ventral hippocampal projections also reach the hypothalamus, a region involved in stress regulation. These connections play a role in modulating anxiety responses. Disruption of this pathway can lead to increased stress hormone levels and overactivity of the HPA axis [61]. In contrast, the ventral SUB projects to the paraventricular nucleus (PVN) of the hypothalamus and helps to buffer stress responses [62, 63]. This highlights the complex interplay between the vHPC and the hypothalamus in regulating anxiety and stress (Fig. 14.2).

In experiments, vCA1 neurons showed increased activity in anxiogenic environments, like the open arms of the EPM, correlating with the baseline anxiety levels of individual animals. Conversely, dCA1 neurons did not show significant changes in activity. In addition, these neurons in the vCA1 were selective for anxiety-inducing zones (like the center zone of the Open-Field Test), suggesting the existence of cells that exhibit stable representations of anxiety-related information or

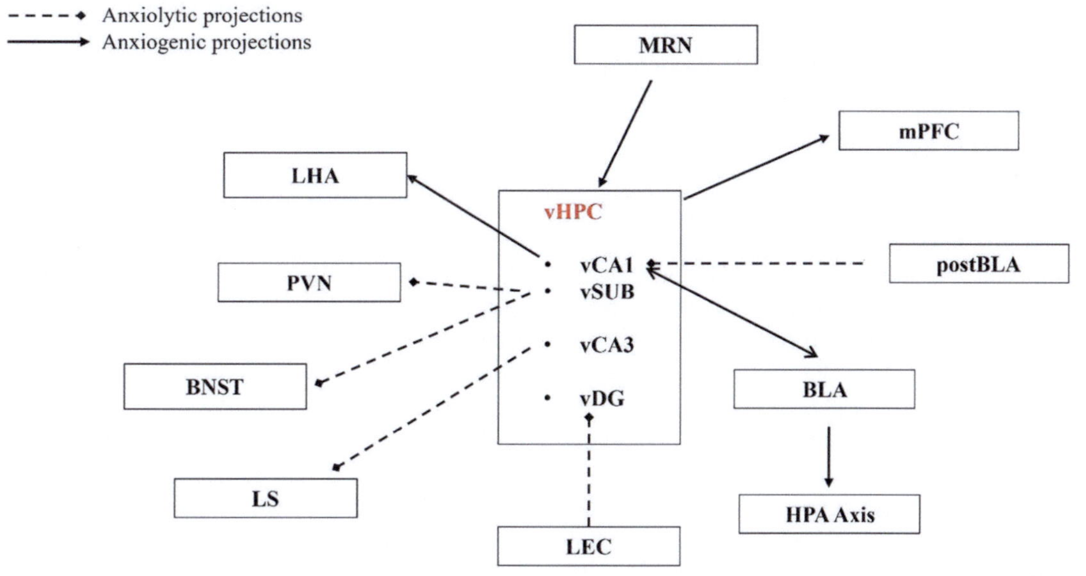

Fig. 14.2 Ventral hippocampal networks involved in anxiety-related behaviors. *vCA1* ventral CA1, *vSUB* ventral subiculum, *vCA3* ventral CA3, *vDG* ventral dentate gyrus, *vHPC* ventral hippocampus, *BLA* basolateral amygdala, *BNST* bed nucleus of the stria terminalis, *HPA* hypothalamo-pituitary-adrenal axis, *LHA* lateral hypothalamic area, *LEC* lateral entorhinal cortex, *LS* lateral septum, *mPFC* medial prefrontal cortex, *MRN* median raphe nucleus, *postBLA* posterior basolateral amygdala, *PVN* paraventricular nucleus

"anxiety cells" that are more abundant in vCA1 than in dCA1 [47].

Optogenetic and pharmacological studies show that manipulating the vHPC, its inputs, and cortical outputs directly impacts anxiety-related behavior [34, 64]. "Anxiety cells," enriched within the vCA1, carry representations of anxiogenic stimuli essential for avoidance behavior. These neurons project to the LHA, respond to anxiogenic environments, and affect avoidance behavior, suggesting the vCA1–LHA projection as another key route for controlling anxiety-like behaviors [47]. Jimenez et al. suggested that this pathway may function similarly to the thalamic "low road" and cortical "high road" in fear conditioning, providing rapid avoidance signals and slower, higher-order representations through cortical integration [47].

The vHPC also projects to the basolateral amygdala (BLA), which is essential for contextual fear expression and regulating generalized fear in non-threatening environments [6]. Inhibiting these projections reduces fear generalization [65]. Theta rhythm synchronization between these areas during fear memory retrieval and extinction further links the vHPC to anxiety.

In a study, it was shown that the activation of vCA1–LHA axonal terminals, but not vCA1–BLA axonal terminals, results in avoidance behavior in anxiety tasks. Kim et al. further investigated, and projection-specific imaging studies revealed a higher concentration of anxiety cells within vCA1–LHA projections than vCA1–BLA projections, with about 80% of the vCA1–LHA population representing the open arms of the EPM. The authors discussed this result as the LHA integrates these representations with inputs from other areas like the BNST, producing anxiolytic effects [66]. According to Jimenez et al., the vHPC projections to the LHA appear to be more implicated in anxiety-related behaviors, despite not affecting fear memory [47]. Thus, routing anxiety representations to the LHA is critical for generating avoidance behaviors.

Research shows that ventral hippocampal projections induce significant feedforward inhibition of pyramidal neurons in the infralimbic region of the mPFC, leading to the relapse of extinguished fear. Blocking these projections or GABAergic receptors in this region reduces fear relapse [67]. Another study highlighting the vHPC's role in anxiety and fear regulation reveals that the ventral hippocampal projections to the PFC are involved in adrenergic-mediated plasticity of glutamatergic neurotransmission during development. Neonatal lesions of these projections in rats disrupt the fear extinction and adrenergic control of glutamatergic synaptic plasticity in the PFC [68]. Theta-frequency synchronization between the vHPC and mPFC strengthens during anxiety, suggesting a collaborative role in generating behavioral inhibition suggesting a collaborative role in generating behavioral inhibition [42, 46].

Overall, the vHPC's projections to the mPFC, amygdala, and LHA are crucial in generating and modulating anxiety responses. Understanding these pathways and their interactions and identifying the underlying causes of ventral hippocampal hyperactivity in neural circuits involved in fear/anxiety and stress response provide insight into potential treatments for anxiety disorders.

14.3.3 Theta Rhythm Synchronization

Brain oscillations are believed to regulate cognitive functions by linking different brain regions and separating distinct neuronal activities. Hippocampal theta rhythm, which is a large-amplitude near-sinusoidal rhythm resulting from synchronous phasic firing of cells, having a fundamental peak frequency ranging from 4 to 14 Hz [69]. Influenced by serotonin from the raphe nuclei, hippocampal theta is associated with various brain and behavioral functions, including voluntary movement, sensorimotor integration, spatial navigation, learning, memory, and affective responses [69, 70].

Increased hippocampal theta observed in anxiety-provoking situations like the open arms of the EPM [7]. Additionally, studies by Gray and McNaughton showed that anxiolytics consistently reduce hippocampal theta frequency in both anesthetized and awake animals, linking hippocampal activity to anxiety regulation [7]. In anesthetized animals, theta oscillations, typically recorded via the dHPC, occur at lower frequencies and are sensitive to atropine or scopolamine, indicating their role in sensory processing rather than movement [71]. Despite their indirect connection, some studies have shown that neural activity in the mPFC synchronizes with theta-frequency oscillations in the dHPC [72]. In contrast to this result, a study showed that the theta-frequency activity in the mPFC and vHPC, but not dHPC, is highly correlated at baseline and increases in anxiogenic environments. This synchronization mechanism suggests cooperation between the vHPC and mPFC during anxiety, allowing the mPFC to recognize threatening environments and modulate the amygdala to produce defensive and anxiety-related behaviors; in other words, it inhibits exploratory behavior [46]. In line with this study, freely moving 5-HT1A knockout mice exhibit increased anxiety, and theta-frequency activity is elevated in the vHPC in novel environments. However, it remains unchanged in the dHPC [42].

One possibility is that it reflects a state of heightened information processing within the anxiety network. The increased synchrony between the vHPC and mPFC could represent an attempt to better understand and respond to potential threats. Further research is needed to investigate whether different tasks modulate mPFC-vHPC synchrony in other frequency ranges. Future research can explore whether manipulating theta activity through targeted interventions might help regulate anxiety.

14.3.4 Hippocampal Neurogenesis and Anxiety

Neural plasticity in the adult brain, particularly within the subgranular zone of the DG [73], where stem cells produce progenitors that divide into immature neurons, allows for adaptation to environmental changes through the generation of adult-born GCs [74]. Newborn GCs, about 10% of the GC population, are more excitable and plastic than mature GCs [6] and are linked to anxiolytic and antidepressant effects via their activation in the ventral DG [75, 76]. These new neurons decrease the excitability of mature GCs and mainly respond to stress or anxiogenic environments [15, 77]. Both mature and immature GCs are involved in anxiety regulation [15, 78], with mature GCs in the ventral DG activated in anxiogenic environments and adult-born neurons inhibiting these stress-responsive cells, thus protecting against anxiety [15]. Mossy cells in the DG regulate this process by balancing direct (glutamatergic) and indirect (GABAergic) pathways, influencing neural stem cell activation and quiescence, respectively [6].

Newborn GCs, mossy cells, and local GABAergic interneurons, particularly parvalbumin (PV)-expressing interneurons, modulate GC activity and influence pattern separation, which is essential for encoding contextual information [6]. These cells play a role in fear extinction, and PV-expressing interneurons regulate neurogenesis and anxiety by controlling the survival of newborn GCs and inhibiting quiescent neural stem cells via GABA release [79]. This suggests that PV-expressing interneurons may regulate aberrant neurogenesis and excessive anxiety.

The hippocampal pattern separation process, supported by the DG-CA3 circuit, relies on neurogenesis to create selective activation patterns, reducing overlap between similar stimuli [9, 80] and helping manage fear and anxiety responses. Reducing mature GC excitability increases sparse coding in the GC layer, improving pattern separation and cognitive flexibility, which helps reduce anxiety-like behavior [81, 82]. Additionally, granule cell output is crucial for cognitive functions like context discrimination and precision in fear retrieval [83]. However, the role of neurogenesis in anxiety regulation is complex, with studies showing mixed results. For instance, reducing immature GCs increases anxiety-like behavior in the novelty-suppressed feeding test [84]. At the same time, neurogenesis has also been shown to increase fear generalization and anxiety in response to unpredictable threats [85]. Knockdown of newborn neurons has been shown to impair fear extinction [86, 87], while Kirby et al.'s study increased activation of newborn neurons after acute stress, which has been shown to enhance fear extinction memory [88], suggesting these neurons help the brain adapt to stress and prevent excessive anxiety.

Stress hormone receptors are absent in neuronal progenitor cells. However, both MR and glucocorticoid receptors (GR) are present in new neurons and may mediate the effects of stress on neurogenesis [6]. Stress hormones influence neurogenesis, with low glucocorticoid levels activating MRs to induce an anxiolytic phenotype [89]. In contrast, high levels activate GRs, reducing progenitor cell proliferation [90] and altering neuron integration [91]. Glucocorticoids have been shown to enhance fear generalization by increasing the size of the DG cell population activated by aversive stimuli. Lesuis et al. showed that inhibiting this neuronal population reduces fear memory generalization [92]. An imbalance in hippocampal excitation-inhibition affects DG excitability and contributes to anxiety behavior [93].

Chronic stress, like repeated social defeat, reduces neurogenesis and increases anxiety-like behavior [94]. Some anxiolytics enhance neurogenesis and normalize DG activity [95], suggesting the importance of neurogenesis in anxiety regulation [96]. Chronic antidepressant treatment enhances MR density in the hippocampus [6], and SSRIs like fluoxetine prevent fear generalization and enhance extinction [97] by accelerating GC maturation, potentially reducing DG plasticity and activity [98], which may influence anxiety behavior. According to Sah et al., normalization of hippocampal hypoactivity could be a neurobiological marker for effective behavioral remission [96].

In summary, understanding the intricate interplay between hippocampal plasticity, neurogenesis, and local inhibitory circuits offers a fascinating window into anxiety-related behavior regulation. Future research can explore how to harness this knowledge to develop more targeted treatments for anxiety disorders.

14.4 Clinical Implications

14.4.1 Pharmacological Studies and Therapeutic Interventions Targeting the Hippocampus

Pharmacological studies and therapeutic interventions targeting the hippocampus highlight its role in anxiety regulation. Anxiolytic drugs generally reduce hippocampal theta frequency induced by reticular formation stimulation in animals [70]. Suppression of evoked hippocampal theta frequency is considered a reliable marker of anxiolytic drug action [69]. Clinically proven anxiolytics like diazepam, buspirone, and fluoxetine suppress the frequency of reticularly elicited hippocampal theta in anesthetized rats, while procognitive drugs increase the power of reticular-elicited theta [99]. Exceptions include bilateral histamine infusion into the lateral septum, which reduces anxiety-like responses while increasing theta frequency, and septal muscimol infusion, which increases theta frequency while reducing anxiety-like behaviors [100].

Studies show aberrant hippocampal activity in patients with anxiety disorders, with treatments like SSRIs deactivating the hippocampus. The activation of 5-HT neurons in the dHPC is linked to anxiety-related behaviors, with the effect varying based on the origin of these neurons [20]. SSRIs increase neurogenesis via 5-HT1A, abundant in the hippocampal CA1, and 5-HT2B receptors, while ablation of serotonergic neurons reduces neurogenesis, and blocking 5-HT2A/C receptors reduces cell proliferation [20]. In the DG, 5-HT1A receptors enhance adult neurogenesis and mediate granule cell proliferation induced by SSRIs [101]. Chronic SSRI treatment increases mossy cell expressing Sonic hedgehog activity [102]. Their ablation increases anxiety-like behavior [31, 103]. New antipsychotics target 5-HT2A and 5-HT1A receptors on glutamatergic pyramidal neurons, GABAergic interneurons, and 5-HT1A receptors in the raphe nuclei [20]. 5-HT4 receptors also promote neurogenesis and facilitate neurotransmission, contributing to SSRI effects [36]. 5-HT4 receptor knockout mice exhibit reduced activity, anhedonia, and anxiety linked to changes in BDNF, Arc, and trkB mRNA levels in the hippocampus and desensitization of 5-HT1A autoreceptors in the DRN [14].

Beyond the 5-HT system, research is exploring other potential targets within the hippocampus for anxiety treatment. Mossy cell activity can be increased by activating dopamine D2 receptors, which are involved in the anxiolytic effects of the dopamine system following chronic stress [104]. This suggests that drugs targeting the dopamine system within the hippocampus may offer a novel avenue for treating anxiety, particularly in individuals with stress-related disorders.

Hippocampal GABAergic deficits are associated with rodent anxiety behavior, mainly decreases in α1- and α2-containing GABA type A receptor (GABA-A) subtypes. A decrease in the α1-GABA-A subtype in the hippocampus induced by early-life stress is linked to anxiety-like behavior in adult male rats [105]. Conversely, increasing GABA-A receptor function via α2-containing GABA-A receptors in DG and CA3 neurons reduces anxiety in mice [106].

Physical exercise enhances local inhibitory mechanisms in the hippocampus, leading to anxiolytic effects. Blocking GABA-A receptors in the ventral DG reverses the anxiolytic effects of physical exercise [48]. Additionally, serotonergic median raphe fibers primarily target GABAergic interneurons in the DG, suppressing the firing of GCs. This suggests that SSRIs may exert therapeutic effects in anxiety disorders partly by decreasing DG activity via GABAergic neurotransmission [6].

The research on the hippocampus as a target for anxiety treatment is still in its initial stages, but it holds immense promise. Future studies can explore how to best manipulate specific pathways within the hippocampus, such as those involving mossy cells and the dopamine system, to achieve optimal therapeutic effects.

14.5 Hippocampus Aging and Alzheimer Disease

The hippocampus is a very important region of the brain involved in memory and cognitive processes. This structure shows severe changes at both a structural and functional level with age. More specifically, the volume of the hippocampus decreases dramatically after 50 years of age, parallel to changes at the cellular level. Although the degree of atrophy attributed to hippocampal aging has been controversial, it is widely accepted that aging is responsible for the shrinkage of this brain structure. Major factors underlying this atrophy include neuronal loss, decreased neurogenesis, neurovascular changes, increased neuroinflammation, and oxidative stress [107, 108].

Aging leads to significant changes in brain size, vascular structure, and cognition. Brain volume decreases, and changes occur at all levels, from molecular alterations to morphological transformations. It is known that there is a decrease in volume as we age, especially in the frontal cortex of the brain. However, volume reduction has also been found in the temporal lobe, cerebellar vermis, cerebellar hemispheres, and hippocampus. Some studies suggest that the hippocampus shows the greatest effect of aging. While the most affected areas in men are the frontal and temporal lobes, the most affected areas in women are the hippocampus and parietal lobes [109].

In human and animal experimental studies, it has been reported that the hippocampus undergoes changes at both the macro and cellular levels with aging. In one study, a correlation was observed between age-related decline in hippocampal-dependent learning and memory in female rats and a decrease in hippocampal volume [110]. In a longitudinal study in humans, an age-related decrease in hippocampal volume and activity was detected on MRI images and was associated with decreased performance on episodic memory tasks over a 20-year period [111].

Although some studies have shown that hippocampal volume decreases with age [112], some studies have reported no significant change in hippocampal volume with age [113], and one study reported a significant decrease in hippocampal volumes only in men [114]. Some studies failed to show a clear relationship between structural and functional hippocampal changes and memory decline with age [111].

In a longitudinal study, individuals aged between 26 and 82 years were included in the study, and MR images were monitored over a 5-year period, resulting in a modest decrease in hippocampal volume in individuals under 50 years of age, while a dramatic volume loss of approximately 1.2% per year was detected in individuals over 50 years of age [115]. In a different longitudinal study, several hippocampal subregions, including Cornu Ammonis (CA) 1–4, subiculum, and DG, were found to undergo changes over a 30-month period in participants older than 49 years [116].

It is thought that one of the underlying causes of the age-related decrease in hippocampal volume may be the loss of neurons in this brain structure during the aging process. A potential decrease in neuronal production or neurogenesis with age is also known to be a cause of the age-related decrease in hippocampal volume [117].

There is a clearly documented consequence of aging on cognitive decline, in particular, episodic memory. This drop in normal aging is clear, although it is much more dramatic in a number of neurodegenerative diseases, such as Alzheimer's disease (AD). In AD, these cognitive deficits, particularly the memory-related ones, are not only more pronounced, but they also worsen progressively with the development of the disease. Specifically, in AD, the hippocampus has been mentioned to undergo extreme atrophy. This atrophy is closely related to the accumulation of beta-amyloid plaques and tau protein tangles, which are the hallmarks of the pathology of AD. Such pathological changes confer considerable neuronal loss and synaptic dysfunction, potentially directly responsible for the profound impairments in memory seen in AD [118, 119].

The atrophy of the hippocampus due to AD is much more severe when contrasted with normal aging, and such atrophy is one of the hallmarks for the diagnosis and progression of AD. MRI can detect hippocampal atrophy, making it an important imaging biomarker for the early diagnosis of AD and for predicting conversion from MCI to AD. The degree of hippocampal atrophy, as measured by MRI, is strongly correlated with the severity of cognitive impairment in AD patients [120, 121].

Factors that are neurotrophic—that is, support neurogenesis and neuron survival—also decline with aging and are further reduced in the context of AD. To be specific, BDNF and vascular endothelial growth factor (VEGF) levels are decreased in aging and further decreased in AD. It is these very neurotrophic factors that are lost with aging and contribute to the compromised neurogenesis and neuronal health seen in the aging hippocampus, further driving cognitive decline [122, 123].

Those risk factors associated with an exaggeration of hippocampal dysfunction due to aging include obesity, chronic stress, depression, and genetic background. Such risk factors do not only have an effect on hippocampal atrophy but also enhance vulnerability to AD. On the other hand, regular physical exercise and a balanced diet are recommended to reduce hippocampal atrophy and promote healthy cognition with aging [124, 125].

Imaging methods, particularly MRI, have helped a great deal in discovering the extent of atrophy in aging and Alzheimer's disease. All these MRI-related studies have continually demonstrated that atrophy of the hippocampus is one of the strong predictors of cognitive decline, particularly during the early stages of AD. The ability to document the process of hippocampal shrinkage over time has provided vital information regarding the progression of the disease and has been helpful in assessing the efficacy of therapeutic interventions aimed at slowing such cognitive decline [126, 127].

In Alzheimer's disease, the hippocampus is the first region that undergoes pathological changes, such as the deposition of beta-amyloid plaques and tau protein tangles. These markers incite an unknown set of events, which result in synaptic loss and neuronal death, leading to severe memory loss and cognitive deficits. The degree of hippocampal atrophy in AD contrasts sharply with the degree occurring under normal aging, suggesting that early detection and intervention are required if cognitive capacity is to be retained to the greatest degree and for as long as possible [128, 129].

This effect is also very strongly dependent upon lifestyle factors in age-related changes within the hippocampus. Physical activity enhances neurogenesis and improves hippocampal function, while, on the other hand, a diet rich in antioxidants and omega-3 fatty acids confers benefits on brain health through actions that diminish oxidative stress and inflammation. These protective factors delay the onset of cognitive decline and reduce the risk of developing AD, underscoring the importance of preventive strategies in maintaining cognitive health during aging [130, 131].

In summary, aging alone significantly affects the hippocampus, and even more so in AD. While normal aging is associated with volume, neurogenic, and synaptic decline of the hippocampus, these effects have a much greater magnitude in AD. Indeed, the atrophy in the hippocampus of AD subjects is closely interlinked with the buildup of beta-amyloid plaques and tau tangles, ultimately leading to serious cognitive impairments, particularly in memory. MRI studies play a critical role in detecting and tracking atrophy of the hippocampus. In fact, MRI has been a very valuable tool in the diagnosis of AD and in monitoring disease progression. On the other hand, lifestyle interventions—including regular physical exercise and a healthy diet—may moderate the age-related effects on the hippocampus, slowing down the decline in cognition and thus reducing the risk for AD. Genetic predisposition, environmental factors, and lifestyle features interplay in a way that underscores the importance of a holistic approach toward maintaining healthy cognition with age.

14.6 Conclusion

In conclusion, the hippocampus emerges as a pivotal structure in the modulation of anxiety-related behaviors, with its diverse subregions and neurotransmitter systems playing distinct roles. The interplay between hippocampal circuits and external stressors underscores the complexity of anxiety disorders and the necessity for targeted therapeutic approaches. Advances in our understanding of hippocampal neurogenesis and plasticity offer promising avenues for developing treatments aimed at normalizing hippocampal function. Overall, while the neural basis of differences in anxiety-related behaviors and anxiety disorders remains mostly unresolved, it may involve differences in fear extinction, hippocampal responses to stress, and neurogenesis. Further research is essential to unravel the precise mechanisms by which the hippocampus influences anxiety, which will inform more effective interventions for anxiety disorders. By deepening our comprehension of hippocampal contributions to anxiety, we move closer to mitigating the burden of these pervasive conditions on affected individuals.

References

1. Akiskal HS. Anxiety: definition, relationship to depression, and proposal for an integrative model. In: Anxiety and the anxiety disorders. Routledge; 2019. p. 787–98.
2. Davis M, et al. Phasic vs sustained fear in rats and humans: role of the extended amygdala in fear vs anxiety. Neuropsychopharmacology. 2010;35(1):105–35.
3. Endler NS, Kocovski NL. State and trait anxiety revisited. J Anxiety Disord. 2001;15(3):231–45.
4. Rosen JB, Schulkin J. From normal fear to pathological anxiety. Psychol Rev. 1998;105(2):325.
5. Stein DJ, Nesse RM. Threat detection, precautionary responses, and anxiety disorders. Neurosci Biobehav Rev. 2011;35(4):1075–9.
6. Ghasemi M, et al. Anxiety and hippocampal neuronal activity: relationship and potential mechanisms. Cogn Affect Behav Neurosci. 2022;22(3):431–49.
7. McNaughton N, Gray JA. Anxiolytic action on the behavioural inhibition system implies multiple types of arousal contribute to anxiety. J Affect Disord. 2000;61(3):161–76.
8. Bernstein EE, McNally RJ. Exploring behavioral pattern separation and risk for emotional disorders. J Anxiety Disord. 2018;59:27–33.
9. Kheirbek MA, et al. Neurogenesis and generalization: a new approach to stratify and treat anxiety disorders. Nat Neurosci. 2012;15(12):1613–20.
10. Thibaut F. Anxiety disorders: a review of current literature. Dialogues Clin Neurosci. 2017;19(2):87–8.
11. Koen N, Stein DJ. Core dimensions of anxiety disorders. In: Geddes JR, Andreasen NC, Goodwin GM, editors. New Oxford textbook of psychiatry. USA: Oxford University Press; 2020. p. 897–904.
12. Wittchen H-U, Beesdo-Baum K. Epidemiology of anxiety disorders. In: Geddes JR, Andreasen NC, Goodwin GM, editors. New Oxford Textbook of Psychiatry. USA: Oxford University Press; 2020. p. 917–27.
13. McLean CP, et al. Gender differences in anxiety disorders: prevalence, course of illness, comorbidity and burden of illness. J Psychiatr Res. 2011;45(8):1027–35.
14. Amigó J, et al. The absence of 5-HT4 receptors modulates depression- and anxiety-like responses and influences the response of fluoxetine in olfactory bulbectomised mice: adaptive changes in hippocampal neuroplasticity markers and 5-HT1A autoreceptor. Neuropharmacology. 2016;111:47–58.
15. Anacker C, et al. Hippocampal neurogenesis confers stress resilience by inhibiting the ventral dentate gyrus. Nature. 2018;559(7712):98–102.
16. Muir J, et al. Ventral hippocampal afferents to nucleus accumbens encode both latent vulnerability and stress-induced susceptibility. Biol Psychiatry. 2020;88(11):843–54.
17. Lüscher C, et al. G protein-coupled inwardly rectifying K+ channels (GIRKs) mediate postsynaptic but not presynaptic transmitter actions in hippocampal neurons. Neuron. 1997;19(3):687–95.
18. Turgut YB, Turgut M. A mysterious term hippocampus involved in learning and memory. Childs Nerv Syst. 2011;27:2023–5.
19. Shi H-J, et al. Hippocampus: molecular, cellular, and circuit features in anxiety. Neurosci Bull. 2023;39(6):1009–26.
20. Bombardi C, et al. Serotonin modulation of hippocampal functions: from anatomy to neurotherapeutics. Prog Brain Res. 2021;261:83–158.
21. Insausti R, Amaral DG. Hippocampal formation. In: The human nervous system, vol. 2. Elsevier; 2004. p. 871–914.
22. Schultz C, Engelhardt M. Anatomy of the hippocampal formation. Front Neurol Neurosci. 2014;34:6–17.
23. Zeidman P, Maguire EA. Anterior hippocampus: the anatomy of perception, imagination and episodic memory. Nat Rev Neurosci. 2016;17(3):173–82.
24. Bannerman D, et al. Regional dissociations within the hippocampus—memory and anxiety. Neurosci Biobehav Rev. 2004;28(3):273–83.
25. Fanselow MS, Dong H-W. Are the dorsal and ventral hippocampus functionally distinct structures? Neuron. 2010;65(1):7–19.
26. Amaral DG, Witter MP. The three-dimensional organization of the hippocampal formation: a review of anatomical data. Neuroscience. 1989;31(3):571–91.
27. Pitkänen A, et al. Reciprocal connections between the amygdala and the hippocampal formation, perirhinal cortex, and postrhinal cortex in rat: a review. Ann N Y Acad Sci. 2000;911(1):369–91.
28. Bernier BE, et al. Dentate gyrus contributes to retrieval as well as encoding: evidence from context fear conditioning, recall, and extinction. J Neurosci. 2017;37(26):6359–71.
29. Heigele S, et al. Bidirectional GABAergic control of action potential firing in newborn hippocampal granule cells. Nat Neurosci. 2016;19(2):263–70.
30. Alvarez DD, et al. A disynaptic feedback network activated by experience promotes the integration of new granule cells. Science. 2016;354(6311):459–65.
31. Jinde S, et al. Hilar mossy cell degeneration causes transient dentate granule cell hyperexcitability and impaired pattern separation. Neuron. 2012;76(6):1189–200.

32. Myers CE, Scharfman HE. A role for hilar cells in pattern separation in the dentate gyrus: a computational approach. Hippocampus. 2009;19(4):321–37.
33. Miguelez C, et al. L-DOPA elicits non-vesicular releases of serotonin and dopamine in hemiparkinsonian rats in vivo. Eur Neuropsychopharmacol. 2016;26(8):1297–309.
34. Kjaerby C, et al. Serotonin 1B receptors regulate prefrontal function by gating callosal and hippocampal inputs. Cell Rep. 2016;17(11):2882–90.
35. Di Giovanni G, De Deurwaerdère P. New therapeutic opportunities for 5-HT2C receptor ligands in neuropsychiatric disorders. Pharmacol Ther. 2016;157:125–62.
36. Segi-Nishida E. The effect of serotonin-targeting antidepressants on neurogenesis and neuronal maturation of the hippocampus mediated via 5-HT1A and 5-HT4 receptors. Front Cell Neurosci. 2017;11:142.
37. Gray JA. Précis of the neuropsychology of anxiety: an enquiry into the functions of the septo-hippocampal system. Behav Brain Sci. 1982;5(3):469–84.
38. Levita L, et al. The behavioural inhibition system, anxiety and hippocampal volume in a non-clinical population. Biol Mood Anxiety Disord. 2014;4:1–10.
39. Morellini F, et al. Impaired fear extinction due to a deficit in Ca^{2+} influx through L-type voltage-gated Ca^{2+} channels in mice deficient for tenascin-C. Front Integr Neurosci. 2017;11:16.
40. Dunsmoor JE, Paz R. Fear generalization and anxiety: behavioral and neural mechanisms. Biol Psychiatry. 2015;78(5):336–43.
41. Lacagnina AF, et al. Distinct hippocampal engrams control extinction and relapse of fear memory. Nat Neurosci. 2019;22(5):753–61.
42. Adhikari A, Topiwala MA, Gordon JA. Single units in the medial prefrontal cortex with anxiety-related firing patterns are preferentially influenced by ventral hippocampal activity. Neuron. 2011;71(5):898–910.
43. Hogg S. A review of the validity and variability of the elevated plus-maze as an animal model of anxiety. Pharmacol Biochem Behav. 1996;54(1):21–30.
44. Shah AA, Treit D. Infusions of midazolam into the medial prefrontal cortex produce anxiolytic effects in the elevated plus-maze and shock-probe burying tests. Brain Res. 2004;996(1):31–40.
45. Kjelstrup KG, et al. Reduced fear expression after lesions of the ventral hippocampus. Proc Natl Acad Sci. 2002;99(16):10825–30.
46. Adhikari A, Topiwala MA, Gordon JA. Synchronized activity between the ventral hippocampus and the medial prefrontal cortex during anxiety. Neuron. 2010;65(2):257–69.
47. Jimenez JC, et al. Anxiety cells in a hippocampal-hypothalamic circuit. Neuron. 2018;97(3):670–83.e6
48. Schoenfeld TJ, et al. Physical exercise prevents stress-induced activation of granule neurons and enhances local inhibitory mechanisms in the dentate gyrus. J Neurosci. 2013;33(18):7770–7.
49. Schoenfeld TJ, et al. Anxiolytic actions of exercise in absence of new neurons. Hippocampus. 2016;26(11):1373–8.
50. Tye KM, et al. Amygdala circuitry mediating reversible and bidirectional control of anxiety. Nature. 2011;471(7338):358–62.
51. Bannerman DM, et al. Double dissociation of function within the hippocampus: spatial memory and hyponeophagia. Behav Neurosci. 2002;116(5):884.
52. Peper JS, et al. Genetic influences on human brain structure: a review of brain imaging studies in twins. Hum Brain Mapp. 2007;28(6):464–73.
53. Barrós-Loscertales A, et al. Behavioral inhibition system activity is associated with increased amygdala and hippocampal gray matter volume: a voxel-based morphometry study. NeuroImage. 2006;33(3):1011–5.
54. Cherbuin N, et al. Hippocampal volume is positively associated with behavioural inhibition (BIS) in a large community-based sample of mid-life adults: the PATH through life study. Soc Cogn Affect Neurosci. 2008;3(3):262–9.
55. Gross C, et al. Serotonin1A receptor acts during development to establish normal anxiety-like behaviour in the adult. Nature. 2002;416(6879):396–400.
56. Gross C, et al. Altered fear circuits in 5-HT1A receptor KO mice. Biol Psychiatry. 2000;48(12):1157–63.
57. Bergami M, Berninger B, Canossa M. Conditional deletion of TrkB alters adult hippocampal neurogenesis and anxiety-related behavior: the relationship between visual long-term memory and change blindness. Commun Integr Biol. 2009;2(1):14–6.
58. Li C-Q, et al. Development of anxiety-like behavior via hippocampal IGF-2 signaling in the offspring of parental morphine exposure: effect of enriched environment. Neuropsychopharmacology. 2014;39(12):2777–87.
59. Vertes RP. Differential projections of the infralimbic and prelimbic cortex in the rat. Synapse. 2004;51(1):32–58.
60. Kjelstrup KB, et al. Finite scale of spatial representation in the hippocampus. Science. 2008;321(5885):140–3.
61. Terpstra AR, et al. Higher anxiety symptoms predict progressive hippocampal atrophy in the chronic stages of moderate to severe traumatic brain injury. Neurorehabil Neural Repair. 2017;31(12):1063–71.
62. Ulrich-Lai YM, Herman JP. Neural regulation of endocrine and autonomic stress responses. Nat Rev Neurosci. 2009;10(6):397–409.
63. Tafet GE, Nemeroff CB. Pharmacological treatment of anxiety disorders: the role of the HPA axis. Front Psych. 2020;11:529446.
64. Felix-Ortiz AC, et al. BLA to vHPC inputs modulate anxiety-related behaviors. Neuron. 2013;79(4):658–64.
65. Ortiz S, et al. Anterior cingulate cortex and ventral hippocampal inputs to the basolateral amygdala selectively control generalized fear. J Neurosci. 2019;39(33):6526–39.

66. Kim S-Y, et al. Diverging neural pathways assemble a behavioural state from separable features in anxiety. Nature. 2013;496(7444):219–23.
67. Marek R, et al. Hippocampus-driven feed-forward inhibition of the prefrontal cortex mediates relapse of extinguished fear. Nat Neurosci. 2018;21(3):384–92.
68. Bhardwaj SK, et al. Impaired adrenergic-mediated plasticity of prefrontal cortical glutamate synapses in rats with developmental disruption of the ventral hippocampus. Neuropsychopharmacology. 2014;39(13):2963–73.
69. Yeung M, Dickson CT, Treit D. Intrahippocampal infusion of the Ih blocker ZD7288 slows evoked theta rhythm and produces anxiolytic-like effects in the elevated plus maze. Hippocampus. 2013;23(4):278–86.
70. Horváth J, et al. Anxious and nonanxious mice show similar hippocampal sensory evoked oscillations under urethane anesthesia: difference in the effect of buspirone. Neural Plast. 2015;2015:186323.
71. Kramis R, Vanderwolf CH, Bland BH. Two types of hippocampal rhythmical slow activity in both the rabbit and the rat: relations to behavior and effects of atropine, diethyl ether, urethane, and pentobarbital. Exp Neurol. 1975;49(1):58–85.
72. Jones MW, Wilson MA. Theta rhythms coordinate hippocampal–prefrontal interactions in a spatial memory task. PLoS Biol. 2005;3(12):e402.
73. Catavero C, Bao H, Song J. Neural mechanisms underlying GABAergic regulation of adult hippocampal neurogenesis. Cell Tissue Res. 2018;371:33–46.
74. Zhao C, Deng W, Gage FH. Mechanisms and functional implications of adult neurogenesis. Cell. 2008;132(4):645–60.
75. Schmidt-Hieber C, Jonas P, Bischofberger J. Enhanced synaptic plasticity in newly generated granule cells of the adult hippocampus. Nature. 2004;429(6988):184–7.
76. Wu MV, Hen R. Functional dissociation of adult-born neurons along the dorsoventral axis of the dentate gyrus. Hippocampus. 2014;24(7):751–61.
77. Marín-Burgin A, et al. Unique processing during a period of high excitation/inhibition balance in adult-born neurons. Science. 2012;335(6073):1238–42.
78. Samuels BA, et al. 5-HT1A receptors on mature dentate gyrus granule cells are critical for the antidepressant response. Nat Neurosci. 2015;18(11):1606–16.
79. Song J, et al. Neuronal circuitry mechanism regulating adult quiescent neural stem-cell fate decision. Nature. 2012;489(7414):150–4.
80. Sahay A, et al. Increasing adult hippocampal neurogenesis is sufficient to improve pattern separation. Nature. 2011;472(7344):466–70.
81. Anacker C, Hen R. Adult hippocampal neurogenesis and cognitive flexibility—linking memory and mood. Nat Rev Neurosci. 2017;18(6):335–46.
82. Wilson CG, et al. Trait anxiety impairs cognitive flexibility when overcoming a task acquired response and a preexisting bias. PLoS One. 2018;13(9):e0204694.
83. Nakashiba T, et al. Young dentate granule cells mediate pattern separation, whereas old granule cells facilitate pattern completion. Cell. 2012;149(1):188–201.
84. Deng W, Gage FH. The effect of immature adult-born dentate granule cells on hyponeophagial behavior is related to their roles in learning and memory. Front Syst Neurosci. 2015;9:34.
85. Glover LR, et al. Ongoing neurogenesis in the adult dentate gyrus mediates behavioral responses to ambiguous threat cues. PLoS Biol. 2017;15(4):e2001154.
86. Deng W, et al. Adult-born hippocampal dentate granule cells undergoing maturation modulate learning and memory in the brain. J Neurosci. 2009;29(43):13532–42.
87. Stone SS, et al. Functional convergence of developmentally and adult-generated granule cells in dentate gyrus circuits supporting hippocampus-dependent memory. Hippocampus. 2011;21(12):1348–62.
88. Kirby ED, et al. Acute stress enhances adult rat hippocampal neurogenesis and activation of newborn neurons via secreted astrocytic FGF2. elife. 2013;2:e00362.
89. Xing X, et al. Mineralocorticoid receptors in the ventral hippocampus are involved in extinction memory in rats. Psych J. 2014;3(3):201–13.
90. Anacker C, et al. Glucocorticoid-related molecular signaling pathways regulating hippocampal neurogenesis. Neuropsychopharmacology. 2013;38(5):872–83.
91. Fitzsimons C, et al. Knockdown of the glucocorticoid receptor alters functional integration of newborn neurons in the adult hippocampus and impairs fear-motivated behavior. Mol Psychiatry. 2013;18(9):993–1005.
92. Lesuis SL, et al. Glucocorticoids promote fear generalization by increasing the size of a dentate gyrus engram cell population. Biol Psychiatry. 2021;90(7):494–504.
93. Roshan-Milani S, et al. Prenatal stress and increased susceptibility to anxiety-like behaviors: role of neuroinflammation and balance between GABAergic and glutamatergic transmission. Stress. 2021;24(5):481–95.
94. Aoki Y, et al. Selective attenuation of electrophysiological activity of the dentate gyrus in a social defeat mouse model. J Physiol Sci. 2017;67:507–13.
95. Lempel AA, et al. Chronic pregabalin treatment decreases excitability of dentate gyrus and accelerates maturation of adult-born granule cells. J Neurochem. 2017;140(2):257–67.
96. Sah A, et al. Anxiety-rather than depression-like behavior is associated with adult neurogenesis in a female mouse model of higher trait anxiety- and comorbid depression-like behavior. Transl Psychiatry. 2012;2(10):e171.

97. Pedraza LK, et al. Chronic fluoxetine prevents fear memory generalization and enhances subsequent extinction by remodeling hippocampal dendritic spines and slowing down systems consolidation. Transl Psychiatry. 2019;9(1):53.
98. Åmellem I, et al. A critical period for antidepressant-induced acceleration of neuronal maturation in adult dentate gyrus. Transl Psychiatry. 2017;7(9):e1235.
99. McNaughton N, Kocsis B, Hajos M. Elicited hippocampal theta rhythm: a screen for anxiolytic and procognitive drugs through changes in hippocampal function? Behav Pharmacol. 2007;18(5–6):329–46.
100. Chee SSA, Menard JL, Dringenberg HC. Behavioral anxiolysis without reduction of hippocampal theta frequency after histamine application in the lateral septum of rats. Hippocampus. 2014;24(6):615–27.
101. Santarelli L, et al. Requirement of hippocampal neurogenesis for the behavioral effects of antidepressants. Science. 2003;301(5634):805–9.
102. Oh S-J, et al. Hippocampal mossy cell involvement in behavioral and neurogenic responses to chronic antidepressant treatment. Mol Psychiatry. 2020;25(6):1215–28.
103. Gonzalez-Reyes LE, et al. Sonic Hedgehog is expressed by hilar mossy cells and regulates cellular survival and neurogenesis in the adult hippocampus. Sci Rep. 2019;9(1):17402.
104. Choi M-H, et al. Role of dopamine D2 receptor in stress-induced myelin loss. Sci Rep. 2017;7(1):11654.
105. Mahmoodkhani M, et al. Long-term decreases in the expression of calcineurin and GABAa receptors induced by early maternal separation are associated with increased anxiety-like behavior in adult male rats. Dev Neurosci. 2021;42(2–4):135–44.
106. Engin E, et al. Modulation of anxiety and fear via distinct intrahippocampal circuits. elife. 2016;5:e14120.
107. Smith A, Jones B. Neurodegenerative changes in aging. J Neurosci Res. 2020;35(4):567–75.
108. Brown C, Green D. Synaptic plasticity and aging. Neurobiol Aging. 2019;28(3):203–10.
109. Peters R. Ageing and the brain. Postgrad Med J. 2006;82(964):84–8.
110. Driscoll I, et al. The aging hippocampus: a multi-level analysis in the rat. Neuroscience. 2006;139:1173–85.
111. Persson J, et al. Longitudinal structure-function correlates in elderly reveal MTL dysfunction with cognitive decline. Cereb Cortex. 2012;22:2297–304.
112. Walhovd KB, et al. Effects of age on volumes of cortex, white matter and subcortical structures. Neurobiol Aging. 2005;26:1261–70.
113. Du ATN, et al. Age effects on atrophy rates of entorhinal cortex and hippocampus. Neurobiol Aging. 2006;27:733–40.
114. Pruessner JC, Collins DL, Pruessner M, Evans AC. Age and gender predict volume decline in the anterior and posterior hippocampus in early adulthood. J Neurosci. 2001;21:194–200.
115. Raz N, et al. Differential aging of the medial temporal lobe: a study of a five-year change. Neurology. 2004;62:433–8.
116. Raz N, et al. Trajectories of brain aging in middle-aged and older adults: regional and individual differences. NeuroImage. 2010;51:501–11.
117. Luis EB, et al. The effects of aging in the hippocampus and cognitive decline. Neurosci Biobehav Rev. 2017;79:66–86.
118. Doe J, White E. Imaging biomarkers in Alzheimer's disease. Alzheimers Dement. 2021;17(6):1125–33.
119. Lee K, Kim H. Tau pathology and hippocampal atrophy in Alzheimer's disease. J Alzheimers Dis. 2022;30(7):755–62.
120. Wang Z, Zhang L. MRI assessment of hippocampal atrophy in Alzheimer's disease. NeuroImage Clin. 2020;26:102225.
121. Williams RA, Smith PK. Predictive value of hippocampal atrophy in the progression of mild cognitive impairment. J Neurol. 2018;15(2):129–37.
122. Holmes C, Lovestone S. Neurotrophic factors in Alzheimer's disease. Mol Psychiatry. 2017;22(1):1–7.
123. Naylor M, Bullmore E. The role of neurotrophic factors in the aging brain. Nat Rev Neurosci. 2019;20(5):300–9.
124. Erickson KI, Kramer AF. Physical activity and brain plasticity in late adulthood. Annu Rev Psychol. 2018;69(1):29–51.
125. Scarmeas N, Stern Y. Diet, nutrition, and the aging brain. Lancet Neurol. 2021;20(9):735–46.
126. Jack CR, Petersen RC. The role of imaging in the diagnosis of Alzheimer's disease. Brain Imaging Behav. 2019;13(4):1119–25.
127. Whitwell JL, Josephs KA. Hippocampal atrophy in Alzheimer's disease versus aging. Neurobiol Aging. 2020;39:1–9.
128. Sperling RA, Johnson KA. Early detection of Alzheimer's disease pathology using imaging and biomarkers. Brain. 2018;141(4):1265–78.
129. Braak H, Braak E. Neuropathological stageing of Alzheimer-related changes. Acta Neuropathol. 1991;82(4):239–59.
130. Voss MW, Vivar C. Exercise and hippocampal neurogenesis. Hippocampus. 2016;26(7):731–48.
131. Gomez-Pinilla F, Ying Z. Dietary factors and their impact on hippocampal plasticity and cognitive function. Mol Neurobiol. 2010;41(2–3):85–96.

Part II

Role of the Hippocampus in Memory, Cognition and Learning

Role of the Hippocampus in Cognition

15

Enes Akyüz and Feyza Şule Aslan

15.1 Introduction

The hippocampus is a forebrain structure located in the medial temporal lobe and consists of the subiculum, CA1, CA2, CA3, and dentate gyrus regions [1]. This region, which is part of the limbic system, contains neuronal networks underlying learning [2]. Hippocampal activity contributes to long-term memory, learning, decision-making, episodic memory, and cognitive processes [3].

Cognition is a process consisting of the steps of acquiring, processing, storing, and retrieving information [4]. In this process, various skills, including memory, attention, motor skills, verbal and language skills, learning, recalling past experiences, and active decision-making, are acquired [5]. Since the hippocampus is a memory center, it is normal that cognitive skills are acquired in this region. However, the role of the hippocampus in memory and cognition should not be confused. Compared to cognition, memory involves only the storage and retrieval of information. A cognitive map is created by providing a spatial and temporal framework for episodic memory [6].

The hippocampus contributes to cognitive processes through memory formation, information processing, spatial navigation, and learning. The hippocampus, transforming short-term memory into long-term memory, ensures that new information is received and associated with old information [7, 8]. It regulates the planning of behaviors, the decision-making process, and the processing of emotional experiences using learned information [9]. It also controls flexible cognitive functions such as adaptation to new situations and problem-solving. The combination of these roles shows the wide influence and importance of the hippocampus in cognitive processes. The hippocampus needs direct communication with other brain regions to regulate cognitive processes [10]. In this book chapter, to elucidate the complex role of the hippocampus in cognition, its anatomical location and its communication with various brain regions were comprehensively analyzed. The profound impact of these connections on learning, decision-making, and various cognitive scenarios was investigated. Moreover, the cognitive role of hippocampal activity was elucidated by examining its role in spatial cognition and cognitive flexibility.

E. Akyüz (✉)
Department of Biophysics, University of Health Sciences International Faculty of Medicine, İstanbul, Türkiye

F. Ş. Aslan
Marmara University Faculty of Medicine, İstanbul, Türkiye
e-mail: feyza.sule@marun.edu.tr

M. Turgut et al. (eds.), *The Human Hippocampus*, https://doi.org/10.1007/978-3-031-88907-3_15

15.2 Anatomy of Hippocampus and Connected Brain Regions

The hippocampus is a C-shaped, two-layered gray matter structure within the parahippocampal gyrus. It is defined as a curved and convoluted layer of cortex that folds into the central surface of the temporal lobe [2, 11]. The hippocampus consists of the dentate gyrus, subiculum, and cornu ammonis (CA). In the dentate gyrus, there are three layers from the outside to the inside: the molecular layer, granular layer, and polymorphic layer. Granular neurons here receive input from the entorhinal cortex and send mossy fibers to pyramidal cells in the CA. The subiculum section is located between the hippocampus, entorhinal cortex, and other cortices. It is responsible for receiving and processing information from CA1 and the entorhinal cortex and transmitting it through the fornix [12]. The fornix is the main output bundle of the hippocampus, which consists of fimbriae, crus, body, and column. Hippocampal pyramidal cell axons consisting of white fibers form the fornix. It contributes to memory processes by transmitting information from the hippocampus to other regions of the brain and also synapses with the cingulate gyrus, returning to the hippocampus through the connection with the septal region of the fornix. This neuronal intrinsic circuit, known as the Papez circuit, plays a crucial role in memory processing. The cornu ammonis, the most important region in the hippocampus where memory is processed, is divided into four different main regions: CA1, CA2, CA3, and CA4. This region consists of a deep polymorphic layer (stratum oriens) and a pyramidal layer (stratum pyramidale) with a superficial or molecular layer. CA1 is a region that sends outputs to the entorhinal cortex. CA2, a small region between CA1 and CA3, receives input from the entorhinal cortex. CA3 is the largest CA region, receiving input from mossy fibers of granule cells in the dentate gyrus and cells in the entorhinal cortex. The neural connections in CA3 are richer than in other hippocampal regions, and its pyramidal layer is divided into three sublayers: stratum lacunosum-moleculare, stratum radiatum, and stratum lucidum, which receives mossy fiber input from the dentate gyrus (Fig. 15.1). The CA3 region is thought to be the "pacemaker" of the hippocampus. Receiving, sending, and processing of information are mostly located in the CA3 region [13–15].

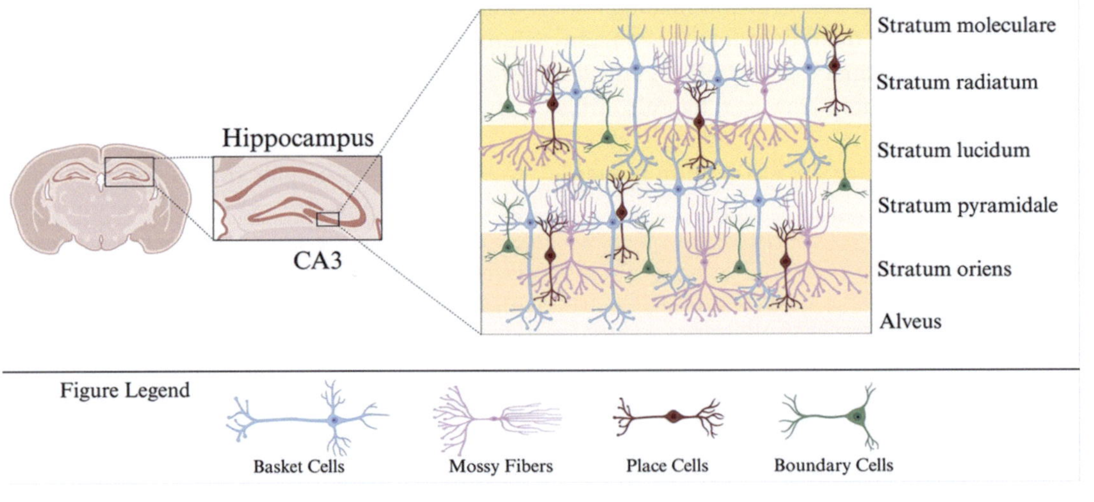

Fig. 15.1 The CA3 region in the hippocampus consists of six different layers. These layers are stratum lacunosum-moleculare, stratum radiatum, stratum lucidum, stratum pyramidale, and stratum oriens. In these layers, there are basket, boundary, place cells, and mossy fibers

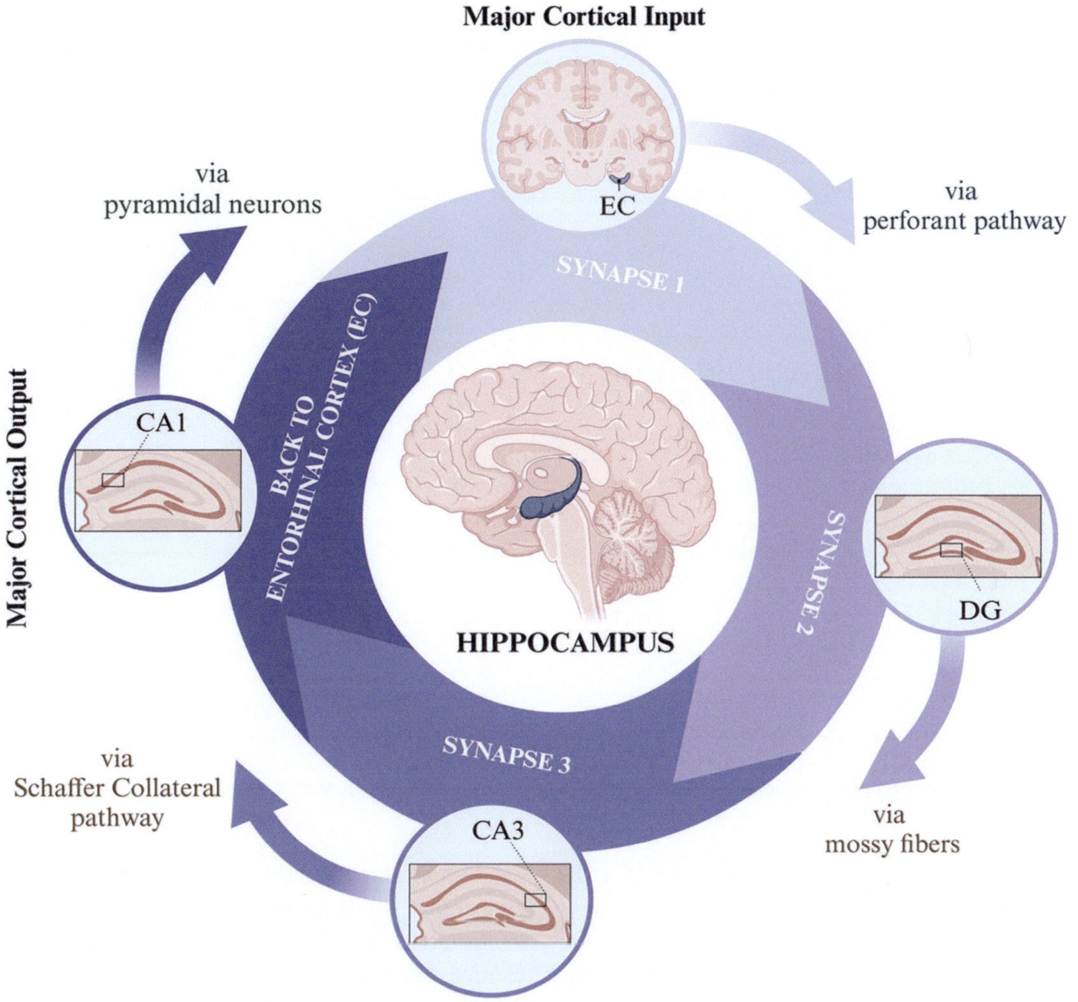

Fig. 15.2 The trisynaptic circuit or trisynaptic loop is an indicator of synaptic transmission in the hippocampus. In this loop, the entorhinal cortex (EC) provides the main cortical input to the hippocampus via the perforant pathway to the dentate gyrus (DG) (Synapse 1). The DG extends to the CA3 region via mossy fibers (Synapse 2). CA3 goes to the CA1 region via Schaffer's collateral (Synapse 3). Finally, pyramidal neurons from the CA1 region return to the EC to complete the loop. The main cortical input from the EC is evaluated in the loop and transmitted as the main cortical output from the CA1 region. Thanks to this loop, the encoding, retrieval, and processing steps of cognitive function are realized

The connection of the hippocampus within its parts is called the "trisynaptic loop." In this connection, the entorhinal cortex provides the main cortical input to the hippocampus via the perforant pathway to the dentate gyrus. The information coming to the dentate gyrus is sent to the CA3 region via mossy fibers. The cycle is completed when CA3 extends to CA1 via the Schaffer Collateral, and the input to CA1 returns to the entorhinal cortex (Fig. 15.2). This loop contributes to memory and cognitive processing. In addition to the main inputs from the entorhinal cortex, the hippocampus also receives direct inputs from the perirhinal cortex and postrhinal cortex. There are also important subcortical inputs from the locus coeruleus, raphe nucleus, nucleus reuniens, and amygdala. The CA3 and CA1 regions send outputs to the nucleus accum-

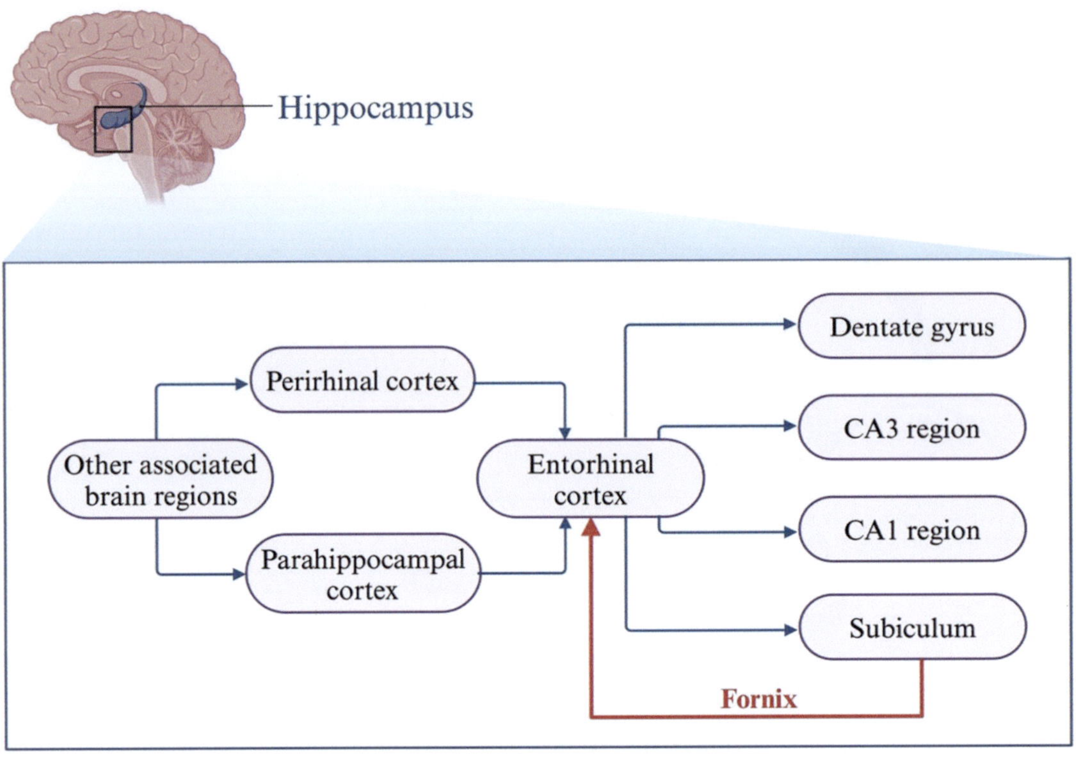

Fig. 15.3 Connection of hippocampus and other brain regions in cognitive functions

bens, amygdala, and prefrontal cortex via the fornix (Fig. 15.3) [16–18].

All this anatomical structure of the hippocampus provides an understanding of its roles in the evaluation of incoming information, sending it to the relevant region, and memory, learning, decision-making, and cognitive processes. The cortical and subcortical connections of the dorsal and ventral hippocampus differ, and accordingly, region-specific roles emerge. The dorsal hippocampus receives inputs from parts of the medial entorhinal cortex with spatial cells, while the ventral hippocampus is closely connected to the prefrontal cortex and amygdala. The dorsal hippocampus appears to be more important than ventral areas for encoding spatial and non-spatial memory [19]. A dorsal hippocampal connection extends to the presubiculum, which connects to the hippocampus by connecting the parietal cortex to the cingulate cortex and the parahippocampal cortex. The ventral hippocampal connection is located between the perirhinal, parahippocampal, temporal lobe, and entorhinal cortex. A third connection is between the orbitofrontal and ventromedial-prefrontal cortex and the hippocampal, entorhinal, and perirhinal cortex. The dorsal connection regulates the "where," the ventral connection regulates the "what," and the tertiary connection regulates the "reward" system and ensures that this information is integrated into the hippocampus. This suggests that the dorsal hippocampus and ventral hippocampus are differentially involved in spatial-emotional memory and cognition. These different connections are critical for important brain functions, including memory formation, emotional processing, and other cognitive processes [20].

As a result, each hippocampal subregion contributes uniquely to the encoding, consolidation, and retrieval of memories. The connectivity of the hippocampus within itself and with other regions is important in elucidating its role in cognitive functions. The input-output mechanism in regions connected through pathways ensures the correct evaluation of information. An example of a neuroanatomical pathway in which the hippocampus contributes is the Papez loop, which returns from the hippocampus through the fornix to the mammillary body, thalamus, cingulate gyrus, and parahippocampal region and back to the hippocampus. These pathways manifest different functions through dorsal, ventral, or subregions of the hippocampus. It is through these connections that complex networks and memory processes in the brain are understood (Fig. 15.3).

15.3 Learning, Decision-Making, and Associative Processes

Learning occurs as a result of the interaction between neurons and synapses. During learning, the brain modifies synapses to improve behavior. Physiologically, the learning process starts with electrical and chemical communication between neurons. This communication is carried out through neurotransmitters and electrical waves. Neurotransmitters associated with learning, such as dopamine, serotonin, and glutamate, contribute to the creation of memory, which is the physiological basis of learning. Memory is achieved through synaptic strengthening and repetition of neural activation that increases neuronal connections. Strengthening synapses through synaptic plasticity enables encoding and remembering new information in the learning process [21, 22].

The learning process takes place in many parts of the brain, including the neocortex, hippocampus, and basal ganglia. The hippocampus contributes to the learning process in cognitive activity through the medial temporal lobe (MTL). The MTL, including the hippocampus, has anatomical and functional connections that enable rapid learning to take place. Through these connections, information from the temporal, parietal, and frontal cortices is sent to the entorhinal cortex. The incoming stimulus is sent from the entorhinal cortex to the dentate gyrus, CA3, and CA1 area and back to the entorhinal cortex. Through this cycle, the encoding, retrieval, and processing steps of cognitive function are realized (Fig. 15.3) [23–25]. Moreover, the role of the hippocampus in learning is explained by neurogenesis. Neurogenesis is defined as the process of generating neurons, and it is suggested that this process continues in the hippocampus. It has been found that the hippocampus provides excitability and flexibility to the learning process with neurogenesis, and that neurons at different maturation stages make significant contributions to learning and memory [26]. Findings such as the decrease in neurogenesis over time and hippocampal atrophy suggest that hippocampus-dependent spatial learning decreases with age. Increasing hippocampal-dependent spatial learning may preserve overall cognitive function [27].

In summary, the learning process represents the accumulation of knowledge and experiences that an individual has, and this knowledge and these experiences help in the decision-making process. Adaptive behaviors provide an increase in quality of life and fulfillment of cognitive functions [28]. The prefrontal cortex (PFC) is recognized as a key brain region in the decision-making process. The decision-making process is realized as a result of the interaction of the PFC with cortical and subcortical structures. These connections are provided by the PFC, orbitofrontal cortex, anterior cingulate cortex, amygdala, thalamus, and cerebellum. Moreover, as in learning, it is observed that the MTL and hippocampus work together with the PFC to accept or reject the proposed plan in the decision-making process [29–31].

PFC and hippocampus cooperate in memory-based decision making. In a study, theta oscillation cycles in the hippocampus and behavioral sequences in the PFC were found to be involved in spatial decision-making [32]. While the hippocampus is linked to spatial navigation and memory encoding, the PFC is associated with cognitive functions such as decision-making. A dynamic interaction between these two regions ensures that the decision-making process is carried out correctly [33]. The hippocampus contributes to the decision-making process with flexible decision-making roles based on memory, inference, generalization, and thinking [34]. Due to the inferential and prospective functions of the hippocampus, it is necessary in the decision-making process, especially in the evaluation of decisions that need to be made under uncertain conditions [35]. By creating cognitive maps of relational information, the hippocampus strengthens the spatial map and weakens the predictive map in the decision-making process.

It provides an understanding of spatial locations and relationships by encoding the experienced transition structure as well as spatial relationships [36]. In this process, spatial memory contributes significantly to the decision-making process not only through the encoding of experiences but also through social learning. "Social learning," defined as learning from the experiences of the social environment, facilitates learning. The hippocampus also contributes to social learning through its cognitive functions and neural and behavioral mechanisms. This suggests that the hippocampus plays an important role in cognitive functions not only spatially but also through the processing of social experiences [37].

The hippocampus contributes to cognitive functions such as learning, memory, and decision-making through the interaction of multiple brain systems. It supports a flexible concept-learning mechanism that combines adaptation to spatial situations, rapid decision-making, and evaluation of information from the environment.

15.4 The Role of the Hippocampus in Spatial Cognition and Cognitive Flexibility

Spatial cognition is a representation of orientation and action with respect to the location of oneself, others, objects, or the environment. This cognitive process consists of the steps of understanding, extending, and communicating spatial information. Developmental, neural, and cognitive research has identified how people acquire and use information about their environment to determine where they are. It is evaluated based on scales such as memory, language, motor control, and attention [38–41]. Recent data show that the complex structure of spatial cognition is interpreted in different specialized subdivisions of the brain. Interaction between these different brain parts is necessary to enhance spatial cognition [42].

The hippocampus plays a central role in the processing of spatial cognition. Because of its critical role in memory, navigation, and cognition, the spatial role of the hippocampus is being investigated for cognitive mapping. The role of the hippocampus in spatial cognition is being elucidated by investigating other brain regions that receive hippocampal input and output. Through its memory and extensive connections, it generates peripheral cognition based on movement information, enabling navigation and positioning [7, 43]. During these connections, the hippocampus makes map-like spatial encodings with the entorhinal cortex and sends outputs to parahippocampal and hindbrain regions to link these maps to fixed environmental cues. The spatial codes generated by the entorhinal cortex work with frontal lobe areas to plan routes during navigation [44].

Investigation of the relationship between the hippocampus and the entorhinal cortex in spatial cognition has revealed the cells involved in this process [45]. Synaptic integrative mechanisms contribute to the encoding, storage, and retrieval of information in these cells [46]. Providing a better understanding of the mechanism of spatial cognition, these cells provide a highly detailed picture of current location and direction. The main cells involved in the construction of cognitive maps are place and boundary cells in the hippocampus and grid cells in the entorhinal cortex [47]. Place cells are a type of pyramidal neuron that is activated when a person enters a specific place. Encoding of place information, formation of place-related memory, and wayfinding ability are realized by place cells [48]. Boundary cells determine the boundaries of place and space in spatial cognition [49]. Grid cells in the entorhinal cortex integrate information about location, distance, and direction and create a coordinate system to understand location in space [50]. The hippocampus, which acts as *an important bridge* in the processes of processing new information and remembering old information through these cells, provides a person's ability to understand

and navigate the environmental world—that is, to form spatial memory. The person also needs to adapt to environmental variability while processing spatial information.

Cognitive flexibility is a dynamic process involving the appropriate regulation of behavior and thoughts in response to changing environmental stimuli [51, 52]. It provides complex responses such as finding new, adaptive solutions to changing demands and multiple tasks. The generation of appropriate behavioral responses is developed in a long-term manner through the ability to selectively switch between mental processes. Cognitive flexibility emerges through the interaction of cognitive and sensory-motor mechanisms [53, 54]. Among these interactions, the frontal lobes and hippocampus are prominently involved in successful behavior. The interaction between the orbitofrontal cortex and hippocampus is important in providing functional characteristics to advanced cognitive processes and cognitive flexibility [55]. The hippocampus contributes to social behavior and flexible cognition through memory. Its connections with other brain regions enable the flexible use of information. It is involved in the production, recombination, and flexible use of information through different roles such as language use, decision-making, social bonding, memory, and orientation [56]. Language use is the production of complex and novel expressions in a fast yet fluent manner and is not only based on vocabulary but is also associated with cognitive flexibility. In this process, the hippocampus contributes to the integration of multiple sources of information with the associative binding and representational flexibility provided by the declarative memory system [57].

The role of the hippocampus in cognitive flexibility is particularly evident in cases of hippocampal impairment. Studies have shown that hippocampal impairments lead to reduced cognitive flexibility in post-traumatic stress disorder and autism spectrum disorder [58, 59]. The dentate gyrus region can reduce memory through neurogenesis-mediated inhibition. This may help reduce anxiety and depressive-like behaviors by providing improved cognitive flexibility. Thus, the dentate gyrus in the hippocampus may enable the development of new treatment strategies for mood regulation and anxiety disorders [60]. Furthermore, ongoing neurogenesis in the hippocampus modulates network excitability and synaptic plasticity. Neurogenesis supports cognitive flexibility by disrupting established memories and promoting relearning. As a result, hippocampal neurogenesis contributes to cognitive flexibility by altering the response to evoked memory [61, 62].

Overall, the hippocampus contributes to the formation of spatial cognition and cognitive flexibility by interacting with other brain regions. Cognitive flexibility involves a person's ability to adapt to changing conditions, learn new information, and solve problems, while spatial memory involves a person's ability to understand and navigate the environmental world. The hippocampus helps in processing new information and remembering old information. In turn, it regulates cognitive flexibility through the processes of evaluating information from the environment, associating it with existing knowledge, and determining appropriate behaviors. Spatial cognition and cognitive flexibility function properly thanks to the hippocampus, a brain region with both functional and structural flexibility and in which neurogenesis is constantly ongoing.

15.5 Conclusion

The hippocampus is a brain region primarily associated with memory. It provides a spatial and temporal framework for episodic memory, creating a cognitive map. The dentate gyrus, cornu ammonis, and subiculum provide connections between other brain regions to store, process, and retrieve information from the environment. It helps in cognitive processes such as memory, learning, spatial cognition, cognitive flexibility, language skills, decision-making, and attention. Structurally and functionally, the hippocampus contributes to cognition through complex mechanisms such as the Papez circuit and the trisynaptic loop. It facilitates learning by processing incoming information in the cognitive process and helps to make decisions by choosing among learned alternatives. The hippocampus enables the formation of spatial cognition through the steps of understanding, expanding, and communicating the information received from the environment. It provides cognitive flexibility by regulating behaviors and thoughts appropriately in changing environmental conditions. All these cognitive activities are carried out by the hippocampus, which integrates the inputs from various brain regions and sends the output to the relevant region. A comprehensive understanding of its anatomical connections provides an understanding of the role of the hippocampus in the complex structure of cognition and cognitive processes.

References

1. van Strien NM, Cappaert NL, Witter MP. The anatomy of memory: an interactive overview of the parahippocampal-hippocampal network. Nat Rev Neurosci. 2009;10(4):272–82. https://doi.org/10.1038/nrn2614.
2. Fogwe LA, Reddy V, Mesfin FB. Neuroanatomy, hippocampus. StatPearls Publishing; 2023.
3. Ambrogioni L, Ólafsdóttir HF. Rethinking the hippocampal cognitive map as a meta-learning computational module. Trends Cogn Sci. 2023;27(8):702–12. https://doi.org/10.1016/j.tics.2023.05.011.
4. Rao YL, Ganaraja B, Murlimanju BV, Joy T, Krishnamurthy A, Agrawal A. Hippocampus and its involvement in Alzheimer's disease: a review. 3 Biotech. 2022;12(2):55. https://doi.org/10.1007/s13205-022-03123-4.
5. Khera T, Rangasamy V. Cognition and pain: a review. Front Psychol. 2021;12:673962. https://doi.org/10.3389/fpsyg.2021.673962.
6. Gellman M, RickTurner J, editors. Encyclopedia of behavioral medicine. New York: Springer; 2013. https://doi.org/10.1007/978-1-4419-1005-9.
7. Lisman J, Buzsáki G, Eichenbaum H, Nadel L, Ranganath C, Redish AD. Viewpoints: how the hippocampus contributes to memory, navigation and cognition. Nat Neurosci. 2017;20(11):1434–47. https://doi.org/10.1038/nn.4661.
8. Olsen RK, Moses SN, Riggs L, Ryan JD. The hippocampus supports multiple cognitive processes through relational binding and comparison. Front Hum Neurosci. 2012;6:146. https://doi.org/10.3389/fnhum.2012.00146.
9. Hartley T, Bird CM, Chan D, Cipolotti L, Husain M, Vargha-Khadem F, Burgess N. The hippocampus is required for short-term topographical memory in humans. Hippocampus. 2007;17(1):34–48. https://doi.org/10.1002/hipo.20240.
10. Tyng CM, Amin HU, Saad MNM, Malik AS. The influences of emotion on learning and memory. Front Psychol. 2017;8:1454. https://doi.org/10.3389/fpsyg.2017.01454.
11. Dekeyzer S, De Kock I, Nikoubashman O, Vanden Bossche S, Van Eetvelde R, De Groote J, Acou M, Wiesmann M, Deblaere K, Achten E. "Unforgettable"—a pictorial essay on anatomy and pathology of the hippocampus. Insights Imaging. 2017;8(2):199–212. https://doi.org/10.1007/s13244-016-0541-2.
12. Chauhan P, Jethwa K, Rathawa A, Chauhan G, Mehra S. The anatomy of the hippocampus. In: Pluta R, editor. Cerebral ischemia. Exon Publications; 2021.
13. Witter MP. Connectivity of the hippocampus. In: Cutsuridis V, Graham B, Cobb S, Vida I, editors. Hippocampal microcircuits, Springer series in computational neuroscience. Cham: Springer; 2018. https://doi.org/10.1007/978-3-319-99103-0_1.
14. Mercer A, Thomson AM. Cornu Ammonis regions-antecedents of cortical layers? Front Neuroanat. 2017;11:83. https://doi.org/10.3389/fnana.2017.00083.
15. Cherubini E, Miles R. The CA3 region of the hippocampus: how is it? What is it for? How does it do it? Front Cell Neurosci. 2015;9:19. https://doi.org/10.3389/fncel.2015.00019.
16. Knierim JJ. The hippocampus. Curr Biol. 2015;25(23):R1116–21. https://doi.org/10.1016/j.cub.2015.10.049.
17. Basu J, Siegelbaum SA. The corticohippocampal circuit, synaptic plasticity, and memory. Cold Spring Harb Perspect Biol. 2015;7(11):a021733. https://doi.org/10.1101/cshperspect.a021733.
18. Bayer H. Deciphering CA2 connectivity. Nat Neurosci. 2014;17(2):152. https://doi.org/10.1038/nn0214-152.
19. Moser MB, Moser EI. Functional differentiation in the hippocampus. Hippocampus. 1998;8(6):608–19. https://doi.org/10.1002/(SICI)1098-1063(1998)8.
20. Huang CC, Rolls ET, Hsu CH, Feng J, Lin CP. Extensive cortical connectivity of the human hippocampal memory system: beyond the "what" and "where" dual stream model. Cereb Cortex (New York, N.Y.: 1991). 2021;31(10):4652–69. https://doi.org/10.1093/cercor/bhab113.
21. Boss BJ. The neuroanatomical and neurophysiological basis of learning. J Neurosci Nurs. 1986;18(5):256–64. https://doi.org/10.1097/01376517-198610000-00004.
22. Lillicrap TP, Santoro A, Marris L, Akerman CJ, Hinton G. Backpropagation and the brain. Nat Rev Neurosci. 2020;21(6):335–46. https://doi.org/10.1038/s41583-020-0277-3.
23. Seger CA, Miller EK. Category learning in the brain. Annu Rev Neurosci. 2010;33:203–19. https://doi.org/10.1146/annurev.neuro.051508.135546.
24. Becker S, Wojtowicz JM. A model of hippocampal neurogenesis in memory and mood disorders. Trends Cogn Sci. 2007;11(2):70–6. https://doi.org/10.1016/j.tics.2006.10.013.

25. Green CS, Bavelier D. Exercising your brain: a review of human brain plasticity and training-induced learning. Psychol Aging. 2008;23(4):692–701. https://doi.org/10.1037/a0014345.
26. Deng W, Aimone JB, Gage FH. New neurons and new memories: how does adult hippocampal neurogenesis affect learning and memory? Nat Rev Neurosci. 2010;11(5):339–50. https://doi.org/10.1038/nrn2822.
27. Konishi K, Mckenzie S, Etchamendy N, Roy S, Bohbot VD. Hippocampus-dependent spatial learning is associated with higher global cognition among healthy older adults. Neuropsychologia. 2017;106:310–21. https://doi.org/10.1016/j.neuropsychologia.2017.09.025.
28. Lee D. Decision making: from neuroscience to psychiatry. Neuron. 2013;78(2):233–48. https://doi.org/10.1016/j.neuron.2013.04.008.
29. Broche-Pérez Y, Herrera Jiménez LF, Omar-Martínez E. Neural substrates of decision-making. Bases neurales de la toma de decisiones. Neurologia (Barcelona, Spain). 2016;31(5):319–25. https://doi.org/10.1016/j.nrl.2015.03.001.
30. Herd S, Krueger K, Nair A, Mollick J, O'Reilly R. Neural mechanisms of human decision-making. Cogn Affect Behav Neurosci. 2021;21(1):35–57. https://doi.org/10.3758/s13415-020-00842-0.
31. Zeithamova D, Mack ML, Braunlich K, Davis T, Seger CA, van Kesteren MTR, Wutz A. Brain mechanisms of concept learning. J Neurosci. 2019;39(42):8259–66. https://doi.org/10.1523/JNEUROSCI.1166-19.2019.
32. Tang W, Shin JD, Jadhav SP. Multiple time-scales of decision-making in the hippocampus and prefrontal cortex. eLife. 2021;10:e66227. https://doi.org/10.7554/eLife.66227.
33. Tavares LCS, Tort ABL. Hippocampal-prefrontal interactions during spatial decision-making. Hippocampus. 2022;32(1):38–54. https://doi.org/10.1002/hipo.23394.
34. Biderman N, Bakkour A, Shohamy D. What are memories for? The hippocampus bridges past experience with future decisions. Trends Cogn Sci. 2020;24(7):542–56. https://doi.org/10.1016/j.tics.2020.04.004.
35. Attaallah B, Petitet P, Zambellas R, Toniolo S, Maio MR, Ganse-Dumrath A, Irani SR, Manohar SG, Husain M. The role of the human hippocampus in decision-making under uncertainty. Nat Hum Behav. 2024;8:1366. https://doi.org/10.1038/s41562-024-01855-2.
36. Garvert MM, Saanum T, Schulz E, Schuck NW, Doeller CF. Hippocampal spatio-predictive cognitive maps adaptively guide reward generalization. Nat Neurosci. 2023;26(4):615–26. https://doi.org/10.1038/s41593-023-01283-x.
37. Leblanc H, Ramirez S. Linking social cognition to learning and memory. J Neurosci. 2020;40(46):8782–98. https://doi.org/10.1523/JNEUROSCI.1280-20.2020.
38. Waller D, Nadel L. Introduction: frameworks for understanding spatial thought (or wrapping our heads around space). In: Waller D, Nadel L, editors. Handbook of spatial cognition. American Psychological Association; 2013. p. 3–12. http://www.jstor.org/stable/j.ctv1chs7vk.5.
39. Klencklen G, Després O, Dufour A. What do we know about aging and spatial cognition? Reviews and perspectives. Ageing Res Rev. 2012;11(1):123–35. https://doi.org/10.1016/j.arr.2011.10.001.
40. Burgess N. Spatial cognition and the brain. Ann N Y Acad Sci. 2008;1124:77–97. https://doi.org/10.1196/annals.1440.002.
41. Marshall JC, Fink GR. Spatial cognition: where we were and where we are. NeuroImage. 2001;14(1 Pt 2):S2–7. https://doi.org/10.1006/nimg.2001.0834.
42. Landau B. Spatial cognition. In: Ramachandran VS, editor. Encyclopedia of the human brain. Academic; 2002. p. 395–418. https://doi.org/10.1016/B0-12-227210-2/00326-5.
43. Yuan J, Guo W, Zha F, Wang P, Li M, Sun L. A bionic spatial cognition model and method for robots based on the hippocampus mechanism. Front Neurorobot. 2022;15:769829. https://doi.org/10.3389/fnbot.2021.769829.
44. Epstein RA, Patai EZ, Julian JB, Spiers HJ. The cognitive map in humans: spatial navigation and beyond. Nat Neurosci. 2017;20(11):1504–13. https://doi.org/10.1038/nn.4656.
45. Moser EI, Moser MB, McNaughton BL. Spatial representation in the hippocampal formation: a history. Nat Neurosci. 2017;20(11):1448–64. https://doi.org/10.1038/nn.4653.
46. Schmidt-Hieber C, Nolan MF. Synaptic integrative mechanisms for spatial cognition. Nat Neurosci. 2017;20(11):1483–92. https://doi.org/10.1038/nn.4652.
47. Hartley T, Lever C, Burgess N, O'Keefe J. Space in the brain: how the hippocampal formation supports spatial cognition. Philos Trans R Soc Lond Ser B Biol Sci. 2013;369(1635):20120510. https://doi.org/10.1098/rstb.2012.0510.
48. Welberg L. A goal for place cells. Nat Rev Neurosci. 2007;8:164. https://doi.org/10.1038/nrn2101.
49. Barry C, Lever C, Hayman R, Hartley T, Burton S, O'Keefe J, Jeffery K, Burgess N. The boundary vector cell model of place cell firing and spatial memory. Rev Neurosci. 2006;17(1–2):71–97. https://doi.org/10.1515/revneuro.2006.17.1-2.71.
50. Moser EI, Roudi Y, Witter MP, Kentros C, Bonhoeffer T, Moser MB. Grid cells and cortical representation. Nat Rev Neurosci. 2014;15(7):466–81. https://doi.org/10.1038/nrn3766.
51. Uddin LQ. Cognitive and behavioural flexibility: neural mechanisms and clinical considerations. Nat Rev Neurosci. 2021;22(3):167–79. https://doi.org/10.1038/s41583-021-00428-w.
52. Hohl K, Dolcos S. Measuring cognitive flexibility: a brief review of neuropsychological, self-report, and neuroscientific approaches. Front Hum Neurosci. 2024;18:1331960. https://doi.org/10.3389/fnhum.2024.1331960.

53. Ionescu T. Exploring the nature of cognitive flexibility. New Ideas Psychol. 2012;30(2):190–200. https://doi.org/10.1016/j.newideapsych.2011.11.001.
54. Dajani DR, Uddin LQ. Demystifying cognitive flexibility: implications for clinical and developmental neuroscience. Trends Neurosci. 2015;38(9):571–8. https://doi.org/10.1016/j.tins.2015.07.003.
55. Thonnard D, Callaerts-Vegh Z, D'Hooge R. Effects of orbitofrontal cortex and ventral hippocampus disconnection on spatial reversal learning. Neurosci Lett. 2021;750:135711. https://doi.org/10.1016/j.neulet.2021.135711.
56. Rubin RD, Watson PD, Duff MC, Cohen NJ. The role of the hippocampus in flexible cognition and social behavior. Front Hum Neurosci. 2014;8:742. https://doi.org/10.3389/fnhum.2014.00742.
57. Duff MC, Brown-Schmidt S. The hippocampus and the flexible use and processing of language. Front Hum Neurosci. 2012;6:69. https://doi.org/10.3389/fnhum.2012.00069.
58. Zabag R, Deri O, Gilboa-Schechtman E, Richter-Levin G, Levy-Gigi E. Cognitive flexibility in PTSD individuals following nature adventure intervention: is it really that good? Stress (Amsterdam, Netherlands). 2020;23(1):97–104. https://doi.org/10.1080/10253890.2019.1645113.
59. Uddin LQ. Brain mechanisms supporting flexible cognition and behavior in adolescents with autism spectrum disorder. Biol Psychiatry. 2021;89(2):172–83. https://doi.org/10.1016/j.biopsych.2020.05.010.
60. Anacker C, Hen R. Adult hippocampal neurogenesis and cognitive flexibility—linking memory and mood. Nat Rev Neurosci. 2017;18(6):335–46. https://doi.org/10.1038/nrn.2017.45.
61. Burghardt NS, Park EH, Hen R, Fenton AA. Adult-born hippocampal neurons promote cognitive flexibility in mice. Hippocampus. 2012;22(9):1795–808. https://doi.org/10.1002/hipo.22013.
62. Martínez-Canabal A, López-Oropeza G, Sotres-Bayón F. Hippocampal neurogenesis facilitates cognitive flexibility in a fear discrimination task. Front Behav Neurosci. 2024;17:1331928. https://doi.org/10.3389/fnbeh.2023.1331928.

Role of the Hippocampus in Time and Memory

16

Bircan Aslan

16.1 Introduction

The relationship between memory, a higher cognitive function, and the hippocampus has been known since Henry Molaison [1]. It has been proposed that memory has six components: space, sensation/perception, time, reaction, and affect [2]. It has been thought since Aristotle that memory is temporally organized and that we remember experiences organized in time, and Tulving stated that time is the basic organizing dimension of episodic memory [3]. The concept of time is also important for discriminating similar events that happened in the same places at different times [4]. With the statement that the hippocam pus has a key role in episodic memory, the temporal role of the hippocampus in memory began to be investigated [5, 6]. In recent years, the relationship of the hippocampus with space and time, which are among the components of memory, has begun to be examined, and many studies have been conducted especially on its relationship with space. The relationship with time has gained importance in recent years, and the focus of this article is the relationship of the hippocampus with time and memory.

B. Aslan (✉)
Department of Physiology, Yozgat Bozok University Faculty of Medicine, Yozgat, Türkiye
e-mail: bircan.aslan@bozok.edu.tr

16.1.1 Hippocampus and Memory

16.1.1.1 Episodic Memory

The hippocampus is the center of declarative memory, a type of long-term memory, and a component of the cortico-hippocampal circuit necessary for declarative memory processes. Declarative memory in other mammals is typically defined as spatial, temporal, associative, or rapidly acquired memory, whereas in humans, it is usually expressed by the conscious recall of facts and events. The hippocampus is a central structure in episodic memory, one of the two subtypes of declarative memory [7–9]. Tulving defined episodic memory and argued that it is unique to humans. On the other hand, later researchers argued that rodents also have episodic memory and that rodent studies can help in understanding human episodic memory [10]. The events we see while watching a movie are tempo rally continuous. However, in our memory (episodic memory), when we remember that movie, we draw boundaries between the events according to the differences between them and remember each part as an episode [11]. Episodic memory, which constitutes our memories, consists of spatial, temporal, and semantic information related to each other. The job of the hippocampus in episodic memory formation is to combine this spatial, temporal, or semantic information to form meaningful wholes. While memory acquisition requires an intact hippocampus,

M. Turgut et al. (eds.), *The Human Hippocampus*, https://doi.org/10.1007/978-3-031-88907-3_16

as we know from Henry Molaison, the hippocampus has no function in memory recall [7]. Animal and human studies have also clearly demonstrated the critical role of the hippocampus in episodic memory [12, 13].

It has been suggested that the hippocampus maps experiences according to when and where they occurred [10]. When we recall an event in our memory, we also remember information such as the place, time, and sequence of the event. The hippocampus cooperates with the entorhinal, parahippocampal, and perirhinal cortices in the medial temporal cortex to answer the questions of what, where, and when, which are the components of episodic memory. The what information received from the neocortex is carried from the perirhinal cortex to the lateral entorhinal cortex and then to the hippocampus, and the where and when information is carried from the parahippocampal cortex to the medial entorhinal cortex and then to the hippocampus. The hippocampus integrates this information in a context and forms a coherent and detailed memory (Fig. 16.1). This task of the hippocampus is important as it encodes not only the event but also the context and time of the event, enabling different experiences with overlapping memories to be distinguished [3, 7, 14].

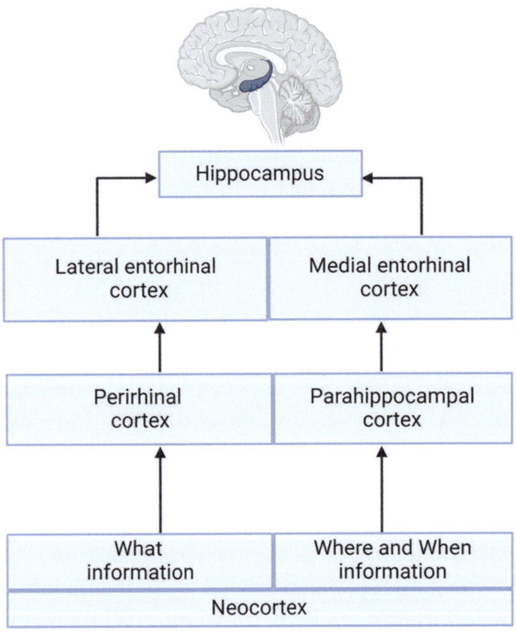

Fig. 16.1 Representations of what information and where and when information in the medial temporal lobe. (Adapted from Howard Eichenbaum's article [14])

The parahippocampal and perirhinal cortex select and filter inputs to the hippocampus and information to be stored in memory. The entorhinal cortex is closely related to the hippocampus as it sends cortical input to the hippocampus and receives output from the hippocampus. Animal studies have shown that the hippocampus is important in recognition memory; the perirhinal cortex is sufficient for remembering simple objects, while the hippocampus is needed as the object becomes more complex [7]. Human studies have reported that recognition is of two types: either by familiarity of the stimulus or by recollection of the stimulus with old memories [15]. It has been shown that the hippocampus may especially support recollection-related recognition and that people with hippocampus damage have impaired recollection-related recognition memory [10]. Human fMRI studies have reported activation in both the hippocampus and parahippocampal region during temporal sequence memory tasks [16]. Furthermore, studies have shown that hippocampal activity during recall of object sequences reflects identity and temporal location for all objects. In an fMRI study, Hsieh et al. showed that during retrieval of learned object sequences, hippocampal activity patterns include the temporal location of the objects in the sequences, the parahippocampal cortex generates neuronal patterns encoding temporal location, and the perirhinal cortex encodes object information [10, 17].

16.1.1.2 Spatial Memory and Navigation

The terms spatial and non-spatial appear frequently in hippocampal research, but the distinction is not sharply demarcated. When we speak of space, we speak in terms of a reference point. We refer to the positioning of objects in space as egocentric if we refer to our body or allocentric if we refer to the position of objects relative to each other. The hippocampus is important in rodent allocentric spatial memory. O'Keefe and Nadel reported that the hippocampus supports episodic memory in temporal and spatial contexts [11] and organizes our memories in episodic memory with temporal and spatial frames [18]. When we remember an event, it appears in our memory as a temporally ordered sequence. Navigation is like a sequence of events and, like episodic memory, consists of sequential representations of events or locations. In memory-related navigation, the hippocampus is considered to encode information about time intervals and sequencing [10, 19]. Navigation was critical for the survival of primitive animals and humans. O'Keefe and Nadel put forward the theory that the hippocampus forms a cognitive map and stated that the hippocampus provides a spatial map that helps animals organize and remember environmental information. In studies, it was observed that hippocampal neurons fired when experimental animals passed through places in their living environment [20, 21].

Studies have shown that the hippocampus is important in spatial memory and that the CA3 region is critical and that lesions of this region impair memory with a spatial component [22, 23]. O'Keefe and Dostrovsky discovered place cells, pyramidal neurons in the CA1 region of the rat dorsal hippocampus that encode the animal's location within a given environment and are thought to represent a cognitive map in navigation. Due to the role of the hippocampus in spatial navigation, it has been suggested that cues in the external environ ment influence the firing of place cells [24]. Hippocampal recordings from subsequent rodent experiments showed that if the animal follows a specific route in a maze, place cells encode the spatial location as the steps of a sequence, and the sequence of these steps is represented in the firing patterns of place cells [4, 6, 18]. Moreover, it has been reported that place cells exhibit pathway-specific firing sequences prior to correct memory choices, while their firing decreases or does not fire at all prior to incorrect memory choices, thus predicting the accuracy of the subsequent memory [3].

16.1.2 Hippocampus and Time

Since the hippocampus represents spatial and temporal contexts in episodic memory, it might be thought to encode space and time together, but findings show that they are encoded by the same neurons but independently and mixed [25]. Space and time are not seen as separate entities but are referred to as spacetime [26]. When we talk about hippocampal spatial memory, we include the fourth dimension, time, in addition to the three-dimensional spatial memory [7].

16.1.2.1 Source of Time Information

With the support of different regions of the brain, including prefrontal and parietal regions, the hippocampus receives temporal information from the medial entorhinal cortex and uses temporal information from multiple sources in the temporal organization of memories [2, 3, 27]. Although the hippocampus is associated with time, the mechanism of this relationship is not yet understood. Two views have been proposed for the neural mechanisms underlying time processing; the first view suggests that timing is centrally generated and distributed to areas of the brain. The second view claims that each subsystem generates its own timing. Although there is evidence to support that each neuronal system generates its own timing and uses it for processing [28], it is still unclear whether the actual mechanism is one of these two views or a combination of both [3].

16.1.2.2 Representation of Time in Sub-hippocampal Areas

In studies on the representation of time in sub-areas of the hippocampus recorded from single neurons, it has been reported that temporal representations are encoded in CA1, CA2, and CA3 areas. However, there is no consensus between computational models and experimental data in the literature, and lesion studies are also variable [10]. While computational models point to the CA3 region due to the connectivity properties of the cells it contains, studies on the cells encoding the passage of time have focused on the CA1 region, as you will see from the articles cited in this article, and the first evidence came from CA1 neurons recorded in a task in which rats had to remember sequences of events. Although the fact that most of the studies were conducted in the CA1 region may be a reason for the discrepancies between the computational model and experimental data in the literature, there is no doubt that the CA1 region is important [3, 4]. In an experiment in which rats were asked to learn to associate objects with a certain amount of time between them, rats with CA3 lesions were as successful as normal rats, while rats with CA1 lesions were impaired in learning, and it was stated that the CA3 region is important in spatial and the CA1 region in temporal representation. In another study, it was reported that the CA3 region was effective in a 3-s short-term sequential memory task in an odor task, and when the duration was extended to 10 s, the CA1 region became as important as CA3 [22, 29]. Temporal and spatial coding of CA3 and CA1 neurons were examined in rats performing a maze task, and similar temporal modulation was observed in the activities of CA3 and CA1 neurons in the task requiring memory, and it was stated that the CA3 and CA1 regions were jointly responsible for temporal coding [27]. Evidence has been reported that the CA2 region may also be important for time. It has been reported that the activities of neurons in the CA2 region can be utilized to understand and distinguish the time difference between events, that neuronal activity patterns change as the time difference between repeated events increases, and that these changes help animals to understand time intervals on longer time scales [3, 12, 23].

16.1.2.3 Temporal Organization

The events that make up episodic memory have temporal organization, and for this, the hippocampus helps connect consecutive events by bridging time gaps in order to remember sequences of events and form sequential memories [12]. The temporal function of the hippocampus includes the order and duration of events in memory [26]. Although the strict chaining model and temporal context models have been proposed to explain how the hippocampus performs temporal organization, the evidence favors the temporal context model. This model states that neurons fire in sequential order to encode temporal context. It is the duration of firing that indicates the passage of time. This model suggests that the brain carries a representation of recent events and a progressively weaker representation of older events. In temporal organization, the hippocampus is like a structure in which related events are placed. For this, it helps to process and remember the sequence of two events, when they occurred, and the time between events [3, 6].

Time and space, which cannot be measured, are handled as duration and distance in order to be measurable [30]. Duration refers to the interval between the beginning and end of an event or the interval between two events. The hippocampus is critical in retrospective timing, where time is encoded in the background and randomly to support organizing sequences of events in memory. Duration information becomes accessible when recalling event sequences and is inferred from the sequential firing of time cells that encode the temporal representation of different event sequences [13]. People with hippocampal damage have severe problems remembering the sequence of events but less so with distance and duration estimation. There are also studies in the literature on the perception of time intervals in rats showing that the hippocampus has a function in time perception and duration memory. Hippocampal-damaged rats have difficulty remembering the sequence of odors, whereas they recognize recent odors [2, 30, 31]. Studies have shown that the hippocampus is particularly involved in bridging temporal gaps that lack specific external cues to connect disconnected events that form sequential memories [32].

16.1.2.4 Sequential Memory

Sequential memory is responsible for sequential memory for spatial and non-spatial events [22], and one of the main tasks of the hippocampus is to bridge successive events and link them together in time [31]. Establishing relationships between temporally close events is important in the formation of episodic memory in humans, and in experiments, subjects showed better recall performance when they recalled using time. Animal and human experiments have shown that the hippocampus is necessary for recalling the sequence of memories, but not for remembering a specific event. In human studies, it has been clearly demonstrated that the hippocampus performs temporal encoding in memory [3, 18, 26].

There is a relationship between episodic memory and odor [33], and hippocampal neurons have been found to be sensitive to odor recognition, even though hippocampal lesions do not impair odor recognition memory [10]. Kesner pointed to the critical role of the hippocampus in sequential learning beyond odor recognition [4] and designed list learning experimental models parallel to human experiments for use in rats. Since then, these models have been used in research on the encoding of temporal sequences and temporal ordering. To test the ability of normal and hippocampus-damaged rats to remember odors and their temporal order in an experiment investigating the role of the hippocampus in sequential memory, rats sampled and were rewarded for each of the odors in a list, and then, a few minutes later, in an order test, they were presented with two non-consecutive odors from the list and asked to choose the less experienced odor for reward. While normal rats performed well in temporal order memory, hippocampal-damaged rats had impaired temporal order memory at all delays and performed poorly in sequential memory tests [31].

When a group of healthy and hippocampal-damaged patients were asked what they remembered after walking around a university campus, hippocampal-damaged subjects remembered fewer details such as objects seen, events, and locations of events than the healthy group. While the healthy subjects remembered the sequence of consecutive events correctly, the patients partially remembered the events but failed to remember the sequence of events. These results are in line with animal experiments, and the function of the hippocampus to form the sequence of events is supported by both lesion and imaging data [6]. In fMRI studies, hippocampal activity has been shown to increase during the learning and retrieval of nonspatial temporal sequence information. Jenkins and Ranganath, in their fMRI study, after subjects were shown a series of objects, asked them to recall the order in which the objects were shown using working memory. The hippocampal activity of the subjects who remembered correctly increased compared to those who remembered incorrectly, thus demonstrating the role of the hippocampus in the temporal context of sequential memory [10, 34]. Temporal encoding not only represents elapsed time but also helps to represent the sequence of events in experiences. Knowing the sequence of events is important because it allows us to use our experiences to predict and plan what will happen in the future [30, 31].

16.1.2.5 Time Cells

It has been discovered that in the hippocampus, there are neurons that fire at specific moments of an event and create temporal maps of experiences. It has been found that the firing patterns of these neurons differ at time intervals bridging different object-odor pairs, indicating that these neurons uniquely encode memory and organize it over time. Since the properties of these neurons are similar to place cells and they can encode time, these cells are called time cells and their firing fields are called time fields, inspired by place fields [18, 31, 35, 36]. Time cells have been reported to be present in the medial entorhinal cortex in addition to the hippocampus, and it has been shown that timing is impaired in lesions of this region [27]. It has been suggested that these cells are important in the sequential organization of episodic memories, and time cells are thought to work according to the temporal context memory model. According to this model, neurons fire in sequential order to encode the temporal context, with the firing time indicating the passage of time. This model explains the re-timing of time cells discussed below and the different firing patterns that are generated to distinguish overlapping memories [3].

In free-ranging animals, time cells were found to be activated at specific locations. In rats performing a spatial alternation task involving right-left turns and running on a treadmill between turns, neurons that fired at certain moments while the animal was on the treadmill also fired when the animal passed through certain points of the maze, indicating that place cells are also time cells. Time cells are activated sequentially and overlap their firing sites, bridging the entire latency between two events in the sequence [3, 13]. Time cells bridge the delay only when animals need to use memory, not

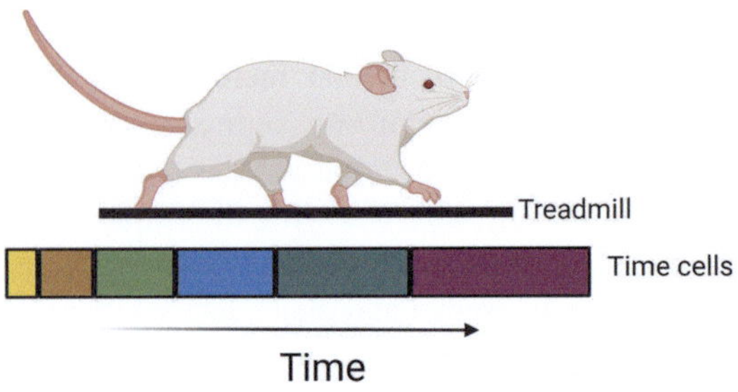

Fig. 16.2 Time cells and time fields. Each colored box represents a time cell. As time progresses and time passes, the time fields grow. (Adapted from Howard Eichenbaum's article [6])

when memory is not required [18, 24]. Time cells indicate the passage of time in a time interval, such as the latency in a task. In studies examining the firing of these cells, changes in the firing pattern when delay times increase have shown that their firing is a representation of elapsed time [12]. In a maze task experiment, the size of the place fields remained the same, whereas the size of the time fields increased with latency on the treadmill, showing that the encoding of time is non-linear [6, 27]. That is, time cells fired later in a sequence fired for a longer period of time, suggesting that time is encoded as a scalar (Fig. 16.2). Although time cells encode very short temporal intervals, they can also encode longer time intervals [10]. Time cells have been found to be associated with the theta rhythm, which is thought to be critical in declarative memory [9]. While gamma oscillations encode items within a temporal frame during episodic memory formation, the theta rhythm has been reported to regulate the sequential flow of information and the temporal ordering of items [37]. It has also been reported that theta oscillations fine-tune the timing of place cell firing during wakefulness and play a critical role in the encoding and consolidation of spatial memory [38].

Time and space are processed by similar mechanisms in the hippocampus. In one experiment, it was observed that the activity patterns of most time cells were reorganized when a temporal parameter was changed, similar to the reorganization of the activity patterns of place cells when a spatial cue was changed [3]. Just as place cells are remapped when the spatial context of the animal changes, time cells are also rescheduled when the temporal structure or behavioral context of a task changes [11]. Just as the activity of place cells is affected by non-spatial variables, the firing patterns of time cells are affected by variables such as location and behavior [36].

16.2 Discussion

Hippocampal neurons can encode time independent of behavior and space. It organizes this encoding and temporal ordering of events with time cells. Experiments reporting that the hippocampus encodes time have been challenged by saying that time is confused with the animal's position and distance and that the activities seen belong to spatial features, not time. This is because the hippocampus can encode not only time and space but also information that is important in the behavioral context. In certain behavioral models, place cells can encode spatial (distance and direction) and non-spatial (smell) features and actions in different situations, while time cells can encode location, distance, and ongoing behavior in free-moving animals. However, studies have shown that when the effects of variable factors are removed, the temporal signal generated by the short duration and successive firing and sequencing of time cells is independent of position and movement [3, 11, 18, 25, 31]. When we look at the studies to which we owe what we know today about time cells, neurons encoding the passage of time were first observed by Manns et al. in an experiment in which they aimed to investigate how rat CA1 neurons behave in encoding and retrieval processes in an odor sequence memory task and their relationship with the theta rhythm. In the experiment, which included two different recognition memory tasks, rats received a reward for the odors they sampled in one of the tasks, and after a delay period, they were asked to identify which of the odors presented for testing was presented first. In all recordings, the firing patterns of CA1 ensembles gradually changed even within a few minutes of encoding the sequences. Neural activity was correlated with time on correct test trials but not on incorrect trials. These results are the first evidence that the temporal context represented in the hippocampus changes gradually [2, 9, 10].

It has been proposed that hippocampal neurons can produce sequential firing patterns through environmental cues or spontaneous internal mechanisms. To test these hypotheses, Pastalkova et al. examined the activity patterns of CA1 neurons in different environments, such as a wheel and a maze, in a spatial task in which rats rotated alternately left and right in a T-maze, including a fixed duration wheel run separating trials. The animal had to run on the wheel during the delay between trials and then turn left or right for a reward. Although the animal's position relative to the maze remained constant during the run [10], it was reported for the first time that a number of neurons fired at specific moments during the wheel run, with all moments of each run represented by a series of short neuronal activations. The researchers called these neurons episodic cells because they fired transiently and reliably at specific moments in the wheel task. The mechanisms by which these episodic cells form episodic fields were found to be similar to those of place cells. These neuron arrays not only encoded the time elapsed during the delay time in the wheel, but it was also reported that the path of the rat after the delay time could be predicted with these arrays. These sequence patterns represented a unique array of neurons that were activated in a specific time pattern from the time the animal entered the wheel until it won the reward. Because the firing sequences were different but internally consistent across trials in which the rat turned left or right, it was thought that the sequences could be linked to the context of the trial. It was found that idiothetic cues related to movement did not play a critical role in the firing of neurons. Researchers have suggested that in learning, external stimuli play a critical role in determining the temporal sequence of events in neuronal representations, while in recall, imagination, and planning, the sequences may be determined by the internal dynamics of the network [24, 31]. Manns et al. could not determine whether different temporal context representations were produced for certain memories since the task involved unique memories in each trial. Pastalkova et al. observed that the different hippocampal neuron arrays they saw were associated with subsequent memory choices and that these arrays were disrupted when animals made errors [2]. Itskov et al. analyzed Pastalkova's data and reported that the firing patterns of rat hippocampal CA1 neurons can generate reliable and long-lasting sequences in the absence of variable external stimuli and that the sequences recorded from these neurons can accurately estimate the time elapsed during wheel running [28]. In Pastalkova et al.'s study, a wheel-running task was used for theta activity associated with time-related firing patterns. Gill et al. examined the firing patterns of CA1 neurons during the delay between trials in a rewarded plus maze task in rats to determine whether continuous motion associated with theta activity is necessary for the generation of firing patterns that occur during the delay period. In the first part of the experiment, trials began in one of the non-reward arms of a plus maze, and rats traveled to the reward arm; in the second part of the experiment, the reward arm was switched, and trials began in one of the other arms. Between trials, rats were moved to a platform adjacent to the maze where they could move freely without having to run continuously. During the time between trials, time-specific firing patterns emerged, and firing sequences varied between the two sessions. Because the rats moved freely during the delay, cells known to be place fields were excluded from the analysis, thus demonstrating that the measured activity encoded time. Delay-related temporal firing patterns were different for left-right turns and were able to predict the direction of the turn. This suggests that delay-related temporal firing patterns may be important in the maintenance of memory across delay [31, 39].

The findings on the relationship between hippocampal neurons and time shown in rodent studies can be generalized to primates. Naya and Suzuki found that during a short-term memory test in which monkeys with head immobilization were shown two images with a time delay between them, most neurons represented time, although few neurons encoded what the images were during the time delay [2, 10]. However, in this study, it was observed that time was generally encoded instead of the temporal order of objects. This was thought to be due to high interference, which confused the information and prevented reliable encoding of objects [4, 16].

MacDonald et al. investigated whether CA1 neurons encode sequential events and how they encode time connecting separate events in a non-spatial task [36]. In the experiment, rats were exposed to an odor after sampling one of two objects followed by a fixed latency period, using associative memory and digging in the sand for reward if the odor matched the object. Neurons encoded the passage of time by firing sequentially at specific points in time during the delay period according to a specific pattern. These neurons are called time cells because they generate firing patterns that reflect the passage of time [18]. In addition to the passing of time, time cells were also found to encode the animal's position, behavior, objects it was exposed to, and odors. Despite the possibility of confusion of temporal signals with behavior, position, speed, and head direction, the firing patterns of neurons were analyzed, and it was found that although neurons also represented behavior and position during the delay, these variables did not explain the time signal generated by neurons. Moreover, when the delay was extended, the firing patterns of most neurons were different and re-timed. Moreover, as the end of the delay period approached, the firing neurons remained active longer and their time fields became larger, indicating that time was encoded as a scalar. Since the animal did not move freely during the delay period, it was suggested that movement during running could not be effective in neuronal firing patterns [2, 31, 36]. Path integration is one of the main tasks of the hippocampus and refers to the fact that the firing of neurons during movement is affected by the distance traveled. To exclude the effect of path integration in the encoding of time in the hippocampus, Kraus et al. recorded from CA1 neurons while rats ran in place at variable speeds on a treadmill in an eight-shaped maze. The rats' location, behavior, and head position remained constant, while running time and distance were varied so that time and distance were variable and other parameters were constant. It was shown that CA1 neurons fired independently of time and distance on the treadmill and that the hippocampus encodes both spatial and temporal information. Time and distance were encoded simultaneously in the hippocampus; some neurons encoded time better and some neurons encoded distance better, while the effect of fixed position was found to be weak. The same neurons fired in patterns indicating elapsed time on the treadmill and space in the maze. However, it could not be determined whether time and space were encoded simultaneously in the same neuron [24, 31, 36, 39, 40]. Previous studies have suggested that the time signal observed in the hippocampus may be partially influenced by movement-related cues or path integration of neurons during the delay [32].

Macdonald and colleagues wanted to test whether time encoding would be seen in delay times recorded from hippocampal neurons in the completely still state because the animals were moving [24, 36, 39, 40]. By fixing the rats' heads, they recorded the activity of CA1 neurons during latency in a memory test in a rewarded odor pattern matching task while the rats were motionless in the same position. This eliminated the possibility that the firing of time cells could be affected by movement. On each trial, the test odor was presented after a fixed delay following the sample odor. Thirty percent of the hippocampal neurons fired in short bursts during the delay period, encoding the elapsed time. When the animal's head was still and stationary and the same behavior was maintained, time cells were found to generate time signals independent of location and distance, encoding the passage of time independent of path integration. Just as place cells exhibit different firing patterns in different spatial contexts, different firing patterns of time cells were observed in trials with different odor combinations. While some neurons signaled for either time or distance, most neurons represented both variables in different proportions, indicating that most of these neurons encode both time and distance. That is, most hippocampal neurons were found to be both place cells and time cells [2, 10, 31, 32]. The relationship of the hippocampus to time is also seen in classical conditioning. When Modi et al. immobilized the head of mice, subjected them to classical conditioning, and examined CA1 neurons, they reported that neurons formed time-related patterns [3, 31, 41].

Kraus et al. wondered whether medial entorhinal cortex grid cells, like hippocampal place cells, could encode time as well as being responsible for spatial encoding. In their experiments, Kraus et al. examined the firing properties of medial entorhinal cortex grid cells while rats ran on a treadmill. On the treadmill, the grid cells precisely encoded intervals and many of the neurons represented the time spent and distance traveled on the treadmill. Thus, it was found that temporal information was also encoded in the medial entorhinal cortex [10, 42]. It has been suggested that place cells and medial entorhinal cortex grid cells can also represent both place fields and time fields in different situations [43, 44]. To investigate the role of the medial entorhinal cortex in the encoding of time in CA1 neurons, Robinson et al. found that memory and the encoding of time in CA1 neurons were severely impaired when medial entorhinal cortex neurons were inactivated. On the other hand, the encoding of location was not affected, which strengthened the idea that spatial and temporal encoding in the hippocampus can be separated and that the medial entorhinal cortex has significant importance in temporal encoding and that the hippocampus receives information from the medial entorhinal cortex [10, 45]. Abramson et al. wondered whether the coding of hippocampal neurons is flexible according to behavior and task; they used the experimental data of Kraus et al. [40] and analyzed it according to behavior. They found that when rats ran for a fixed time, time coding dominated in neurons, while distance coding dominated when rats ran a fixed distance. They found that CA1 neurons were able to encode distance, location, or time according to the experiment or task, and that the coding of the hippocampus was adaptive and adjusted according to the most appropriate parameters for the task [46].

16.2.1 Hippocampus in Memory Consolidation

In the experiments, it was found that the firing patterns of place cells observed in rats performing the maze task were replayed in an accelerated manner during periods of sleep and quiet wakefulness. These replay patterns are important for memory consolidation, and when they are disrupted, animals fail memory tests. It has also been reported that hippocampal replay is synchronized with cortical replay, which is critical for memory retention [3, 18, 26, 31].

The hippocampus plays an active role in the retrieval of newly acquired information into short-term memory. During sleep, when there is no external information flow, the brain has the opportunity to organize and reinforce newly learned and encoded information. Experiences are quickly recorded in neuronal circuits in the hippocampus, where temporary engrams are created and stored. These temporary engrams are then gradually transferred to the neocortex and made permanent. The brain replays the information acquired during the day and transfers it to the cortex for long-term memory. In this way, engrams are formed. When we want to recall information, these engrams are activated and support our memory retrieval. Different brain oscillations are involved in the consolidation process. Theta waves are recorded in the hippocampus as well as in the cortex and subcortical regions during active wakefulness and REM sleep, particularly during spatial exploration. During new experiences, CA3 and CA1 neurons coordinate with theta oscillations to form groups of neurons that encode new information. These CA3 ensembles then trigger SWR (Sharp-Wave Ripples) events for consolidation during sleep. SWR is a signal that occurs during NREM sleep, originating from the CA3 and CA1 subregions of the hippocampus, and is involved in the flow of information between the hippocampus and the cortex [7, 30, 38]. The input from CA3 constitutes the "Sharp wave," and the interaction between CA1 pyramidal neurons and interneurons constitutes the "Ripple" part. During sleep, CA3 neurons are spontaneously activated in synchronized bursts, which trigger CA1 neurons to be activated as well. SWRs reactivate CA1 ensembles and strengthen synapses between these neurons. Sequences created during novel experiences are consolidated through SWR-dominated replay during NREM sleep following the experience. In other words, SWRs are important for consolidating memory through replay. Replays have also been observed in humans. Additionally, a study found that the duration of SWR activity associated with consolidation lasted much longer than previously thought. SWRs also occur outside of sleep, during quiet wakeful states, such as while performing tasks like cleaning or cooking, though their content differs from that during sleep. These wakefulness replays have been associated with remembering recent past experiences and making decisions based on this information [38, 47–49].

It was unclear how neuronal activity in the brain restores balance after SWR bursts in the hippocampus during NREM sleep. In a new study, Karaba et al. recorded from the CA1, CA2, and CA3 regions during memory tasks and sleep in mice and rats to understand how the hippocampus maintains this balance. They identified an activity called Barrage of Action Potentials (BARRs), characterized by the firing of neurons and Synuclein Gamma (+) cells (Cholecystokinin (+) basket cells) in the CA2 region, with a longer duration than SWRs. They found that BARRs selectively inhibit CA1 pyramidal neurons that are reactivated during and after learning in sleep. This inhibition of cells that are highly active during SWRs—when learned information is replayed in the memory consolidation process—prevents pathological synchronization. It has been reported that cellular activity should be balanced for memory consolidation and that both under- or over-reactivation can lead to memory impairment. In addition, memory problems were observed when BARRs were disrupted by optogenetic methods. Thus, BARRs help maintain a homeostatic balance in both memory and neuronal firing rates in the hippocampus during sleep [50, 51].

16.3 Conclusion

Studies have shown that the hippocampus encodes time, space, and task-related behaviors. It has been reported that temporal representation can be affected by many parameters and that hippocampal neurons encode time even when these parameters are eliminated in experiments. On the other hand, in the experiments, recordings were taken from hippocampal neuron ensembles, and there is still no known study in which recordings were taken directly from only time cells. In this context, studies with neurons directly encoding time instead of general hippocampal neurons are also necessary. While more studies are needed to better understand the subject, it is crucial to elucidate the encoding of temporal information in hippocampal subfields and the mechanisms of memory-related temporal representation in the hippocampus, as well as its relationship with neighboring hippocampal structures. On the other hand, the consolidation and preservation of memory are just as important as its formation. As the activities and changes during consolidation—closely associated with the hippocampus—are better understood, our knowledge of memory, which has played a crucial role in humanity's advancement of knowledge, including the development of artificial intelligence, will continue to progress.

References

1. Squire LR. The legacy of patient H.M. for neuroscience. Neuron. 2009;61(1):6–9.
2. Eichenbaum H, Komorowski R, MacDonald CJ, Kraus BJ, Robitsek J. How does the hippocampus support the spatial and temporal attributes of memory. In: Jackson PA, Chiba AA, Berman RF, Ragozzino ME, editors. The neurobiological basis of memory a system, attribute, and process analysis. Springer; 2016. p. 39–58.
3. Eichenbaum H. Time cells in the hippocampus: a new dimension for mapping memories. Nat Rev Neurosci. 2014;15(11):732–44.
4. Ranganath C, Hsieh LT. The hippocampus: a special place for time. Ann N Y Acad Sci. 2016;1369(1):93–110.
5. Vargha-Khadem F, et al. Differential effects of early hippocampal pathology on episodic and semantic memory. Science. 1997;277(5324):376–80.

6. Eichenbaum H. Time (and space) in the hippocampus. Curr Opin Behav Sci. 2017;17:65–70.
7. Unal G. The cortico-hippocampal circuit: the brain's center for mapping and declarative memory. J Ankara Univ Faculty Med. 2019;72(1):13–24.
8. Kesner RP. In: Clark RE, Martin SJ, editors. Behavioral neuroscience of learning and memory. Springer; 2018. p. 334.
9. Manns JR, et al. Hippocampal CA1 spiking during encoding and retrieval: relation to theta phase. Neurobiol Learn Mem. 2007;87(1):9–20.
10. Ranganath C. Time, memory, and the legacy of Howard Eichenbaum. Hippocampus. 2019;29(3):146–61.
11. Ekstrom AD, Ranganath C. Space, time, and episodic memory: the hippocampus is all over the cognitive map. Hippocampus. 2018;28(9):680–7.
12. Sabariego M, et al. In the temporal organization of episodic memory, the hippocampus supports the experience of elapsed time. Hippocampus. 2021;31(1):46–55.
13. MacDonald CJ. Prospective and retrospective duration memory in the hippocampus: is time in the foreground or background? Philos Trans R Soc Lond Ser B Biol Sci. 2014;369(1637):20120463.
14. Dickerson BC, Eichenbaum H. The episodic memory system: neurocircuitry and disorders. Neuropsychopharmacology. 2010;35(1):86–104.
15. Eichenbaum H. In: Clark RE, Martin SJ, editors. Behavioral neuroscience of learning and memory. Springer; 2018. p. 101–18.
16. Naya Y, Suzuki WA. Integrating what and when across the primate medial temporal lobe. Science. 2011;333(6043):773–6.
17. Hsieh LT, et al. Hippocampal activity patterns carry information about objects in temporal context. Neuron. 2014;81(5):1165–78.
18. Eichenbaum H. Memory on time. Trends Cogn Sci. 2013;17(2):81–8.
19. Terada S, et al. Temporal and rate coding for discrete event sequences in the hippocampus. Neuron. 2017;94(6):1248–1262.e4.
20. Scoville WB, Milner B. Loss of recent memory after bilateral hippocampal lesions. J Neurol Neurosurg Psychiatry. 1957;20(1):11–21.
21. O'Keefe J, Nadel L. The hippocampus as a cognitive map. Oxford University Press; 1978.
22. Farovik A, Dupont LM, Eichenbaum H. Distinct roles for dorsal CA3 and CA1 in memory for sequential nonspatial events. Learn Mem. 2010;17(1):12–7.
23. Mankin EA, et al. Hippocampal CA2 activity patterns change over time to a larger extent than between spatial contexts. Neuron. 2015;85(1):190–201.
24. Pastalkova E, et al. Internally generated cell assembly sequences in the rat hippocampus. Science. 2008;321(5894):1322–7.
25. Eichenbaum H. On the integration of space, time, and memory. Neuron. 2017;95(5):1007–18.
26. Rangel LM, Quinn LK, Chiba AA. Space, time, and the hippocampus. In: Jackson PA, Chiba AA, Berman RF, Ragozzino ME, editors. The neurobiological basis of memory. Springer; 2016. p. 59–76.
27. Salz DM, et al. Time cells in hippocampal area CA3. J Neurosci. 2016;36(28):7476–84.
28. Itskov V, et al. Cell assembly sequences arising from spike threshold adaptation keep track of time in the hippocampus. J Neurosci. 2011;31(8):2828–34.
29. Kesner RP, Hunsaker MR, Gilbert PE. The role of CA1 in the acquisition of an object-trace-odor paired associate task. Behav Neurosci. 2005;119(3):781–6.
30. Buzsáki G, Tingley D. Space and time: the hippocampus as a sequence generator. Trends Cogn Sci. 2018;22(10):853–69.
31. Eichenbaum H. Elements of information processing in hippocampal neuronal activity: space, time, and memory. In: Hannula DE, Duff MC, editors. The hippocampus from cells to systems structure, connectivity, and functional contributions to memory and flexible cognition. Springer; 2017. p. 69–94.
32. MacDonald CJ, et al. Distinct hippocampal time cell sequences represent odor memories in immobilized rats. J Neurosci. 2013;33(36):14607–16.
33. Eichenbaum H, Otto T. Odor-guided learning and memory in rats: is it 'special'? Trends Neurosci. 1993;16(1):22–4.
34. Jenkins LJ, Ranganath C. Distinct neural mechanisms for remembering when an event occurred. Hippocampus. 2016;26(5):554–9.
35. Banquet JP, et al. Time as the fourth dimension in the hippocampus. Prog Neurobiol. 2021;199:101920.
36. MacDonald CJ, et al. Hippocampal "time cells" bridge the gap in memory for discontiguous events. Neuron. 2011;71(4):737–49.
37. Heusser AC, et al. Episodic sequence memory is supported by a theta-gamma phase code. Nat Neurosci. 2016;19(10):1374–80.
38. Girardeau G, Lopes-Dos-Santos V. Brain neural patterns and the memory function of sleep. Science. 2021;374(6567):560–4.
39. Gill PR, Mizumori SJ, Smith DM. Hippocampal episode fields develop with learning. Hippocampus. 2011;21(11):1240–9.
40. Kraus BJ, et al. Hippocampal "time cells": time versus path integration. Neuron. 2013;78(6):1090–101.
41. Modi MN, Dhawale AK, Bhalla US. CA1 cell activity sequences emerge after reorganization of network correlation structure during associative learning. Elife. 2014;3:e01982.
42. Kraus BJ, et al. During running in place, grid cells integrate elapsed time and distance run. Neuron. 2015;88(3):578–89.
43. Montchal ME, Reagh ZM, Yassa MA. Precise temporal memories are supported by the lateral entorhinal cortex in humans. Nat Neurosci. 2019;22(2):284–8.
44. Shapiro ML. Time is just a memory. Nat Neurosci. 2019;22(2):151–3.
45. Robinson NTM, et al. Medial entorhinal cortex selectively supports temporal coding by hippocampal neurons. Neuron. 2017;94(3):677–688.e6.

46. Abramson S, et al. Flexible coding of time or distance in hippocampal cells. Elife. 2023;12:e83930.
47. Buzsáki G. Two-stage model of memory trace formation: a role for "noisy" brain states. Neuroscience. 1989;31(3):551–70.
48. Giri B, et al. Hippocampal reactivation extends for several hours following novel experience. J Neurosci. 2019;39(5):866–75.
49. Klinzing JG, Niethard N, Born J. Mechanisms of systems memory consolidation during sleep. Nat Neurosci. 2019;22(10):1598–610.
50. Mou X, Ji D. A BARRage of firing while asleep. Science. 2024;385(6710):710–1.
51. Karaba LA, et al. A hippocampal circuit mechanism to balance memory reactivation during sleep. Science. 2024;385(6710):738–43.

Epilepsy and the Hippocampus

17

Amal Satte and Ali Akhaddar

17.1 Introduction

Though small in size, the hippocampus is one of the brain's most complex structures and plays a significant role in memory formation and spatial navigation but also contributes to the pathogenesis of the most frequent focal epileptic syndromes—mesial temporal lobe epilepsy (MTLE). The particular cytoarchitecture and connectivity of the hippocampus make it vulnerable to epileptogenesis. Hippocampal sclerosis is by far the most frequent and the most studied type of MTLE, where the hippocampus, its nearby structures, and even remote cortical areas are involved. Data on epilepsies involving the hippocampus per se are limited due to the complex, intricate connections of this structure. These connections favor the spread of epileptic activity with different clinical manifestations. This chapter provides an updated review of the state-of-the-art knowledge on hippocampal epilepsies, including their etiologies, pathophysiology, clinical features, and treatment paradigms. In addition, we will discuss recent progress in neuroimaging and electrophysiology, as well as the range of therapeutic options from pharmacological approaches through cutting-edge surgical methods to neuromodulation strategies.

17.2 Epidemiology

Epidemiological data specifically focused on "hippocampal epilepsy" are less common in the literature, as this term is not as widely used as MTLE, which typically involves the hippocampus and surrounding structures: the amygdala, entorhinal cortex, perirhinal cortex, and parahippocampal cortex.

MTLE is the most common form of focal epilepsy in adults. However, precise prevalence figures are challenging to establish due to diagnostic variability and population differences.

MTLE typically presents in late childhood, adolescence, or early adulthood, with a peak onset in the second decade of life.

Some studies suggest a slight female predominance, but this has not been confirmed by all studies.

MTLE occurs worldwide, but its relative frequency may vary between regions due to differences in risk factors and genetic backgrounds. It is probably underdiagnosed in some regions of the globe.

A. Satte (✉)
Department of Neurophysiology, Mohammed V Military Teaching Hospital, Rabat, Morocco

Mohammed V University in Rabat, Rabat, Morocco
e-mail: satteamal@gmail.com

A. Akhaddar
Mohammed V University in Rabat, Rabat, Morocco

Department of Neurosurgery, Avicenne Military Hospital of Marrakech, Marrakech, Morocco

M. Turgut et al. (eds.), *The Human Hippocampus*, https://doi.org/10.1007/978-3-031-88907-3_17

Risk factors comprise a history of febrile seizures, particularly complex febrile seizures, central nervous system infections, traumatic brain injury, perinatal hypoxia, and status epilepticus.

The main pathological finding of MTLE is hippocampal sclerosis (HS) . Other causes include tumors, vascular malformations, and developmental abnormalities.

MTLE often follows a progressive course, with increasing seizure frequency and potential cognitive decline over time if left untreated [1–3].

17.3 Pathophysiology of Hippocampal Epilepsies

17.3.1 General Considerations

The hippocampus comprises several distinct subregions: The dentate gyrus (DG) is a densely packed granule cell layer, serving as the primary input region. It acts as a "gate" controlling input to the hippocampus. The Cornu Ammonis (CA) fields comprise three regions:

- CA1, which contains pyramidal neurons crucial for output processing and is often involved in seizure propagation and generalization.
- CA2, a small, specialized region with unique connectivity and a potential role in seizure initiation and propagation.
- CA3, notable for its recurrent collateral connections; it plays an important role in pattern completion and rapid encoding of novel information.

The subiculum, the main output structure of the hippocampus, is involved in information integration and output regulation [4–8].

17.3.2 Causes and Pathophysiology of Hippocampal Epilepsies

17.3.2.1 Hippocampal Sclerosis

According to radiological, surgical, and postmortem studies, HS is the leading lesion found in mesial temporal lobe epilepsy as well as other epilepsy syndromes. The cause of hippocampal sclerosis (HS) remains incompletely understood and is probably multifactorial. Current research suggests a complex interplay of potential contributing factors, including but not limited to a history of febrile seizures, genetic predisposition, neuroinflammatory processes, and aberrant neurodevelopmental mechanisms. The precise pathogenic pathways and the relative significance of each factor in the development of HS are still subjects of ongoing scientific investigation [1, 2].

The hippocampus is particularly susceptible to hyperexcitability and the development of seizures due to its intrinsic neuronal properties and circuitry. The disruption of the excitation-inhibition balance is a key mechanism underlying the development of hippocampal epilepsy [9, 10].

Two predominant hypotheses have been proposed to elucidate the mechanisms of epileptogenesis in hippocampal sclerosis (HS) :

- The "Dormant Basket Cell" Hypothesis: This theory posits that inhibitory basket cells become hypofunctional due to the loss of excitatory input from damaged dentate mossy cells and other neurons in the affected region. This results in a reduction of GABAergic inhibition, leading to hyperexcitability of principal neurons, potentially culminating in seizure activity and further neuronal damage. On the other hand, the hippocampus is characterized by a relatively poor vascular supply, with inhibitory interneurons located in deep intraparenchymal regions. This anatomical configuration renders these neurons particularly susceptible to pathophysiological insults such as hypoxia, ischemia, and oxidative stress [7, 8].
- The "Mossy Fiber Sprouting" Hypothesis: This model suggests that aberrant sprouting of mossy fibers, which originate from dentate granule cells, occurs in the supragranular region of the inner molecular layer of the fascia dentata. This pathological reorganization is hypothesized to form recurrent excitatory circuits, leading to hyperexcitability and epileptiform discharges [9].

However, the relative contributions of these mechanisms to epileptogenesis in HS remain a subject of debate, and other additional factors are probably involved.

Another subject of debate is whether HS is a cause or a consequence of epilepsy. While some authors highlight the role of seizures as precipitating factors for hippocampal sclerosis, other studies have demonstrated that this relationship is not universally applicable [10–13].

Other causes of hippocampal epilepsy include all the different lesions that can be observed in other cerebral lobes, including malformation of cortical development, perinatal injury, arteriovenous malformations, tumors, infections, and autoimmune encephalitis [14, 15].

Among the various etiologies of encephalitis, limbic encephalitis demonstrates a predilection for the mesial temporal region. These recently delineated nosological entities are characterized by inflammation of the limbic system, particularly affecting the hippocampus, amygdala, and other medial temporal lobe structures. The condition can be triggered by various factors, involving autoantibodies targeting neuronal cell surface or intracellular antigens, and can be paraneoplastic. The inflammatory process leads to neuronal dysfunction or neuronal death, depending on the type of antibody, and results in a range of cognitive, behavioral, and neurological symptoms, including memory impairment, seizures, and psychiatric disturbances [16, 17].

Certain infections are known to involve the limbic system. The most well known is herpes simplex virus (HSV) type 1. The predilection of HSV for the mesiotemporal region is not fully understood, but several theories exist:

- Neuroanatomical pathway: The olfactory and trigeminal nerves may serve as entry routes, providing direct access to the temporal lobes.
- Cellular receptors: The presence of specific cellular receptors in the limbic system may facilitate viral entry and replication.
- Blood-brain barrier: The mesiotemporal region may have areas of relative blood-brain barrier vulnerability.
- Neuronal cell type susceptibility: Certain neuronal populations in the limbic system might be more susceptible to viral infection.
- Latency and reactivation: The ability to establish latency in sensory ganglia with subsequent reactivation may contribute to their tropism [18–20].

Other infections can affect the limbic system, including West Nile virus, Japanese encephalitis virus, Toscana virus, enterovirus, COVID-19, and neurosyphilis [21–23].

Familial mesial temporal lobe epilepsy (FMTLE) is a form of epilepsy with a genetic component. Most patients have rare seizures while those with frequent seizures often have hippocampal atrophy progression and may require surgery for improvement. Inheritance patterns in FMTLE families suggest a complex mode of inheritance, similar to idiopathic generalized epilepsies [24–26].

Finally, it is noteworthy that the hippocampus may be implicated in epileptogenic processes beyond its anatomical boundaries. Hippocampal involvement has been documented not only in focal extrahippocampal epilepsies but also in certain generalized epilepsies such as juvenile myoclonic epilepsy [27, 28].

17.4 Histopathology of HS

The International League Against Epilepsy (ILAE) has delineated a classification system for hippocampal sclerosis (HS) in temporal lobe epilepsy (TLE). HS ILAE type 1 is characterized by severe neuronal depletion and gliosis, predominantly affecting the CA1 and CA4 regions. In contrast, HS ILAE type 2 exhibits neuronal loss and gliosis primarily in the CA1 region, while HS ILAE type 3 demonstrates these pathological changes predominantly in the CA4 region. Additionally, some surgical hippocampal specimens from TLE patients may present with preserved neuronal populations but exhibit reactive gliosis, a condition termed "no hippocampal sclerosis"(noHS). HS ILAE type 1 has been associated with a higher incidence of initial precipitating injuries occurring before five years of age, earlier seizure onset, and a more favorable post-surgical seizure control prognosis. The clinical and prognostic implications of HS ILAE types 2 and 3 have been less extensively investigated. However, preliminary reports suggest these sub-types may be linked to less favorable outcomes and distinct epilepsy histories, including variations in the age of seizure onset. The implementation of this international consensus classification aims to facilitate the characterization of specific clinicopathologic syndromes. Furthermore, it is expected to elucidate variability in neuroimaging and electrophysiological findings, as well as postsurgical seizure control outcomes across these subtypes [29].

17.5 Semiology of Hippocampal Seizures

17.5.1 General Considerations

Hippocampal seizures exhibit a diverse array of clinical manifestations, intricately linked to the precise topography and propagation patterns of epileptic activity within the hippocampal formation. This semiological heterogeneity is attributed to the complex anatomical and functional architecture of the hippocampus [30]. The symptomatology may encompass psychic, mnemonic, emotional, autonomic, and sensory phenomena, reflecting the hippocampus's involvement in multiple cognitive and physiological processes. The specific manifestations are contingent upon several factors:

- Initial localization: The precise origin of the epileptic discharge within the hippocampal subregions (e.g., CA1, CA3, dentate gyrus) influences the nature of the inaugural symptoms.
- Propagation pattern: The diffusion of ictal activity through intrahippocampal circuits and to adjacent limbic structures determines the sequential evolution of clinical manifestations.
- Lateralization: Involvement of the right or left hippocampus can engender notable differences, particularly in mnemonic disturbances (verbal vs. visuospatial memory).
- Implicated neural networks: The activation or inhibition of specific connections with other cerebral regions (e.g., amygdala, entorhinal cortex, temporal neocortex) modulates the observed semiology.
- Electrophysiological characteristics: The frequency, amplitude, and synchronization of neuronal discharges influence the intensity and duration of clinical manifestations.
- Individual factors: Cerebral plasticity, comorbidities, and the patient's epileptic history can modulate the clinical expression of seizures.

Adhering to the most recent International League Against Epilepsy (ILAE) classification, focal seizures are now systematically categorized into distinct subtypes. These include focal (previously termed partial) motor or non-motor seizures, further delineated by the absence or presence of impaired awareness (previously termed "complex seizures"). Additionally, the classification acknowledges the potential for focal seizures to evolve into bilateral tonic-clonic seizures, previously termed secondary generalization [31].

17.5.2 Semiology of MTLE Seizures, Including Hippocampal Seizures

Aura is a subjective symptom perceived by the patient. Typically, when awareness is preserved, the ictal phase commences with an aura, which varies depending on the subregion where the seizures start.

The most reported and characteristic symptom of a hippocampal seizure is a visceral or ascending epigastric aura. It accounts for around 80% of cases. In a study aiming to further describe visceral aura in temporal lobe epilepsies, Kuan found that abdominal aura was reported in both MTLE and non-MTLE; however, patients with MTLE more frequently experienced a rising sensation, originating in the epigastric or periumbilical area and often extending to the chest or the abdomen. Moreover, according to the same study, epigastric sensations were associated with a left-side seizure onset, while non-epigastric sensations were linked to a right-sided onset [28, 32].

The dreamy state is also a frequently reported aura in mesiotemporal lobe epilepsy. It is not specific to the hippocampus and can also be produced by stimulation of the amygdala and parahippocampal gyrus. It manifests as "déjà vu" or "jamais vu" sensations, complex hallucinations, or an out-of-body feeling [33].

Ictal fear can be seen in around 30% of cases and is not pathognomonic of hippocampal topography. Less frequently, patients may report gustatory or olfactory hallucinations. Visual hallucinations have been reported but are more suggestive of an occipital topography [34, 35].

Aura may remain isolated or be followed by impaired awareness manifesting as motionless staring. It is noteworthy that consciousness alteration in temporal lobe seizures is characterized by a delayed onset, typically manifesting in the mid-to-late ictal phase. This pattern contrasts with some other seizure types where consciousness may be impaired from the outset. The underlying mechanism is thought to involve pathological hypersynchronization between thalamic structures and widespread associative cortical networks. This thalamocortical dysrhythmia disrupts the normal patterns of neural activity necessary for maintaining conscious awareness. Another possible mechanism is an inhibitory effect of bilateral temporal lobe seizures on subcortical arousal systems [36, 37].

Automatisms that are typically oro-alimentary, in the form of chewing or swallowing, then take place. Alarcon highlights that automatisms can occur with minimal loss of consciousness. They are pathognomonic of mesiotemporal epilepsy when they follow an epigastric aura. Gestural automatisms reflect neocortical propagation. Verbal automatisms are less frequent; they can be in the native or a foreign language but are typically unintelligible and manifest in the nondominant hemisphere. Autonomic manifestations such as cardiovascular and respiratory changes, piloerection, pallor, or flushing are common in mesiotemporal seizures and may contribute to sudden unexpected death in epilepsy (SUDEP).

The duration of the seizures tends to be longer compared to frontal lobe or lateral temporal lobe seizures, up to 3 min, with a longer postictal confusion. They less frequently evolve to bilateralization [32, 38–43].

17.5.3 Lateralizing Clinical Symptoms

Certain semiological features of temporal lobe seizures provide valuable lateralizing information. The early head version is often ipsilateral, while the late deviation is contralateral. Unilateral automatisms are ipsilateral; conversely, dystonic posturing is contralateral and is considered as the most prevalent and dependable indicator of seizure lateralization. Seizures with ictal spitting often stem from the right hemisphere. Postictal aphasia or dysphasia is indicative of seizure onset in the language-dominant hemisphere. Motion arrest and fear have been reported more frequently in left MTLE. Ipsilateral postictal nose wiping, contralateral clonus, and hypokinesia correctly localize the focus in a significant proportion of patients but these signs are less commonly observed [32, 39, 44].

17.5.4 Hippocampal Status Epilepticus

Hippocampal status epilepticus can manifest as various symptoms, including cognitive dysfunction in different cognitive domains, causing confusion, aphasia, amnesia, and psychiatric symptoms suggesting psychosis. Most of the time, clinical features are insufficient to establish the diagnosis. Remote risk factors for seizures, severely impaired mental state, and ocular movement abnormalities may help, if present, suggest the diagnosis [45, 46].

17.5.5 Clinical Semiology of Hippocampal Seizures: A Subregional Analysis

Hippocampal seizures present a complex and diverse array of clinical manifestations, reflecting the intricate functional architecture of the hippocampal formation. The semiology can be delineated based on the involvement of specific hippocampal subregions, each contributing unique features to the ictal phenomenology. In this paragraph, we endeavor to elucidate the clinical semiology through the physiological characteristics of each subregion. However, it is important to highlight that seizures affecting exclusively one subregion remain rare and that the clinical presentation often involves a complex interplay between hippocampal subregions and extrahippocampal structures. Understanding these subregion-specific manifestations aids in more precise localization of seizure onset, potentially improving surgical outcomes and guiding the development of targeted therapies. However, it is important to interpret these correlations cautiously, always considering the complex and dynamic nature of hippocampal seizures.

17.5.5.1 Dentate Gyrus

The DG serves as the primary input gate to the hippocampus. It plays a crucial role in pattern separation, which is the ability to discriminate between similar sensory inputs or experiences. Impaired pattern separation causes perceptual distortions or difficulty distinguishing between similar experiences, either during the ictal or the postictal period. This would manifest as a derealization or depersonalization phenomenon. Patients may also experience confusion about the familiarity of surroundings that can manifest as "déjà vu": an inappropriate feeling of familiarity in a novel situation caused by inappropriate activation of existing memory traces for new stimuli, or "jamais vu": a sense of unfamiliarity in a known environment caused by failure to properly match a familiar sensory input with its stored representation [47, 48].

It has been demonstrated that neurogenesis continues in DG throughout adulthood. This phenomenon contributes to cognitive flexibility and allows adaptation to novel environments. A disruption of this process in the postictal period may contribute to perseverative behaviors and difficulty adapting to environmental changes in the postictal period [49–51].

As the DG is connected to the entorhinal cortex and CA3, it plays an important role in encoding contextual and spatial information in episodic memories. Patients can experience anterograde amnesia for events during and shortly after the seizure and or spatial disorientation [52, 53].

The DG is also connected with limbic structures involved in emotional processing. Thus, patients may experience various emotional and affective symptoms such as fear, anxiety, or sudden changes of mood [54].

Altered consciousness and automatisms reflect the spread of seizure activity to other hippocampal regions and the disruption of hippocampal-cortical interactions.

Finally, the DG plays a key role in controlling seizures and preventing their spread throughout the brain. Under normal conditions, it has strong inhibitory mechanisms that help regulate neural activity and prevent excessive excitation. In epilepsy, these inhibitory functions can become compromised, reducing the dentate gyrus's ability to control abnormal electrical activity. This can lead to a higher likelihood of seizures, as the balance between excitation and inhibition is disrupted, allowing seizures to occur more easily and spread more widely [54, 55].

17.5.5.2 CA3 Region

The CA3 region is known for its auto-associative network (rapid encoding of new information and association with existing memories), and its involvement in short-term memory, pattern completion, and spatial navigation. CA3 seizure involvement often presents with déjà vu or jamais vu sensation, temporary amnesia, due to inappropriate activation of memory circuits. Abnormal smells or tastes are explained by the connections to the amygdala and olfactory cortex. Rising epigastric sensation can be reported at the beginning of the seizure, and is explained by the connections to the insula or activation of the autonomic nervous system. Fear or anxiety and emotional changes occur due to the connection to the amygdala and other limbic structures. Motor automatisms are explained by the connection with the amygdala, frontal, and insular cortex. Altered consciousness: confusion, unresponsiveness, staring, as CA3 is involved in consciousness [56, 57].

Visual hallucinations or illusions are less common than other symptoms. They tend to be more complex compared to patterns seen in occipital seizures. They may be explained by erroneous pattern completion processes but also by the connections of CA3 to visual processing regions [58].

17.5.5.3 CA1 Region

As a major output node of the hippocampus, CA1 is critical for temporal association memory. It is involved in memory consolidation, spatial and temporal processing, and novelty detection. It also contributes to pattern completion and separation. Seizures involving CA1 can manifest with experiential phenomena, such as "déjà vu" or "jamais vu", autobiographical memory recall, often vivid and involuntary, feelings of fear or anxiety, olfactory and gustatory hallucinations, due to connections with the primary olfactory cortex, autonomic symptoms, and emotional changes, due to connections with the autonomic nervous system and amygdala and other limbic structures. Automatisms, speech disturbances, and altered consciousness can also be seen in other subregions [47, 59, 60].

17.5.5.4 CA2 Region

Though historically understudied, CA2 is increasingly recognized for its role in social memory. The role, of CA2 in epilepsy is still being elucidated: it may have a facilitating role in seizure spread between CA3 and CA1. The disruption of CA2 function may also contribute to social memory deficit that can be observed in temporal lobe epilepsy [61–63].

17.5.5.5 Subiculum

As the main output structure of the hippocampus, the subiculum is involved in different processes, including spatial processing, memory consolidation, executive function, emotional processing, and information integration. It also contributes through its inhibitory control to regulate the excitability of the hippocampus. Subicular involvement in seizures often leads to various symptoms. Autonomic signs are prominent due to the connection with the hypothalamus. They include cardiovascular changes, gastrointestinal sensations, and pilomotor responses. Emotional changes, "déjà vu" or "jamais vu" experiences, memory disturbances, olfactory and gustatory hallucinations, and disorientation may also occur. Automatisms can also be seen, and complex visual hallucinations are less common. Due to its extensive connections, it is involved in the bilateralization of hippocampal seizures [64–66].

17.6 Complications and Comorbidities

Psychiatric troubles are a frequently reported comorbidity in mesiotemporal lobe epilepsy, especially depression, anxiety, aggression, psychosis, and interictal dysphoria. Psychiatric complications should therefore be systematically investigated, especially when certain treatments that may exacerbate them are being considered, such as Levetiracetam or deep brain stimulation.

Cognitive impairment can also be associated with MTLE, including memory deficits, learning difficulties, executive function deficits, and attention problems. They can be further aggravated by certain antiseizure medications, especially with polytherapies [38, 67–69].

There's a strong comorbidity between autism spectrum disorder and mesiotemporal epilepsy. Neuroimaging studies have revealed hippocampal morphological abnormalities in pediatric autism spectrum disorder (ASD) populations that are similar to adult patients with temporal epilepsy. The identified structural modifications in the hippocampus may be implicated in the social and cognitive impairments characteristic of ASD. These findings suggest a potential neuroanatomical link between ASD and hippocampal epilepsy [70–72].

Many studies have shown that epilepsy and migraine are frequently associated. No significant correlation was observed between the lateralization of cephalgia and the ipsilateral occurrence of hippocampal sclerosis in individuals diagnosed with both migraine and temporal lobe epilepsy. In cases of familial mesiotemporal epilepsy, the association between epilepsy and migraine probably stems from shared genetic factors that may be part of a broader category of neurological conditions, "Primary Brain Channelopathies" [73–75].

Patients with temporal lobe epilepsy (TLE) exhibit a higher prevalence of reproductive disorders. This comorbidity may be attributed to multiple factors: seizure effects may directly impact reproductive function, long-term neurological changes associated with TLE, and antiseizure medication use [76–78].

Sleep disturbances are common in MTLE manifesting as increased sleep latency and fragmentation. On another hand, patients may have a higher risk of obstructive sleep apnea [79–81].

Seizure-related injuries are underreported in clinical settings, potentially underestimating the true impact of epilepsy-associated morbidity. Patients with pharmacoresistant temporal lobe epilepsy appear to be at higher risk for experiencing injuries compared to those with pharmacoresistant extratemporal lobe epilepsy [82].

There are significant associations between mesiotemporal epilepsy and SUDEP risk factors. Mueller demonstrated a correlation between temporal lobe epilepsy and brainstem atrophy, particularly in regions crucial for autonomic control. This structural alteration may predispose patients to fatal autonomic dysregulation during seizures. Other studies have revealed that ictal hypoventilation contributes to cardiac arrhythmia and subsequent mortality in SUDEP cases. The current hypothesis posits that SUDEP results from postictal cardiorespiratory failure, typically triggered by generalized tonic-clonic seizures, through a complex, multifactorial pathophysiological process. Epidemiological data indicate that young adults with poorly controlled seizures are at heightened risk for SUDEP [83–85].

17.7 Differential Diagnosis

Hippocampal seizures must be differentiated from other types of epileptic seizures as well as from non-epileptic conditions.

17.7.1 Non-epileptic Conditions

Transient ischemic attacks: typically manifest as a focal neurological deficit.

Migraine with aura: the aura here is typically visual, lasts longer, and is followed by headache.

Transient global amnesia is a condition characterized by sudden anterograde and retrograde amnesia. It occurs in adults aged 50–70. Patients report no aura, and no automatisms are observed. EEG during the episode does not show any ictal abnormalities. The episodes are longer (hours) compared to hippocampal seizures. An increased risk of epilepsy after transient global amnesia has been reported in a study [86, 87].

Panic attacks manifest as sudden fear with autonomic symptoms. They last longer than a typical seizure and are not followed by confusion. The diagnosis can be challenging. Long-term video-EEG may help elucidating the cause of the symptoms [88].

Psychogenic Non-epileptic Seizures (PNES) Differentiating PNES from hippocampal seizures involves evaluating clinical history, seizure semiology, and psychiatric background, with video-EEG monitoring as the gold standard for diagnosis [89].

Syncope and epileptic seizures share many clinical features, including automatisms and vocalizations. Some features can be suggestive of a hippocampal seizure, such as postictal confusion and tongue bites if the seizure had evolved into a bilateral tonic-clonic seizure. Electroencephalography, tilt test, creatine, and prolactin levels may help differentiate the two conditions [90].

Hippocampal seizures, especially when manifesting as isolated epigastric aura can be misdiagnosed as ***gastrointestinal disorders*** [91, 92].

17.7.2 Other Epilepsies

17.7.2.1 Temporal Non-hippocampal Epilepsies

In a retrospective analysis of patients with focal seizures of temporal lobe origin who achieved seizure freedom after surgery, two groups were compared: hippocampal temporal lobe seizures and extrahippocampal temporal lobe seizures (ETS). The study examined historical information, ictal semiology, and ictal EEG patterns. Significant differences were observed in the prevalence of febrile convulsions history, types of auras, and early ictal manifestations between the two groups. HTS patients more frequently reported a history of febrile convulsions and presented more frequent epigastric auras and early oral automatisms. In contrast, ETS patients more commonly reported experiential auras and early motor involvement of the contralateral upper extremity without oral automatisms. Later ictal and postictal features were similar in both groups. These findings highlight distinct early clinical features in HTS and ETS, potentially aiding in their differentiation and localization [41].

17.7.2.2 Frontal Lobe Epilepsies

In frontal localizations, early motor manifestations are more prominent, with sometimes a more abrupt alteration of awareness. Frontal seizures usually occur during sleep, their duration is shorter and patients present no or less postictal confusion, compared to hippocampal seizures [15, 32].

17.7.2.3 Occipital Lobe Seizures

Hippocampal seizures rarely include visual hallucinations that may be misinterpreted as ictal hallucinations stemming from the occipital lobe. MTLE visual hallucinations are often complex, involving scenes, people, or recognizable objects. They are usually accompanied by emotional disturbances, vivid memories, or déjà vu sensations. Ictal visual hallucinations in occipital seizures are typically simple, such as flashes of light, colored spots, or geometric shapes. They may include transient visual field losses or eye movement abnormalities and are typically limited to the visual phenomenon without significant emotional or memory components [32, 34, 93].

17.7.2.4 Absence Seizures

Absence seizures, which are the main clinical feature of some genetic generalized epilepsies, are characterized by sudden and brief lapses in consciousness, typically lasting for a few seconds. Absences can be accompanied by subtle motor signs such as eyelid fluttering, lip-smacking, or minor automatisms. These “focal symptoms” can lead to the misdiagnosis of generalized genetic epilepsy as focal epilepsy. In complex clinical scenarios, careful evaluation of clinical features and electroencephalographic (EEG) recording is crucial for resolving ambiguities in seizure localization and characterization [94, 95].

17.8 Diagnostic Workup

17.8.1 Electroencephalography (EEG)

17.8.1.1 Scalp EEG

From a technical point of view, the 10–20 system is used for standardized positioning of electroencephalography (EEG) electrodes on the scalp. This system allows for a uniform distribution of electrodes and ensures consistency in recording and interpreting brain signals across different patients and medical centers. Extra EEG sensors are positioned between the standard 10–20 system electrodes. Other non-standard electrodes, such as sphenoidal, nasopharyngeal, or anterior temporal, can improve the precision of ictal and interictal abnormalities in MTLE.

Interictal EEG in hippocampal epilepsy may show a normal background, an asymmetry

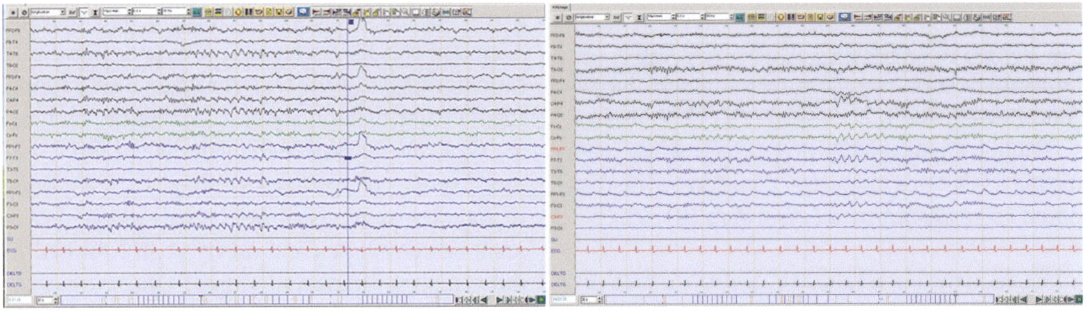

Fig. 17.1 A 38-year-old patient with temporal epilepsy. EEG shows bilateral TIRDA (temporal intermittent rhythmic delta activity): bilateral temporo-occipital delta rhythm that is sometimes predominant in the left temporal regions

of background activity, or focal slowing in temporal regions. Interictal abnormalities consist of epileptiform and non-epileptiform abnormalities. Non-epileptiform abnormalities present as focal temporal dysrhythmia that can be intermittent or persistent. This pattern is non-specific to hippocampal epilepsies and can also be recorded in patients with tumors or stroke. Temporal intermittent rhythmic delta activity (TIRDA) consists of trains of rhythmic delta. TIRDA is more associated with temporal lobe epilepsy (Fig. 17.1). Epileptiform abnormalities are spikes or sharp waves localized on temporal basal electrodes (Fig. 17.2). They can be unilateral or bilateral. When routine EEG is normal, a prolonged recording reveals abnormalities in most patients. Interictal discharges are more prevalent during NREM sleep and better localized during REM sleep.

Ictal EEG: During the aura, EEG may be normal. An initial focal, regional, or more diffuse background attenuation can then be seen. The typical pattern is the build-up of rhythmic theta activity that can be lateralized or not (Fig. 17.3). It is important to note that even ictal rhythmic theta can be contralateral to the side of seizure origin.

EEG in mesiotemporal status epilepticus: EEG usually remains the only means of diagnosing a status epilepticus of mesiotemporal origin, particularly when the patient exhibits confusion, cognitive or psychotic disturbances with subtle or absent signs that could guide the diagnosis (e.g., subtle finger or eyelid clonus, etc.). Mesiotemporal status epilepticus typically shows repetitive electrographic seizures (Fig. 17.4), continuous spike-and-wave activity, or lateralized periodic discharges.

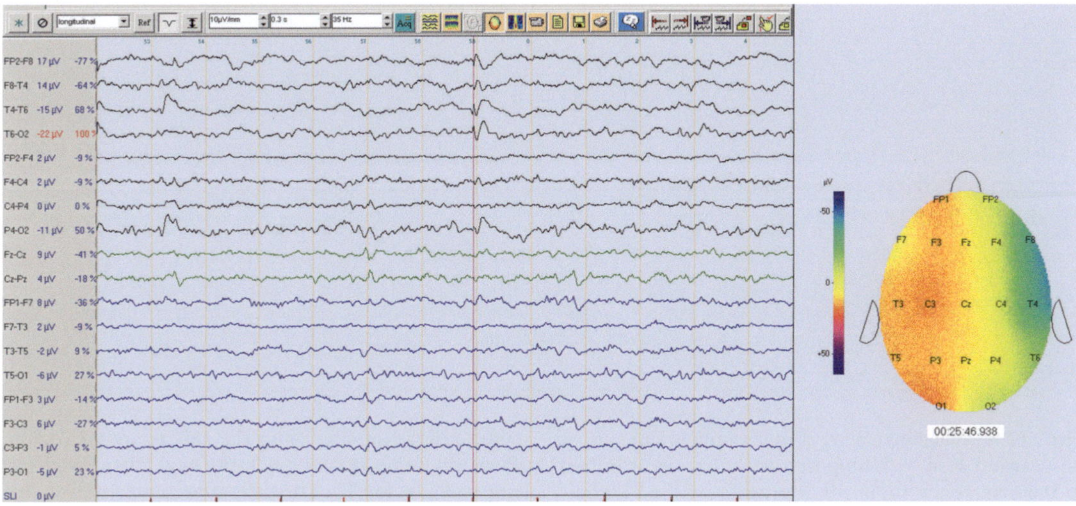

Fig. 17.2 Sleep EEG, 38-year-old patient with right hippocampus sclerosis and typical mesiotemporal seizures (see Fig. 17.3). Interictal EEG shows spike and wave in the temporal region, with phase reversal on T4. Voltage map: maximum negativity right temporal

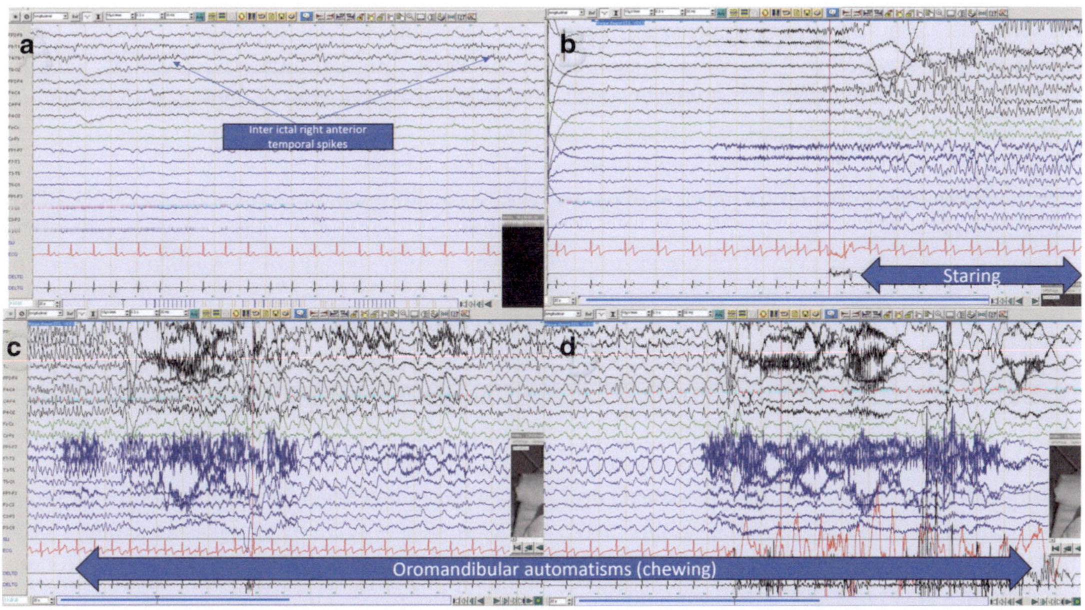

Fig. 17.3 A 38-year-old patient with hippocampal sclerosis revealed by typical mesiotemporal seizures: typical ascending digestive aura, followed by motionless staring, then oromandibular automatisms (chewing). Video-EEG of one of these seizures showing an initial background activity attenuation (**a**), followed by rapid activity (**b**) and a build-up of rhythmic theta activity that is diffuse but more prominent in the right hemisphere (**b**, **c**, and **d**)

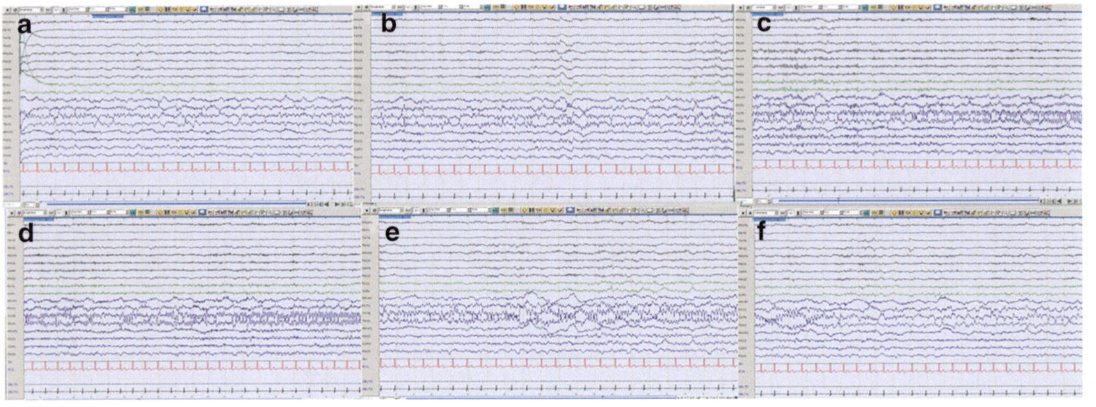

Fig. 17.4 Nonconvulsive status epilepticus in a 42-year-old man with hippocampal sclerosis, admitted to the emergency room for confusion with no associated symptoms. EEG shows focal left temporal slowing and left temporal spikes (**a**, **b**), followed by a nonmotor seizure arising from the left temporal region, where rhythmic spikes are seen (**b**–**e**), then focal left temporal slowing (**f**). The patient presented during this recording with repetitive seizures similar to this one, every 50–60 s

It is important to emphasize that some nonepileptic EEG patterns can sometimes be mistaken for temporal lobe abnormalities. These mimics of temporal EEG abnormalities encompass artifacts, breech rhythm, and benign EEG variants such as rhythmic midtemporal discharges or wicket spikes (Fig. 17.5). A careful analysis of artifacts and physiological variants in EEG recordings can prevent misinterpretation of normal brain activity as pathological [14, 32, 39, 96–101].

In most cases, the diagnosis of hippocampal epilepsy is made based on clinical and noninvasive EEG.

17.8.1.2 Intracranial EEG in Hippocampal Seizures

Intracranial EEG monitoring plays a vital role in presurgical assessment for drug-resistant epilepsy patients, especially when noninvasive methods prove insufficient for pinpointing seizure origins. This approach is recommended for cases involving suspected bilateral temporal lobe seizures, possible associated extratemporal foci, or when seizure activity occurs near functionally important brain areas.

The technique employs various electrode types, such as subdural grids, strips, and depth electrodes, tailored to each patient's specific needs.

By enabling precise identification of seizure onset zones and delineation of the epileptogenic region, intracranial EEG significantly contributes to successful surgical outcomes.

Despite certain limitations, it remains the gold standard for localizing epileptic brain tissue and informing surgical decisions, particularly in complex cases.

Normal hippocampal intracranial EEG (iEEG) recordings typically show occasional delta waves, which should not be mistaken as signs of epilepsy or other disorders. However, several features indicate abnormal hippocampal iEEG: almost constant delta activity, absence of higher-frequency background waves, decreased occurrence of spindles and sharp waves, and frequent interictal discharges.

Seizures can begin in two distinct ways: rapid, low-amplitude spike discharges exceeding 10 Hz, and slow, high-amplitude hypersynchronous spikes below 2 Hz. Both patterns eventually develop into the typical rhythmic, high-amplitude ictal activity. It remains unclear whether these different onset types correlate with specific neuropathological changes, such as neuronal loss, gliosis, or other biological alterations. Additionally, the potential prognostic significance of these two patterns remains unclear [14, 96, 97, 102–105].

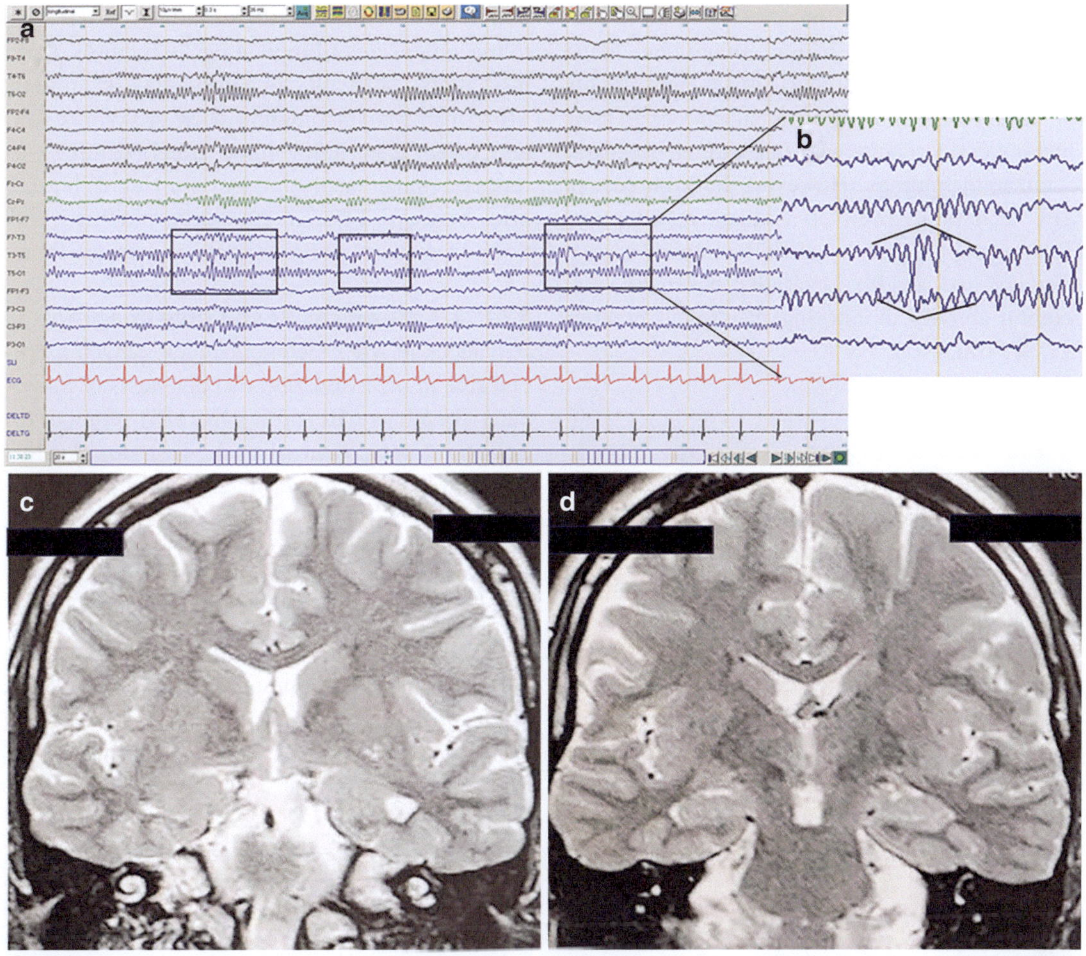

Fig. 17.5 A 37-year-old woman was referred for presurgical evaluation for "mesiotemporal epilepsy with hippocampus sclerosis." She reported non-stereotyped episodes, beginning either with paresthesia of the whole body or dizziness; she then fell down, didn't lose consciousness but was unable to respond. The episodes were always followed by crying and lasted at least 30 min. EEG was interpreted as "interictal focal left temporal spikes" and was confirmed by the MRI, which showed an enlarged left ventricle temporal horn (**b**). The semiology of the non-stereotyped, long episodes was more suggestive of PNES rather than temporal epilepsy. EEG shows brief bursts of sharply contoured waves (**a**) but without slow after-wave and with a diamond shape, which is typical of wicket spikes (**b**). MRI shows an enlarged left temporal horn (**c**), but the shape, signal, and volume of the homolateral hippocampus are normal and similar to the right one (**d**)

17.8.2 Magnetoencephalography

Magnetoencephalography (MEG) is a functional noninvasive technique that measures the magnetic field generated by the electrical activity of neurons. MEG has emerged as a significant adjunct in the detection and localization of epileptiform activity originating from deep brain structures, including the hippocampus. Contemporary research has challenged previous assumptions regarding MEG's limitations, confirming its ability to detect signals from deep limbic structures, especially the hippocampus and amygdala. In clinical epileptology, MEG serves as a complementary modality to electroencephalography (EEG) for interictal spike detection while offering unique insights into nonlesional cases and extratemporal epilepsies. Furthermore, MEG contributes to the delineation of the eloquent cortex and holds prognostic value in predicting postoperative seizure outcomes. MEG demonstrates the potential to reduce the necessity for invasive monitoring in focal epilepsy cases, showing concordance with intracranial electrode localization findings. However, the high cost of MEG technology limits its widespread adoption. Electroencephalography (EEG) remains a cost-effective and readily available alternative for cerebral activity monitoring in clinical practice [37, 97, 106, 107].

17.8.3 Neuroimaging

17.8.3.1 Computed Tomography (CT)

A CT scan offers the advantages of being easily accessible and cost-effective. It is also useful in emergencies, especially in agitated patients or when magnetic resonance imaging (MRI) is not available or contraindicated. A CT scan can detect pathologies like hemorrhage (Fig. 17.6) and calcifications. However, the importance of a CT scan in exploring hippocampal lesions has markedly diminished with the advent of MRI due to the small size of hippocampal structures, making them difficult to visualize accurately. Moreover, the surrounding bony structures usually cause a partial volume effect, resulting in reduced image resolution and blurring of boundaries between different structures, making it difficult to accurately differentiate or measure them.

17.8.3.2 Magnetic Resonance Imaging

Advances in MRI techniques have played a crucial role in the evaluation and diagnosis of hippocampal epilepsy. Regarding the MRI protocol for epilepsy, the International League Against Epilepsy (ILAE) Neuroimaging Task Force recommended a set of MRI acquisition sequences known as the HARNESS-MRI protocol that is regarded as the optimal standard for imaging in epilepsy. According to this protocol, mandatory sequences are high-resolution 3D T1-weighted MRI for optimal evaluation of brain morphology. High-resolution 3D fluid-attenuated inversion recovery (FLAIR) is the best sequence for detecting signal anomalies. High in-plane resolu-

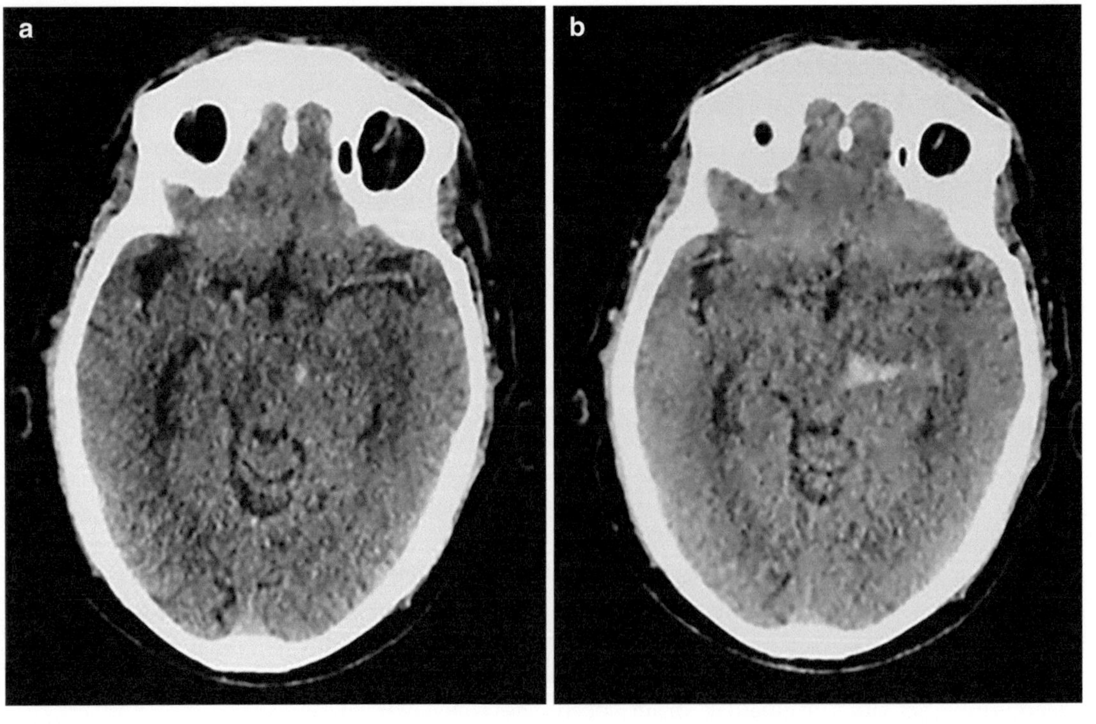

Fig. 17.6 Axial CT scan showing left hippocampal spontaneous hematoma in a 56-year-old man (**a**, **b**)

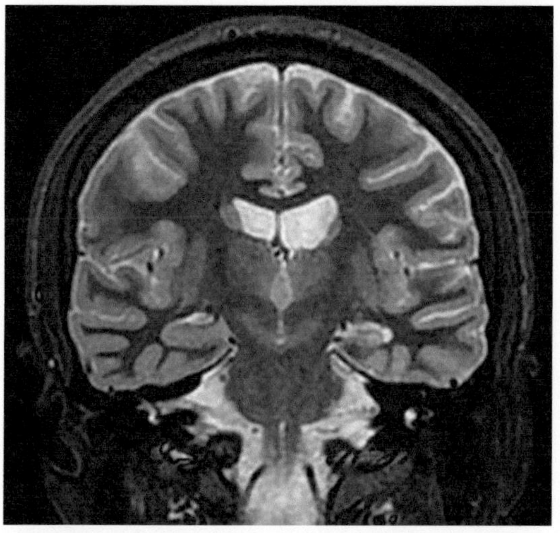

Fig. 17.7 MRI, coronal T2: hippocampal sclerosis: hippocampal asymmetry. The left hippocampus is smaller than the right one; it is hyperintense and has lost its internal architecture. Slight enlargement of the left temporal horn. (Courtesy of Prof. En-Nouali H.)

tion 2D coronal T2-weighted MRI is the best sequence for assessing the hippocampal internal structure. Optional sequences can be added, such as T1-weighted MRI with gadolinium in cases of tumors, infectious or inflammatory processes, or vascular malformations. T2* is hemosensitive and thus can be useful in cases of hemorrhage, iron deposits, and calcifications. Slices of 1 mm or less reduce the partial volume effect.

The most frequent cause of hippocampal epilepsy is hippocampal sclerosis (HS) . It is typically characterized on MRI by atrophy, loss of internal structure, increased T2 signal, and decreased T1 signal (Fig. 17.7). Automated volumetric segmentation techniques for hippocampal analysis have been developed for hippocampal analysis; however, manual segmentation performed by experienced radiologists remains the gold standard for hippocampal volume estimation. Ipsilateral atrophy of the temporal lobe, fornix, mammillary body, insula, and thalamus may be associated with hippocampal sclerosis.

Spectroscopy-MR helps identify metabolic alterations in epileptic tissues. In HS, it shows a mean reduction in *N*-acetyl-aspartate signal, an increase in creatine and choline peaks, and a low choline/creatine ratio. This noninvasive technique can be useful in lateralizing seizure foci when HS is bilateral.

When interpreting MRI in hippocampal epilepsy, careful consideration must be given to various confounding factors that may lead to misinterpretation of imaging findings: FLAIR imaging can produce false-positive results due to an enhanced hyperintense signal. Hippocampal asymmetry and hyperintense signal in T2-weighted images can also be found in the normal population. A unilateral enlarged temporal horn occurs in about one-third of healthy controls (Fig. 17.5). Unilateral or bilateral hippocampal hyperintensities are found in about one-third of normal controls without concomitant atrophy. It is estimated that standard MRI evaluation performed by nonspecialized radiologists may result in misdiagnosis in approximately 60% of cases.

Functional MRI (fMRI) primarily uses the blood oxygen level-dependent (BOLD) effect to infer brain activity by detecting changes in magnetic susceptibility due to blood oxygenation. This method allows for the visualization of neural activity through increased cerebral blood flow and resultant changes in T2* signals. Task-based fMRI is crucial in presurgical workups for language dominance and motor or sensory mapping, offering a noninvasive alternative to procedures like the Wada test. Additionally, EEG-fMRI combines electrical and hemodynamic data to localize epileptic foci, enhancing clinical decision-making. Resting-state fMRI further maps functional networks, identifying atypical connectivity related to seizure generation, thereby holding potential for clinical application in epilepsy management.

Diffusion-weighted MRI (DWI) measures how water molecules move within brain tissue, helping to understand the structure of white matter. DWI data helps create detailed maps of white matter tracts in the brain through tractography. Tractography can help visualize brain networks and predict outcomes after surgery [38, 39, 108–112].

MRI also enables the diagnosis of other lesion-related causes of hippocampal epilepsies and allows for the study of their relationships, such as benign (Fig. 17.8) and malignant tumors

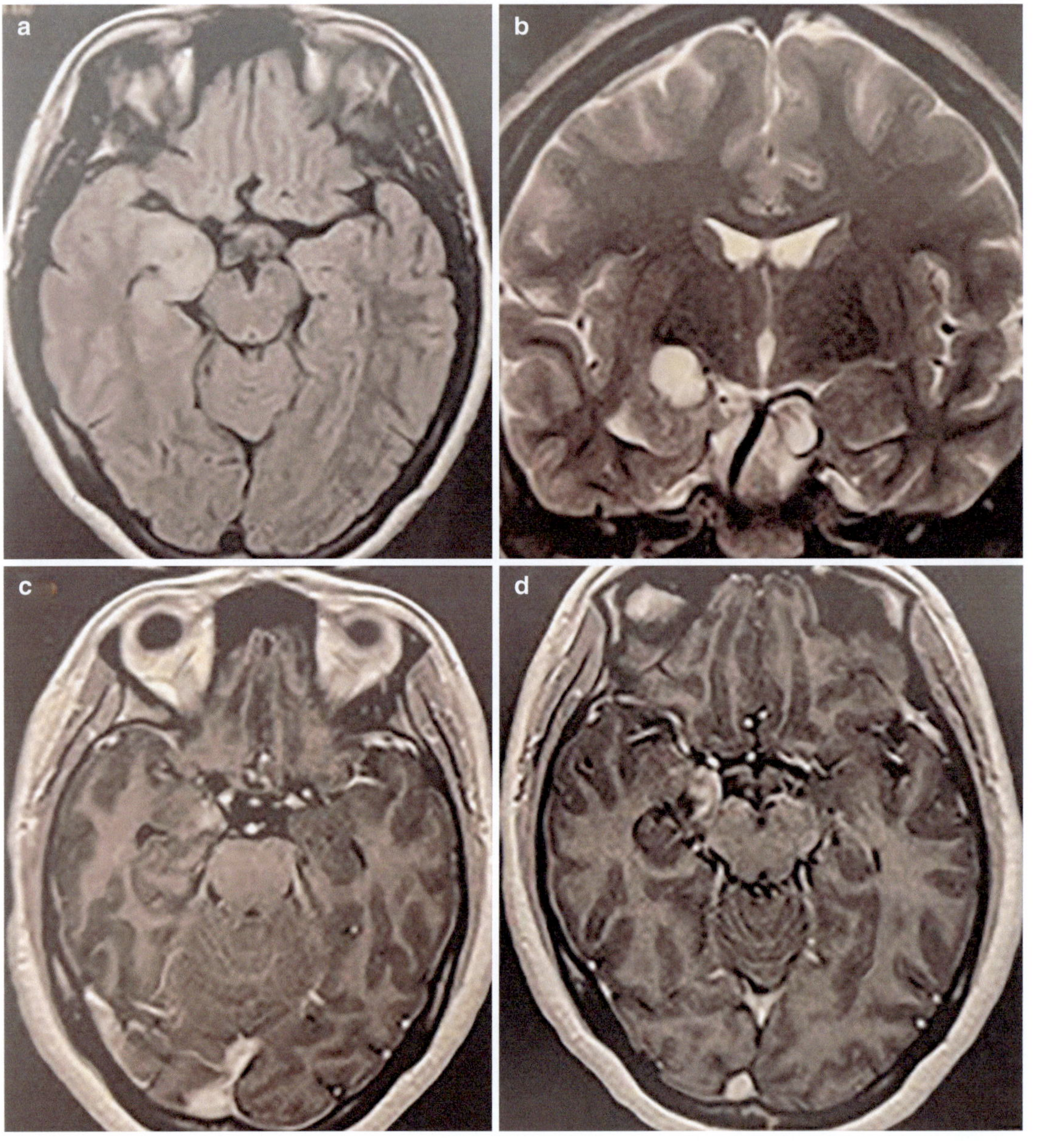

Fig. 17.8 A 20-year-old woman presented, since she was 10, with brief episodes of ascending digestive aura, followed by ictal spitting, then loss of consciousness and agitation. MRI shows an intra-axial expansive process involving the uncus and hippocampus, hyperintense on T2 and FLAIR sequences, hypointense on T1, with heterogeneous contrast enhancement. The process is not accompanied by cerebral edema and does not cause compression. The combination of clinical presentation and MRI findings is suggestive of a low-grade glial tumor

(Fig. 17.9), brain malformations, and arteriovenous malformations. In cases of limbic encephalitis, MRI can be normal or can show hyper T2 signals in the medial temporal lobe, typically bilateral (Fig. 17.10) [32, 113].

17.8.3.3 Functional Imaging Modalities

Functional imaging techniques, such as PET scan and ictal single-photon emission computed tomography (SPECT), can help identify the specific regions of hypometabolism or hypermetabo-

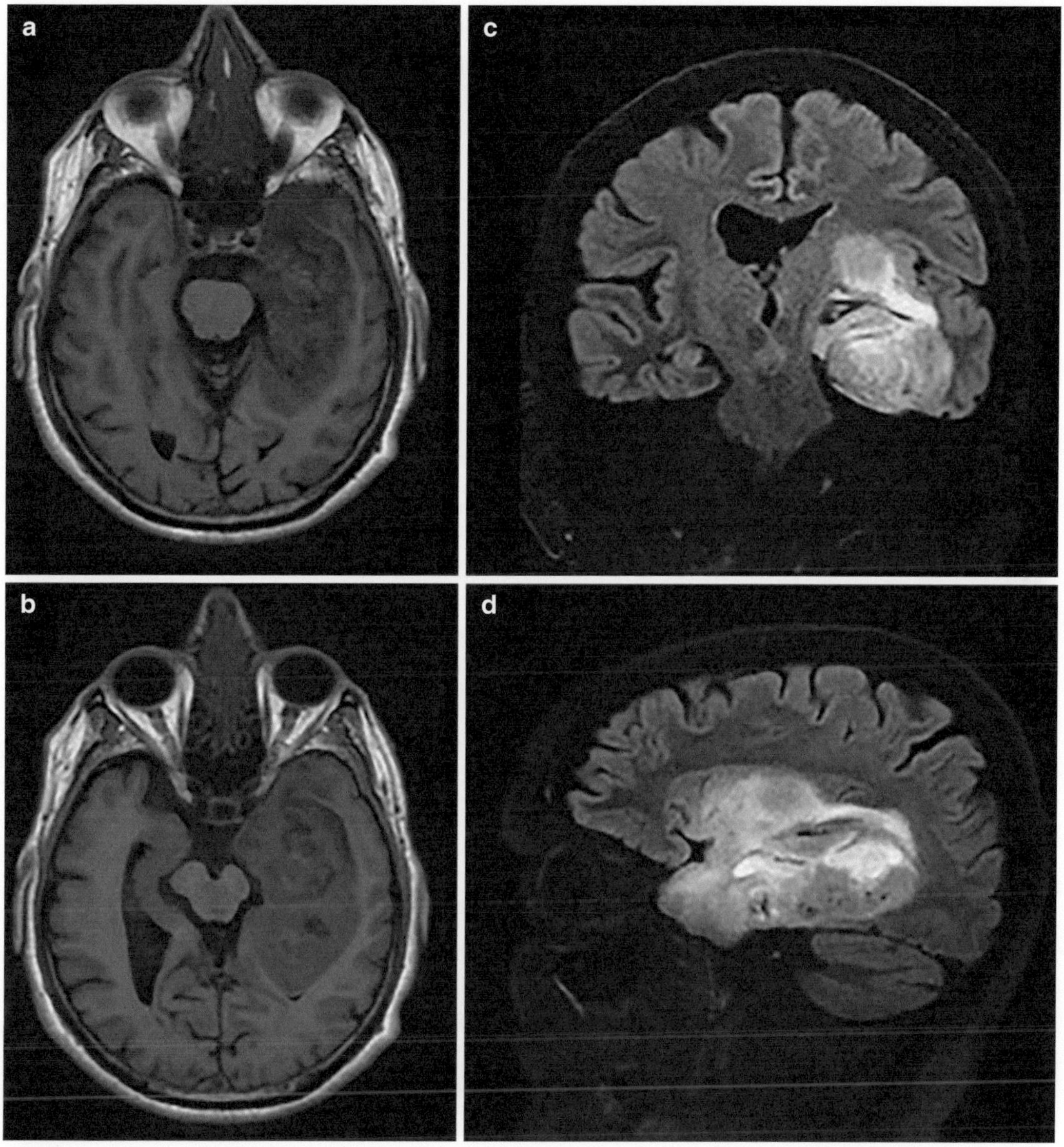

Fig. 17.9 Left mesial temporal lobe glioma in a 40-year-old woman as seen on axial T1-weighted magnetic resonance imaging (**a**, **b**) and on coronal (**c**) and sagittal (**d**) FLAIR sequences

lism within the hippocampus, further refining the localization of the epileptic focus.

Fluorodeoxyglucose (FDG)-PET frequently demonstrates temporal hypometabolism in temporal lobe epilepsy patients, correlating with HS identified through MRI. Occasionally, PET scans reveal hippocampal hypermetabolism, potentially indicative of subclinical ictal activity. The synergistic use of PET and MRI provides complementary data, with 95% of patients exhibiting either MRI-detected HS or PET-detected temporal hypometabolism. PET imaging can also elucidate multiregional hypometabolism, predominantly in the ipsilateral temporal lobe. The underlying pathophysiology of this widespread hypometabolism remains incompletely elucidated (Fig. 17.9). Flumazenil (FMZ) is another radiotracer that can be used to image the distribu-

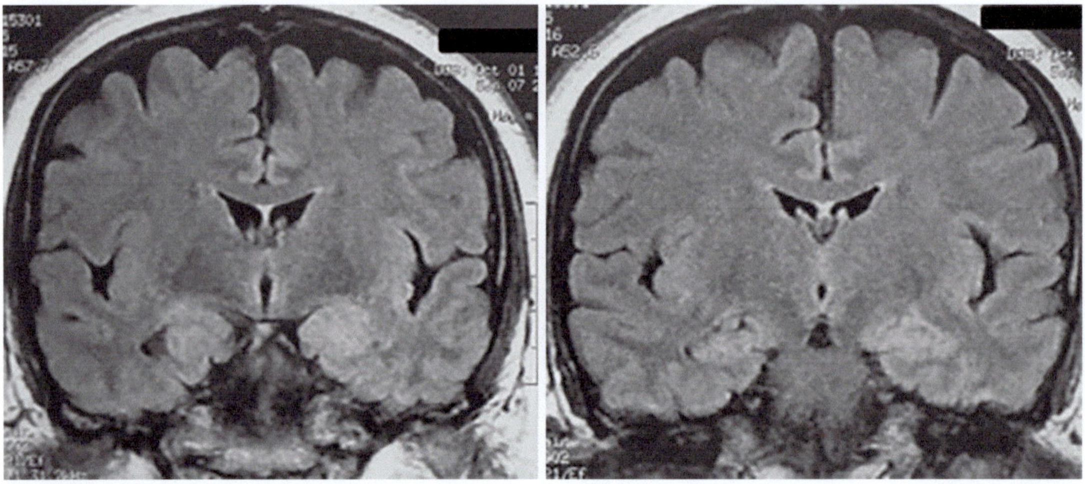

Fig. 17.10 A 39-year-old woman presented with repeated, stereotyped, brief episodes of "déjà vu," followed by motionless staring. MRI showed bilateral symmetrical FLAIR hyperintensity in the hippocampus and amygdala. Laboratory workup revealed positive anti-GAD

tion and density of GABA-A receptors in the brain. FMZ-PET demonstrates superior sensitivity, specificity, and localization precision compared to FDG-PET in the evaluation of hippocampal epilepsy. Nevertheless, FMZ's high cost and its limited accessibility are significant barriers to its utilization by many medical centers worldwide [114–117] (Fig. 17.11).

Single-photon emission computed tomography data demonstrates high lateralization accuracy (91%) in mesial temporal lobe epilepsy, comparable to FLAIR MRI. Ictal SPECT typically reveals ipsilateral temporal hyperperfusion with frontal hypoperfusion. It achieves the highest sensitivity (97%) for temporal lobe seizure localization [118–120].

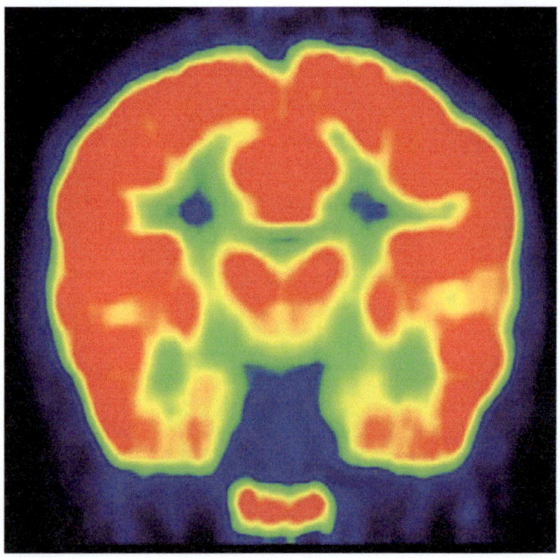

Fig. 17.11 FDG-PET of a patient presenting mesiotemporal epilepsy with no clinical or electroencephalographic lateralizing signs and bilateral HS on MRI. FDG-PET shows mesiotemporal hypometabolism, associated with ipsilateral insular hypometabolism. (Courtesy of Prof. Biyi A.)

17.8.4 Neuropsychological Testing in Hippocampal Epilepsies

Neuropsychological evaluation plays a crucial role in the assessment of hippocampal epilepsy, offering valuable insights into cognitive deficits, particularly memory impairments indicative of hippocampal damage. Neuropsychological tests, including the intracarotid amobarbital test (Wada test), are effective in documenting dysfunction associated with lateralized temporal lobe seizure onset and predicting seizure outcomes post-surgery, especially concerning verbal memory decline in cases involving the language-dominant hemisphere. The intracarotid amobarbital test (Wada test), while still used in some cases, has not demonstrated superiority to a comprehensive evaluation combining neuropsychological testing and fMRI for assessing language lateralization and contralateral hippocampal memory function potential.

Neuropsychological tests also allow an objective evaluation of cognitive impairments, and psychological and behavioral alterations in hippocampal epilepsies. The hallmark cognitive deficit in mesial temporal lobe epilepsy with hippocampal sclerosis (MTLE-HS) is impairment of long-term memory consolidation or retrieval of newly acquired information. The severity of memory dysfunction is influenced by multiple factors, including seizure frequency and severity, age of epilepsy onset, medication, degree of hippocampal atrophy, and epilepsy duration. Structural abnormalities detected through neuroimaging techniques correlate with cognitive impairments in TLE patients, supporting the use of combined neuroimaging and neuropsychological assessments for a comprehensive evaluation [38, 121–125].

17.8.5 Other Tests

In non-HS epilepsies, depending on the suspected etiology, further investigation may include CSF analysis, PCR, viral serologies, or neural antibodies, among others.

17.9 Management of Hippocampal Epilepsies

17.9.1 Antiseizure Medication

Antiseizure medication (ASM) choice should take into account, on one hand, the type of seizures and epileptic syndrome, the patient's age, sex, and comorbidities, and on the other hand, the pharmacological profile of ASMs, including mechanism of action, pharmacokinetics, efficacy spectrum, adverse effect profile, drug interactions, therapeutic index, formulations, dosing regimen, special population considerations, and pharmacoeconomics. The choice should thus be tailored according to the patient's and ASM's specificities.

The study SANAD II compared Levetiracetam, Zonisamide, and Lamotrigine as first-line treatments for patients with focal epilepsy. It concluded that Lamotrigine should remain a first-line treatment for patients with focal epilepsy and should be the standard treatment in future trials.

Another study found that Carbamazepine was superior in short-term seizure control but had a higher rate of adverse events and treatment failure due to adverse events, while Lamotrigine was better tolerated with fewer adverse events and lower withdrawal rates, making it a suitable alternative, especially for long-term treatment.

Oxcarbazepine is effective in reducing seizure frequency and improving quality of life, with a tolerability profile similar to Lamotrigine.

Thus, as a first-line treatment, Lamotrigine or Carbamazepine should be chosen, bearing in mind that Carbamazepine is more effective in short-term seizure control, and Lamotrigine is better tolerated. Levetiracetam can be considered if there is no history of psychiatric disorder, as it is associated with a higher risk of inducing psychotic disorders, especially in females and those with temporal lobe involvement, while Carbamazepine has a protective effect against such disorders.

No definitive strategy exists for managing focal epilepsy uncontrolled by the first ASM. Clinical studies show no significant difference in efficacy or tolerability between alternative monotherapy and adjunctive therapy. In animal studies, the combination of ASMs with different or multiple mechanisms of action was the most effective, while the association of drugs with similar mechanisms of action was associated with more toxicity. According to clinical studies, these medications can be favorably combined: Levetiracetam with Lacosamide or Lamotrigine, Lamotrigine with Valproate and possibly with Levetiracetam and Topiramate, Carbamazepine with Gabapentin and Levetiracetam.

Among third-generation ASMs for adjunctive treatment of focal seizures, Cenobamate is the most effective, while Brivaracetam and Lacosamide are the best tolerated.

However, the probability of achieving pharmacological control becomes negligible following the failure of one or two primary antiepileptic drugs (AEDs). The implementation of polypharmacy, involving more than two or three AEDs, even when rationally conceived, is likely to exacerbate adverse effects such as cognitive impairment and somnolence, without conferring significant therapeutic benefit. These patients, warrant prompt evaluation for neurosurgical intervention, especially those with HS who are the optimal candidates for surgical treatment among symptomatic focal epilepsies, with the highest likelihood of achieving seizure freedom post-operatively [5, 32, 126–135].

17.9.2 Surgical Treatment

Surgical intervention for hippocampal epilepsy, particularly mesial temporal lobe epilepsy, encompasses a variety of techniques such as amygdalohippocampectomy and hippocampal transection to minimize complications and preserve memory function. Research indicates that individuals suffering from drug-resistant epilepsy experience notable benefits from epilepsy surgery, with resective surgery proving to be more effective than medical treatment in terms of managing seizures and enhancing quality of life. A comparison of surgical methods reveals that anterior temporal lobectomy (ATL) leads to a higher rate of complete seizure remission and satisfactory seizure management in contrast to selective amygdalohippocampectomy (SAH). ATL emerges as the preferred surgical option for drug-resistant hippocampal epilepsy, demonstrating a significant success rate in achieving long-term seizure remission among patients and enhancing overall quality of life. Temporal lobectomy involves the resection of a portion of the temporal lobe to curtail the propagation of epileptic discharges [136–138].

To achieve comparable seizure control while reducing surgical morbidity and preserving cognitive function, innovative less invasive surgical techniques for epilepsy have been developed. These minimally invasive approaches encompass Stereotactic Radiosurgery (SRS): Utilizing platforms such as Gamma Knife and CyberKnife, SRS delivers focused ionizing radiation to targeted mesial temporal structures in MTLE. Stereotactic Radiofrequency Thermocoagulation (SRT): employs radiofrequency energy for the precise ablation of epileptogenic foci. Laser Interstitial Thermal Therapy (LITT): LITT involves the application of laser-generated thermal energy for targeted tissue ablation. MRI-guided Focused Ultrasound Ablation (FUS): uses high-intensity focused ultrasound for noninvasive thermal ablation of epileptogenic tissue. These emerging approaches present promising alternatives to conventional surgical interventions. Their implementation is associated with reduced complication rates, enhanced neuropsychological outcomes, and increased patient acceptability. However, it is noteworthy that traditional surgical procedures, such as ATL and SAH, continue to demonstrate superior efficacy in terms of long-term seizure freedom rates [110, 139–142].

17.9.3 Neuromodulation Therapies

In select cases where resective surgery is contraindicated or ineffective, neuromodulation therapies can be considered. Vagus nerve stimulation (VNS) is the most well-known technique; it acts on hippocampus by modulating neuro-inflammatory responses, enhancing cholinergic activity, and upregulating $\alpha 7$ nicotinic acetylcholine receptor expression. These mechanisms contribute to reduced seizure frequency and duration, decreased pro-inflammatory cytokine levels, and activation of anti-inflammatory pathways in the hippocampus. VNS is also effective in treating depression, a frequent comorbidity of hippocampal epilepsy. Responsive neural stimulation (RNS) has emerged as a promising therapeutic approach for hippocampal epilepsy, demonstrating efficacy in reducing seizure duration through targeted stimulation of specific brain regions such as the subiculum and CA1, with frequency-dependent effects. Deep brain stimulation (DBS) targets stimulation of regions such as the parahippocampal cortex, hippocampus, and anterior thalamic nucleus. The most effective DBS targets for treating temporal lobe epilepsy include the anterior nucleus of the thalamus (ANT), the hippocampus (both unilateral and bilateral), and the ventral hippocampal commissure (VHC). Each of these targets has shown significant potential in reducing seizure frequency, and cognitive improvement in MTLE patients [143–147].

17.9.4 Dietary Therapies and Hippocampal Epilepsies

Despite being an age-old therapy, dietary treatments like the ketogenic diet (KD) continue to play a significant role in managing pharmacoresistant epilepsies. In the particular case of hippocampal epilepsies, KD has shown promising neuroprotective effects. Studies have demonstrated that KD improves seizures and cognitive impairments and prevents hippocampal neuronal injury and apoptosis. It has also been found to ameliorate the balance between excitatory and inhibitory neurotransmitters, reduce oxidative stress, and improve inflammatory responses in the hippocampus post-status epilepticus. Less restrictive diets have been developed to improve the palatability [148–150].

17.9.5 Other Treatments

Depending on the cause of epilepsy, specific treatment should be initiated, such as antiviral treatment. Autoimmune-associated seizure disorders require immunotherapy and the treatment of the associated neoplasm when they are paraneoplastic.

Treatment of comorbidities is important, especially depression and cognitive impairment, highlighting the importance of a multidisciplinary, tailored management approach for patients with hippocampal epilepsy.

17.10 Conclusion

The relationship between the hippocampus and its surrounding structures poses both challenges and opportunities in understanding and treating hippocampal epilepsy. Our knowledge of this condition has advanced significantly, thanks to technological progress and a deeper understanding of the complexities of the mesiotemporal network. It is crucial to integrate insights from basic science, clinical research, and new technologies to develop more effective, tailored treatments for patients with MTLE. By continuing to explore how the hippocampus functions and malfunctions within the broader mesiotemporal network, we aim not only to better control seizures but also to enhance the overall quality of life for those affected by this complex form of epilepsy.

References

1. Grattan-Smith JD, Harvey AS, Desmond PM, Chow CW. Hippocampal sclerosis in children with intractable temporal lobe epilepsy: detection with MR imaging. Am J Roentgenol. 1993;161(5):1045–8.
2. Thom M. Review: Hippocampal sclerosis in epilepsy: a neuropathology review. Neuropathol Appl Neurobiol. 2014;40(5):520–43.
3. Shariff S, Nouh HA, Inshutiyimana S, Kachouh C, Abdelwahab MM, Nazir A, et al. Advances in understanding the pathogenesis of epilepsy: unraveling the molecular mechanisms. Health Sci Rep. 2024;7(2):e1896.
4. Huberfeld G, Blauwblomme T, Miles R. Hippocampus and epilepsy: findings from human tissues. Rev Neurol (Paris). 2015;171(3):236–51.
5. Akdogan I, Goksin N. Experimental epilepsy models and morphologic alterations of experimental epilepsy models in brain and hippocampus. In: Shad Kaneez F, editor. Underlying mechanisms of epilepsy. IntechOpen; 2011. http://www.intechopen.com/books/underlying-mechanisms-of-epilepsy/experimental-epilepsy-models-and-morphologic-alterations-of-experimental-epilepsy-models-in-brain-an.
6. Zhu X, Yao Y, Yang J, Ge Q, Niu D, Liu X, et al. Seizure-induced neuroinflammation contributes to ectopic neurogenesis and aggressive behavior in pilocarpine-induced status epilepticus mice. Neuropharmacology. 2020;170:108044.

7. Ulitin AY, Vasilenko AV, Ivanenko AV, Bubnova PD, Rasulov ZM, Sokolov IA, et al. Neurophysiological and morphological features of the formation of the pathological hippocampal system in structural epilepsy (Literature review). Russ J Pers Med. 2022;2(1):83–92.
8. Bekenstein JW, Lothman EW. Dormancy of inhibitory interneurons in a model of temporal lobe epilepsy. Science. 1993;259(5091):97–100.
9. Cavazos J, Golarai G, Sutula T. Mossy fiber synaptic reorganization induced by kindling: time course of development, progression, and permanence. J Neurosci. 1991;11(9):2795–803.
10. Harvey AS, Grattan-Smith JD, Desmond PM, Chow CW, Berkovic SF. Febrile seizures and hippocampal sclerosis: frequent and related findings in intractable temporal lobe epilepsy of childhood. Pediatr Neurol. 1995;12(3):201–6.
11. Thompson SM. Consequences of epileptic activity *in vitro*. Brain Pathol. 1993;3(4):413–9.
12. Davies KG, Hermann BP, Dohan FC, Foley KT, Bush AJ, Wyler AR. Relationship of hippocampal sclerosis to duration and age of onset of epilepsy, and childhood febrile seizures in temporal lobectomy patients. Epilepsy Res. 1996;24(2):119–26.
13. Koepp MJ. Hippocampal sclerosis: cause or consequence of febrile seizures. J Neurol Neurosurg Psychiatry. 2000;69(6):716–7.
14. Jan MM, Sadler M, Rahey SR. Electroencephalographic features of temporal lobe epilepsy. Can J Neurol Sci. 2010;37(4):439–48.
15. Blair RDG. Temporal lobe epilepsy semiology. Epilepsy Res Treat. 2012;2012:1–10.
16. Dalmau J, Graus F. Antibody-mediated encephalitis. N Engl J Med. 2018;378(9):840–51.
17. Smith KM, Budhram A, Geis C, McKeon A, Steriade C, Stredny CM, et al. Autoimmune-associated seizure disorders. Epileptic Disord. 2024;26:415–34.
18. Damasio AR, Van Hoesen GW. The limbic system and the localisation of herpes simplex encephalitis. J Neurol Neurosurg Psychiatry. 1985;48(4):297–301.
19. Nagashima K, Kobayashi Y, Kojima H, Hasegawa H, Kurata T. Herpes encephalitis and paraneoplastic limbic encephalitis. Neuropathology. 1998;18(2):215–21.
20. Bradshaw MJ, Venkatesan A. Herpes simplex virus-1 encephalitis in adults: pathophysiology, diagnosis, and management. Neurotherapeutics. 2016;13(3):493–508.
21. Tyler KL. Emerging viral infections of the central nervous system: Part 1. Arch Neurol. 2009;66(8) https://doi.org/10.1001/archneurol.2009.153.
22. Shnayder NA, Sirbiladze TK, Demko IV, Petrova MM, Nasyrova RF. Limbic encephalitis associated with COVID-19. Encyclopedia. 2021;2(1):26–35.
23. Shrimpton M, Malhotra A. Neurosyphilis presenting as limbic encephalitis. BMJ Case Rep. 2023;16(12):e258794.
24. Gambardella A, Labate A, Giallonardo A, Aguglia U. Familial mesial temporal lobe epilepsies: clinical and genetic features. Epilepsia. 2009;50(s5):55–7.
25. Liu C, Qiao XZ, Wei ZH, Cao M, Wu ZY, Deng YC. Molecular typing of familial temporal lobe epilepsy. World J Psychiatry. 2022;12(1):98–107.
26. Crompton DE, Scheffer IE, Taylor I, Cook MJ, McKelvie PA, Vears DF, et al. Familial mesial temporal lobe epilepsy: a benign epilepsy syndrome showing complex inheritance. Brain. 2010;133(11):3221–31.
27. Tae WS, Hong SB, Joo EY, Han SJ, Cho JW, Seo DW, et al. Structural brain abnormalities in juvenile myoclonic epilepsy patients: volumetry and voxel-based morphometry. Korean J Radiol. 2006;7(3):162.
28. Kuan YC, Shih YH, Chen C, Yu HY, Yiu CH, Lin YY, et al. Abdominal auras in patients with mesial temporal sclerosis. Epilepsy Behav. 2012;25(3):386–90.
29. Blümcke I, Thom M, Aronica E, Armstrong DD, Bartolomei F, Bernasconi A, et al. International consensus classification of hippocampal sclerosis in temporal lobe epilepsy: a Task Force report from the ILAE Commission on Diagnostic Methods. Epilepsia. 2013;54(7):1315–29.
30. Schwartzkroin PA. Role of the hippocampus in epilepsy. Hippocampus. 1994;4(3):239–42.
31. Fisher RS, Cross JH, French JA, Higurashi N, Hirsch E, Jansen FE, et al. Operational classification of seizure types by the International League Against Epilepsy: Position Paper of the ILAE Commission for Classification and Terminology. Epilepsia. 2017;58(4):522–30.
32. Panayiotopoulos CP. Symptomatic and cryptogenic (probably symptomatic) focal epilepsies. In: A clinical guide to epileptic syndromes and their treatment. London: Springer London; 2010. p. 435–96. https://doi.org/10.1007/978-1-84628-644-5_15.
33. Vignal JP, Maillard L, McGonigal A, Chauvel P. The dreamy state: hallucinations of autobiographic memory evoked by temporal lobe stimulations and seizures. Brain. 2006;130(1):88–99.
34. Hatano K, Shimizu T, Matsumoto H, Suzuki I, Hashida H. Dreamy state, delusions, audiovisual hallucinations, and metamorphopsia in a lesional lateral temporal lobe epilepsy followed by ipsilateral hippocampal sclerosis. Case Rep Neurol. 2019;11(2):209–16.
35. Beniczky S, Tatum WO, Blumenfeld H, Stefan H, Mani J, Maillard L, et al. Seizure semiology: ILAE glossary of terms and their significance. Epileptic Disord. 2022;24(3):447–95.
36. Arthuis M, Valton L, Régis J, Chauvel P, Wendling F, Naccache L, et al. Impaired consciousness during temporal lobe seizures is related to increased long-distance cortical–subcortical synchronization. Brain. 2009;132(8):2091–101.
37. Englot DJ, Yang L, Hamid H, Danielson N, Bai X, Marfeo A, et al. Impaired consciousness in temporal lobe seizures: role of cortical slow activity. Brain. 2010;133(12):3764–77.

38. Chatzikonstantinou A. Epilepsy and the hippocampus. In: Szabo K, Hennerici MG, editors. Frontiers of neurology and neuroscience. S. Karger AG; 2014. p. 121–42. https://karger.com/books/book/245/chapter/5169796.
39. Baulac M. MTLE with hippocampal sclerosis in adult as a syndrome. Rev Neurol (Paris). 2015;171(3):259–66.
40. Alarcon G, Elwes RDC, Polkey CE, Binnie CD. Ictal oroalimentary automatisms with preserved consciousness: implications for the pathophysiology of automatisms and relevance to the international classification of seizures. Epilepsia. 1998;39(10):1119–27.
41. Gil-Nagel A. Ictal semiology in hippocampal versus extrahippocampal temporal lobe epilepsy. Brain. 1997;120(1):183–92.
42. INSERM. Mesial temporal lobe epilepsy with hippocampal sclerosis. In: Definitions. Qeios; 2020. https://www.qeios.com/read/definition/103459.
43. Feindel W. Localization of discharge in temporal lobe automatism. Arch Neurol Psychiatr. 1954;72(5):605.
44. Dupont S, Samson Y, Nguyen-Michel VH, Zavanone C, Navarro V, Baulac M, et al. Lateralizing value of semiology in medial temporal lobe epilepsy. Acta Neurol Scand. 2015;132(6):401–9.
45. Krishnan V, Drislane FW, Benatar MG. Cognitive manifestations of focal status epilepticus. In: Drislane FW, Kaplan PW, editors. Status epilepticus. Cham: Springer International Publishing; 2018. p. 259–73. https://doi.org/10.1007/978-3-319-58200-9_21.
46. Husain AM, Horn GJ, Jacobson MP. Non-convulsive status epilepticus: usefulness of clinical features in selecting patients for urgent EEG. J Neurol Neurosurg Psychiatry. 2003;74(2):189–91.
47. Bartsch T, Deuschl G. Transient global amnesia: functional anatomy and clinical implications. Lancet Neurol. 2010;9(2):205–14.
48. Kesner RP, Rolls ET. A computational theory of hippocampal function, and tests of the theory: new developments. Neurosci Biobehav Rev. 2015;48:92–147.
49. Benarroch EE. Adult neurogenesis in the dentate gyrus: general concepts and potential implications. Neurology. 2013;81(16):1443–52.
50. Boldrini M, Fulmore CA, Tartt AN, Simeon LR, Pavlova I, Poposka V, et al. Human hippocampal neurogenesis persists throughout aging. Cell Stem Cell. 2018;22(4):589–599.e5.
51. Jessberger S, Parent JM. Epilepsy and adult neurogenesis, vol. 2015. Cold Spring Harb Perspect Biol; 2015. p. a020677.
52. Naik AA, Brodovskaya A, Subedi S, Akram A, Kapur J. Extrahippocampal seizure and memory circuits overlap. eNeuro. 2022;9(4):ENEURO.0179-22.2022.
53. Bui AD, Nguyen TM, Limouse C, Kim HK, Szabo GG, Felong S, et al. Dentate gyrus mossy cells control spontaneous convulsive seizures and spatial memory. Science. 2018;359(6377):787–90.
54. Dengler CG, Coulter DA. Normal and epilepsy-associated pathologic function of the dentate gyrus. In: Progress in brain research. Elsevier; 2016. p. 155–78. https://linkinghub.elsevier.com/retrieve/pii/S0079612316300097.
55. Buckmaster PS, Abrams E, Wen X. Seizure frequency correlates with loss of dentate gyrus GABAergic neurons in a mouse model of temporal lobe epilepsy. J Comp Neurol. 2017;525(11):2592–610.
56. Engel J. Mesial temporal lobe epilepsy: what have we learned? Neuroscientist. 2001;7(4):340–52.
57. Rolls ET. The mechanisms for pattern completion and pattern separation in the hippocampus. Front Syst Neurosci. 2013;7:74. https://doi.org/10.3389/fnsys.2013.00074/abstract.
58. Bien CG, Benninger FO, Urbach H, Schramm J, Kurthen M, Elger CE. Localizing value of epileptic visual auras. Brain. 2000;123(2):244–53.
59. Ang CW, Carlson GC, Coulter DA. Massive and specific dysregulation of direct cortical input to the hippocampus in temporal lobe epilepsy. J Neurosci. 2006;26(46):11850–6.
60. Smith BN, Dudek FE. Short- and long-term changes in CA1 network excitability after kainate treatment in rats. J Neurophysiol. 2001;85(1):1–9.
61. Tzakis N, Holahan MR. Social memory and the role of the hippocampal CA2 region. Front Behav Neurosci. 2019;13:233.
62. Lisgaras CP, Oliva A, Mckenzie S, LaFrancois J, Siegelbaum SA, Scharman HE. Hippocampal area CA2 controls seizure dynamics, interictal EEG abnormalities and social comorbidity in mouse models of temporal lobe epilepsy; 2023. https://doi.org/10.1101/2023.01.15.524149.
63. Boehringer R, Polygalov D, Huang AJY, Middleton SJ, Robert V, Wintzer ME, et al. Chronic loss of CA2 transmission leads to hippocampal hyperexcitability. Neuron. 2017;94(3):642–655.e9.
64. Stafstrom CE. The role of the subiculum in epilepsy and epileptogenesis. Epilepsy Curr. 2005;5(4):121–9.
65. Fei F, Wang X, Xu C, Shi J, Gong Y, Cheng H, et al. Discrete subicular circuits control generalization of hippocampal seizures. Nat Commun. 2022;13(1):5010.
66. Fei F, Wang X, Wang Y, Chen Z. Dissecting the role of subiculum in epilepsy: research update and translational potential. Prog Neurobiol. 2021;201:102029.
67. Vinti V, Dell'Isola GB, Tascini G, Mencaroni E, Cara GD, Striano P, et al. Temporal lobe epilepsy and psychiatric comorbidity. Front Neurol. 2021;12:775781.
68. Novak A, Vizjak K, Rakusa M. Cognitive impairment in people with epilepsy. J Clin Med. 2022;11(1):267.
69. Devinsky O, Bear D. Varieties of aggressive behavior in temporal lobe epilepsy. Am J Psychiatry. 1984;141(5):651–6.
70. García-Peñas JJ. Autism, epilepsy and temporal lobe pathology. Rev Neurol. 2009;48(Suppl 2):S35–45.
71. Banker SM, Gu X, Schiller D, Foss-Feig JH. Hippocampal contributions to social and cog-

nitive deficits in autism spectrum disorder. Trends Neurosci. 2021;44(10):793–807.
72. Dager SR, Wang L, Friedman SD, Shaw DW, Constantino JN, Artru AA, et al. Shape mapping of the hippocampus in young children with autism spectrum disorder. AJNR Am J Neuroradiol. 2007;28(4):672–7.
73. Garg D, Tripathi M. Borderlands of migraine and epilepsy. Neurol India. 2021;69(7):91.
74. Valenti MP, Cretin B, Rudolf G, Dylgjeri S, De Saint-Martin A, Hirsch E. Migraines et épilepsies temporales mésiales familiales: quel est le lien? Rev Neurol (Paris). 2009;165(10):774–81.
75. Güven B, Çilliler AE. Headache in patients with mesial temporal lobe epilepsy with hippocampal sclerosis. Epilepsy Behav. 2021;121:108081.
76. Atli Eklioglu O, Ilgin S. Adverse effects of antiepileptic drugs on hormones of the hypothalamic-pituitary-gonadal axis in males: a review. Toxicology. 2022;465:153043.
77. Cutia CA, Christian-Hinman CA. Mechanisms linking neurological disorders with reproductive endocrine dysfunction: insights from epilepsy research. Front Neuroendocrinol. 2023;71:101084.
78. Fawley JA, Pouliot WA, Dudek FE. Epilepsy and reproductive disorders: the role of the gonadotropin-releasing hormone network. Epilepsy Behav. 2006;8(3):477–82.
79. Lin K, Scarlatelli-Lima A, Sukys-Claudino L. Disturbed subjective and objective parameters of sleep in patients with drug-resistant mesial temporal lobe epilepsy (P7.021). Neurology. 2015;84(14_Suppl):P7.021.
80. Scarlatelli-Lima AV, Sukys-Claudino L, Watanabe N, Guarnieri R, Walz R, Lin K. How do people with drug-resistant mesial temporal lobe epilepsy sleep? A clinical and video-EEG with EOG and submental EMG for sleep staging study. eNeurologicalSci. 2016;4:34–41.
81. Yildiz FG, Tezer FI, Saygi S. Temporal lobe epilepsy is a predisposing factor for sleep apnea: a questionnaire study in video-EEG monitoring unit. Epilepsy Behav. 2015;48:1–3.
82. Friedman DE, Gilliam FG. Seizure-related injuries are underreported in pharmacoresistant localization-related epilepsy. Epilepsia. 2010;51(1):43–7.
83. Leitner DF, Mills JD, Pires G, Faustin A, Drummond E, Kanshin E, et al. Proteomics and transcriptomics of the hippocampus and cortex in SUDEP and high-risk SUDEP patients. Neurology. 2021;96(21):e2639. https://doi.org/10.1212/WNL.0000000000011999.
84. Mueller SG, Bateman LM, Laxer KD. Evidence for brainstem network disruption in temporal lobe epilepsy and sudden unexplained death in epilepsy. NeuroImage Clin. 2014;5:208–16.
85. Bateman LM, Spitz M, Seyal M. Ictal hypoventilation contributes to cardiac arrhythmia and SUDEP: report on two deaths in video-EEG–monitored patients. Epilepsia. 2010;51(5):916–20.
86. Lee SJ, Lee TK, Bae YJ, Kim M. Increased risk of epilepsy after transient global amnesia: a population-based study in South Korea. Clin Neurol Neurosurg. 2024;243:108357.
87. Nehring SM, Spurling BC, Kumar A. Transient global amnesia. In: StatPearls. Treasure Island, FL: StatPearls Publishing; 2024. http://www.ncbi.nlm.nih.gov/books/NBK442001/.
88. Castellana F, D'Onofrio G, Ciccone F, Di Claudio MT, Pugliatti M, Popolizio T, et al. Ictal fear or panic attack, this is the question—a video–EEG study. Brain Sci. 2024;14(6):594.
89. Mostacci B, Bisulli F, Alvisi L, Licchetta L, Baruzzi A, Tinuper P. Ictal characteristics of psychogenic nonepileptic seizures: what we have learned from video/EEG recordings—a literature review. Epilepsy Behav. 2011;22(2):144–53.
90. Lempert T. Synkopen. Nervenarzt. 1997;68(8):620–4.
91. Carbonari G, Tonti G, Di Pisa V, Franzoni E, Cordelli DM. Pediatric epilepsies misdiagnosed as gastrointestinal disorders. Epilepsy Behav. 2018;83:137–9.
92. Bayram AK, Canpolat M, Karacabey N, Gumus H, Kumandas S, Doğanay S, et al. Misdiagnosis of gastroesophageal reflux disease as epileptic seizures in children. Brain Dev. 2016;38(3):274–9.
93. Revdal E, Arntsen V, Doan TP, Kvello-Alme M, Kvistad KA, Bråthen G, et al. Experiential seizures related to the hippocampal-parahippocampal spatial representation system. Epilepsy Behav Rep. 2020;14:100386.
94. Koren J, Lang C, Gritsch G, Mayer L, Hartmann M, Hafner S, et al. Idiopathic generalized epilepsies in the epilepsy monitoring unit: systematic quantification of focal EEG and semiological signs. Clin Neurophysiol. 2024;162:82–90.
95. Seneviratne U, Cook M, D'Souza W. Focal abnormalities in idiopathic generalized epilepsy: a critical review of the literature. Epilepsia. 2014;55(8):1157–69.
96. Williamson PD, French JA, Thadani VM, Kim JH, Novelly RA, Spencer SS, et al. Characteristics of medial temporal lobe epilepsy: II. Interictal and ictal scalp electroencephalography, neuropsychological testing, neuroimaging, surgical results, and pathology. Ann Neurol. 1993;34(6):781–7.
97. Javidan M. Electroencephalography in mesial temporal lobe epilepsy: a review. Epilepsy Res Treat. 2012;2012:1–17.
98. Binnie CD, Marston D, Polkey CE, Amin D. Distribution of temporal spikes in relation to the sphenoidal electrode. Electroencephalogr Clin Neurophysiol. 1989;73(5):403–9.
99. Giovannini G, Kuchukhidze G, McCoy MR, Meletti S, Trinka E. Neuroimaging alterations related to status epilepticus in an adult population: definition of MRI findings and clinical-EEG correlation. Epilepsia. 2018;59(S2):120–7.
100. Gélisse P, Benbadis SR, Crespel A, Tatum WO. Overcoming traps and pitfalls leading to

misinterpretation of normal EEG variants and variation of the background activity. J Neurol. 2024;271(7):3869–78.
101. Lowenstein DH, Aminoff MJ. Clinical and EEG features of status epilepticus in comatose patients. Neurology. 1992;42(1):100.
102. Frauscher B, Mansilla D, Abdallah C, Astner-Rohracher A, Beniczky S, Brazdil M, et al. Learn how to interpret and use intracranial EEG findings. Epileptic Disord. 2024;26(1):1–59.
103. Kokkinos V. Interpretation of the intracranial electroencephalogram of the human hippocampus. Neurosurg Clin N Am. 2024;35(1):73–82.
104. Kovac S, Vakharia VN, Scott C, Diehl B. Invasive epilepsy surgery evaluation. Seizure. 2017;44:125–36.
105. Blount JP, Cormier J, Kim H, Kankirawatana P, Riley KO, Knowlton RC. Advances in intracranial monitoring. Neurosurg Focus. 2008;25(3):E18.
106. Baumgartner C, Pataraia E. Revisiting the role of magnetoencephalography in epilepsy. Curr Opin Neurol. 2006;19(2):181–6.
107. Ebersole JS, Squires KC, Eliashiv SD, Smith JR. Applications of magnetic source imaging in evaluation of candidates for epilepsy surgery. Neuroimaging Clin N Am. 1995;5(2):267–88.
108. Isnard J, Bourdillon P. Morphological imaging of the hippocampus in epilepsy. Rev Neurol (Paris). 2015;171(3):298–306.
109. Bernasconi A, Cendes F, Theodore WH, Gill RS, Koepp MJ, Hogan RE, et al. Recommendations for the use of structural magnetic resonance imaging in the care of patients with epilepsy: a consensus report from the International League Against Epilepsy Neuroimaging Task Force. Epilepsia. 2019;60(6):1054–68.
110. Wang I, Bernasconi A, Bernhardt B, Blumenfeld H, Cendes F, Chinvarun Y, et al. MRI essentials in epileptology: a review from the ILAE Imaging Taskforce. Epileptic Disord. 2020;22(4):421–37.
111. Hammen T, Schwarz M, Doelken M, Kerling F, Engelhorn T, Stadlbauer A, et al. ^{1}H-MR spectroscopy indicates severity markers in temporal lobe epilepsy: Correlations between metabolic alterations, seizures, and epileptic discharges in EEG. Epilepsia. 2007;48(2):263–9.
112. Cohen-Gadol AA, Pan JW, Kim JH, Spencer DD, Hetherington HH. Mesial temporal lobe epilepsy: a proton magnetic resonance spectroscopy study and a histopathological analysis. J Neurosurg. 2004;101(4):613–20.
113. Zoccarato M, Valeggia S, Zuliani L, Gastaldi M, Mariotto S, Franciotta D, et al. Conventional brain MRI features distinguishing limbic encephalitis from mesial temporal glioma. Neuroradiology. 2019;61(8):853–60.
114. Gaillard WD, Bhatia S, Bookheimer SY, Fazilat S, Sato S, Theodore WH. FDG-PET and volumetric MRI in the evaluation of patients with partial epilepsy. Neurology. 1995;45(1):123–6.
115. Salanova V, Markand O, Worth R, Smith R, Wellman H, Hutchins G, et al. FDG-PET and MRI in temporal lobe epilepsy: relationship to febrile seizures, hippocampal sclerosis and outcome. Acta Neurol Scand. 2009;97(3):146–53.
116. Hodolic M, Topakian R, Pichler R. ^{18}F-Fluorodeoxyglucose and ^{18}F-flumazenil positron emission tomography in patients with refractory epilepsy. Radiol Oncol. 2016;50(3):247–53.
117. Ryvlin P. Clinical utility of flumazenil-PET versus [18F]fluorodeoxyglucose-PET and MRI in refractory partial epilepsy. A prospective study in 100 patients. Brain. 1998;121(11):2067–81.
118. Duncan R, Patterson J, Roberts R, Hadley DM, Bone I. Ictal/postictal SPECT in the pre-surgical localisation of complex partial seizures. J Neurol Neurosurg Psychiatry. 1993;56(2):141–8.
119. Jafari-Khouzani K, Elisevich K, Karvelis KC, Soltanian-Zadeh H. Quantitative multi-compartmental SPECT image analysis for lateralization of temporal lobe epilepsy. Epilepsy Res. 2011;95(1–2):35–50.
120. Kim JH, Im KC, Kim JS, Lee S, Lee JK, Khang SK, et al. Ictal hyperperfusion patterns in relation to ictal scalp EEG patterns in patients with unilateral hippocampal sclerosis: a SPECT study. Epilepsia. 2007;48(2):270–7.
121. DeFelipe J, DeFelipe-Oroquieta J, Furcila D, Muñoz-Alegre M, Maestú F, Sola RG, et al. Neuroanatomical and psychological considerations in temporal lobe epilepsy. Front Neuroanat. 2022;16:995286.
122. Loring DW. Neuropsychological evaluation in epilepsy surgery. Epilepsia. 1997;38:s4. https://doi.org/10.1111/j.1528-1157.1997.tb04535.x.
123. Kim H, Yi S, Son EI, Kim J. Lateralization of epileptic foci by neuropsychological testing in mesial temporal lobe epilepsy. Neuropsychology. 2004;18(1):141–51.
124. Allone C, Lo Buono V, Corallo F, Pisani LR, Pollicino P, Bramanti P, et al. Neuroimaging and cognitive functions in temporal lobe epilepsy: a review of the literature. J Neurol Sci. 2017;381:7–15.
125. Garcia-Ramos C, Struck AF, Cook C, Prabhakaran V, Nair V, Maganti R, et al. Network topology of the cognitive phenotypes of temporal lobe epilepsy. Cortex. 2021;141:55–65.
126. Kanner AM, Bicchi MM. Antiseizure medications for adults with epilepsy: a review. JAMA. 2022;327(13):1269.
127. Brigo F, Marson A. Approach to the medical treatment of epilepsy. Contin Lifelong Learn Neurol. 2022;28(2):483–99.
128. Zaccara G, Lattanzi S, Brigo F. Which treatment strategy in patients with epilepsy with focal seizures uncontrolled by the first anti-seizure medication? Epilepsy Behav. 2021;121:108031.
129. Lattanzi S, Trinka E, Zaccara G, Striano P, Russo E, Del Giovane C, et al. Third-generation antiseizure medications for adjunctive treatment of focal-onset

seizures in adults: a systematic review and network meta-analysis. Drugs. 2022;82(2):199–218.
130. Marson AG, Al-Kharusi AM, Alwaidh M, Appleton R, Baker GA, Chadwick DW, et al. The SANAD study of effectiveness of carbamazepine, gabapentin, lamotrigine, oxcarbazepine, or topiramate for treatment of partial epilepsy: an unblinded randomised controlled trial. Lancet. 2007;369(9566):1000–15.
131. Nevitt SJ, Sudell M, Weston J, Tudur Smith C, Marson AG. Antiepileptic drug monotherapy for epilepsy: a network meta-analysis of individual participant data. Cochrane Database Syst Rev. 2017; https://doi.org/10.1002/14651858.CD011412.pub2.
132. Brodie MJ, Besag F, Ettinger AB, Mula M, Gobbi G, Comai S, et al. Epilepsy, antiepileptic drugs, and aggression: an evidence-based review. Pharmacol Rev. 2016;68(3):563–602.
133. Klein S, Bankstahl M, Löscher W. Inter-individual variation in the effect of antiepileptic drugs in the intrahippocampal kainate model of mesial temporal lobe epilepsy in mice. Neuropharmacology. 2015;90:53–62.
134. Chen Z, Lusicic A, O'Brien TJ, Velakoulis D, Adams SJ, Kwan P. Psychotic disorders induced by antiepileptic drugs in people with epilepsy. Brain. 2016;139(10):2668–78.
135. Rogawski MA, Löscher W. The neurobiology of antiepileptic drugs. Nat Rev Neurosci. 2004;5(7):553–64.
136. Seferi A, Dashi F, Lilamani E, Seferi L, Rroji A, Alimehmeti R. Surgical treatment for mesial temporal lobe epilepsy associated with hippocampal sclerosis. Our Experience. Albanian J Trauma Emerg Surg. 2023;7(1):1101–3.
137. Pitskhelauri DI, Kudieva ES, Melikyan AG, Vlasov PA, Kamenetskaya MI, Zaitsev OS, et al. Surgical treatment of drug-resistant epilepsy following hippocampal sclerosis. Vopr Neirokhirurgii Im NN Burdenko. 2021;85(5):31.
138. Sonmezturk HH. Surgical treatment of epilepsy. In: Atlas of EEG, seizure semiology, and management. 3rd ed. Oxford University Press; 2022. p. 350–7. https://academic.oup.com/book/40146/chapter/341504666.
139. Kohlhase K, Zöllner JP, Tandon N, Strzelczyk A, Rosenow F. Comparison of minimally invasive and traditional surgical approaches for refractory mesial temporal lobe epilepsy: a systematic review and meta-analysis of outcomes. Epilepsia. 2021;62(4):831–45.
140. Iranmehr A, Dabbagh Ohadi MA, Chavoshi M, Jahanbakhshi A, Slavin KV. Minimally invasive procedures for hypothalamic hamartoma-related epilepsy: a systematic review and meta-analysis. Neurosurg Focus. 2022;53(4):E8.
141. Youngerman BE, Save AV, McKhann GM. Magnetic resonance imaging-guided laser interstitial thermal therapy for epilepsy: systematic review of technique, indications, and outcomes. Neurosurgery. 2020;86(4):E366–82.
142. Losarcos NG, Miller J, Fastenau P, Amina S, Bailey C, Deveraux M, et al. Stereotactic-EEG-guided radiofrequency multiple hippocampal transection (SEEG-guided-RF-MHT) for the treatment of mesial temporal lobe epilepsy: a minimally invasive method for diagnosis and treatment. Epileptic Disord. 2021;23(5):682–94.
143. Wang R, Beg U, Padmanaban V, Abel TJ, Lipsman N, Ibrahim GM, et al. A systematic review of minimally invasive procedures for mesial temporal lobe epilepsy: too minimal, too fast? Neurosurgery. 2021;89(2):164–76.
144. Wang S, Zhao M, Li T, Zhang C, Zhou J, Wang M, et al. Long-term efficacy and cognitive effects of bilateral hippocampal deep brain stimulation in patients with drug-resistant temporal lobe epilepsy. Neurol Sci. 2021;42(1):225–33.
145. McLachlan RS, Pigott S, Tellez-Zenteno JF, Wiebe S, Parrent A. Bilateral hippocampal stimulation for intractable temporal lobe epilepsy: impact on seizures and memory. Epilepsia. 2010;51(2):304–7.
146. Saucedo-Alvarado PE, Velasco AL, Aguado-Carrillo G, Cuellar-Herrera M, Trejo-Martínez D, Márquez-Franco R, et al. Optimizing deep brain stimulation for the treatment of drug-resistant temporal lobe epilepsy: a pilot study. J Neurosurg. 2022;137(3):768–75.
147. Zhang F, Yang Y, Zheng Y, Zhu J, Wang P, Xu K. Combination of matching responsive stimulations of hippocampus and subiculum for effective seizure suppression in temporal lobe epilepsy. Front Neurol. 2021;12:638795.
148. Qiao Q, Qu Z, Tian S, Cao H, Zhang Y, Sun C, et al. Ketogenic diet alleviates hippocampal neurodegeneration possibly via ASIC1a and the mitochondria-mediated apoptotic pathway in a rat model of temporal lobe epilepsy. Neuropsychiatr Dis Treat. 2022;18:2181–98.
149. Kawamura M Jr, Ruskin DN, Masino SA. Metabolic therapy for temporal lobe epilepsy in a dish: investigating mechanisms of ketogenic diet using electrophysiological recordings in hippocampal slices. Front Mol Neurosci. 2016;9:112. https://doi.org/10.3389/fnmol.2016.00112/full.
150. Shehata NI, Abdelsamad MA, Amin HAA, Sadik NAH, Shaheen AA. Ameliorating effect of ketogenic diet on acute status epilepticus: insights into biochemical and histological changes in rat hippocampus. J Food Biochem. 2022;46(9):e14217. https://doi.org/10.1111/jfbc.14217.

18 Impact of Exercise on Hippocampal Structure and Function

Zekine Pündük

18.1 Introduction

The hippocampus is a part of the limbic system and is located in the medial temporal lobes. This system plays an important role in the processing of memory events, the learning process, and also regulates behaviors. Intrinsic and extrinsic factors can change the function and structure of the hippocampus. These can make it particularly susceptible to neurodegenerative diseases; however, neuroprotective effects can be induced by physical exercise [1]. Physical exercise can influence rates of hippocampal neurogenesis, which is a potent stimulus [2, 3], and this can have a significant effect on cognitive performance. An inactive lifestyle, for example, is related to cognitive problems [4]. On the other hand, regular physical exercise reduces the early risk of Alzheimer's disease due to improvement in hippocampus volume, function, and cognitive events of the brain [5, 6]. Several studies have recently shown the positive effects of exercise on neurons in the brain, resulting in the improvement of cognitive processes in the older population [7, 8].

In part, the impact of physical exercise on neurogenesis is related to an alteration in several neurotransmitter systems of the brain, involving the endocrine system, such as serotonin, dopamine, acetylcholine, and norepinephrine [9]. These hormones are essential for metabolic functions, tissue regeneration, cardiovascular health, and other metabolic activities [10–13]. Also, estrogens and sex hormones [10, 11], somatotrope cells [3, 11, 14, 15], vascular endothelial growth factor [7], irisin/FNDC5 [16, 17], and cortisol [18] play an important role in brain health mechanisms. All these hormones can penetrate the human blood-brain barrier and influence neurogenesis and the cognitive system. Accordingly, regular physical exercise reduces depression and protects against atrophy and age-related mental disorders such as Alzheimer's or dementia [19]. In this matter, researchers have now shown that even minimal physical activity has positive effects on brain mechanisms, such as improved memory, state of mind, cognitive processing, plasticity, and learning capacity [20, 21]. This review discusses the changes in molecular pathways and the possible mechanisms that modulate neurogenesis in the brain through different intensities and types of training regimes.

18.1.1 Physical Exercise Modulates BDNF and Other Physiological Mechanisms

Physical exercise activates a series of signaling mechanisms in the brain. The moving muscle communicates with the central nervous system (CNS), and a number of chemical messages are

Z. Pündük (✉)
Balıkesir University Faculty of Sport Sciences, Balıkesir, Türkiye
e-mail: zkn1938@gmail.com

M. Turgut et al. (eds.), *The Human Hippocampus*, https://doi.org/10.1007/978-3-031-88907-3_18

released into the bloodstream from other organs. One of these is brain-derived neurotrophic factor (BDNF), which is secreted during exercise and communicates with the CNS. Additionally, research results have reported that other physiological mediators, such as insulin-like growth factor (IGF-1), cytokines, and vascular endothelial growth factor (VEGF), also play an important role in neurogenesis [22–24]. Especially, cytokines, which are a group of low-molecular-weight glycoproteins, are known as intercellular messengers and act as pro-inflammatory and anti-inflammatory tools. In particular, levels of the Interleukin-6 (IL-6) family of cytokines are significantly increased in response to exercise [25] and produce biological effects. It seems these effects are highly related to increased BDNF expression and interleukin response to exercise. Also, proteins and metabolic products released from the periphery of the skeletal muscles with exercise, such as Cathepsin B and Irisin/FNDC5, regulate BDNF expression in the hippocampus [26]. Mostly, BDNF expression in the brain occurs by binding to the receptor Tropomyosin Kinase B (TrkB); this complex of BDNF-TrkB stimulates many signaling pathways in the secondary system, such as Cascades and Protein Phosphorylation Cascades [27]. All these mechanisms, such as the neuroendocrine system, neurotransmitters, neuronal insulin mechanism, irisin mechanism, anti-neuronal inflammation and anti-neuronal oxidative response mechanism, and anti-apoptosis effect, are regulated with exercise and improve brain functions.

Moreover, BDNF is a family of neurotrophins identified in the hippocampus, participates in processes linked to cognitive systems, and plays an important role in many psychiatric diseases such as mood disorders (Alzheimer's, schizophrenia, dementia, anxiety disorders) [28]. Experimental and observational studies have shown that physical exercise improves cognitive activities; it functionally improves brain structure and decreases the risk of neurological degenerative diseases associated with the secretion of BDNF factors [29]. Previous work by Neeper et al. has shown that the level of BDNF increases with exercise in animals [30]. Animal studies have shown that long-term exercise upregulates BDNF gene and protein expression, especially in the hippocampus and other brain areas, such as the amygdala, cerebellum, and perirhinal cortex [31]. Raichlen and Gordon have reported that a favorable relationship exists between brain volume and endurance exercise performance in humans, and this helps in the development of human movement cognitive functions [32]. A study by Mattson et al. showed that endurance exercise training upmodulated the level of BDNF in circulation, resulting in an improvement in brain capacity in the hippocampus [33]. Mattson's study showed that BDNF expression increased significantly due to endurance exercise, improving exercise capacity and promoting brain growth, especially in the hippocampus. Although the exact biochemical mechanism of exercise-induced BDNF expression is unknown, its positive effects on the brain are well established. In addition, research supports that increasing BDNF levels with regular exercise exerts a beneficial effect against Alzheimer. Fahimi et al. reported that in mice subjected to a 4-week treadmill exercise, BDNF mRNA and protein levels increased, and brain morphology and cellular structure also changed positively [34]. In another study, it was reported that an increased BDNF level with exercise also affects dopamine release, which is necessary for the life of neurons and improvements in learning and memory [35].

18.1.2 Different Type and Intensity Exercise and BDNF Expression

Exercise training modalities are performed with different aims to improve heart/respiratory system fitness, musculoskeletal, and metabolic activity or for general health. The most common methods of exercise training are aerobic endurance exercise and/or weight lifting/resistance exercise (i.e., strength training) [36]. Aerobic exercise/training (AE) is a moderately intensive continuous training (MICT), and it generally involves long-duration or distance running, cycling, or jogging [36]. High-intensity or very high-intensity interval training (HIIT) is also a running strategy and is performed at high intensity with short distances (according to maximal heart rate, 80–100%) with a maximum of 60 s × maximum repetition, with 3–4 min rests [37].

The effects of AE or HIIT exercise on brain health have been reported, specifically BDNF expression. The moderate or high intensity of physical exercise, daily or intermittent, is also important in the BDNF expression of the hippocampus. One study showed that the levels of BDNF proteins after running (about 100–160%), in both daily and occasional exercise training protocols, were elevated in the rats' hippocampus. Nonetheless, at the end of the 1-week exercise training session, the BDNF protein levels showed no more significant increase in either the everyday exercise or occasional exercise period (about 100–115%, respectively) [38]. These experimental results indicate that the rate of BDNF protein levels (daily exercise versus occasional exercise) is dependent on the absolute amount of exercise days. The amount of BDNF increase is affected by acute exercise and regular exercise training (chronic exercise). BDNF protein levels may be an adaptive response to regular exercise training. However, research on the effects of exercise training on BDNF levels in the brain is still unclear.

Few long-term research studies involving animals and normal healthy participants have shown that moderate-intensity, long-duration aerobic exercise increases the ratio of plasticity synapses while also increasing the number of neurons and angiogenesis via upregulation of BDNF levels, with improvements in volume and white-matter integrity in the hippocampus [39–44]. One study also reported that the impact of exercise and training performed at different types, intensities, and durations (aerobic, resistance, and strength exercise) can increase BDNF and brain function. Moreover, a 12-week moderate-intensity (70% Maximal Oxygen Uptake) exercise training program with different durations (1–4, 5–8, and 9–12 weeks of exercise intervention) indicated a significant increase in left hippocampal volume and hippocampal *N*-acetylaspartate (NAA) levels by ≈16%, using a new model of MRI systems [45]. This result demonstrates that prolonged aerobic exercise at the same intensity improves neuronal integrity. Previous studies have demonstrated that volumetric changes [46], hippocampal volume [40], NAA concentration [47], and memory performance in the hippocampus [40] are related to high physical performance in humans.

Physical acute exercise affects dopaminergic, serotonergic, and noradrenergic physiological mechanisms. Physiological adaptations develop in these systems due to the stimulating effect of exercise as many neurotransmitters are increasingly released during exercise. According to the study by Lin and Kuo, the three main monoamine neurotransmitters (noradrenaline, dopamine, serotonin, or hydroxytryptamine (5-HT)) are released and regulated during exercise [48]. According to this study, regular exercise increases norepinephrine levels in the spinal cord pons-medulla compared to sedentary patients; this may support the improvement of cognitive features in patients [48]. The regulation of this system with exercise varies based on the brain area and the exercise/training intensity and its duration. For example, it has been reported that 4 weeks of moderately intensive treadmill exercise upregulated these neurotransmitters (dopamine, serotonin, and norepinephrine) in the hippocampus [49], and another study reported that 1 week of high-intensity running treadmill exercise also enhanced the number of neurons in the hippocampus [50].

Insulin signaling is essential for the protection of neuronal mechanisms, and it is also required to prevent defects in the BDNF transfusion process in the brain. Insulin regulation disorder is related to the formation of metabolic diseases, such as diabetes, obesity, cardiovascular diseases, and hypertension. Also, an impairment in insulin metabolism signaling causes cognitive disorders in the brain [51]. Cellular insulin receptors are intensely secreted from pyramidal cell axons in the hippocampal part of the brain and are localized in the predominant regions of the brain involved in cognitive functions such as learning and memory. This receptor is found in various regions of the brain; however, it is present in high concentration in the hypothalamus, hippocampus, and cerebral cortex area. In these regions, the hippocampus has a key role in the memory storage of newly learned information. Under normal physiological conditions, insulin and insulin-like growth factors (IGFs) send signals to the neurons of the brain and hippocampal region to maintain their physiological functions. However, when these growth factors begin to get suppressed, the functionality of this region is disrupted and initiates diseases such as Alzheimer's. For example, it has been reported that damage to a part of memory in old mouse brains is related to increased inflammation and decreased anti-inflammatory markers, but these damages regress with 20 min of daily strenuous exercise [52]. Therefore, in the brain, physical exercise has an anti-inflammatory action by increasing insulin signaling in hippocampal neurons. This effect is probably seen in the central and peripheral nervous systems and acts in the neuroprotective mechanism (protection of neurons).

Physical activity also affects the production of irisin, which is a myokine released from the skeletal muscles during exercise [53]. Irisin increases the metabolic rate as a myokine in the circulation and is known to have a protective effect against obesity by maintaining glucose balance [54]. The increased irisin level in the circulation with exercise increases the action of BDNF levels from cortical neurons. It has also been found to increase neuroprotective genes in the hippocampus [55]. The irisin level, which increases especially with endurance exercise, is probably associated with dopamine release, and it activates BDNF expression. Thus, irisin and BDNF are increased by exercise and reduce ischemia-induced neuronal damage and contribute to increasing neuroprotective effects [56]. It is also thought that exercise-induced irisin release has a therapeutic effect against ischemia; thus, it has a beneficial effect in reducing neuropathic-related pain. As a response to physical exercise, the autonomic nervous system (ANS) is activated to maintain homeostasis in the brain hypothalamic-pituitary-adrenal axis, and as a result, plasma cortisol and catecholamine levels are increased. Physical exercise influences the immune system by stimulating the secretion of hormones; the growth hormone and prolactin are secreted by the pituitary gland. Thus, exercise stimulates the antioxidant and anti-inflammatory mechanisms, and this mechanism protects the cardiovascular system as well as neurons.

18.1.3 How Much Exercise Should Be Done for Brain Health? Exercise and Training Approaches

For exercise to be beneficial to brain health, the question "how much exercise is enough?" needs to be answered. Although meta-analysis results from the last 37 studies (involving 1300 young individuals and elderly participants) explain that exercise has a positive effect, they are insufficient to explain these benefits through mechanisms of improvement in cognitive performance and cardiovascular fitness.

In order for the physical fitness level to have a beneficial impact on cognitive functions, it requires a certain intensity level to improve cardiovascular endurance. Together with this, some studies argue that high-intensity exercise improves the functional performance of the brain by increasing neurotransmitter levels more than moderate-intensity exercise [57]. These contradictory results show that the effects of exercise on brain tissue are dependent on exercise intensity and duration. Studies showed that exercises performed in moderate-intensity running and walking modes are generally monitored for 30–60 min. The common result of these studies is that although running at moderate intensity for 60 min increases hippocampal BDNF activity, low-intensity long-distance running (2 h) also increases BDNF activity [58]. Thus, it has been reported that long-term exercise increases hippocampal BDNF levels better than acute exercise and provides a positive "neuroplastic environment" for the recovery of the brain, especially for diseases such as stroke.

Although exercise increases protein levels, it may also be beneficial for brain functions as it also increases stress hormones and corticosterone, depending on exercise intensity. Since the increase in corticosterone with high-intensity exercise can suppress BDNF activity, this may cause concern in terms of brain health. There's an old saying for clinicians: "everything should be in moderation." This also seems to apply to exercise prescription's positive effects on brain health. All studies state that physical training, especially aerobic training, has an effect on the generation of new neuron cells, new blood vessels, synapses, and the production of necessary protein molecules. According to the guide published by the World Health Organization, regular exercise is generally aerobic muscle exercise that improves muscle strength and endurance (long-term exercise for large muscle groups), balance exercises (back walking, turning and walking, single-leg balance exercises), muscle-strengthening exercises, and resistance exercises. It has been reported as (muscle exercises against load and resistance) and flexibility (muscle flexibility exercises that extend joint range of motion and muscle length) [59]. However, in studies where different types of exercise and training affect different brain regions, Vossa et al. reported that 12-week aerobic moderate walking training created more white matter in the brain's frontal and temporal lobes compared to flexibility training, thus increasing hippocampal volume [60]. Correlational studies have shown that individuals with higher physical athletic condition levels have larger hippocampal volumes in preadolescents and adolescents [61]. Also, similar results have been shown in elderly people who performed regular 1-year moderate aerobic exercise, and this resulted in an elevated hippocampus volume with better memory functions [40]. Killgore et al. have investigated the effects of exercise on the hippocampus with gray matter volume in healthy adults (18–45 years old). The findings of this study indicate that there is a positive correlation between total weekly exercise time and the volume of the right side of the hippocampus [62]. Moreover, 16 weeks of progressive aerobic exercise significantly improved the left hippocampal volume, including the hippocampal subfield (CA4_DG), during exercise sessions compared to the control group [63]. However, random-effects meta-analysis results have demonstrated that aerobic exercise had no significant effect on total hippocampal volume in some of the participants [64].

In studies on cognitive features, it has been concluded that aerobic running and muscle strength training may differ. For example, after aerobic training, a significant improvement was noted in memory-related functions, and this is similar to the improvement that was also recorded after training against resistance [65]. While an increase in volume was noted in the hippocampal region even after walking, a similar change was not noted in the stretching exercise [40]. In this case, combined training may be better than uniform training. It has been reported that coordination training shows similar developments in different parts of the brain. In studies conducted on physical training, it is possible to say that the impact of exercise training on brain knowledge systems varies by the type of physical activity. For instance, it was determined that entrepreneurship (planning, organization, etc.) among cognitive features increased after aerobic training [66]. Another study reported improvements in memory functions [67]. In this case, it is possible to say that combined training is more ideal in terms of affecting different parts and features of the brain.

18.2 Conclusion

In general, a literature search showed that regular aerobic physical exercise training had positive beneficial effects on the hippocampus of the left and right side volumes in the human brain. High-intensity interval training may pose a risk to those with heart health problems. However, an analysis of 23 studies, in which 1117 patients participated and 227 training hours were recorded, applied a high-intensity interval training protocol, including 4 × 4 min interval training, 3 min rest between intervals, 30 s to 3 min interval training protocol, as well as a moderate-intensity interval protocol (30/60). It has been reported that high-intensity intervals are more effective in neuroprotection compared to (min training season, 7134 training season, 5606 training hours). However, since there are scarce studies on this subject, there is a need to conduct more detailed studies to determine the mechanism of action. As a result, even if aerobic and different types of combined training have a more positive effect on brain structure and functions, it should be considered that HIIT training is also significant.

Acknowledgments We gratefully acknowledge Dr. Khalid Rahman, School of Pharmacy and Biomolecular Sciences, Liverpool John Moores University, for English language editing of this chapter.

References

1. Bartsch T, Wulff P. The hippocampus in aging and disease: from plasticity to vulnerability. Neuroscience. 2015;309:1–16. https://doi.org/10.1016/j.neuroscience.2015.07.084.
2. Trinchero MF, Herrero M, Schinder AF. Rejuvenating the brain with chronic exercise through adult neurogenesis. Front Neurosci. 2019;13:1000. https://doi.org/10.3389/FNINS.2019.01000.
3. Bednarczyk MR, Aumont A, Décary S, Bergeron R, Fernandes KJ. Prolonged voluntary wheel-running stimulates neural precursors in the hippocampus and forebrain of adult CD1 mice. Hippocampus. 2009;19(10):913–27. https://doi.org/10.1002/hipo.20621.
4. Maasakkers CM, Thijssen DH, Knight SP, Newman L, O'Connor JD, Scarlett S, Carey D, et al. Hemodynamic and structural brain measures in high and low sedentary older adults. J Cereb Blood Flow Metab. 2021;41(10):2607–16. https://doi.org/10.1177/0271678X211009382.
5. Cooper C, Moon HY, Van Praag H. On the run for hippocampal plasticity. Cold Spring Harb Perspect Med. 2018;8(4):a029736. https://doi.org/10.1101/CSHPERSPECT.A029736.
6. Ben-Zeev T, Shoenfeld Y, Hoffman JR. The effect of exercise on neurogenesis in the brain. Isr Med Assoc J. 2022;24(8):533–8.
7. Hötting K, Röder B. Beneficial effects of physical exercise on neuroplasticity and cognition. Neurosci Biobehav Rev. 2013;37:2243–57. https://doi.org/10.1016/J.NEUBIOREV.2013.04.005.
8. Deslandes A, Moraes H, Ferreira C, Veiga H, Silveira H, Mouta R, et al. Exercise and mental health: many reasons to move. Neuropsychobiology. 2009;59(4):191–8. https://doi.org/10.1159/000223730.
9. Kraemer RR, Kilgore JL, Kraemer GR, Daniel Castracane V. Growth hormone, IGF-i, and testosterone responses to resistive exercise. Med Sci Sports Exerc. 1992;24:1346–52. https://doi.org/10.1249/00005768-199212000-00007.
10. Kraemer RR, Durand RJ, Acevedo EO, Johnson LG, Kraemer GR, Hebert EP, Castracane VD. Rigorous running increases growth hormone and insulin-like growth factor-I without altering ghrelin. Exp Biol Med (Maywood). 2004;229(3):240–6. https://doi.org/10.1177/153537020422900304.
11. Kraemer RR, Francois MR, Sehgal K, Sirikul B, Valverde RA, Castracane VD. Amylin and selective glucoregulatory peptide alterations during prolonged exercise. Med Sci Sports Exerc. 2011;43:1451–6. https://doi.org/10.1249/MSS.0B013E3182114AB9.
12. Acevedo EO, Kraemer RR, Kamimori GH, Durand RJ, Johnson LG, Castracane VD. Stress hormones, effort sense, and perceptions of stress during incremental exercise: an exploratory investigation. J Strength Cond Res. 2007;21:283–8. https://doi.org/10.1519/00124278200702000-00050.
13. Durand RJ, Castracane VD, Hollander DB, Tryniecki JL, Bamman MM, O'Neal S, et al. Hormonal responses from concentric and eccentric muscle contractions. Med Sci Sports Exerc. 2003;35(6):937–43. https://doi.org/10.1249/01.MSS.0000069522.38141.0B.
14. Kraemer RR, Heleniak RJ, Tryniecki JL, Kraemer GR, Okazaki NJ, Castracane VD. Follicular and luteal phase hormonal responses to low-volume resistive exercise. Med Sci Sports Exerc. 1995;27:809–17. https://doi.org/10.1249/00005768-199506000-00004.
15. Kraemer RR, Shockett P, Webb ND, Shah U, Castracane VD. A transient elevated irisin blood concentration in response to prolonged, moderate aerobic exercise in young men and women. Horm Metab Res. 2014;46:150–4. https://doi.org/10.1055/S-0033-1355381.
16. Kraemer RR, Goldfarb AH, Reeves GV, Meachum WA, Daniel Castracane V. Effects of partial vascular occlusion on irisin responses to loaded muscle contractions. Appl Physiol Nutr Metab. 2016;41:332–4. https://doi.org/10.1139/APNM-2015-0464.
17. Kraemer RR, Blair S, Kraemer GR, Castracane VD. Effects of treadmill running on plasma beta-endorphin, corticotropin, and cortisol levels in male and female 10K runners. Eur J Appl Physiol Occup Physiol. 1989;58:845–51. https://doi.org/10.1007/BF02332217.
18. Laurin D, Verreault R, Lindsay J, MacPherson K, Rockwood K. Physical activity and risk of cognitive impairment and dementia in elderly persons. Arch Neurol. 2001;58:498–504. https://doi.org/10.1001/archneur.58.3.498.
19. Spalding KL, Bergmann O, Alkass K, Bernard S, Salehpour M, Huttner HB, et al. Dynamics of hippocampal neurogenesis in adult humans. Cell. 2013;153:1219–27. https://doi.org/10.1016/j.cell.2013.05.002.
20. Phillips C, Baktir MA, Srivatsan M, Salehi A. Neuroprotective effects of physical activity on the brain: a closer look at trophic factor signaling. Front Cell Neurosci. 2014;8:170. https://doi.org/10.3389/fncel.2014.00170.
21. Smith CD, Chebrolu H, Wekstein DR, Schmitt FA, Jicha GA, Cooper G, Markesbery WR. Brain

structural alterations before mild cognitive impairment. Neurology. 2007;68(16):1268–73. https://doi.org/10.1212/01.wnl.0000259542.54830.34.
22. Sofi F, Valecchi D, Bacci D, Abbate R, Gensini GF, Casini A, Macchi C. Physical activity and risk of cognitive decline: a meta-analysis of prospective studies. J Intern Med. 2011;269(1):107–17. https://doi.org/10.1111/j.1365-2796.2010.02281.x. Epub 2010 Sep 10
23. Kohman RA, Bhattacharya TK, Wojcik E, Rhodes JS. Exercise reduces activation of microglia isolated from hippocampus and brain of aged mice. J Neuroinflammation. 2013;18(10):114. https://doi.org/10.1186/1742-2094-10-114.
24. Pedersen BK, Febbraio MA. Muscle as an endocrine organ: focus on muscle derived interleukin6. Physiol Rev. 2008;88:1379–406.
25. Moon HY, Becke A, Berron D, Becker B, Sah N, Benoni G, et al. Running-induced systemic cathepsin B secretion is associated with memory function. Cell Metab. 2016;24:332–40. https://doi.org/10.1016/j.cmet.2016.05.025.
26. Yoshii A, Constantine-Paton M. Post-synaptic BDNF-TrkB signalling in synapse maturation, plasticity and disease. Dev Neurobiol. 2010;70:304–22. https://doi.org/10.1002/dneu.20765.
27. von Bohlen Und Halbach O, von Bohlen Und Halbach V. The BDNF effects on dendritic spine morphology and hippocampal function. Cell Tissue Res. 2018;373:729–41. https://doi.org/10.1007/s00441-017-2782-x.
28. Forlenza OV, Diniz BS. Effect of brain-derived neurotrophic factor Val66Met polymorphism and serum levels on the progression of mild cognitive impairment. World J Biol Psychiatry. 2010;11:774–80.
29. Stillman CM, Cohen J, Lehman ME, Erickson KI. Mediators of physical activity on neurocognitive function: a review at multiple levels of analysis. Front Hum Neurosci. 2016;8(10):626. https://doi.org/10.3389/fnhum.2016.00626.
30. Neeper SA, Gómez-Pinilla F, Choi J, Cotman C. Exercise and brain neurotrophins. Nature. 1995;373(6510):109. https://doi.org/10.1038/373109a0.
31. Hopkins ME, Bucci DJ. BDNF expression in perirhinal cortex is associated with exercise-induced improvement in object recognition memory. Neurobiol Learn Mem. 2010;94:278–84.
32. Raichlen DA, Gordon AD. Relationship between exercise capacity and brain size in mammals. PLoS One. 2011;6:e20601. https://doi.org/10.1371/journal.pone.0020601.
33. Mattson MP. Evolutionary aspects of human exercise—born to run purposefully. Ageing Res Rev. 2012;11:347–52.
34. Fahimi A, Baktir MA, Moghadam S, Mojabi FS, Sumanth K, McNerney MW, Ponnusamy R, Salehi A. Physical exercise induces structural alterations in the hippocampal astrocytes: exploring the role of BDNF-TrkB signaling. Brain Struct Funct. 2017;222(4):1797–808. https://doi.org/10.1007/s00429-016-1308-8.
35. Zsuga J, Tajti G, Papp C, Juhasz B, Gesztelyi R. FNDC5/irisin, a molecular target for boosting reward-related learning and motivation. Med Hypotheses. 2016;90:23–8. https://doi.org/10.1016/j.mehy.2016.02.020.
36. Kang J, Ratamess N. Which comes first? Resistance before aerobic exercise or vice versa? ACSM's Health Fit J. 2014;18:9–14. https://doi.org/10.1249/FIT.0000000000000004.
37. Gibala MJ, McGee SL. Metabolic adaptations to short-term high intensity interval training: a little pain for a lot of gain? Exerc Sport Sci Rev. 2008;36:58–63. https://doi.org/10.1097/JES.0b013e318168ec1f.
38. Berchtold NC, Chinn G, Chou M, Kesslak JP, Cotman CW. Exercise primes a molecular memory for brain-derived neurotrophic factor protein induction in the rat hippocampus. Neuroscience. 2005;133:853–61.
39. Voss MW, Vivar C, Kramer AF, van Praag H. Bridging animal and human models of exercise-induced brain plasticity. Trends Cogn Sci. 2013;17(10):525–44. https://doi.org/10.1016/j.tics.2013.08.001.
40. Erickson KI, Voss MW, Prakash RS, Basak C, Szabo A, Chaddock L, Kim JS, et al. Exercise training increases size of hippocampus and improves memory. Proc Natl Acad Sci USA. 2011;108(7):3017–22. https://doi.org/10.1073/pnas.1015950108.
41. Kleemeyer MM, Kühn S, Prindle J, Bodammer NC, Brechtel L, Garthe A, Kempermann G, Schaefer S, Lindenberger U. Changes in fitness are associated with changes in hippocampal microstructure and hippocampal volume among older adults. NeuroImage. 2016;1(131):155–61. https://doi.org/10.1016/j.neuroimage.2015.11.026.
42. Maass A, Düzel S, Goerke M, Becke A, Sobieray U, Neumann K, Lövden M, et al. Vascular hippocampal plasticity after aerobic exercise in older adults. Mol Psychiatry. 2015;20(5):585–93. https://doi.org/10.1038/mp.2014.114.
43. Thomas AG, Dennis A, Rawlings NB, Stagg CJ, Matthews L, Morris M, Kolind SH, et al. Multimodal characterization of rapid anterior hippocampal volume increase associated with aerobic exercise. NeuroImage. 2016;1(131):162–70. https://doi.org/10.1016/j.neuroimage.2015.10.090.
44. Sexton CE, Betts JF, Demnitz N, Dawes H, Ebmeier KP, Johansen-Berg HA. Systematic review of MRI studies examining the relationship between physical fitness and activity and the white matter of the ageing brain. NeuroImage. 2016;1(131):81–90. https://doi.org/10.1016/j.neuroimage.2015.09.071.
45. Den Ouden L, Kandola A, Suo C, Hendrikse J, Costa RJS, Watt MJ, Lorenzetti V, et al. The influence of aerobic exercise on hippocampal integrity and function: preliminary findings of a multi-modal imaging analysis. Brain Plast. 2018;4(2):211–6. https://doi.org/10.3233/BPL-170053.
46. Pajonk FG, Wobrock T, Gruber O, Scherk H, Berner D, Kaizl I, Kierer A, et al. Hippocampal plastic-

ity in response to exercise in schizophrenia. Arch Gen Psychiatry. 2010;67(2):133–43. https://doi.org/10.1001/archgenpsychiatry.2009.193.
47. Erickson KI, Weinstein AM, Suttoni BP, Prakash RS, Voss MW, Chaddock L, Szabo AN, et al. Beyond vascularization: aerobic fitness is associated with N-acetylaspartate and working memory. Brain Behav. 2012;2(1):32–41. https://doi.org/10.1002/brb3.30.
48. Lin TW, Kuo YM. Exercise benefits brain function: the monoamine connection. Brain Sci. 2013;3:39–53.
49. Chen HI, Lin LC, Yu L, Liu YF, Kuo YM, Huang AM, Chuang JI, Wu FS, Liao PC, Jen CJ. Treadmill exercise enhances passive avoidance learning in rats: the role of down-regulated serotonin system in the limbic system. Neurobiol Learn Mem. 2008;89(4):489–96. https://doi.org/10.1016/j.nlm.2007.08.004.
50. Chennaoui M, Grimaldi B. Effects of physical training on functional activity of 5-HT1B receptors in rat central nervous system: role of 5-HT-moduline. Naunyn-Schmiedeberg Arch Pharmacol. 2000;361:600–4.
51. Dodd GT, Tiganis T. Insulin action in the brain: roles in energy and glucose homeostasis. J Neuroendocrinol. 2017;29(10) https://doi.org/10.1111/jne.12513.
52. Lovatel GA, Elsner VR, Bertoldi K, Vanzella C, Moysés Fdos S, Vizuete A, Spindler C, et al. Treadmill exercise induces age-related changes in aversive memory, neuroinflammatory and epigenetic processes in the rat hippocampus. Neurobiol Learn Mem. 2013;101:94–102. https://doi.org/10.1016/j.nlm.2013.01.007.
53. Dun SL, Lyu RM, Chen YH, Chang JK, Luo JJ, Dun NJ. Irisin-immunoreactivity in neural and non-neural cells of the rodent. Neuroscience. 2013;14(240):155–62. https://doi.org/10.1016/j.neuroscience.2013.02.050.
54. Wrann CD, White JP, Salogiannnis J, Laznik-Bogoslavski D, Wu J, Ma D, Lin JD, Greenberg ME, Spiegelman BM. Exercise induces hippocampal BDNF through a PGC-1α/FNDC5 pathway. Cell Metab. 2013;18(5):649–59. https://doi.org/10.1016/j.cmet.2013.09.008.
55. Boström P, Wu J, Jedrychowski MP, Korde A, Ye L, Lo JC, Rasbach KA, Boström EA, et al. A PGC1-α-dependent myokine that drives brown-fat-like development of white fat and thermogenesis. Nature. 2012;481(7382):463–8. https://doi.org/10.1038/nature10777.
56. Li DJ, Li YH, Yuan HB, Qu LF, Wang P. The novel exercise-induced hormone irisin protects against neuronal injury via activation of the Akt and ERK1/2 signaling pathways and contributes to the neuroprotection of physical exercise in cerebral ischemia. Metabolism. 2017;68:31–42. https://doi.org/10.1016/j.metabol.2016.12.003.
57. Winter B, Breitenstein C, Mooren FC, Voelker K, Fobker M, Lechtermann A, Krueger K, et al. High impact running improves learning. Neurobiol Learn Mem. 2007;87(4):597–609. https://doi.org/10.1016/j.nlm.2006.11.003.
58. Ploughman M, Granter-Button S, Chernenko G, Tucker BA, Mearow KM, Corbett D. Endurance exercise regimens induce differential effects on brain-derived neurotrophic factor, synapsin-I and insulin-like growth factor I after focal ischemia. Neuroscience. 2005;136(4):991–1001. https://doi.org/10.1016/j.neuroscience.2005.08.037.
59. WHO. Making a difference. The World Health Report 1999. Health Millions. 1999;25(4):3–5.
60. Voss MW, Heo S, Prakash RS, Erickson KI, Alves H, Chaddock L, Szabo AN, et al. The influence of aerobic fitness on cerebral white matter integrity and cognitive function in older adults: results of a one-year exercise intervention. Hum Brain Mapp. 2013;34(11):2972–85. https://doi.org/10.1002/hbm.22119.
61. Herting MM, Nagel BJ. Aerobic fitness relates to learning on a virtualMorris Water Task and hippocampal volume in adolescents. Behav Brain Res. 2012;233:517–25. https://doi.org/10.1016/j.bbr.2012.05.012.
62. Killgore WD, Olson EA, Weber M. Physical exercise habits correlate with gray matter volume of the hippocampus in healthy adult humans. Sci Rep. 2013;3(1):3457.
63. Frodl T, Strehl K, Carballedo A, Tozzi L, Doyle M, Amico F, et al. Aerobic exercise increases hippocampal subfield volumes in younger adults and prevents volume decline in the elderly. Brain Imaging Behav. 2020;14:1577–87.
64. Firth J, Stubbs B, Vancampfort D, Schuch F, Lagopoulos J, Rosenbaum S, Ward PB. Effect of aerobic exercise on hippocampal volume in humans: a systematic review and meta-analysis. NeuroImage. 2018;166:230–8.
65. Voss MW, Prakash RS, Erickson KI, Basak C, Chaddock L, Kim JS, Alves H, et al. Plasticity of brain networks in a randomized intervention trial of exercise training in older adults. Front Aging Neurosci. 2010;2:32. https://doi.org/10.3389/fnagi.2010.00032.
66. Colcombe SJ, Kramer AF, Erickson KI, Scalf P, McAuley E, Cohen NJ, Webb A, et al. Cardiovascular fitness, cortical plasticity, and aging. Proc Natl Acad Sci USA. 2004;101(9):3316–21. https://doi.org/10.1073/pnas.0400266101.
67. Liu-Ambrose T, Donaldson MG, Ahamed Y, Graf P, Cook WL, Close J, Lord SR, Khan KM. Otago home-based strength and balance retraining improves executive functioning in older fallers: a randomized controlled trial. J Am Geriatr Soc. 2008;56(10):1821–30. https://doi.org/10.1111/j.1532-5415.2008.01931.x.

Role of the Hippocampus in Learning

19

Mehmet Akif Baktır

19.1 Introduction

Although we may not possess medical expertise, most of us are familiar with the hippocampus. The brain is the subject of extensive individual and clinical study, making it the most extensively studied portion of the body. Initial research has identified the hippocampus as a key component involved in olfactory investigations. Later research demonstrated that the hippocampus functions as a central point for memory. The hippocampus, located in the innermost part of the medial temporal lobe of the brain, is acknowledged as a vital element in the formation of complex long-term learning and memory. Many studies have provided evidence that the hippocampus plays a crucial role in many processes, such as spatial orientation and episodic memory [1].

The hippocampus is characterized by its remarkable plasticity, which refers to its capacity to undergo both structural and functional modifications in response to experiences and environmental stimuli [2]. Neuronal plasticity is considered a crucial foundation for the hippocampus's involvement in learning and memory. Deficits in spatial and contextual memory have been linked to impairments in hippocampal long-term potentiation (LTP), which is a process that enhances synaptic connections [3]. Furthermore, the suppression of the formation of new neurons in the hippocampus, known as hippocampal neurogenesis, has been demonstrated to negatively impact the ability of experimental animals to remember spatial and contextual information.

19.1.1 The Role of the Hippocampus in Learning

Studies conducted on both rats and humans have demonstrated that the hippocampus plays a crucial role in the creation of spatial and contextual memory [4]. Individuals who have damage to the hippocampus, medial temporal lobe, or frontal lobe frequently experience difficulties in remembering spatial and contextual information. This emphasizes the crucial significance of the hippocampus in several cognitive domains [5].

Extensive studies are being conducted worldwide to investigate the constituents of the hippocampus's involvement in learning. By comprehending the function of the hippocampus in the process of learning, we may significantly advance our understanding of the development of several disorders, including depression, dementia, and Alzheimer's, which have reached epidemic proportions in modern society [2].

Learning is often characterized as the process of obtaining new knowledge or abilities through

M. A. Baktır (✉)
Department of Physiology, Erciyes University Faculty of Medicine, Kayseri, Türkiye
e-mail: drbaktir@yahoo.com

M. Turgut et al. (eds.), *The Human Hippocampus*, https://doi.org/10.1007/978-3-031-88907-3_19

individual experiences [2]. Learning occurs when new synaptic connections are created in neurons due to chemical, electrical, and structural changes, which are triggered by external inputs from the individual's surroundings. Learning, a process assisted by the intricate connections of neurons, is enhanced through the repetition and accumulation of experiences [3]. An organism's capacity to gain knowledge and skills from experiences and to learn by encoding, storing, and retrieving information is vital to adapting to changing situations and maintaining survival [6–10].

There are many different kinds of learning:

1. Associative learning
   (a) Conditioned response
   (b) Instrumental conditioning
2. Non-associative learning
   (a) Habituation
   (b) Sensitization

1. **Associative Learning**: It is a form of learning that occurs when a connection is made between two stimuli.
   (a) **Classical conditioning** refers to a type of learning in which a reflex response is elicited by a previously neutral stimulus through repeated association. This is the quintessential illustration of associative learning. When a stimulus that does not produce a response on its own (conditioned) is repeatedly paired with another stimulus that does produce a response (unconditioned), eventually, a response will occur when the conditioned stimulus is presented alone. Put simply, it refers to the organism's ability to associate one type of stimulus with another. The time period between the stimulus that was conditioned and its unconditioned counterpart is important in the traditional conditioning process.

   The duration of this time interval should be precisely 0.5 s on average, and it is crucial that the conditioned stimulus be presented prior to the unconditioned stimulus [7].

   Russian physiologist Ivan Pavlov was the first to uncover the phenomenon of conditioned reflex. In this experiment, Pavlov demonstrated that a dog, which had been trained to associate the sound of a bell with receiving meat and subsequently salivating, eventually began salivating at the sound of the bell alone after a specific number of repetitions. Ingesting food into the oral cavity serves as a physiological trigger for the production of saliva. Over time, it becomes evident that the sole auditory perception of a bell triggers salivation as a conditioned response.
   (b) **Operant conditioning**, also referred to as instrumental conditioning, refers to the process of learning to perform certain actions in exchange for a reward or punishment. In classical conditioning, a connection between two distinct stimuli is acquired. In contrast, in this case, the link between a specific behavior and its corresponding response is acquired. Once again, the time period between the behavior and its corresponding response holds significance [9]. Offering a reward as a form of positive reinforcement encourages the recurrence of a behavior, while administering punishment or inducing suffering serves as negative reinforcement that deters the behavior [10]. This conditioning strategy is employed in a medical setting to correct an unfavorable circumstance. For instance, individuals with persistent alcoholism are given an emetic with the use of alcohol, resulting in the association of symptoms of nausea and vomiting with alcohol intake. This approach aims to discourage the use of alcohol [7].
2. **Non-associative Learning**:
   (a) **Habituation**: Refers to the gradual decrease or total absence of a motor response to stimuli that are regarded as unimportant. In this type of learning, a mild or low-frequency repetitive stimulus that does not have a beneficial or detrimental effect on the human being results in a gradual reduction in the amount of opening of Ca^{2+} channels at the presynap-

tic junction. Because of the low frequency of the stimulus, the channels that do open ultimately close, leading to a decline in the release of brain chemicals. This leads to a steady decrease in the motor neuron response.

(b) **Sensitization**: The organism exhibits a response to a stimulus that is typically neutral and would not elicit a response but does so when another sensory stimulation is introduced. An instance of sensitization can be observed when a woman perceives her child's cry while asleep, despite being unable to detect other sounds.

19.1.2 The Correlation Between LTP in the Hippocampus and Learning

Synaptic plasticity describes the modification in the functioning of synapses according to previous activity. This alteration can manifest as either a brief, transient occurrence (spanning a few seconds or minutes) or a prolonged, enduring state (persisting for hours or even days in certain in vivo models). Modulating the activity of a synapse is regarded as crucial in the creation of learning and memory. There are a multitude of synaptic plasticity variations. The outcome is contingent upon which particular neurons are involved, the length of time the alteration persists, the direction of the modification, and the specific brain region implicated. The most common type of synaptic plasticity that has been thoroughly studied is known as LTP, which was first documented in 1973 [11]. The initial study utilized recordings from the anesthetized rabbit hippocampus. It was observed that recurrent stimulation paradigms led to an amplification in synaptic responses, which exhibited an average duration ranging from 30 min to 10 h. LTP, or long-term potentiation, is commonly defined as a persistent enhancement in synaptic response that happens after the application of certain stimulation paradigms [12]. When LTP is triggered by applying a high-frequency tetanic shock to the presynaptic cell, it results in a permanent enhancement of the cells' response to specific inputs. Nerves have the ability to retain information for extended durations by modifying their synaptic plasticity [13]. The diverse temporal phases of LTP are driven by various cellular and molecular processes. The formation of LTP necessitates the influx of calcium ions (Ca^{2+}) into the postsynaptic neuron and the activation of protein kinases. Subsequent phases of LTP, on the other hand, rely on enduring alterations in gene transcription and protein synthesis. Ion channels are activated, resulting in the motion of ions through both receptor-mediated and voltage-controlled pathways. This activation enables intracellular signaling proteins to undergo phosphorylation, which in turn promotes the beginning of gene expression [14].

Previous investigations indicate that the hippocampus was associated with olfaction or emotion [15, 16]. Dr. William Scoville [17] conducted the inaugural surgical procedure that provided valuable understanding of the role of the hippocampus. He performed surgery on a patient who presented with a complaint of recurring seizures. During the surgical procedure, he removed several brain regions, including the hippocampus, amygdala, collateral sulcus, and entorhinal cortex [18]. After undergoing this surgical procedure, the patient experienced a cessation of seizures but developed significant mental deficits. Specifically, there were impairments in the capacity to store and retain new factual information (semantic memory) as well as recall past events (episodic memory). However, cognitive abilities such as reasoning, perception, and working memory remained unaffected.

During the 1980s, Dr. Richard Morris [19] devised the Morris water tank test to teach mice how to identify the geographic location of a concealed platform in murky water. This test was used to assess the functioning of the hippocampus, and it was shown that impairments in learning performance were substantially associated with hippocampal lesions. At the moment, many studies have clarified the function of the hippocampus in the creation and upkeep of spatial memory [20, 21]. The hippocampus is additionally implicated in the process of long-term potentiation, a neuronal mechanism that acts as the foundation for learning and memory. Long-term potentiation is the process by which synaptic connections between neurons are strengthened due to repeated activation. This process is thought to be a mechanism within cells that is connected to the acquisition of knowledge and the development of memories. The hippocampus plays an essential function in long-term potentiation and synaptic plasticity, which are basic mechanisms underlying learning and memory. Research has shown that inhibiting or altering hippocampal function can hinder the process of solidifying new memories, while enhancing hippocampal activity can improve learning and memory abilities. The hippocampus is essential for learning and memory, namely in the creation and recall of spatial memory.

19.1.2.1 Neurobiological Mechanisms Associated with Learning in the Hippocampus

The learning process in the hippocampus is controlled by a multitude of biochemical mechanisms. These pathways involve the stimulation of glutamate receptors, adjustment of calcium signaling, and involvement of other intracellular signaling molecules. One example is the NMDA receptor, which is a specific type of glutamate receptor that has an essential role in long-term potentiation and the process of learning. The activation of the NMDA receptor initiates a cascade of intracellular processes that enables the entry of calcium ions into the postsynaptic neuron, ultimately resulting in the improvement of synaptic connections. Furthermore, the stimulation of intracellular signaling molecules, such as cAMP and protein kinases, may regulate synaptic plasticity and enhance the processes of learning and memory in the hippocampus [12, 22].

19.1.3 The Function of Neurotransmitters in the Process of Learning in the Hippocampus

Multiple neurotransmitters have been shown to exert a substantial influence on the cognitive processes of learning and memory in the hippocampus. The neurotransmitters in question are acetylcholine, glutamate, and dopamine. Acetylcholine has been recognized for its impact on improving synaptic plasticity and aiding in the formation and recall of memories in the hippocampus. Glutamate, the main neurotransmitter responsible for boosting neuronal activity, plays a vital role in synaptic plasticity and the creation of new memories in the hippocampus. Dopamine, a neurotransmitter engaged in the regulation of reward and motivation, also plays a role in the acceleration of learning and memory functions in the hippocampus. The hippocampus plays a role in the process of consolidating and recalling memories [23, 24].

The hippocampus, an essential part of the brain, has a crucial role in the processes of learning and the creation of memories. The core of this process involves a complex relationship between different neurotransmitters that regulate the intricate communication between neurons responsible for our capacity to encode, store, and recall information.

Glutamate is a crucial neurotransmitter in the process of hippocampal learning. Glutamate, the main neurotransmitter that promotes brain activity, is essential for initiating long-term potentiation, a key process of synaptic plasticity thought to be responsible for learning and memory. Troubles in the transmission of glutamate, such as excitotoxicity, have been linked to the development of neurodegenerative diseases, emphasizing the crucial role of this neurotransmitter in preserving the normal functioning of the brain [25].

Aside from glutamate, the hippocampus is additionally impacted by inhibitory neurotransmitters like gamma-aminobutyric acid. GABA is crucial for regulating synaptic plasticity and controlling the excitability of neurons in the hippocampal circuit [26]. Precise control of the equilibrium between excitatory and inhibitory neurotransmitters is crucial for proper formation and enhancement of memories [27].

Additional neurotransmitters, including hydrogen sulfide, nitric oxide, and carbon monoxide, have been found to have significant functions in synaptic plasticity and the broader mechanisms associated with learning and memory [27]. Neurotransmitters have the ability to affect the functioning of ion channels, the creation of second messengers, and the regulation of gene expression. As a result, they play a role in creating the neuronal responses that are associated with cognitive activities.

19.1.4 The Impact of Physical Exercise on the Process of Learning

Exercise has numerous impacts on cognitive functions, and the hippocampus, a crucial component for learning and memory, has garnered significant attention in studies on this subject. Multiple studies have demonstrated that engaging in voluntary exercise, such as running on a treadmill, enhances the production of new neurons, the ability of synapses to change and adapt, and cognitive function in rats. The ways in which exercise impacts hippocampal function are complex and varied. Research has demonstrated that exercise enhances the generation and release of brain-derived neurotrophic factor (BDNF) and receptor tyrosine kinase B, both of which play a crucial role in the survival, differentiation, and functioning of neurons in the hippocampus. Furthermore, physical activity can augment LTP, a biological process that underlies the acquisition and retention of information, and promote the proliferation of synapses in the hippocampus [28–30].

Many human studies have additionally examined the mental advantages of exercise on hippocampal learning. Research has demonstrated that engaging in aerobic exercise can lead to an expansion in the size of both gray and white matter in the prefrontal cortex in elderly individuals. Moreover, research has shown that older individuals who participate in more intense physical activities have a lower likelihood of experiencing cognitive decline and possess greater volumes of the hippocampus and the middle temporal lobe [29–31]. To summarize, existing research indicates that engaging in regular physical activity is a promising and easily accessible method to enhance cognitive functions such as learning and memory. Furthermore, it appears that the hippocampus, a region of the brain, is a key focus for these advantageous outcomes. The improvement of the formation of new neurons in the hippocampus, the ability of synapses to change and adapt, and the creation of growth factors like BDNF are presumably important mechanisms that clarify how exercise boosts cognitive ability.

19.1.4.1 The Impact of Stress on Learning in the Human Hippocampus

The hippocampus is essential for various cognitive tasks, including learning and navigation. Nevertheless, it is exceedingly susceptible and delicate to the detrimental consequences of stress.

Acute stress, which involves the quick release of stress hormones like cortisol, has been proven to hinder cognitive flexibility and force the brain to switch from using methods that rely on the hippocampus to using more inflexible, habitual responses controlled by the dorsal striatum [32]. This alteration in brain processing can hinder an individual's capacity to adjust to evolving circumstances and efficiently acquire new knowledge. Chronic stress has been found to result in structural alterations in the hippocampus, such as reduced neuronal branching, decreased dendritic density, and modifications in the formation of synaptic spines [33]. These alterations in the structure and form of the brain may have a negative impact on the functioning of the hippocampus, leading to difficulties in learning and memory.

Curiously, the hippocampus is also a recipient of the effects of estrogen, a hormone that has been demonstrated to have a beneficial impact on cognitive performance [34]. This indicates that adjusting the hormonal environment, especially in relation to stress, has the potential to reduce the harmful impact of stress on the ability to learn and remember in the hippocampus.

Furthermore, recent findings indicate that epigenetic processes, which control gene expression, may have a significant impact on how stress affects the hippocampus [35]. Gaining a comprehensive understanding of the intricate connections among stress, hormones, and epigenetic variables could offer essential knowledge for creating specific therapies aimed at improving learning and memory in those under significant stress.

Ultimately, the hippocampus is an essential area of the brain that is highly responsive to the detrimental impacts of stress, leading to significant impairment in an individual's capacity to acquire and retain knowledge. Through the clarification of the fundamental processes and the investigation of possible interventions, scientists and medical professionals can strive to create successful approaches to reduce the harmful impact of stress on the ability to learn and think.

References

1. Andersen P, et al., editors. The hippocampus book, Oxford neuroscience series. New York: Oxford Academic; 2006.
2. Bartsch T, Bartsch T. Introduction: the hippocampus in the clinical neurosciences. In: The clinical neurobiology of the hippocampus: an integrative view. New York: Oxford Academic; 2012.
3. Anand KS, Dhikav V. Hippocampus in health and disease: an overview. Ann Indian Acad Neurol. 2012;15(4):239–46.
4. Shen Y, Specht SM, Li R. The hippocampus: a biological model for studying learning and memory. Prog Neurobiol. 1994;44(5):485–96.
5. Bird CM, Burgess N. The hippocampus and memory: insights from spatial processing. Nat Rev Neurosci. 2008;9(3):182–94.
6. Ganong WF. Ganong'un Tıbbi Fizyolojisi. In: Gökbel H, editor. Nobel Tıp Kitapevi; 2011. p. 289–99.
7. Carew TJ, Hawkins RD, Kandel ER. Differential classical conditioning of a defensive withdrawal reflex in Aplysia californica. Science. 1983;219(4583):397–400.
8. Yigit R. Kontrol Sistemleri, Sindirim ve Bosaltım Fizyolojisi. İzmir: Güven Kitabevi; 2001. p. 293–8.
9. Kandel ER. Cellular mechanisms of learning and the biological basis of individuality. In: Principles of neural science. 3rd ed. Elsevier; 1991. p. 1009–31.
10. Agras W, editor. Comprehensive textbook of physiology; 2000. p. 133–9.
11. Bliss TV, Lømo T. Long-lasting potentiation of synaptic transmission in the dentate area of the anaesthetized rabbit following stimulation of the perforant path. J Physiol 1973;232:331–56.
12. Malenka RC, Bear MF. LTP and LTD: an embarrassment of riches. Neuron 2004;44:5–21.
13. Rygh LJ, Tjølsen A, Hole K, et al. Cellular memory in spinal nociceptive circuitry. Scand J Psychol 2002;43:153–9.
14. Plátenik J, Kuramoto N, Yoneda Y. Molecular mechanisms associated with long-term consolidation of the NMDA signals. Life Sci 2000;67:335–64.
15. Eichenbaum, Howard, and Tim Otto. The hippocampus and the sense of smell. Chemical Signals in Vertebrates 6. Boston, MA: Springer US. 1992;67–77.
16. Papez JW. A proposed mechanism of emotion. 1937. J Neuropsychiatry Clin Neurosci 1994; 69(7):103–12.
17. Scoville WB, Milner B. Loss of recent memory after bilateral hippocampal lesions. J Neurol Neurosurg Psychiatry 1957;20:11.
18. Corkin S, Amaral DG, González RG, et al. HM's medial temporal lobe lesion: findings from magnetic resonance imaging. J Neurosci Nurs 1997;17:3964–79.
19. Morris R, Garrud P, Rawlins J, et al. Place navigation impaired in rats with hippocampal lesions. Nature 1982;297:681–3.
20. Vargha-Khadem F, Gadian DG, Watkins KE, et al. Differential effects of early hippocampal pathology on episodic and semantic memory. Science 1997;277: 376–80.
21. O'Reilly RC, Norman KA. Hippocampal and neocortical contributions to memory: Advances in the complementary learning systems framework. Trends Cogn Sci 2002;6:505–10.
22. Jones MW, Errington ML, French PJ, Fine A, Bliss TV, Garel S, et al. A requirement for the immediate early gene Zif268 in the expression of late LTP and long-term memories. Nat Neurosci. 2001;4(3):289–96.
23. Collingridge GL. Long term potentiation in the hippocampus: mechanisms of initiation and modulation by neurotransmitters. Trends Pharmacol Sci. 1985;6:407–11.
24. Bairy LK, Kumar S. Neurotransmitters and neuromodulators involved in learning and memory. Int J Basic Clin Pharmacol. 2019;8(12):2777–83.
25. Park JM, Jung SC, Eun SY. Long-term synaptic plasticity: circuit perturbation and stabilization. Korean J Physiol Pharmacol. 2014;18(6):457.
26. Collingridge GL, Isaac JT, Wang YT. Receptor trafficking and synaptic plasticity. Nat Rev Neurosci. 2004;5(12):952–62.
27. Shefa U, Kim D, Kim MS, Jeong NY, Jung J. Roles of gasotransmitters in synaptic plasticity and neuropsychiatric conditions. Neural Plast. 2018;2018(1):1824713.
28. Moon HY, van Praag H. On the run for hippocampal plasticity. Cold Spring Harb Perspect Med. 2018;8(4):a029736.
29. Erickson KI, Voss MW, Prakash RS, Basak C, Szabo A, Chaddock L, et al. Exercise training increases size of hippocampus and improves memory. Proc Natl Acad Sci. 2011;108(7):3017–22.
30. Erickson KI, Gildengers AG, Butters MA. Physical activity and brain plasticity in late adulthood. Dialogues Clin Neurosci. 2013;15(1):99–108.
31. Tarumi T, Zhang R. Cerebral hemodynamics of the aging brain: risk of Alzheimer disease and benefit of aerobic exercise. Front Physiol. 2014;5:6.

32. Seehagen S, Schneider S, Rudolph J, Ernst S, Zmyj N. Stress impairs cognitive flexibility in infants. Proc Natl Acad Sci. 2015;112(41):12882–6.
33. Alkadhi K. Brain physiology and pathophysiology in mental stress. Int Scholarly Res Notices. 2013;(1):806104.
34. McEwen B. Estrogen actions throughout the brain. Recent Prog Horm Res. 2002;57:357–84.
35. Hunter RG. Epigenetic effects of stress and corticosteroids in the brain. Front Cell Neurosci. 2012;6:18.

Relationship of Melatonin with the Hippocampus

20

Kubilay Doğan Kılıç, Ali Çağlar Turgut, and Ayşegül Taşkıran Özyazıcı

Abbreviations

AANAT	*N*-Acetylserotonin by arylalkylamine *N*-acetyltransferase
AD	Alzheimer's disease
AHN	Adult hippocampal neurogenesis
AMPH	Amphetamine
APP	Amyloid precursor protein
ASMT	Acetylserotonin *O*-methyltransferase
Aβ	Amyloid-beta
BACE1	Beta-site amyloid precursor protein cleaving enzyme 1
BDNF	Brain-derived neurotrophic factor
BrdU	Bromodeoxyuridine
CA	Cornu ammonis CA
CSF	Cerebrospinal fluid
DG	Dentate gyrus
HD	Huntington's disease
HIOMT	Hydroxyindole-*O*-methyltransferase
HPA	Hypothalamic-pituitary-adrenal
ISL2	ISL LIM-homeobox 2
LPS	Lipopolysaccharide
LTD	Long-term depression
LTP	Long-term potentiation
mtDNA	Mitochondrial DNA
MTX	Methotrexate
NCAMs	Neural cell adhesion molecules
NPCs	Neural progenitor cells
NSCs	Neural stem cells
PD	Parkinson's disease
PRM	Prolonged-release melatonin
RFP	Red fluorescence protein
RGL	Radial glia-like
ROS	Reactive oxygen species
SCN	Suprachiasmatic nucleus
SGZ	Subgranular zone

K. D. Kılıç (✉)
Department of Histology and Embryology, Ege University Faculty of Medicine, İzmir, Türkiye

Institute for Tissue Engineering and Regenerative Medicine, Helmholtz Zentrum München, Neuherberg, Germany
e-mail: kubilay.dogan.kilic@ege.edu.tr; kubilaydogan.kilic@helmholtz-munich.de

A. Ç. Turgut
Department of Radiology, Ege University Faculty of Medicine, İzmir, Türkiye

Department of Histology and Embryology, Aydın Adnan Menderes University Health Sciences Institute, Aydın, Türkiye

A. T. Özyazıcı
Department of Histology and Embryology, Ege University Faculty of Medicine, İzmir, Türkiye

20.1 Introduction

Studying the connection between melatonin and hippocampal function presents an intriguing opportunity in neuroscience, linking the fields of chronobiology and cognitive neuroscience. This chapter thoroughly introduces the physiological

M. Turgut et al. (eds.), *The Human Hippocampus*, https://doi.org/10.1007/978-3-031-88907-3_20

importance of melatonin, the key functions of the hippocampus in the central nervous system, and the rationale for researching the interaction between these two essential components.

Melatonin, a hormone mostly produced by the pineal gland, is crucial in controlling circadian rhythms, the internal clock of the body that adjusts sleep-wake cycles and other physiological functions. In addition to its main production location, melatonin is generated in several extrapineal tissues, such as the retina, digestive system, and immune cells. In these tissues, melatonin has local autocrine and paracrine effects [1]. The generation of melatonin is achieved by a series of enzymatic processes that begin with the amino acid tryptophan and culminate in the activities of aryl alkylamine *N*-acetyltransferase and acetyl serotonin *O*-methyltransferase [2].

In addition to its role in regulating sleep, melatonin also serves several additional physiological functions:

1. Antioxidant properties: Melatonin possesses strong antioxidant effects, effectively eliminating free radicals and safeguarding cells against oxidative stress. This characteristic is especially crucial in protecting cellular constituents, including DNA, proteins, and lipids, from harm that might result in aging and different ailments [3, 4].
2. Immune system modulation: Melatonin modulates the activity of immune cells, thus influencing immunological function. Research has demonstrated that it can boost the immune response and safeguard against infections and inflammation [5, 6].
3. Regulation of mood: Melatonin modulates neurotransmitter systems in the brain, impacting mood and emotional well-being. Abnormal melatonin levels have been linked to mood disorders such as depression and seasonal affective disorder [7, 8].
4. Reproductive function: Melatonin regulates reproductive hormones and can affect the reproductive cycles of both humans and animals [9, 10].
5. Cancer prevention: Melatonin has been found to possess anti-cancer properties since it can impede the growth of neoplastic cells and improve the effectiveness of specific cancer therapies, according to several studies [11, 12].

Comprehending the process of melatonin's production, control, and many functions lays the foundation for investigating its impact on brain regions like the hippocampus. The hippocampus is a key component of the limbic system, localized in the medial temporal lobe of the human brain. It plays an important role in regulating emotions, memory, and spatial navigation. This curved structure consists of several subregions, including the dentate gyrus, cornu ammonis (CA) fields, subiculum, and entorhinal cortex. Each of these subregions contributes distinctly to the overall function of the structure [13].

The hippocampus is most known for its function in memory consolidation and the conversion of short-term memories into long-term memories. Spatial memory is crucial for creating cognitive maps that represent the environment. In addition, the hippocampus has a role in emotional regulation through its interaction with the amygdala and prefrontal cortex [14]. The intricate structure and interconnections of the hippocampus enable its varied functional responsibilities, rendering it a central area of focus for investigating brain function and disease.

The relationship between melatonin and hippocampal function is a rapidly growing field with important implications for comprehending cognitive processes and mental well-being. Several aspects support the justification for examining this relationship:

1. Circadian influence on cognitive functions: The circadian rhythm influences the hippocampus's cognitive processes of learning and memory, reflecting daily patterns. Melatonin, a crucial controller of circadian rhythms, can influence hippocampal activity and, thus, cognitive function [15].
2. Neuroprotective effects of melatonin: The antioxidant capabilities of melatonin provide neuroprotection against oxidative stress, a factor involved in several neurodegenerative disorders. Studying the impact of melatonin on hip-

pocampal neurons might offer valuable information on reducing cognitive decline [16].

3. Regulation of neurogenesis: The hippocampus is among the few brain areas where the process of neurogenesis continues throughout an individual's lifespan. Studies have revealed that melatonin can stimulate the growth of new neurons in animals, indicating that it may benefit the brain's ability to change and improve cognitive performance [17, 18].
4. Implications for mental health: Disruptions in melatonin levels and circadian rhythms have significant effects on mental health, such as mood disorders, depression, and anxiety. Gaining insight into the effect of melatonin on hippocampal function might clarify the underlying mechanisms of these diseases and guide the development of treatment strategies [8].

By investigating the relationship between melatonin and hippocampal function, researchers aim to uncover mechanisms that underpin cognitive health and disease, potentially leading to novel interventions for enhancing brain function and treating neuropsychiatric disorders. This introductory chapter has outlined melatonin's physiological significance, the hippocampus's critical roles in the central nervous system, and the rationale for exploring their relationship. The following sections will delve deeper into the biosynthesis and regulation of melatonin (Sect. 20.2) and the anatomical and functional overview of the hippocampus (Sect. 20.3), setting the stage for a comprehensive understanding of their interplay and implications for cognitive health.

20.2 Biosynthesis and Regulation of Melatonin

20.2.1 Melatonin Synthesis in the Pineal Gland and Other Regions

Melatonin, a hormone crucial for regulating circadian rhythms, is synthesized primarily in the pineal gland, a small endocrine organ in the brain. Synthesis begins with the amino acid tryptophan, which is converted by tryptophan hydroxylase to 5-hydroxytryptophan [19]. Then, this intermediate is decarboxylated to serotonin by aromatic L-amino acid decarboxylase. Serotonin is then acetylated to *N*-acetylserotonin by arylalkylamine *N*-acetyltransferase (AANAT), the rate-limiting enzyme in the production of melatonin. Finally, *N*-acetylserotonin is methylated by acetylserotonin *O*-methyltransferase (ASMT), formerly known as hydroxyindole-*O*-methyltransferase (HIOMT), to produce melatonin [19–21] (Fig. 20.1).

Synthesis of melatonin is not limited to the pineal gland; it also occurs at various extrapineal sites, such as the retina, digestive system, and immune cells. Melatonin usually acts locally at these sites as an autocrine or paracrine signal. However, the regulatory mechanisms in these regions differ and are less affected by light exposure compared to the pineal gland [19, 21]. Recent studies have shown that melatonin is produced in the mitochondria of various cells and contributes significantly to the body's total melatonin levels. This extrapineal melatonin primarily regulates metabolic processes and maintains redox homeostasis in cells. In contrast to pineal melatonin, its production does not follow a circadian rhythm and is not released into the bloodstream but acts locally [21] (Fig. 20.1).

Synthesis of melatonin in the retina follows a biochemical pathway similar to that in the pineal gland and involves the same key enzymes: AANAT and HIOMT. However, the regulation of melatonin synthesis in the retina is unique due to its direct exposure to light and involvement in visual processing [22]. Photoreceptors primarily synthesize retinal melatonin, and there is evidence to suggest that it contributes to the production of rods and cones. Melatonin in the retina influences several functions, including regulating dopamine levels, which affects photoreceptor connectivity and electroretinogram responses. Melatonin also provides neuroprotection to photoreceptors, protecting them from oxidative stress and apoptosis [22] (Fig. 20.1).

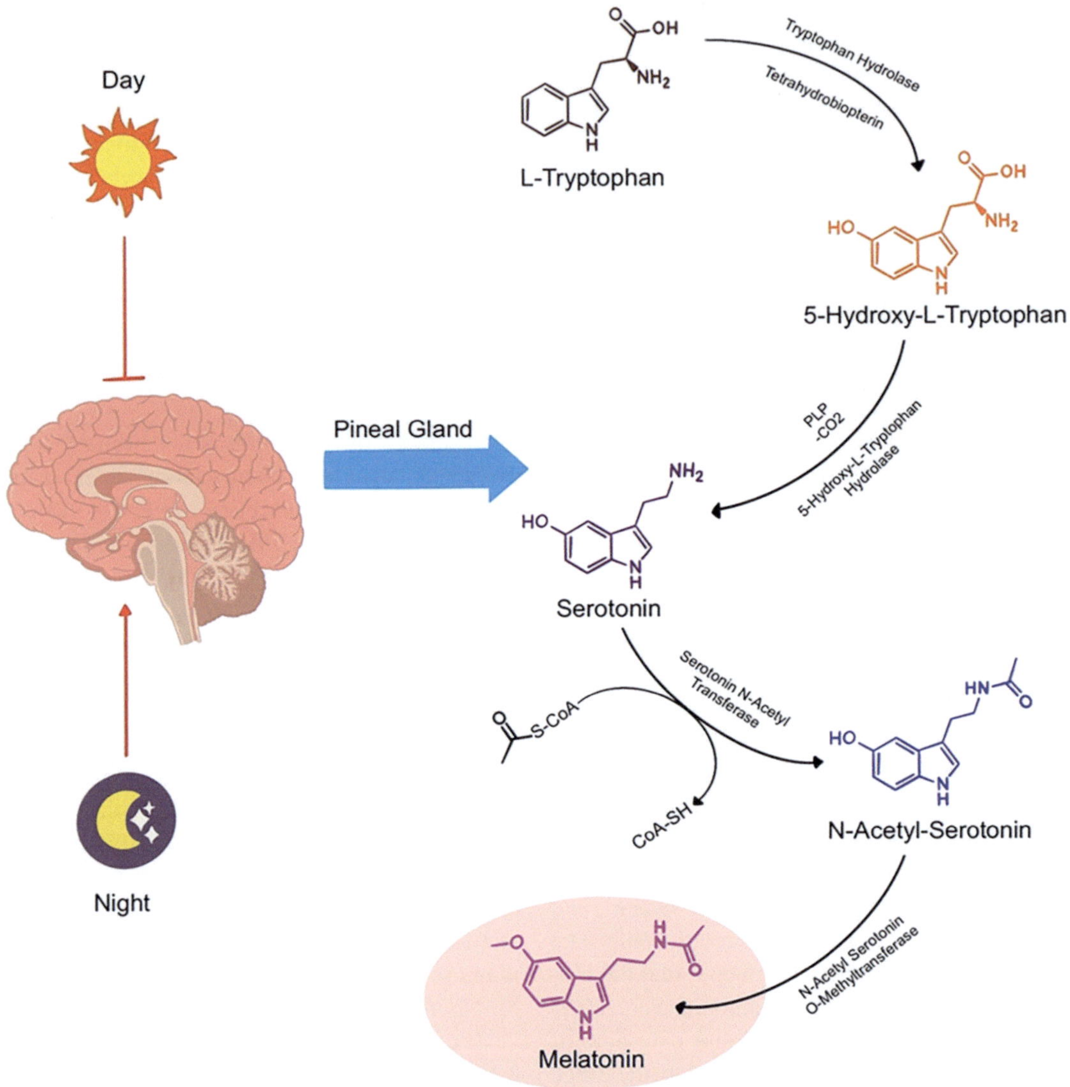

Fig. 20.1 Biosynthesis of melatonin

20.2.2 Regulatory Mechanisms Controlling Melatonin Levels

Melatonin synthesis and secretion regulation are complex processes involving environmental lighting and intrinsic circadian rhythms. The suprachiasmatic nucleus (SCN) of the hypothalamus, which acts as the central circadian clock, controls melatonin production in the pineal gland [21]. The SCN receives photic stimulation directly from the retina via the retinohypothalamic pathway. Light exposure during daylight inhibits melatonin production by suppressing SCN activity, reducing sympathetic outflow to the pineal gland. Conversely, in the absence of light, the SCN stimulates the release of norepinephrine from sympathetic nerves, which innervate the pineal gland. Norepinephrine binds to β-adrenergic receptors on pinealocytes and activates the cAMP-PKA-CREB and PLC-Ca^{2+}-PKC pathways, thereby increasing melatonin synthesis [19, 23, 24].

Circadian rhythms play a vital role in the regulation of melatonin production. The SCN maintains an intrinsic rhythmicity that persists even in perpetual darkness, providing a roughly 24-h cycle of melatonin synthesis. External light-dark cycles can entrain this endogenous rhythm, allowing the organism to adapt to seasonal changes in day length. Besides light and circadian control, melatonin synthesis is modulated by other factors such as temperature, electromagnetic fields, and pharmacological agents [19, 24].

The role of clock genes in the regulation of melatonin production is critical. Clock genes expressed in the pineal gland exhibit rhythmic patterns adjusted by the SCN. Studies have demonstrated that the rhythmic expression of AANAT driven by these clock genes is essential for the nocturnal production of melatonin. SCN lesions disrupt these rhythms, emphasizing the crucial role of the SCN in maintaining circadian regulation of melatonin synthesis [25]. Further investigation into the genetic and molecular mechanisms governing melatonin synthesis has revealed the involvement of several transcription factors and regulatory proteins. For example, the CLOCK-BMAL1 complex regulates the transcription of AANAT, and disruptions in these genes can significantly affect melatonin production. In addition, other transcription factors, such as homeobox genes and microRNAs, have been shown to modulate AANAT expression and melatonin synthesis [22, 26]. Recent studies have highlighted the ISL LIM-homeobox 2 (ISL2) transcription factor as a key player in regulating AANAT expression. ISL2 is rhythmically expressed in pinealocytes, present at high levels during the day and significantly reduced at night. This pattern suggests that ISL2 acts as a repressor of AANAT during the day, preventing melatonin synthesis and contributing to the rhythmic production of melatonin at night. The knockdown of ISL2 results in increased expression of AANAT, further supporting its role as a negative regulator [26].

Another critical factor in melatonin regulation is RP58, a zinc finger protein. It has been revealed that RP58 has an important role in the development and differentiation of nerve cells [27]. Recent studies have found that RP58 is highly expressed in the pineal gland and is downregulated after hypoxia-ischemia. The reduction of RP58 decreases the expression of enzymes related to melatonin production, such as TPH1, ASMT, and AANAT, and clock genes, such as CLOCK and BMAL1, leading to reduced melatonin production and disruption of circadian rhythms. It has been shown that exogenous melatonin administration has a role in the reversion of the harmful effects of RP58 knockdown, highlighting the important role of RP58 in melatonin regulation and pineal gland function [27].

Furthermore, an evolutionary perspective provided by cross-species single-cell analysis highlights the diversity and conservation of pineal gland function among vertebrates. This study has disclosed some transcriptional changes and controls that underline the functional diversity of the gland from zebrafish to mammals. Ligand-receptor interactions, particularly those regarding MDK and PTN, significantly influenced the synthesis of melatonin across species [28].

20.2.3 Extrapineal Melatonin

First identified in the pineal gland, melatonin is recognized to be produced in many other tissues, including the gastrointestinal tract, liver, retina, lymphocytes, and skin melanocytes. However, only pineal and retinal melatonin production depends on the light-dark cycle, while production in other tissues appears to be regulated independently of circadian rhythms. This suggests that melatonin evolved as a multifunctional molecule early in the evolution of life, as it is found in species ranging from bacteria to mammals. In humans and other mammals, melatonin performs several functions, including the regulation of the biological clock, protection against free radicals, regulation of reproduction and sexual maturation, sleep regulation, immunomodulation, and regulation of body weight and energy metabolism [29].

Moreover, recent findings suggest that extrapineal melatonin has an important role in redox homeostasis. It is produced in response to cellular redox states and stress conditions and is a potent antioxidant. This form of melatonin is crucial for protecting cells from oxidative stress and maintaining cellular function, especially in mitochondria, where reactive oxygen species (ROS) are produced [21].

20.2.4 Health Effects of Melatonin and Its Treatment Potential

The potential neuroprotective effects of melatonin have been investigated in various models, such as male BTBR mice, a well-known model for autism spectrum disorder, and these studies demonstrated that melatonin has a role in the reduction of oxidative stress and inflammation in the hippocampus. This study highlighted the ability of melatonin to adjust the expression of key enzymes related to the oxidative stress response, such as Nrf2 and SOD1, as well as inflammation markers, such as NFkB and NLRP3 [30]. Findings suggest that melatonin supplementation may provide neuroprotective benefits in conditions characterized by high oxidative stress and neuroinflammation. Prolonged-release melatonin (PRM) is another area of interest, especially in circadian medicine. It is used for the simulation of the natural physiological pattern of melatonin production and the management of a variety of sleep and circadian rhythm disorders. It effectively restores biological rhythms and improves sleep quality, especially in older adults and those with neurodevelopmental or neurocognitive disorders [31]. Melatonin is also notable for its role in various health conditions, including its potential as an analgesic in pain-related diseases. Melatonin, an antioxidant, clears excess free radicals produced in the body with its anti-stimulating and anti-inflammatory features. These actions make melatonin relevant to conditions such as cancer, mood disorders, and neurodegenerative diseases, including Alzheimer's and Parkinson's disease [32].

20.2.5 Melatonin and Environmental Factors

Exposure to light at night, especially in night shift workers, has shown that it significantly inhibits melatonin synthesis, causing disruptions in circadian rhythms. A comprehensive meta-analysis found that night shift workers had lower levels of the urinary melatonin metabolite 6-sulphatoxymelatonin compared to day shift workers [33]. This indicates that melatonin production is suppressed. Inhibition of melatonin secretion due to light exposure at night has been associated with various adverse health effects, including increased metabolic disorders and cancer risk.

20.3 Comprehensive Neuroarchitectural Analysis of the Hippocampus: Functional Mechanisms and Neurogenesis

20.3.1 Anatomical, Functional, and Structural Overview of the Hippocampus

The hippocampus, an important player in the limbic system of the human brain, is central to converting short-term memories into long-term ones and has a pivotal role in spatial navigation and emotional regulation. This title explores the hippocampus's intricate architecture and multifunctional roles, highlighting ongoing research that sheds light on its complex capabilities.

The hippocampus is a multifaceted, curved structure in the medial temporal lobe of the human brain. It is an important limbic system component, playing a vital role in forming and organizing memories. The hippocampus consists of various parts, including the dentate gyrus, four CA subregions from CA1 to CA4, the subiculum, the presubiculum, the parasubiculum, and the entorhinal cortex [34] (Fig. 20.2).

20.3.1.1 Dentate Gyrus

The dentate gyrus (DG) is the primary structure for information entering the hippocampus, essential for storing new memories. It receives information from the entorhinal cortex through the perforant pathway. Within the DG, granule cells produce new neuronal cells, a process known as "neurogenesis." The DG is a slender, jagged band

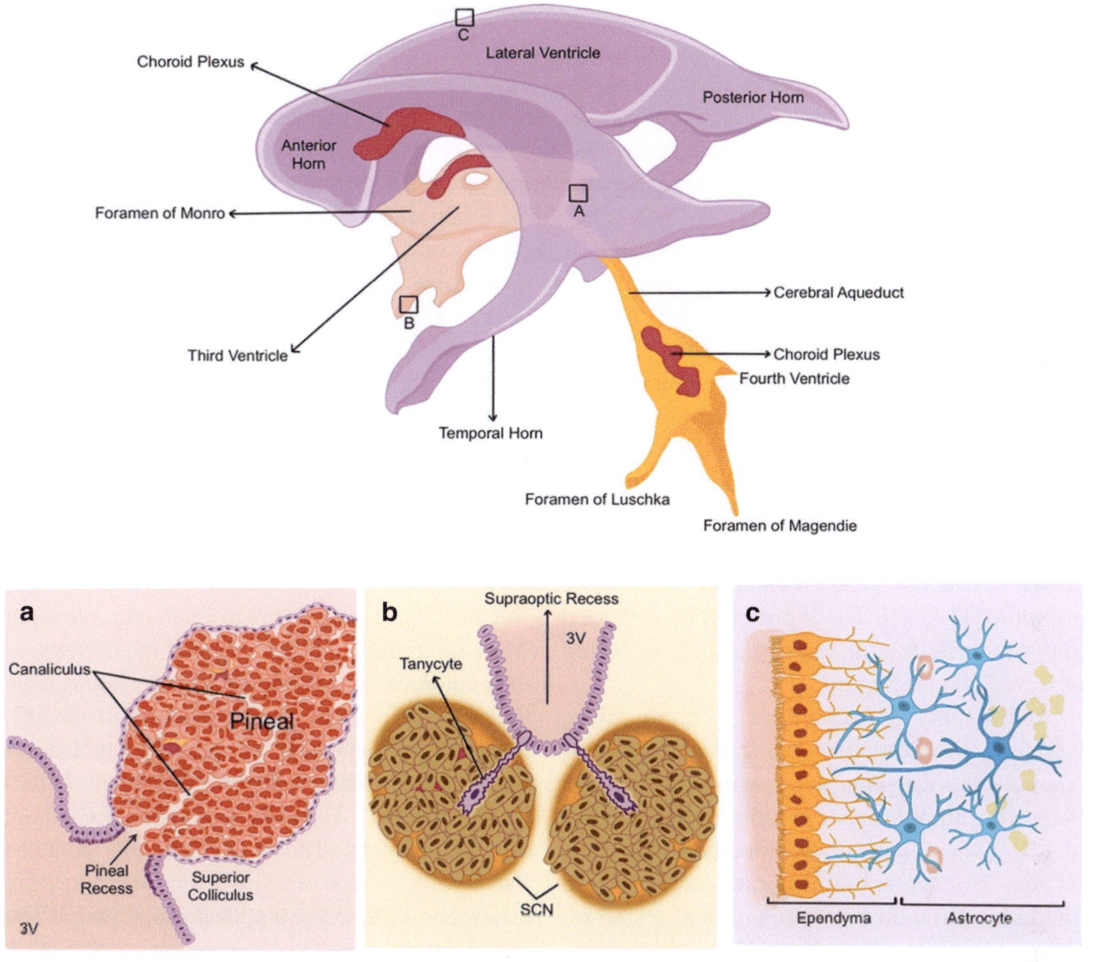

Fig. 20.2 Anatomical, functional, and structural overview of the hippocampus. The diagram provides an overview of the ventricular system of the brain, illustrating the flow of cerebrospinal fluid (CSF) and the role of melatonin in this system. CSF is continuously produced by the choroid plexuses of the lateral, third, and fourth ventricles. At night, pineal-derived melatonin is present in the CSF, which moves through the ventricles and the subarachnoid space via the foramina of Luschka and the foramen of Magendie. From the subarachnoid space, melatonin reaches the periarterial spaces (Virchow-Robin spaces) of the cerebrum and spinal cord, facilitating its entry into the neural parenchyma

of neural tissue that runs from the gyrus fasciolaris on the superior aspect of the parahippocampal gyrus [34, 35]. The structure consists of the fascia dentata, which is divided into three histologically distinct layers: the surface molecular layer, which contains the dendrites of granule cells and various types of interneurons and is a major site of synaptic inputs from the entorhinal cortex; the intermediate granular layer, which is densely packed with granule cells that are the principal neurons of the DG and play a crucial role in processing and transmitting information to the CA3 region; and the deep multiform layer, containing various types of neurons, including mossy cells, which provide excitatory inputs to granule cells and other hilar interneurons [36]. The multiform layer is linked to the CA4 region, known as the hilus region, within the CA. The hilus and the fascia dentata converge to form the DG. The rear segment of the DG, known as the Band of Giacomini, extends backward and medially, dividing the uncus into two separate areas:

the anterior uncinate gyrus and the posterior intralimbic gyrus [36–38]. The DG is anatomically distinct from the parahippocampal gyrus by the presence of the hippocampal sulcus. In addition, the fimbria of the fornix, which surrounds the DG, is separated from it by the fimbrio-dentate sulcus [39, 40].

The hippocampus creates a bulging area at the bottom of the inferior horn of the lateral ventricle. It comprises the archipallial cortex, formed by the inward folding of the DG, CA, and the subiculum. The subiculum connects the six-layer neocortex of the parahippocampal gyrus all the way through. During development, the CA and DG fold and migrate towards the inferior horn of the lateral ventricle at the hippocampal sulcus, bringing the outer molecular layers of the DG and subiculum into closer proximity [39, 40]. The hippocampus is named thus due to its resemblance to a seahorse when observed at a larger scale.

Additionally, it resembles a ram's horn; therefore, its alternative name is the horn of Ammon, the Egyptian deity with a ram's head [41]. The hippocampus histologically consists of six distinct layers, specifically the alveus, stratum oriens, stratum pyramidale, stratum lucidum, stratum radiatum, and stratum lacunosum-moleculare. Neuron growth persists in the subgranular zone (SGZ) of the DG throughout an individual's lifespan. The ongoing generation of new neuronal cells in the hippocampus is vital for several cognitive functions, including learning, long-term memory, spatial memory, and emotional regulation [42]. The alveus is situated within white matter covering the ventricular surface of the hippocampus. It is made up of axons from hippocampal pyramidal cells. These axons unite in the hippocampus and create the fimbria hippocampi. This structure extends backward, overlaying the DG and reaching the splenium of the corpus callosum before continuing as the fornix. The fimbrio-dentate sulcus marks the limit between the fimbria and the DG. The fimbria extends towards the fornix, surrounding the thalamus and separating the two choroidal fissures. The choroidal plexus is within the choroidal fissure [39].

20.3.1.2 Subfields of CA and Their Functions

The subfields of the CA are designated according to their histological characteristics and play distinct roles in different aspects of memory processing. The CA1 region, which is the biggest area of the hippocampus, is primarily responsible for the process of consolidating long-term memories and facilitating spatial navigation. The structure is next to the presubiculum on its outer side and CA2 on its inner side [43, 44]. The major components of the structure are pyramidal cells, which are glutamatergic projection neurons. These cells comprise approximately 90% of the neurons in the structure, while the other 10% are interneurons. The CA2 area, responsible for social memory and recognition, is located next to the DG, with CA1 on its lateral side and CA3 on its medial side [45]. CA2, in contrast to CA3, receives input from the supramammillary area of the hypothalamus but does not receive mossy fiber input from the DG. CA3, located next to the hilus of the DG, is positioned between CA2 and CA4 and has a crucial function in memory recall, namely in pattern completion. The outermost cells of CA3 are often called CA4. The hilus of the DG contains these cells, which are involved in processing information within the hippocampal network. The apical dendrites in the CA3 region receive input from mossy fibers, which originate from the granule cells of the DG [46]. The fibers of the CA3 pyramidal cells participate in the formation of the alveus, fimbria, and fornix. Additionally, some axons in the CA3 are known as Schaffer collaterals, which connect with the dendrites of CA1 pyramidal cells from which axons form synapses with subiculum neurons that are involved in the development of the fimbria and fornix through the alveolar pathway [47].

20.3.1.3 Subiculum

The subiculum is located near the hippocampus, in the CA1 region, and at the edge of the entorhinal cortex. It functions as a critical outflow structure of the hippocampal formation. Histologically, the subiculum exhibits a laminar organization, comprising distinct layers like the molecular layer, pyramidal cell layer, and polymorphic layer.

Functionally, the subiculum is a hub for the hippocampus, transmitting processed information to various brain regions, including the neocortex and thalamus [48]. Additionally, it contributes to spatial navigation, the recall of memories, and the process of learning. It is a significant intermediary hub, conveying information from the hippocampus to several brain areas, including the entorhinal cortex, prefrontal cortex, and thalamus. This relay function has an important role in the consolidation and recovery of memories. The subiculum has place cells similar to those present in the hippocampus. These cells facilitate the encoding of spatial information, assisting in navigation and spatial memory [49]. Studies indicate that the subiculum has a role in controlling emotional reactions and behaviors related to anxiety, stress, and the processing of rewards. The subiculum regulates stress and anxiety levels by interacting with the hypothalamus and amygdala. It functions in the hypothalamic-pituitary-adrenal (HPA) axis, affecting the body's reaction to stress. It plays a role in behaviors connected to reward and motivation, connecting with the nucleus accumbens and other brain regions involved in reward [48, 50]. The subiculum is clinically significant because it is involved in a range of neurological and psychiatric disorders, such as epilepsy, Alzheimer's disease (AD), schizophrenia, and depression. Studies have demonstrated the continuous formation of new neurons in the subiculum, a region of the human brain that has a role in learning and memory, like the DG.

20.3.2 Functional Roles In-Memory Processing, Learning, and Emotional Regulation

The hippocampus has a role in memory and learning but also contributes to emotional regulation. It is involved in memory processing, specifically in the formation, organization, and retrieval of memories. One of its main functions is episodic memory, which entails recalling certain events, settings, and experiences [51]. The hippocampus is vital in converting short-term memories into long-term memories, called "memory consolidation." During sleep, the hippocampus replays daily events, strengthening and integrating new memories with existing knowledge. The innate network of the hippocampus is responsible for unidirectional synaptic connections, which assist in the conversion of short- to long-term memories. The "Papez circuit" functions as a memory circuit, which is necessary for both the integration of emotions and the formation of recent memories [42]. The main components of the circuit include the hippocampus, fornix, mammillary bodies, anterior thalamic nuclei, and cingulate gyrus [42]. The conventional route within the "Papez circuit" involves the transmission of signals from the hippocampus to the mammillary bodies via the fornix, which then establishes connections with the anterior thalamic nuclei. The anterior thalamic nuclei send projections to the cingulate gyrus, which in turn reciprocates by transmitting information to the parahippocampal gyrus, thereby forming a closed loop that connects back to the hippocampus [51–54].

Long-term potentiation (LTP) is a neurophysiological mechanism that increases the effectiveness of synaptic transmission, promoting the fast establishment of new memories [55]. This potentiation happens upon intense stimulation of the presynaptic terminal and is related to the "Schaffer collaterals" and "mossy fibers" in the hippocampus. LTP leads to an extended period of heightened activity in the neurons that receive signals after the synapse, which can last for many days—the high-frequency stimulation results in the buildup of calcium ions in the postsynaptic neurons, which initiates LTP [56]. Even after the original external stimulation stops, impulses continue to be regularly transferred across the synapses in the hippocampus formation because of LTP [57, 58].

The hippocampus is crucial for spatial memory, enabling humans to navigate and comprehend spatial relationships within their environment. This capability is facilitated by specialized neurons known as place cells, which activate when a person is in a particular location, forming a mental map of the surroundings [59]. These cells assist in recalling specific places and the paths to reach them. The hippocampus plays an important role in the formation of memories, and any harm to this region can interfere with the capacity to generate new memories, resulting in anterograde amnesia [60, 61]. The hippocampus also plays a key role in facilitating many learning processes. In associative learning, the formation of connections between diverse stimuli is of special importance. This form of learning is essential for acquiring new abilities and adjusting to environmental changes. The CA3 region of the hippocampus has an important role in this type of learning by facilitating the creation of intricate associative networks through its large recurrent collateral connections [62].

The hippocampus plays a role in pattern separation, distinguishing between similar yet separate events. This function is crucial for reducing interference between memories and ensuring precise recall. The granule cells in the DG have a vital function in this process since they encode distinct elements of each event, therefore facilitating the accuracy of learning and memory [63, 64]. In addition to its functions in memory and learning, the hippocampus has a critical role in regulating emotions and mood. It engages with other brain areas, like the amygdala and the prefrontal cortex, to regulate emotional reactions. The hippocampus has a role in modulating the HPA axis, which is necessary for the regulation of the physiological response of the human brain to stress. Prolonged stress can impact the functioning of the hippocampus, resulting in memory deficits and heightened feelings of anxiety. Neurogenesis in the hippocampus, especially in the DG, has been associated with enhancements in mood. Depression is frequently accompanied by decreased neurogenesis, whereas increased neurogenesis is linked to antidepressant benefits and stress resistance. This condition highlights the hippocampus's significance in maintaining emotional balance and mental wellness [65–67].

20.3.3 Neurogenesis Within the Hippocampus

The journey of understanding adult hippocampal neurogenesis (AHN) in both animals and humankind reflects a significant evolution in scientific thought. Initially, the work of Joseph Altman in the 1960s using 3H-thymidine to label proliferating cells in the adult rat hippocampus faced skepticism, as the prevailing belief was that adult brains were incapable of generating new neurons [68, 69]. This skepticism began to shift with further research, notably from Fernando Nottebohm's group in the 1980s, which demonstrated seasonal changes in the control nuclei of canaries, suggesting a physiological role for neurogenesis linked to the birds' singing behavior [70]. Building on this foundation, a pivotal study by Eriksson et al. in 1998 provided direct evidence of AHN in humans [71]. They used bromodeoxyuridine (BrdU), a synthetic nucleoside that integrates into the DNA of dividing cells, to trace new neuron development in the human brain. The study involved cancer patients who had received BrdU as part of their treatment regimen. Post-mortem examinations revealed that several months after BrdU administration, the labeled cells in the DG not only survived but also expressed various neuronal markers, confirming the presence of adult-born dentate granule cells in humans. Together, these studies have progressively unraveled the complexities of neurogenesis, underscoring its potential implications for brain plasticity, learning, and memory, both in research contexts and possible therapeutic applications.

Neurogenesis, the process of generating new neuronal cells, occurs throughout human life in the adult hippocampus, specifically in the DG. This process is crucial for maintaining hippocampal function and has several important implications. This phenomenon primarily occurs in the SGZ of the DG [72]. In this region, neural progenitor cells (NPCs) proliferate and differentiate into granule cells, which integrate into the existing hippocampal circuitry. This integration is crucial for maintaining and enhancing hippocampal function. Recent studies emphasize the multi-stage process of neurogenesis, which includes NPC proliferation, neuronal differentiation, migration, and maturation [73].

Stem cell populations in tissues, including neural stem cells (NSCs) in the adult mammalian brain, are often heterogeneous. Different models suggest varying identities and activities of NSCs.

Radial glia-like (RGL) cells expressing GFAP, Nestin, and Sox2 (type 1 cell) show NSC properties, such as self-renewal and multipotent capacities. However, studies indicate heterogeneity among RGLs, with some RGLs exhibiting different properties [74, 75]. Non-radial Sox2-expressing precursors may also have multipotent characteristics, and other increasing cell populations could function as NSCs under certain conditions. Single-cell gene expression analysis has revealed limited gene specificity in quiescent NSCs, suggesting a complex developmental sequence in the hippocampal lineage [76]. Recent research indicates that single NSCs do not self-renew long term, implying NSC persistence at the population level [77].

Even today, the origin of adult NSCs is only partly understood, and the prevailing model posits that adult NSCs derive from the dentate neuroepithelium, which produces embryonic granule neurons. An alternative model suggests that adult NSCs derive from sonic hedgehog-responsive cells in the ventral hippocampus during late gestation, which relocates to the dorsal hippocampus [78]. More advanced tracing tools are needed to elucidate the origins of adult NSCs fully.

One of the primary roles of AHN is related to memory formation and cognitive flexibility. New neurons in the DG are particularly important for pattern separation, a process that allows for the discrimination of similar inputs, thus enhancing memory precision and learning capacity [79]. This function is vital for forming distinct memories and adapting to new environments [80]. AHN also contributes to emotional regulation. Enhanced neurogenesis has been associated with improved mood and resilience to stress, while decreased neurogenesis is linked to conditions such as depression and anxiety. The process by which new neurons are integrated into the hippocampal network plays a key role in stress response and emotional health [81]. Studies have shown that antidepressant treatments often stimulate neurogenesis, which is thought to be one of the mechanisms through which these treatments exert their effects [82]. Pathologically, AHN is influenced by various conditions such as epilepsy, stroke, major depression, addiction, and neurodegenerative diseases, including AD. Specific findings include a decrease in progenitor and proliferating cells in major depression, with some evidence suggesting that antidepressants may increase AHN. In AD, there is a physiological decline due to aging and disease-specific reductions in AHN, which could be linked to the early cognitive decline seen in patients [83].

Physical exercise, environmental enrichment, and cognitive stimulation are well-documented factors that enhance hippocampal neurogenesis. These activities increase the proliferation of NPCs and promote the survival and integration of new neurons, thereby improving cognitive functions and emotional health [73]. On the other hand, chronic stress and aging are significant negative regulators of neurogenesis. Prolonged stress exposure leads to elevated levels of glucocorticoids, which can inhibit the proliferation of NPCs and reduce the survival of new neurons. Similarly, aging is associated with a natural decline in neurogenesis, contributing to age-related cognitive decline and increased vulnerability to mood disorders [81].

Understanding the mechanisms of AHN has significant therapeutic implications. Enhancing neurogenesis could be a potential strategy for treating cognitive impairments and psychiatric disorders. For instance, selective serotonin reuptake inhibitors and other antidepressants that boost neurogenesis may help alleviate symptoms of depression and anxiety [82]. Additionally, lifestyle interventions, such as physical exercise and cognitive training, are being explored for their potential to enhance neurogenesis and improve mental health outcomes [80].

Recent research continues to explore the complex dynamics of AHN. Advances in imaging techniques and molecular biology have provided deeper insights into how new neurons contribute to hippocampal function and how they regulate their development [73]. Ongoing studies aim to translate these findings into clinical applications that could help mitigate the effects of neurodegenerative diseases and mental health disorders. These insights underscore the dynamic nature of AHN and its potential contribution to cognitive functions and mood regulation, suggesting that changes in neurogenesis could be a critical factor in the pathology of various brain conditions and normal aging.

To conclude, the hippocampus is a vital brain structure with intricate anatomical components that support its multifaceted roles in memory processing, learning, and emotional regulation. The continuous process of neurogenesis within the hippocampus highlights its dynamic nature and potential for adaptability and resilience throughout life. Understanding these aspects of the hippocampus provides valuable insights into its contributions to cognitive and emotional health.

20.4 Melatonin and Hippocampal Neuroprotection

Melatonin, a neurohormone produced by the pineal gland, has demonstrated significant neuroprotective features in various models of brain injury. This title dives into the mechanisms and effects of melatonin in protecting hippocampal neurons, specifically in the CA1 region, from ischemic damage, oxidative stress, and other neurodegenerative conditions. The hippocampus, a critical structure for memory and cognitive functions, is very sensitive to ischemic injury, making the protective role of melatonin particularly important.

20.4.1 Neuroprotective Mechanisms of Melatonin in Ischemic Injury

Ischemic brain injury involves a complex interplay of excitotoxicity, oxidative stress, and inflammatory responses. During ischemia-reperfusion, over-release of excitatory amino acids and activation of their receptors lead to neuronal injury and death. Additionally, oxygen-derived free radicals contribute significantly to lipid peroxidation and neuronal damage [84].

Research has demonstrated that melatonin administration at the onset of cerebral reperfusion significantly protects CA1 neurons from ischemic damage. In this study, Wistar male rats subjected to 10, 20, and 30 min of transient forebrain ischemia received intraperitoneal injections of melatonin (10 mg/kg) immediately after reperfusion and at 2 and 6 h post-reperfusion [84]. The results indicated a significant increase in CA1 neuronal density in melatonin-treated groups compared to saline-treated controls [84]. This protective effect was not observed when melatonin treatment was delayed by 1 h or administered 30 min prior to ischemia, underscoring the importance of timing in melatonin administration for effective neuroprotection [84].

The antioxidant features of melatonin, which penetrates the central nervous system and acts as a free radical scavenger, reduce lipid peroxidation and other oxidative stress-related damage. Melatonin's efficacy in reducing ischemic injury highlights its potential as a therapeutic agent in conditions involving transient forebrain ischemia, where timely intervention is critical for preventing extensive neuronal damage and preserving cognitive functions [84].

20.4.2 Melatonin's Role in Oxidative Stress Protection

Carneiro and Reiter (1997) explored melatonin's effect on δ-aminolevulinic acid-induced lipid peroxidation in brain regions, including the hippocampus. Their findings showed that melatonin significantly protected against oxidative damage in vitro and in vivo, underscoring its potent antioxidant properties. In vitro experiments demonstrated that melatonin and vitamin E reduced the concentration of lipid peroxidation products in a dose-dependent manner in cortical, cerebellar, and hippocampal homogenates. In vivo, melatonin (10 mg/kg) significantly reduced lipid peroxidation in the cerebellum and hippocampus of rats treated with δ-aminolevulinic acid [85].

Studies on aluminum- and phosphine-induced oxidative stress further highlight melatonin's role as a potent antioxidant that can penetrate the blood-brain barrier and protect neural tissues from oxidative damage. A study demonstrated that prolonged exposure to aluminum, a common environmental toxin linked to various neurodegenerative diseases, significantly increased lipid and protein oxidation in rat synaptosomal membranes. Melatonin reduced this oxidative damage in a concentration-dependent manner, indicating its potential to mitigate neurodegenerative effects associated with aluminum exposure [86]. Similarly, later research explored melatonin's protective effects against phosphine-induced oxidative damage. Phosphine, a toxic gas commonly used in agriculture, induces significant oxidative stress and neuronal damage. The study found that melatonin reduced lipid peroxidation and DNA oxidation in the brain, further emphasizing its antioxidant capabilities. Rats treated with melatonin (10 mg/kg) showed decreased levels of oxidative stress markers and improved neuronal survival compared to controls. These findings underscore the potential role of melatonin as a neuroprotective agent against oxidative damage caused by environmental toxins like phosphine [87].

20.4.3 Melatonin in Perinatal Brain Injury and Hypoxia

Various researchers have explored melatonin's neuroprotective effects in perinatal asphyxia, hypoxia models, and epilepsy. Melatonin combined with erythropoietin enhanced neuroprotection in neonatal encephalopathy, showing a fast recovery of EEG findings and low cell death in piglets subjected to hypoxic-ischemic injury. This study demonstrated that melatonin and erythropoietin, both individually and in combination with hypothermia, enhanced neuroprotection in neonatal encephalopathy. Piglets subjected to hypoxic-ischemic injury were treated with hypothermia, melatonin (20 mg/kg), and/or erythropoietin (3000 U/kg). The results showed faster amplitude-integrated EEG recovery, reduced lactate/*N*-acetyl aspartate peak ratios, and decreased TUNEL-positive cells in the sensorimotor cortex [88]. The study concluded that melatonin and erythropoietin are safe and effective adjunct therapies to hypothermia, with melatonin showing significant protection in reducing cell death and promoting oligodendrocyte survival [88]. The mechanisms of protection in the brain of late-gestation fetal sheep by melatonin in response to hypoxia involved fetal sheep subjected to umbilical cord occlusion, a model that closely mimics human perinatal asphyxia. The results showed that melatonin treatment significantly reduced neuronal death, preserved myelination, and reduced inflammation in the fetal brain. These findings suggest that melatonin has significant therapeutic potential in preventing perinatal brain injury [89].

Melatonin's role in regulating microglial polarization via the RhoA/ROCK signaling pathway in a model of temporal lobe epilepsy was demonstrated; melatonin treatment reduced seizure frequency and severity, improved cognitive functions, and promoted the polarization of microglia from a pro-inflammatory M1 phenotype to an anti-inflammatory M2 phenotype. This shift was mediated by inhibiting the RhoA/ROCK signaling pathway, highlighting a novel mechanism by which melatonin exerts neuroprotective effects in epilepsy [90].

Supportively, research has demonstrated the synergistic effects of melatonin and hypothermia in reducing oxygen-glucose deprivation-induced cell death in hippocampal cultures. This synergy suggests potential therapeutic applications for combined treatments in neonatal hypoxic-ischemic encephalopathy. Treating hippocampal cultures with melatonin (10 μM) and hypothermia (33 °C) after oxygen-glucose deprivation significantly reduced cell death and improved neuronal survival. The findings highlight the potential for combining melatonin and hypothermia as an effective therapeutic strategy for neonatal brain injury [91].

Melatonin's effect on δ-aminolevulinic acid-induced lipid peroxidation in brain regions demonstrated that melatonin, both in vitro and in vivo, protected against oxidative damage induced by δ-aminolevulinic acid. In vitro experiments showed that melatonin and vitamin E reduced the concentration of lipid peroxidation products in a dose-dependent manner in cortical, cerebellar, and hippocampal homogenates. In vivo, melatonin (10 mg/kg) significantly reduced lipid peroxidation in the cerebellum and hippocampus of rats treated with δ-aminolevulinic acid

[85]. This study highlights melatonin's role as a potent antioxidant that can penetrate the blood-brain barrier and protect neural tissues from oxidative damage. The findings suggest that melatonin may be beneficial in conditions such as acute intermittent porphyria, where oxidative stress plays an important role in neuropathology.

Investigations have shown the protective effects of melatonin against oxidative damage induced by aluminum in synaptosomal membranes. Prolonged exposure to aluminum, a common environmental toxin, is linked to various neurodegenerative diseases. A study demonstrated that aluminum, in combination with $FeCl_3$ and ascorbic acid, significantly increased lipid and protein oxidation in rat synaptosomal membranes. Melatonin reduced this oxidative damage in a concentration-dependent manner [86]. Specifically, melatonin (1, 5, and 10 mmol/L) reduced the levels of malondialdehyde and 4-hydroxyalkenals as well as protein carbonyl content, indicating its potent antioxidant properties. These findings suggest that melatonin could be a useful neuroprotective agent against aluminum-induced oxidative stress in the brain [86]. Hsu et al. (2002) explored the protective effects of melatonin against phosphine-induced oxidative damage. The study found that melatonin reduced lipid peroxidation and DNA oxidation in the brain, further emphasizing its antioxidant capabilities. Rats treated with melatonin (10 mg/kg) showed decreased levels of oxidative stress markers and improved neuronal survival compared to controls. These findings underscore melatonin's potential as a neuroprotective agent against oxidative damage caused by environmental toxins like phosphine [87].

The combined studies provide robust evidence that melatonin can mitigate oxidative damage caused by various environmental toxins. This highlights the significant potential of melatonin as a protective agent in environments where exposure to neurotoxic substances is prevalent. The antioxidative properties of melatonin, demonstrated through its ability to reduce lipid peroxidation and DNA damage, affirm its therapeutic promise in counteracting oxidative stress-related neural damage. These findings are crucial for developing strategies to prevent and treat neurodegenerative conditions linked to environmental toxins [86, 87].

20.4.4 Melatonin in Neurotoxicity and Inflammation

Leeboonngam et al. (2017), Pinato et al. (2015), and Hoppe et al. (2010) investigated melatonin's protective effects against amphetamine (AMPH)-induced neurotoxicity, neuroinflammation, and amyloid-β toxicity. Chronic AMPH exposure leads to significant oxidative stress and neuronal damage in the hippocampus. The study demonstrated that melatonin (10 mg/kg) administered prior to AMPH exposure significantly decreased AMPH-induced neuronal degeneration in the hippocampus's dentate gyrus, CA1, and CA3 regions. Melatonin attenuated the reduction in the expression of synaptophysin, PSD-95, α-synuclein, and NMDA receptor proteins caused by AMPH. Furthermore, melatonin prevented the AMPH-induced loss of the dopamine transporter and NMDA receptor and the overexpression of α-synuclein in the hippocampal regions. The study also demonstrated that melatonin reduced AMPH-induced decreases in the levels of NMDA receptor downstream signaling molecules, such as calcium/calmodulin-dependent protein kinase II, and the expressions of melatonin receptors MT1 and MT2 [92].

Local melatonin synthesis in response to neuroinflammation induced by lipopolysaccharide (LPS) provided selective neuroprotection, highlighting the role of melatonin in modulating immune responses and protecting against inflammatory damage. LPS-treated rats showed increased melatonin production in the cerebellum, which correlated with reduced markers of neuroinflammation and preserved neuronal integrity. These findings suggest that the anti-inflammatory properties of melatonin contribute to its neuroprotective effects, making it a potential therapeutic agent for inflammatory neurological conditions [93].

Hoppe et al. (2010) studied melatonin's neuroprotective effects against amyloid-β-induced toxicity. The results showed that melatonin reduced neuroinflammation and tau phosphorylation, supporting its therapeutic potential for AD. Rats treated with melatonin (10 mg/kg) showed decreased markers of neuroinflammation and tau pathology in the hippocampus. These findings suggest that melatonin may offer protective benefits in AD by reducing neuroinflammation and tau pathology [94].

20.4.5 Melatonin in Cognitive Impairment and Post-cardiac Arrest Neuroprotection

Feng et al. (2004) and Yang et al. (2018) examined the effects of melatonin on cognitive impairment in ovariectomized rats and post-cardiac arrest neuroprotection. Melatonin significantly improved memory performance and reduced neuronal apoptosis in ovariectomized rats, highlighting its potential for managing menopause-related cognitive decline. The study involved treating ovariectomized rats with melatonin (10 mg/kg) and assessing their performance in a Morris water maze task. The results demonstrated important improvements in spatial memory retention and reduced neuronal cell death in the hippocampus of melatonin-treated rats compared to controls. These findings confirm that melatonin may have therapeutic applications for managing menopause-related cognitive decline, potentially through its antioxidant properties and ability to reduce neuronal apoptosis [95].

In post-cardiac arrest care, melatonin improved neurological functions and preserved mitochondrial integrity, suggesting its potential as a therapeutic agent. Yang et al. (2018) investigated melatonin's impact on neurological outcomes after cardiac arrest. The study found that melatonin improved neurological functions and preserved mitochondrial integrity, suggesting its potential in post-cardiac arrest care. The study involved inducing cardiac arrest in animal models and treating them with melatonin after resuscitation. Neurological functions and mitochondrial health were assessed. The results indicated that melatonin treatment significantly improved neurological outcomes and preserved mitochondrial function, highlighting its potential as a therapeutic agent in post-cardiac arrest care [96].

20.4.6 Melatonin in Neurodegenerative Diseases

Parisotto et al. (2016), Olcese et al. (2009), Stewart and Leung (2005), and Mei et al. (2016) investigated melatonin's effects on Down syndrome, AD, seizure thresholds, and formaldehyde-induced cognitive decline. Chronic melatonin administration caused a reduction in oxidative damage and cellular senescence in a Down syndrome model, suggesting its potential for mitigating cognitive decline. The study involved administering melatonin chronically to Ts65Dn mice, a well-established model for Down syndrome. Various markers of oxidative stress and cellular aging were assessed in the hippocampus. Melatonin treatment was found to significantly reduce oxidative stress markers and improve the overall health of hippocampal neurons. These results suggest that melatonin could be a valuable therapeutic agent for managing cognitive decline in individuals with Down syndrome [97].

Olcese et al. (2009) examined the long-term effects of melatonin in a transgenic AD model. The study found that melatonin reduced amyloid-beta (Aβ) deposition and improved cognitive functions, supporting its potential as a therapeutic agent for AD. The study involved treating transgenic mice, genetically predisposed to develop Alzheimer's-like pathology, with melatonin over an extended period. Cognitive functions were assessed using behavioral tests, and brain tissues were analyzed for Aβ plaques. The results showed that melatonin-treated mice had significantly fewer Aβ deposits and performed better in memory tasks than untreated controls. These findings confirm that melatonin could be a valuable therapeutic approach for managing AD [98].

Stewart and Leung (2005) studied the potential roles of hippocampal melatonin receptors in modulating seizure thresholds. The study found that melatonin receptor antagonists increased seizure latency, suggesting melatonin's potential in seizure management. The study involved administering melatonin receptor antagonists to animal models and assessing their effects on seizure latency. The results showed that blocking melatonin receptors reduced the seizure threshold, indicating that melatonin plays a protective role in seizure regulation. These findings confirm that melatonin or its analogs may be therapeutic agents for managing seizures [99].

Mei et al. (2016) investigated the effects of formaldehyde exposure on cognitive decline and melatonin's protective role [100]. The study revealed that melatonin mitigated formaldehyde-induced oxidative stress and cognitive impairments, highlighting its potential in occupational health [100]. The study involved exposing mice to formaldehyde and assessing cognitive functions through behavioral tests. Melatonin treatment was administered to evaluate its protective effects [100]. The results demonstrated that melatonin significantly reduced oxidative stress markers and improved cognitive performance in formaldehyde-exposed mice. These findings suggest that melatonin could be used to protect against cognitive decline caused by occupational exposure to formaldehyde [100].

20.4.7 Melatonin in Excitotoxicity-Induced Damage

Lee et al. (2005) and Behan et al. (2009) explored the protective mechanisms of melatonin against kainic acid- and quinolinic acid-induced excitotoxicity. Melatonin reduced neuronal death and promoted survival pathways in kainic acid-induced damage. The study involved administering kainic acid to induce excitotoxic damage in the hippocampus, followed by melatonin treatment. The results showed that melatonin significantly reduced neuronal death and enhanced the expression of survival proteins such as Bcl-2. These findings suggest that melatonin could be used to protect against excitotoxicity-induced neuronal damage, which is common in various neurodegenerative diseases [101].

Behan et al. (2009) explored the protective role of melatonin against quinolinic acid-induced hippocampal damage. The study demonstrated melatonin's ability to reduce oxidative stress and neuronal damage, suggesting its use in excitotoxicity and oxidative stress conditions. The study involved administering quinolinic acid to induce neurotoxicity in the hippocampus, followed by melatonin treatment. The results showed that melatonin significantly reduced oxidative stress markers and neuronal damage. These findings highlight melatonin's potential in protecting against neurotoxic damage caused by excitotoxins like quinolinic acid [102].

The findings provide compelling evidence for the neuroprotective role of melatonin in hippocampal ischemia, hyperbilirubinemia-induced brain damage, aluminum-induced oxidative damage, perinatal asphyxia, amphetamine-induced dopaminergic neurotoxicity, late gestation hypoxia, and electromagnetic radiation-induced oxidative stress. The precise timing of administration is crucial for maximizing its protective effects. These results underscore the potential therapeutic applications of melatonin in conditions involving ischemic and oxidative brain injuries.

Melatonin's ability to cross the blood-brain barrier and act as a free radical scavenger highlights its potential as a therapeutic modality for neuroprotection, as discussed. The varied models of brain injury in these studies demonstrate the wide-ranging protective effects of melatonin across different types of neural damage. Exogenous administration of melatonin in ischemic conditions is critical for reducing neuronal death and preserving cognitive functions. In hyperbilirubinemia and aluminum-induced oxidative stress, melatonin's antioxidant properties significantly mitigate oxidative damage, supporting neural integrity. In conditions like perinatal asphyxia, amphetamine-induced neurotoxicity, electromagnetic radiation exposure, cognitive impairment in ovariectomized rats, Down syndrome models, hydrogen peroxide-induced lipid peroxidation, neuroinflammation, phosphine-induced oxidative damage, hypoxic-ischemic encephalopathy, AD, kainic acid-induced excitotoxicity, quinolinic acid-induced hippocampal damage, amyloid-β neurotoxicity, seizure threshold modulation, formaldehyde-induced cognitive decline, post-cardiac arrest care, epilepsy, and methamphetamine-induced cognitive impairments, melatonin not only reduces cell death but also promotes the recovery of neural functions and reduces inflammation.

20.5 Melatonin, Synaptic Plasticity, and Memory

Beyond its well-known role in the regulation of circadian rhythms, melatonin has been found to significantly affect various neural processes, particularly synaptic plasticity and memory functions. This includes modulating mechanisms such as LTP, long-term depression (LTD), and memory consolidation. These findings are crucial for understanding basic neurophysiology and exploring melatonin's potential therapeutic applications, especially in neurodegenerative diseases like AD [98].

20.5.1 Effect on Synaptic Plasticity

Synaptic plasticity is the foundational mechanism through which the brain adapts and learns. It involves the strengthening (potentiation) or weakening (depression) of synapses based on neuronal activity. Melatonin affects these processes through several pathways, which include receptor-mediated signaling, modulation of intracellular cascades, and antioxidant properties.

Research has shown that melatonin can prevent impairments in synaptic plasticity, often induced by elevated oxidative stress conditions [103–107]. Baydas et al. (2005) demonstrated that melatonin could counteract the adverse effects of hyperhomocysteinemia, a condition associated with increased oxidative stress, by restoring the hippocampus's expression of neural cell adhesion molecules (NCAMs). NCAMs are critical for synaptic stabilization and plasticity, suggesting that melatonin helps maintain the structural and functional integrity of synapses under oxidative stress conditions [103].

However, the effects of melatonin on synaptic plasticity are not uniform and can be context dependent. For example, studies by Soto-Moyano et al. (2006) and Talaei et al. (2009) reported that melatonin inhibited LTP in the neocortex and hippocampus, respectively. This inhibition might be related to melatonin's interaction with NMDA receptors, which play a crucial role in LTP induction and maintenance. The NMDA receptor is a glutamate receptor that allows calcium ions to enter the cell, essential for the synaptic changes underlying learning and memory [104, 105].

Recent information regarding the molecular mechanisms of melatonin's action was provided by Benleulmi-Chaachoua et al. (2015), who identified that melatonin receptors MT1 and MT2 are part of presynaptic protein complexes. These receptors modulate neurotransmitter release by affecting voltage-gated calcium channels, which are crucial for initiating and propagating synaptic signals. By influencing calcium influx, melatonin can regulate the release of neurotransmitters such as glutamate and GABA, thereby impacting synaptic plasticity [106].

Circadian rhythms also influence the role of melatonin in synaptic plasticity. Jilg et al. (2019) highlighted that melatonin receptor signaling is essential for maintaining day-night differences in hippocampal LTP, which is necessary for optimal learning and memory function. Disruptions in circadian rhythms, such as those caused by irregular sleep patterns or shift work, can impair cognitive functions, underscoring the importance of melatonin in synchronizing circadian and cognitive processes [107].

20.5.2 Gene Expression and Neurotrophic Factors

In addition to its direct effects on synaptic plasticity, melatonin influences gene expression, particularly in pathways associated with neurotrophic factors. One of the most well-studied pathways is the CREB-BDNF signaling pathway. CREB is a transcription factor that regulates the expression of BDNF (brain-derived neurotrophic factor). This protein supports the survival of neuronal cells and encourages the growth of new neuronal cells and synapses [19].

Sung et al. (2018) showed that melatonin enhances memory processing by activating the CREB-BDNF pathway. This activation is primarily mediated through the MT1 receptor, suggesting a receptor-specific action of melatonin in promoting neuroplasticity. The upregulation of BDNF is particularly significant because it facilitates synaptic strengthening and plasticity, processes essential for learning and memory. Moreover, BDNF plays a critical role in protecting neurons against damage from oxidative stress and neurotoxins, making it a vital component of neurodegenerative diseases [108].

20.5.3 Antioxidant Properties and Neuroprotection

Melatonin's antioxidant properties are well-documented and critical to its neuroprotective capabilities. Oxidative stress results from an imbalance between the production of free radicals and the detoxification of these reactive species or the repair of damage in the human body. In the human brain, oxidative stress can lead to neuronal damage, synaptic dysfunction, and, ultimately, cognitive decline.

Baydas et al. (2005) highlighted the antioxidant properties of melatonin, which neutralize free radicals and reduce oxidative stress, thereby protecting neurons and synapses from damage. This protective effect is crucial in conditions such as hyperhomocysteinemia and other states associated with increased oxidative stress, which are known to impair cognitive function [103].

The neuroprotective role of melatonin extends beyond its antioxidant properties. It has been revealed to modulate the activity of antioxidant enzymes, enhance mitochondrial function, and inhibit the activation of apoptotic pathways. These actions help preserve neuronal integrity and function, which is particularly important in aging and neurodegenerative diseases.

20.5.4 Melatonin and Alzheimer's Disease

AD is typically diagnosed by progressive cognitive decline, which is characterized by the accumulation of Aβ plaques and tau tangles in the human brain. These pathological features lead to synaptic dysfunction, neuronal death, and, ultimately, loss of memory and cognitive abilities.

Liu et al. (2013) provided compelling evidence for melatonin's protective role against Aβ-induced impairments in synaptic plasticity and memory. Their study showed that chronic melatonin treatment could reverse deficits in LTP induction and theta rhythm in the hippocampus caused by Aβ. Additionally, melatonin improved spatial learning performance in the Morris water maze, a standard test for assessing memory and learning in rodents. These findings suggest that melatonin can mitigate Aβ-induced neurotoxicity, preserve synaptic function, and enhance cognitive performance, highlighting its potential as a therapeutic agent in AD [109].

Melatonin's neuroprotective effects in AD are likely mediated through several mechanisms, including its antioxidant properties, regulation of calcium homeostasis, inhibition of Aβ aggregation, and modulation of neuroinflammatory responses. These multifaceted actions make melatonin a promising candidate for the development of treatments aimed at slowing the progression of AD and improving the quality of life for patients.

Beyond AD, neuroprotective and neurogenic properties of melatonin suggest a broad therapeutic potential for various neurodegenerative conditions characterized by oxidative stress and disrupted neurogenesis. It has been demonstrated that melatonin acts as a protective agent against neurotoxic insults, such as those induced by methotrexate (MTX), a chemotherapeutic agent known to cause cognitive impairments. Suwannakot et al. (2022) demonstrated that melatonin mitigates MTX-induced neurotoxicity by restoring antioxidant enzyme levels and promoting neurogenesis markers like BDNF and doublecortin in the hippocampus. These findings underscore melatonin's potential to alleviate cognitive impairments associated with chemotherapy and other oxidative stress-related conditions [110].

20.5.5 Therapeutic Applications and Considerations

The growing body of evidence supporting melatonin's role in cognitive enhancement and neuroprotection underscores its potential as a therapeutic agent. This is particularly relevant for treating cognitive impairments and neurodegenerative diseases in which traditional treatments are limited or ineffective. Melatonin's safety profile, coupled with its broad physiological actions, makes it an attractive candidate for further research and clinical application.

However, the complexity of melatonin's effects necessitates careful consideration in clinical settings. Factors such as the timing of administration, dosage, and patient-specific characteristics (e.g., age, comorbidities, and concurrent medications) significantly influence its efficacy and safety. For instance, the timing of melatonin administration is critical due to its role in circadian rhythm regulation, which can impact cognitive function and overall health.

Future research should focus on large-scale clinical trials to evaluate the long-term safety and efficacy of melatonin, particularly in populations at risk for cognitive decline and neurodegenerative diseases. Such studies should also explore optimal dosing regimens and investigate the potential benefits of combining melatonin with other therapeutic agents to enhance its neuroprotective effects.

20.6 Clinical Implications and Therapeutic Potential of Melatonin

Melatonin, a hormone mainly produced by the pineal gland, is pivotal in regulating circadian rhythms and sleep-wake cycles. Besides its sleep-related functions, melatonin offers substantial neuroprotective properties, including antioxidant defenses, anti-inflammatory effects, and inhibition of apoptotic pathways. These properties are vital in addressing neurodegenerative diseases, including Alzheimer's, Parkinson's, Huntington's, and mood disorders [111, 112].

20.6.1 Alzheimer's Disease

AD is a degenerative neurological condition characterized by a gradual deterioration in cognitive function, including memory loss and changes in behavior. The pathogenesis of AD is characterized by the buildup of Aβ plaques and neurofibrillary tangles consisting of hyperphosphorylated tau protein. Oxidative stress and mitochondrial dysfunction play crucial roles in causing harm to neurons in AD. Melatonin demonstrates strong antioxidant qualities that are essential in countering oxidative stress and are important in AD's pathogenesis. Melatonin aids in reducing oxidative damage to neurons and protects against neurodegeneration by scavenging ROS and increasing antioxidant enzyme activity [113].

One key feature of AD is the formation of neurotoxic Aβ plaques, which contribute to oxidative stress and inflammation. Research has shown that melatonin can inhibit the accumulation of Aβ and reduce its detrimental effects by obstructing the formation of harmful Aβ oligomers [114]. In addition, melatonin enhances the clearance of Aβ from the brain by modulating the activity of enzymes responsible for its degradation. Melatonin regulates cyclic AMP levels to modulate amyloid precursor protein (APP) gene expression and promotes non-amyloidogenic processing via α-secretase. Activation of melatonin receptors MT1 and MT2 initiates signaling pathways that promote the transcription of ADAM10, resulting in a further reduction in the formation of Aβ [115, 116] (Fig. 20.3).

Neurofibrillary tangles, another feature of AD, occur due to the excessive phosphorylation of tau protein. Melatonin hinders this process by regulating kinases and phosphatases that govern tau phosphorylation. AD pathogenesis is greatly

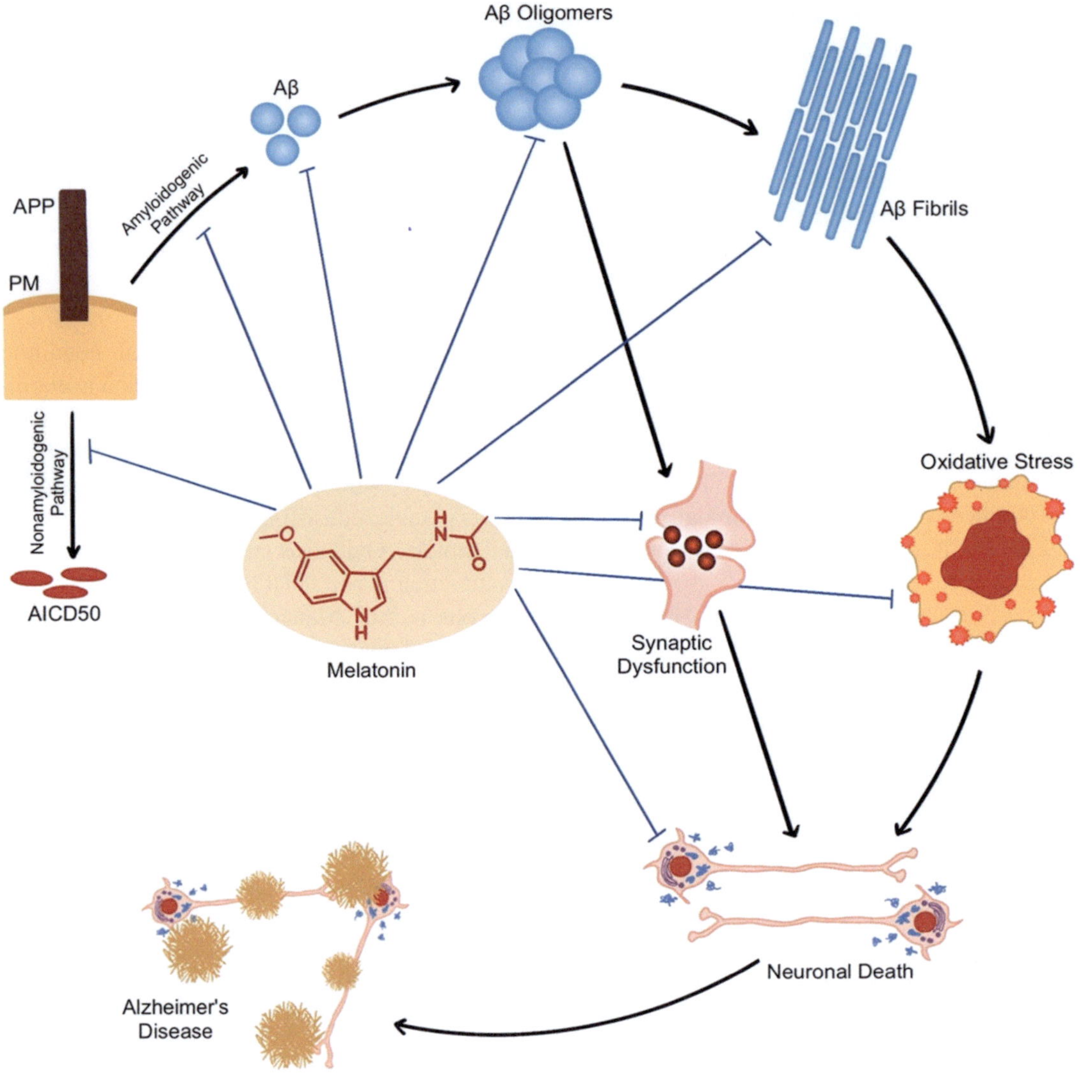

Fig. 20.3 Protective role of melatonin against Alzheimer's disease

influenced by mitochondrial malfunction, and melatonin plays an important role in preserving mitochondrial function by maintaining membrane integrity and inhibiting the release of cytochrome c, a critical stage in apoptosis. Additionally, melatonin improves the efficiency of the mitochondrial electron transport chain, decreasing ROS formation [114] (Fig. 20.3).

AD is characterized by chronic neuroinflammation, and melatonin exerts anti-inflammatory effects by regulating the activity of microglia and astrocytes, lowering the generation of pro-inflammatory cytokines, and blocking the activation of nuclear factor-kappa B (NF-κB). Melatonin, acting as an antioxidant, also inhibits the oxidative inactivation of protein kinase C, an enzyme involved in the non-amyloidogenic processing of APP [113]. Melatonin further enhances the glymphatic system's function, facilitating the clearance of Aβ and tau proteins from the brain, particularly during sleep. Research indicates that persons with AD have reduced levels of melatonin compared to those who are in good health. Furthermore, scientific evidence has demonstrated that the administration of melatonin supplements effectively mitigates moderate cognitive impairment [117]. While experimental animal research offers convincing proof of melatonin's therapeutic advantages in AD, more substantial human trials are required to validate its efficacy in treating AD patients. Research has indicated that the levels of melatonin, a hormone that regulates sleep, decline as individuals get older and are notably reduced in those with AD. The decrease in melatonin levels is linked to disturbed sleep-wake patterns and heightened oxidative stress, both of which play roles in the advancement of AD. Administering melatonin as a supplement to individuals with AD has demonstrated positive effects on sleep quality, a reduction in sundowning (late-day disorientation), and possibly a deceleration in cognitive deterioration [118, 119].

Clinical trials on melatonin's efficacy in AD have shown promising results. Melatonin supplementation has been found to enhance cognitive function and decrease Aβ plaques in animal models by inhibiting β and γ secretase activities and promoting α secretase activity, thus reducing amyloidogenesis [114] (Fig. 20.3). Human studies also have reported benefits, such as improved sleep patterns and cognitive performance. A 6-month randomized, placebo-controlled trial assessing PRM as an add-on therapy to standard AD treatments found that PRM significantly improved sleep quality, onset latency, and morning alertness without negatively affecting cognitive functions [120–123]. These findings underscore the possible roles of PRM in alleviating sleep-related cognitive deficits in AD patients, though further large-scale trials are needed to confirm these effects.

Melatonin's multifaceted neuroprotective properties make it a promising therapeutic agent for AD. Its ability to reduce oxidative stress, inhibit Aβ toxicity, modulate tau protein, preserve mitochondrial function, and exert anti-inflammatory effects positions it as a valuable adjunct in managing AD. Continued research and clinical trials are vital for understanding its therapeutic potential and its optimal use in the management of this debilitating disease (Fig. 20.3).

20.6.2 Parkinson's Disease

Parkinson's disease (PD), one of the most prevalent neurodegenerative disorders, is marked by motor symptoms such as resting tremor, rigidity, bradykinesia, and postural instability [124]. In addition, it exhibits non-motor symptoms such as constipation, cognitive impairment, and sleep disturbances [124]. The disease is defined by the degeneration of dopaminergic neurons and the accumulation of α-synuclein in the substantia nigra pars compacta [125]. The prevalence of PD increases as individuals age, affecting around 0.5–1.0% of persons aged 65–69 and 1.0–3.0% of individuals aged 80 or more. As the global population of aged individuals continues to grow, the occurrence of PD is anticipated to rise significantly. It is projected that the number of PD cases will possibly double from 7 million in 2015 to 13 million by 2040, leading to a substantial societal impact [126]. There is a strong correlation between the disturbance of melatonin levels in the circadian rhythm and a significant decrease in the amount of gray matter in the hypothalamus, which is directly related to the severity of PD. This discovery provides anatomical and physiological proof that supports the existence of circadian rhythm dysfunction in PD [124, 127].

PD patients have substantial changes in their melatonin levels. More precisely, melatonin production diminishes as PD advances, and this drop is connected to the sleep difficulties frequently observed in these patients [128]. Plasma melatonin levels in persons with PD are notably reduced compared to those without the condition, and this decrease is associated with more pronounced sleep disruptions. In addition, there is a significant decrease in the expression of five key clock genes (BMAL1, CLOCK, CRY1, PER1, and PER2) in these individuals' peripheral blood mononuclear cells. The decreased expression of these clock genes is associated with the intensity of sleep-wake disruptions, such as increased daytime drowsiness and worsened sleep quality. The results indicate that changes in melatonin levels and clock gene expressions might be used as possible indicators for identifying and evaluating the development of PD and its related sleep disturbances [129]. Exogenous administration of melatonin has been demonstrated to modulate the expression of the clock gene BMAL1 and enhance motor function [130]. Recent clinical research has shown that melatonin has the potential to be a safer option for treating sleep problems in PD. The experiment found that melatonin had fewer negative effects than other treatments and positively impacted quality and overall quality of life for patients [131].

A 12-week melatonin supplementation significantly improved motor symptoms, reduced C-reactive protein levels in serum, increased total antioxidant capacity and glutathione levels in plasma, and regulated TNF-α, PPAR-γ, and LDLR gene expressions [132]. Another study showed that administering 2 mg of melatonin nightly for 3 months significantly reduced sleep onset latency and improved overall sleep efficiency and quality, indicating a trend toward improvement in motor symptoms [133]. Furthermore, a higher dose of 25 mg of melatonin taken 30 min before bed improved mitochondrial complex I activity and the mitochondrial respiratory control ratio and reduced oxidative stress markers, suggesting potential neuroprotective effects in PD [134].

When assessing melatonin's use in treating neurodegenerative illnesses, examining the possible adverse effects is crucial. Melatonin levels that are not excessive (i.e., less than 5 mg) are typically considered safe for short-term use, although they may cause modest side effects, including headaches and weariness. Nevertheless, there have been reports of more serious adverse effects, such as hypertension, hypotension, and dizziness. However, these problems usually subside after melatonin treatment is stopped [135, 136].

It is hypothesized that changes in melatonin levels have a role in the development of non-motor symptoms of PD, including cognitive decline, depression, and sleep disturbances. The administration of melatonin supplements with prolonged-release properties resulted in enhanced sleep quality, decreased time to fall asleep, and better sleep efficiency, with few negative effects [133]. Gaining a comprehensive understanding of the fluctuations in melatonin levels might provide opportunities for developing novel treatments that target the non-motor symptoms of PD and enhance patients' overall well-being.

20.6.3 Huntington's Disease

Huntington's disease (HD), an inherited neurodegenerative disorder, is characterized by progressive motor deficits, psychiatric symptoms, and cognitive decline [137]. The condition results from an enlarged CAG repeat in the huntingtin gene, which causes a harmful increase in the activity of the mutant huntingtin protein. The clinical signs of HD are predominantly caused by neurodegeneration in the striatum and cortex of the brain [138].

Low levels of melatonin lead to several harmful effects associated with neurodegeneration. Specifically, decreased melatonin levels are associated with higher levels of oxidative stress in mitochondria, leakage of mitochondrial DNA (mtDNA) into the cytosol, and subsequent inflammatory reactions. These processes contribute to the neurological damage seen in diseases like HD, which are notably marked by mtDNA damage [111]. Melatonin has shown neuroprotective properties by preventing the release of mtDNA into the cytosol, which decreases the production of proinflammatory cytokines and affects important inflammatory pathways in the brain, especially in models of HD. This highlights the crucial role of melatonin in regulating mitochondrial dysfunction and its potential as a target for therapy in treating age-related neurodegenerative diseases [139]. In vivo and in vitro research has shown that mtDNA released into the cytosol can activate inflammatory pathways such as the cGAS/STING/IRF3 pathway in HD, leading to synaptic damage and neurodegeneration. Still, melatonin may mitigate these inflammatory reactions, offering insights into potential strategies for managing neuroinflammation in such conditions [140]. In 2014, for the first time, it was shown that HD patients have lower plasma levels of melatonin compared to healthy individuals [141]. Jin-Ho Kim and colleagues reported that HD is a neurodegenerative disorder that affects melatonin production in the brain. A key enzyme called AANAT, necessary for melatonin production, is reduced in HD patients. This reduction is linked to the trapping of AANAT by mutant huntingtin protein, leading to lower melatonin levels in the brain. The study highlights a breakdown in the normal enzyme production process in HD brains [142]. The toxicity caused by mutant huntingtin is linked to the absence of the MT1 receptor. Research conducted in 2011 found that melatonin, through its effects on the MT1 receptor, can delay the onset of HD and reduce mortality. This is achieved by preserving and enhancing the expression of the MT1 receptor and inhibiting the toxic effects caused by mutant huntingtin, suppressing cellular apoptosis, and promoting neuroprotection [143]. Melatonin exhibits protective properties against kainic acid-induced neuronal apoptosis by blocking calpain activation, preventing mitochondrial malfunction, and reducing endoplasmic reticulum stress. Melatonin exhibited neuroprotective properties by restoring mitochondrial function, inhibiting caspase activation, and promoting neuronal survival [139, 144].

In summary, the potential role of melatonin as a neuroprotective agent in HD is particularly promising due to its ability to regulate mitochondrial functions, prevent leakage of mtDNA, and reduce inflammation. Research highlights the decreased levels of melatonin in HD patients and its neuroprotective mechanisms, suggesting melatonin's role in mitigating HD progression and its potential in developing targeted therapies. This enables melatonin to be a major player in the continuous research on treatments for neurodegenerative diseases like HD.

20.6.4 Mood Disorders

Research suggests that melatonin can benefit mood disorders through its actions in the hippocampus. Melatonin receptors are expressed in the hippocampus, and the hormone can influence neuroplasticity, neurogenesis, and synaptic connectivity within this region. These effects are critical because alterations in hippocampal function and structure are often observed in mood disorders such as depression and bipolar disorder [145].

Melatonin has a protective effect on the hippocampus by reducing oxidative stress and inflammation, which can contribute to the neuropathology of mood disorders. Moreover, melatonin may enhance the growth and survival of new neurons in the hippocampus, which are linked to the antidepressant effects observed in various animal studies and preliminary human trials [146].

Additionally, the timing of melatonin secretion plays a role in regulating circadian rhythms, which are frequently disrupted in individuals with mood disorders. By stabilizing circadian rhythms, melatonin can potentially improve mood regulation and overall mental health [147, 148].

Melatonin inhibits cell death in microglia through the Cathepsin B/NLRP3 pathway both in vitro and in vivo. Suppressing the process of microglia cell death due to inflammation reduces the symptoms of depression-like behaviors caused by long-term stress, both in vivo and in vitro [149]. Prolonged and consistent exposure to light causes a decrease in melatonin levels and an increase in oxidative stress in the hippocampus of mice. This leads to the inhibition of glucocorticoid receptors through GSK3β and results in behavior similar to depression. Melatonin reverses this inhibition and improves depression-like behavior in mice [150]. Early-life sleep deprivation causes long-term neurochemical imbalances in the hippocampus, leading to impaired signaling, reduced antioxidant enzyme activity, and neurological dysfunction in adulthood. Administering melatonin during sleep deprivation can prevent cognitive impairments and neuronal damage by protecting hippocampal function and boosting neuronal count. Treatment with melatonin reinstates the expression of Na^+/K^+ ATPase, counteracts the reduced density of GABAergic interneurons, and promotes antioxidant enzyme activity in the hippocampus [146]. The MT2 receptor, which is extensively expressed, has a crucial function in enabling the physiological effects of melatonin in the hippocampus. Suppression of MT1 and/or

MT2 receptors in mice resulted in enhanced anxiety-like behavior, indicating the potential involvement of these receptors in the onset of depression and anxiety. The study suggests that irregularities in these receptors may be a factor in mood disorders [151].

Given these interactions, there is ongoing research into melatonin as a therapeutic agent for treating mood disorders, focusing on its neuroprotective and neuromodulatory effects in the hippocampus. However, more extensive clinical trials are needed to establish its efficacy and safety in this context.

20.7 Methodological Approaches for Studying the Melatonin-Hippocampus Relationship

To fully comprehend the neuroprotective and regulatory roles of this hormone in the brain, it is crucial to possess a thorough awareness of the intricate connection between melatonin and the functioning of the hippocampus. This section provides a comprehensive overview of both in vivo and in vitro models often used in melatonin research and also explores the various methodological strategies employed to examine the correlation between melatonin and the hippocampal area.

Melatonin affects cognition, especially alertness. The MT2 receptor regulates cognitive functions, anxiety, and social behavior. It regulates brain circuits associated with depression, schizophrenia, and bipolar disorder. Studies reveal that mice lacking the melatonin MT2 receptor suffer from focus issues, anxiety, and increased social involvement. Melatonin controls anxiety, social connections, and cognitive processes via the MT2 receptor. Mtnr1b-deficient mice displayed aberrant attention, impulsivity, anxiety, and sociability [152]. MT1 receptor-deficient mice were studied for neurobehavioral effects. The genetic loss of melatonin receptors in mice ($MT1^{-/-}$) leads to depression-like symptoms and sensorimotor gating deficits [153]. The researchers used a "knock-in" procedure to replace the MT1 or MT2 coding sequences with the LacZ reporter, which was then introduced into the mouse brain. This permitted the initial development of cells and neurons with MT1 and MT2 receptors. The study discovered MT2 receptors in the olfactory bulb, forebrain, hippocampus, amygdala, and superior colliculus. These findings suggested that MT2 receptors are more widespread in the brain than previously thought. In contrast, MT1 receptors were only expressed in a few structures [154].

The distribution of the MT1 receptor in cells was determined by utilizing a BAC-transgenic mouse strain that generated red fluorescent protein (RFP) when the natural MT1 promoter was activated. The RFP-MT1 transgenic mouse model offered a distinctive tool for investigating the distribution, cell-specific expression, and in vivo functionality of the MT1 receptor in the mouse brain. The BAC-transgenic mouse line is an invaluable resource for investigating the cellular expression patterns of the MT1 receptor. The study's findings suggested that the MT1 receptor is present in several brain areas involved in behaviors such as pain, fear, and anxiety. Moreover, the research findings indicated that the MT1 receptor is vital in controlling sleep, anxiety, depression, and reproductive functions [155].

Research by Chongyang Zhang and colleagues (2021) showed that endogenous melatonin affects rhythmic behaviors, reproduction, and survival in melatonin-proficient C57BL/6J congenic mice. $AH^{+/+}$ mice with higher amounts of melatonin showed lower body weight, smaller reproductive organ size, and greater daily inactivity. On the other hand, animals who did not have $AH^{-/-}$ exhibited significantly decreased levels. It was shown that $AH^{+/+}$ mice had a higher capacity to adapt their internal circadian rhythm in reaction to a modification in the duration of darkness than $AH^{-/-}$ mice. Melatonin did not impact several processes, including the creation of melatonin receptors or longevity [156]. Different mouse strains are utilized for research on the melatoninergic system. These include C57BL/6J (C57Bl) and C3H/HeN (C3H), which exhibit different levels of melatonin and receptor expressions.

Furthermore, C3H mice are characterized by elevated melatonin levels, and C57Bl mice are characterized by decreased melatonin levels. Moreover, animals with melatonin receptor knockout (KO), namely MT1 KO and MT2 KO mice, are utilized [157]. The study employed mice with melatonin deficiencies (C57Bl) and mice with normal levels (C3H) that had the MT1, MT2, or both receptors removed. Both mice with low melatonin levels and those without receptors had less precise daily locomotion. Movement patterns were affected by the absence of both melatonin receptors. The study found that mice with the MT1 and MT2 genes removed woke up early and had a regular activity pattern. The study found that seasons affected C3H mice's chronotype, activity phase, and daily activity, but not C57Bl mice. Seasonal changes in chronotype, stability, total daily activity, and activity phase length differed in C3H mice but not in C57Bl mice [158]. To evaluate attentional processes, mice lacking the MT2 receptor due to a genetic deletion of the Mtnr1b gene were compared to normal mice of the same litter. An assessment was conducted to evaluate their performance in a translational touchscreen version of the continuous performance task. The study employed Mtnr1b knockout mice crossbred with a C57Bl/6N background strain. Mice that did not have the Mtnr1b gene showed performance levels in the Continuous Performance Test that were largely normal, except for a greater frequency of correct and incorrect alarm responses [152].

Like other organs and tissues in zebrafish, the pineal gland plays a crucial role in melatonin synthesis [159]. The therapeutic properties of melatonin were examined using zebrafish models, yielding valuable insights into the management of many illnesses. The potential of melatonin as a treatment for neurodegenerative, cardiovascular, metabolic, psychiatric, and age-related diseases is emphasized by its ability to reduce oxidative stress, improve mitochondrial function, regulate gene expression, and enhance physiological and cognitive abilities. Conducting more studies is crucial for thoroughly comprehending the fundamental mechanisms and translating these findings into practical treatments for humans [160].

Due to their genetic similarity to humans, the availability of transgenic strains, and their well-characterized physiology, mice and rats are highly versatile rodent models. They are ideal for studying systemic effects, long-term behavioral changes, and complex interactions within the brain. Nevertheless, these models present ethical considerations, substantial expenses, and the intricacy of genetic alterations. Additionally, the dissimilarities in melatonin metabolism between rats and humans might restrict the applicability of the findings. However, zebrafish models provide a cost-effective, genetically manipulable alternative with fast growth and translucent embryos that enable in vivo imaging. They are especially valuable for examining developmental processes and conducting high-throughput drug screening. Although zebrafish have significant benefits, the physiological and anatomical disparities between zebrafish and mammals might constrain certain translational investigations.

Cell culture models are essential for melatonin research, providing controlled environments to study cellular and molecular mechanisms. These models help elucidate melatonin's effects on various cellular processes, from gene expression to neuroprotection. Primary neuronal cultures are derived directly from brain tissues like the hippocampus, cortex, or pineal gland. These cultures retain many of the neurons' in vivo properties, making them ideal for studying neurophysiological processes. In primary cultured hippocampal neurons derived from the hippocampus of adult male C57BL/6 mice, the length and arborization of neurites were significantly augmented by melatonin treatment [161]. AD rat model-derived primary hippocampal cultures were used to assess neuronal toxicity and the neuroprotective effect of melatonin against Aβ1–42-induced toxicity [162]. Melatonin enhanced the GABA-induced current in cultured rat hippocampal neurons in vitro, suggesting a potential pathway for the neuroprotective effects of melatonin [163].

Secondly, immortalized cell lines, such as SH-SY5Y neuroblastoma cells or PC12 pheochromocytoma cells, offer a consistent and replicable model for melatonin research. These cell lines can be cultured indefinitely, providing a stable platform for long-term studies. In a study where 24 melatonin analogs were synthesized as novel monoamine oxidase-B inhibitors to counteract oxidative stress in neuronal PC12 cells, 2 of them showed promising results as multi-targeted candidates for oxidative stress-related PD therapy [164]. Oxaliplatin induces cytotoxicity and apoptosis in SH-SY5Y human neuroblastoma cells, which can be rescued by pre-treatment with melatonin through its antioxidant and free-radical scavenging properties [165]. To investigate the neuroprotective role of melatonin in the modulation of the Sirt1/FoxO1/Rab7 axis in HT22 cells and organotypic hippocampal cultures exposed to oxygen-glucose deprivation followed by reoxygenation, a study was conducted. The study provides novel information regarding the neuroprotective role of melatonin against ischemic injury through the activation of the Sirt1/FoxO1/Rab7 axis in in vitro and ex vivo models [166]. Singrang and colleagues investigated the effects of oxygen-glucose deprivation and reoxygenation on the expression of beta-site amyloid precursor protein-cleaving enzyme 1 (BACE1) and Aβ in SH-SY5Y cells. They explored the potential neuroprotective effects of melatonin in hypoxic environments and concluded that hypoxia regulates the amyloid pathway, while melatonin decreases hypoxia conditions and APP processing [167].

In addition, organotypic brain slices are ex vivo cultures that maintain the three-dimensional architecture and cell diversity of specific brain regions. They bridge the gap between in vitro cell cultures and in vivo animal models. A study evaluated the possible effects of melatonin on microglial cells in organotypic cultures of hippocampal slices prepared from postnatal day 8–10 C57BL/6 wild-type mouse pups, revealing an increase in SIRT1 expression and a shift towards a pro-inflammatory phenotype [168]. Researchers investigated the effect of melatonin on cholinergic neurons in the nucleus basalis of Meynert by treating slices of brain tissue with varying amounts of melatonin. They quantified the number of cholinergic neurons in each slice and found that melatonin increased the number of surviving cholinergic neurons in a dose-dependent manner, meaning that higher doses resulted in greater neuronal survival. Additionally, they identified the presence of two proteins, SIRT1 and MT1A, within the cholinergic neurons, which are implicated in melatonin's protective effects, demonstrating that melatonin promotes the survival of cholinergic neurons in the nucleus basalis of Meynert and confirms the involvement of SIRT1 and MT1A proteins in this process [169].

Exploring the exciting potential of organoids, three-dimensional microtissues were grown from patient cells that mimic in vivo physiology and genetic diversity. These remarkable structures, initiated from cryopreserved material, have been developed for various tissues, including the brain, small intestine, colon, and lung, opening new avenues in neuroscience research [170]. A study was undertaken using in vitro methods to examine the impact of short peptides on the pineal gland. The influence of small peptides on the synthesis of signaling molecules was confirmed by employing an organotypic culture of the pineal gland obtained from 3-month-old rats. The peptides Ala-Glu-Asp-Gly and Lys-Glu-Asp increase the synthesis of the proliferative protein Ki-67 in pineal gland cultures. The peptide Ala-Glu-Asp-Gly stimulates the growth and secretion of pinealocytes, which can restore the pineal gland's functioning at the molecular level [171].

In vitro models are crucial instruments for melatonin research, each with unique advantages and disadvantages. Neuronal cultures obtained directly from brain tissues, known as primary neuronal cultures, maintain several characteristics of neurons as they exist in living organisms. This makes them suitable for investigating neurophysiological processes, synaptic plasticity, and neuroprotection. Nevertheless, preserving these cultures can pose difficulties, have a restricted duration, and fail to completely reproduce the intricate conditions in the whole brain. Immortalized cell lines, such as SH-SY5Y and PC12, provide a robust and reproducible platform for long-term investigations since they are easy to maintain and may be cultured indefinitely. However, they frequently possess only part of the neuronal characteristics and reactions observed in primary cells, which may restrict their ability to mimic in vivo settings accurately. Organotypic brain slices preserve the structural integrity of certain brain areas and cellular heterogeneity, connecting in vitro and in vivo models. These systems are useful in researching neurotoxicity, synapse function, and electrophysiological features. However, they need technical expertise for their preparation, maintenance, and survivability compared to other in vitro systems. Organoids, microtissues generated from patient cells, accurately replicate the physiological and genetic characteristics of living organisms. This makes them highly valuable for investigating human-specific reactions and the causes behind diseases. Nevertheless, the standardization of organoids might pose difficulties, and their growth is characterized by a higher level of complexity and time consumption compared to regular cell cultures.

Technological advancements constantly improve our capacity to investigate the correlation between melatonin and the hippocampus. CRISPR-Cas9 gene editing is an emerging technology that is making important contributions to this field. It enables precise genetic alterations in diverse models, such as creating gene knockouts or insertions. This method is highly advantageous for investigating the functions of melatonin receptors and associated genes, providing a flexible tool for both in vivo and in vitro models. Advanced imaging techniques, including two-photon microscopy and optogenetics, allow direct observation and control of neural activity in live tissues. These methods provide a precise and thorough study of cellular and synaptic processes with great resolution in both space and time. Automated high-throughput screening technologies expedite the quick examination of extensive collections of substances, such as melatonin analogs, hastening the identification of promising medicinal substances and uncovering new pathways affected by melatonin. ScRNA-seq enables the examination of gene expression in individual cells, providing insights into the diversity of cellular reactions to melatonin and identifying distinct alterations in gene expression. Organoid technology, based on stem cells, generates intricate, three-dimensional replicas of human tissues, such as the brain. This provides a biologically accurate environment for investigating disease mechanisms and treatment reactions. Together, these technologies enhance our comprehension of melatonin's influence on the hippocampus, allowing for a more precise simulation of clinical disorders and evaluation of therapeutic approaches.

20.8 Unraveling the Complex Interplay Between Melatonin and Hippocampal Function: Current Insights and Future Directions

The study of the relationship between melatonin and the function of the hippocampus is a multifaceted and varied area of research that has the potential to improve our understanding of cognitive processes significantly. Even though there has been significant progress, it is of the utmost importance to address the numerous unanswered questions and choose the potential direction for further study. Keeping this in mind will ensure that we can completely understand this complex connection and continue to demonstrate our commitment to the development of our region.

A significant amount of research has been conducted on the effects of melatonin on the function of the hippocampus; moreover, the precise molecular and cellular processes at play still need to be determined. Regarding continuing scientific initiatives, the significance of conducting further research that affects the general population must be considered. How melatonin receptors promote various intracellular signaling pathways in hippocampal neurons needs to be further explored. It is important to fully understand melatonin's specific role in regulating the proliferation of new hippocampal neurons throughout the day. Examining the effects of melatonin on the hippocampus's capacity for new neuronal formation, specialization, and survival at different times of the day is crucial. Longitudinal research is crucial for understanding how melatonin affects the development and function of the hippocampus over time. This is particularly relevant for elderly individuals and those with neurological disorders. Such research can enhance our understanding of melatonin's potential therapeutic benefits and risks [18].

Further studies are needed to investigate role of melatonin in regulation of synaptic plasticity in the hippocampus, particularly concerning LTP and LTD. This will provide a more comprehensive understanding of how melatonin impacts learning and memory processes. Exploring the interactions between melatonin and other neurotransmitters and hormones, such as dopamine, serotonin, and cortisol, in modulating hippocampal function is an exciting and promising area of current research. Melatonin is produced by the pineal gland and released into the bloodstream [32]. Gathering extensive data on how hormones affect cognitive function is essential to gain new insights into hormone regulation. Although our understanding has grown, significant gaps still need to be filled before meaningful advancements can be made in this field.

The wide array of potential differences in research designs, techniques for melatonin dispersion, and animal models might lead to conflicting conclusions because of the diversity in research methodology. To improve the reliability and comparability of the study's findings, it is possible to ensure that the methods used throughout the research endeavor are consistent. The translational gap is a significant obstacle because of melatonin's varied physiological properties and the hippocampus's diverse roles in different biological species. However, this obstacle may be overcome. The rationale for this is the pressing need to enhance the methodology for extrapolating research findings from animal studies to human participants. Additional translational research is required to address these knowledge gaps [32, 172].

Most research on how melatonin affects the hippocampus's function has been carried out using animal models. Consequently, this underscores the significance of doing additional research involving human participants, particularly clinical trials, in order to confirm the findings of preclinical studies and evaluate the consequences of such discoveries for the health and well-being of humans. As a result of the complicated nature of the influence that melatonin has on a variety of different physiological systems, there are several challenges that must be overcome to determine the exact impact of melatonin on the hippocampus. Developing and employing biochemically thorough methodologies encompassing melatonin's physicochemical characteristics is essential to achieve this goal. Additionally, the release of melatonin, which may be affected by factors like age, light exposure, and genetic differences, can impact the outcome of conducted investigations. These factors include age, light exposure, and genetic variations. Consequently, it is of the utmost importance to incorporate these aspects into the research design and to implement measures to ensure their control [146, 173].

Various potential avenues for further study have been suggested to tackle lingering inquiries and issues. Researchers may use modern neurodegeneration techniques like functional magnetic resonance imaging and positron emission tomography to study how melatonin affects the connections and activity of the hippocampus. There are alternative legitimate approaches to this investigation, such as clinical trials involving human subjects and use of animal models. By investigating genetic and epigenetic mechanisms that might impact the hippocampus, we can gain a deeper understanding of individual variations and identify potential therapeutic targets. Investigating chronotherapy, which delivers melatonin during certain circadian phases, might enhance its benefits for hippocampal function and cognitive well-being. Multi-omics methodologies, encompassing genomics, proteomics, and metabolomics, might offer valuable insights into the influence of melatonin on hippocampal function across multiple biological levels. Research from many disciplines, such as chronobiology, neurology, psychology, and clinical-based sciences, has contributed to a more profound comprehension of the relationship between melatonin and the hippocampus. To investigate the impact of melatonin supplementation on hippocampal function, cognitive performance, and mental health across diverse populations, it is essential to undertake clinical trials that incorporate many factors. Researchers must investigate the impacts of melatonin due to variations in environmental and lifestyle variables, such as light exposure, sleep habits, and diet, across individuals [30, 146, 173, 174].

Researching the complex and multifaceted relationship between melatonin and the hippocampus is essential for acquiring a more thorough understanding of cognitive processes and developing innovative treatments. The breakthroughs in chronobiology and brain function may expedite the rate of progress in future health concerns in this field.

20.9 Conclusion

Extensive research into the role of melatonin in hippocampal function has yielded profound insights into its multifaceted impact on brain health, linking the fields of chronobiology and cognitive neuroscience. Melatonin, a hormone widely recognized for regulating circadian rhythms and sleep-wake cycles, is increasingly considered critical in various physiological processes beyond its traditional scope. The hormone is produced mainly by the pineal gland and, to a lesser extent, in other body tissues like the retina, digestive system, and immune cells, where it functions in an autocrine or paracrine manner. Its role in the brain, particularly in the hippocampus, has been the focus of significant research, revealing its extensive effects on memory formation, emotion regulation, and spatial navigation. One of the most important findings on this topic is melatonin's powerful neuroprotective properties. As a powerful antioxidant, melatonin effectively scavenges free radicals, reducing oxidative stress, a major contributor to cellular damage and aging. This ability is particularly important in the hippocampus, a brain region highly susceptible to oxidative stress and neurodegeneration. The antioxidant effects of melatonin not only protect neurons from oxidative damage but also play a vital role in reducing the effects of ischemic injury, a common consequence of stroke and other types of brain trauma. This neuroprotective effect is further enhanced by melatonin's ability to modulate the immune response, reduce inflammation, and block apoptotic pathways, thus preserving neuronal integrity and function.

In addition to its antioxidant properties, melatonin has been shown to promote neurogenesis in the hippocampus. Neurogenesis, the process by which new neuronal cells are formed in the brain, is essential for learning, memory, and general cognitive health. This process occurs predominantly in the dentate gyrus, a part of the hippocampus, and is crucial for maintaining brain plasticity and adaptability to new information and experiences. Melatonin's ability to enhance neurogenesis suggests its therapeutic potential in the treatment of neuropsychiatric disorders such as depression and anxiety. These disorders are often associated with reduced neurogenesis. Melatonin may help improve cognitive function and emotional resilience by promoting the growth and integration of new neurons.

Furthermore, melatonin's effects extend to regulating circadian rhythms, critical for synchronizing the body's internal processes with the external environment. This synchronization affects sleep patterns and cognitive functions such as attention, memory consolidation, and mood regulation. Disruptions in circadian rhythms, often caused by shift work, jet lag, or irregular sleep schedules, can increase susceptibility to cognitive impairments and mood disorders. Melatonin's role in regulating these rhythms highlights its importance in maintaining mental well-being and suggests its utility in therapeutic strategies aimed at improving sleep and cognitive function.

The therapeutic potential of melatonin is particularly promising in the context of neurodegenerative diseases, including AD, PD, and other forms of dementia. For example, in AD, melatonin has been found to inhibit the aggregation of Aβ peptides and tau proteins, which are hallmarks of the disease. This inhibition could potentially slow disease progression and improve cognitive outcomes. Furthermore, melatonin's anti-inflammatory and neuroprotective properties further contribute to its potential as a treatment option for these debilitating conditions. Its ability to cross the blood-brain barrier, combined with a favorable safety profile, makes melatonin an attractive candidate for long-term use in managing neurodegenerative diseases.

In conclusion, findings from extensive studies on the potential role of melatonin in hippocampal function highlight the critical importance of the hormone in brain health. Melatonin's multifaceted effects, ranging from antioxidant and neuroprotective properties to promoting neurogenesis and regulating circadian rhythms, highlight its potential as a versatile therapeutic agent. The implications of these findings are far-reaching and suggest new avenues for the treatment of cognitive and emotional disorders, neurodegenerative diseases, and conditions associated with disrupted circadian rhythms. Therefore, continued research is urgently needed to elucidate further the mechanisms underlying melatonin's effects and to develop optimized therapeutic protocols that utilize its full potential. Future studies, particularly large-scale clinical trials, will be crucial in determining the efficacy and safety of melatonin in various clinical settings and paving the way for its broader application in medicine and public health. Through such efforts, melatonin may emerge as a cornerstone in treating and managing a wide range of neurological and psychiatric disorders, ultimately improving the quality of life of many individuals.

Acknowledgements We want to thank Hacer Kerime Taşdemir for her contribution to creating the illustrations for this chapter. Her work has enhanced the presentation of the concepts discussed.

References

1. Savage RA, Zafar N, Yohannan S, Miller J-MM. Melatonin. StatPearls; 2024.
2. Arendt J, Aulinas A. Physiology of the pineal gland and melatonin. Endotext; 2022.
3. Joseph TT, Schuch V, Hossack DJ, Chakraborty R, Johnson EL. Melatonin: the placental antioxidant and anti-inflammatory. Front Immunol. 2024;15:1339304. https://doi.org/10.3389/FIMMU.2024.1339304/BIBTEX.
4. Monteiro KKAC, Shiroma ME, Damous LL, Simões MJ, Simões RS, Cipolla-Neto J, et al. Antioxidant actions of melatonin: a systematic review of animal studies. Antioxidants. 2024;13:439. https://doi.org/10.3390/ANTIOX13040439.
5. Carrillo-Vico A, García-Pergañeda A, Naji L, Calvo JR, Romero MP, Guerrero JM. Expression of membrane and nuclear melatonin receptor mRNA and protein in the mouse immune system. Cell Mol Life Sci. 2003;60:2272–8. https://doi.org/10.1007/S00018-003-3207-4/METRICS.
6. Guerrero J, Reiter R. Melatonin-immune system relationships. Curr Top Med Chem. 2005;2:167–79. https://doi.org/10.2174/1568026023394335.
7. Won E, Na KS, Kim YK. Associations between melatonin, neuroinflammation, and brain alterations in depression. Int J Mol Sci. 2022;23:305. https://doi.org/10.3390/IJMS23010305.
8. Tonon AC, Pilz LK, Markus RP, Hidalgo MP, Elisabetsky E. Melatonin and depression: a translational perspective from animal models to clinical studies. Front Psych. 2021;12:638981. https://doi.org/10.3389/FPSYT.2021.638981/BIBTEX.
9. Olcese JM. Melatonin and female reproduction: an expanding universe. Front Endocrinol (Lausanne). 2020;11:515354. https://doi.org/10.3389/FENDO.2020.00085/BIBTEX.
10. Gelen V, Şengül E, Kükürt A, Gelen V, Şengül E, Kükürt A. An overview of effects on reproductive physiology of melatonin. In: Melatonin—recent updates. IntechOpen; 2022. https://doi.org/10.5772/INTECHOPEN.108101.
11. Li Y, Li S, Zhou Y, Meng X, Zhang J-J, Xu D-P, et al. Melatonin for the prevention and treatment of cancer. Oncotarget. 2017;8:39896–921. https://doi.org/10.18632/ONCOTARGET.16379.
12. Talib WH, Alsayed AR, Abuawad A, Daoud S, Mahmod AI. Melatonin in cancer treatment: current knowledge and future opportunities. Molecules. 2021;26:2506. https://doi.org/10.3390/MOLECULES26092506.
13. Berger JI, Billig AJ, Sedley W, Kumar S, Griffiths TD, Gander PE. What is the role of the hippocampus and parahippocampal gyrus in the persistence of tinnitus? Hum Brain Mapp. 2024;45:e26627. https://doi.org/10.1002/HBM.26627.
14. Pronier É, Morici JF, Girardeau G. The role of the hippocampus in the consolidation of emotional memories during sleep. Trends Neurosci. 2023;46:912–25. https://doi.org/10.1016/j.tins.2023.08.003.
15. Cardinali DP. Melatonin: clinical perspectives in neurodegeneration. Front Endocrinol (Lausanne). 2019;10:447671. https://doi.org/10.3389/FENDO.2019.00480/BIBTEX.
16. Ikram M, Park HY, Ali T, Kim MO. Melatonin as a potential regulator of oxidative stress, and neuroinflammation: mechanisms and implications for the management of brain injury-induced neurodegeneration. J Inflamm Res. 2021;14:6251. https://doi.org/10.2147/JIR.S334423.
17. Leung JWH, Cheung KK, Ngai SPC, Tsang HWH, Lau BWM. Protective effects of melatonin on neurogenesis impairment in neurological disor-

ders and its relevant molecular mechanisms. Int J Mol Sci. 2020;21:1–30. https://doi.org/10.3390/IJMS21165645.
18. Romero A, Morales-García J, Ramos E. Melatonin: a multitasking indoleamine to modulate hippocampal neurogenesis. Neural Regen Res. 2023;18:503. https://doi.org/10.4103/1673-5374.350189.
19. do Amaral FG, Cipolla-Neto J. A brief review about melatonin, a pineal hormone. Arch Endocrinol Metab. 2018;62:472–9. https://doi.org/10.20945/2359-3997000000066.
20. Hagemeister M, Hamilton L, Wandrey N, Hill M, Mounce E, Mosel N, et al. Evaluation of rhodanine indolinones as AANAT inhibitors. ChemMedChem. 2024;19:e202300567. https://doi.org/10.1002/cmdc.202300567.
21. Reiter RJ, Sharma R, Tan D-X, Chuffa LGA, da Silva DGH, Slominski AT, et al. Dual sources of melatonin and evidence for different primary functions. Front Endocrinol (Lausanne). 2024;15:1414463. https://doi.org/10.3389/fendo.2024.1414463.
22. Felder-Schmittbuhl MP, Hicks D, Ribelayga CP, Tosini G. Melatonin in the mammalian retina: synthesis, mechanisms of action and neuroprotection. J Pineal Res. 2024;76:e12951. https://doi.org/10.1111/jpi.12951.
23. Megha KB, Arathi A, Shikha S, Alka R, Ramya P, Mohanan PV. Significance of melatonin in the regulation of circadian rhythms and disease management. Mol Neurobiol. 2024;61:5541–71. https://doi.org/10.1007/s12035-024-03915-0.
24. Martínez Soriano F, Blasco-Serra A, González-Soler EM, Hernández-Sánchez S, Valverde Navarro AA. Topographic morphometry of the pineal gland of the rat. A 24-hours period, light-dark cycle and seasonal study. Eur J Anat. 2023;27:587–99. https://doi.org/10.52083/XDNS9336.
25. Blancas-Velazquez AS, Bering T, Bille S, Rath MF. Role and neural regulation of clock genes in the rat pineal gland: *Clock* modulates amplitude of rhythmic expression of *Aanat* encoding the melatonin-producing enzyme. J Pineal Res. 2023;75:12893. https://doi.org/10.1111/jpi.12893.
26. Mattern KM-J, Blancas-Velázquez AS, Ngo MT, Bille S, Hertz H, Bering T, et al. The ISL LIM-homeobox 2 transcription factor is negatively regulated by circadian adrenergic signaling to repress the expression of *Aanat* in pinealocytes of the rat pineal gland. J Pineal Res. 2023;75:12905. https://doi.org/10.1111/jpi.12905.
27. Yang H, Zhang Z, Ding X, Jiang X, Tan L, Lin C, et al. *RP58* knockdown contributes to hypoxia-ischemia-induced pineal dysfunction and circadian rhythm disruption in neonatal rats. J Pineal Res. 2023;75:e12885. https://doi.org/10.1111/jpi.12885.
28. Zheng J, Song W, Zhou Y, Li X, Wang M, Zhang C. Cross-species single-cell landscape of vertebrate pineal gland. J Pineal Res. 2024;76:e12927. https://doi.org/10.1111/jpi.12927.
29. Gonciarz M, Lombard I, Konecki L, Gietka P, Pajdo R, Brzozowski T. Potential therapeutic role of melatonin in hepatobiliary diseases: current evidence and clinical observations. J Physiol Pharmacol. 2023;74(2):109–30.
30. Bonetti M, Giugno L, Borsani E, Bonomini F. Potential neuroprotective effect of melatonin in the hippocampus of male BTBR mice. Nutrients. 2024;16:1652. https://doi.org/10.3390/nu16111652.
31. Del Casale A, Arena JF, Giannetti F, Minervino A, Biggio G, Girardi P. The use of prolonged-release melatonin in circadian medicine: a systematic review. Minerva Med. 2024;115:125–42. https://doi.org/10.23736/S0026-4806.24.09303-0.
32. Ahmad SB, Ali A, Bilal M, Rashid SM, Wani AB, Bhat RR, et al. Melatonin and health: insights of melatonin action, biological functions, and associated disorders. Cell Mol Neurobiol. 2023;43:2437–58. https://doi.org/10.1007/s10571-023-01324-w.
33. Lu J, Zou R, Yang Y, Bai X, Wei W, Ding R, et al. Association between nocturnal light exposure and melatonin in humans: a meta-analysis. Environ Sci Pollut Res. 2023;31:3425–34. https://doi.org/10.1007/s11356-023-31502-8.
34. Chauhan P, Jethwa K, Rathawa A, Chauhan G, Mehra S. The anatomy of the hippocampus. In: Cerebral ischemia. Exon Publications; 2021. p. 17–30. https://doi.org/10.36255/exonpublications.cerebralischemia.2021.hippocampus.
35. Reiter RJ, Sharma R, Cucielo MS, Tan DX, Rosales-Corral S, Gancitano G, et al. Brain washing and neural health: role of age, sleep, and the cerebrospinal fluid melatonin rhythm. Cell Mol Life Sci. 2023;80:88. https://doi.org/10.1007/s00018-023-04736-5.
36. Palomero-Gallagher N, Kedo O, Mohlberg H, Zilles K, Amunts K. Multimodal mapping and analysis of the cyto- and receptor architecture of the human hippocampus. Brain Struct Funct. 2020;225:881–907. https://doi.org/10.1007/S00429-019-02022-4/FIGURES/14.
37. Hildebrandt-Einfeldt L, Yap K, Paul MH, Stoffer C, Zahn N, Drakew A, et al. Crossed entorhino-dentate projections form and terminate with correct layer-specificity in organotypic slice cultures of the mouse hippocampus. Front Neuroanat. 2021;15:637036. https://doi.org/10.3389/FNANA.2021.637036.
38. Lübbers K, Frotscher M. Differentiation of granule cells in relation to GABAergic neurons in the rat fascia dentata—combined Golgi/EM and immunocytochemical studies. Anat Embryol (Berl). 1988;178:119–27. https://doi.org/10.1007/BF02463645/METRICS.
39. Parmar SK, Pruthi N, Ravindranath R, Ravindranath Y, Somanna S, Philip M. Anatomical variations of the temporomesial structures in normal adult brain—a cadaveric study. J Neurosci Rural Pract. 2018;9:317. https://doi.org/10.4103/JNRP.JNRP_73_18.
40. Fogwe LA, Reddy V, Mesfin FB. Neuroanatomy, hippocampus. StatPearls; 2023.

41. Knierim JJ. The hippocampus. Curr Biol. 2015;25:R1116–21. https://doi.org/10.1016/j.cub.2015.10.049.
42. Rao YL, Ganaraja B, Murlimanju BV, Joy T, Krishnamurthy A, Agrawal A. Hippocampus and its involvement in Alzheimer's disease: a review. 3 Biotech. 2022; https://doi.org/10.1007/s13205-022-03123-4.
43. Hao ZY, Zhong Y, Ma ZJ, Xu HZ, Kong JY, Wu Z, et al. Abnormal resting-state functional connectivity of hippocampal subfields in patients with major depressive disorder. BMC Psychiatry. 2020;20:71. https://doi.org/10.1186/S12888-020-02490-7.
44. González-Arnay E, Pérez-Santos I, Jiménez-Sánchez L, Cid E, Gal B, de la Prida LM, et al. Immunohistochemical field parcellation of the human hippocampus along its antero-posterior axis. Brain Struct Funct. 2024;229:359–85. https://doi.org/10.1007/S00429-023-02725-9.
45. Robert V, Cassim S, Chevaleyre V, Piskorowski RA. Hippocampal area CA2: properties and contribution to hippocampal function. Cell Tissue Res. 2018;373:525–40. https://doi.org/10.1007/S00441-017-2769-7.
46. Cherubini E, Miles R. The CA3 region of the hippocampus: how is it? What is it for? How does it do it? Front Cell Neurosci. 2015;9:126449. https://doi.org/10.3389/FNCEL.2015.00019/BIBTEX.
47. Mercer A, Thomson AM. Cornu ammonis regions-antecedents of cortical layers? Front Neuroanat. 2017;11:83. https://doi.org/10.3389/fnana.2017.00083.
48. O'Mara S. The subiculum: what it does, what it might do, and what neuroanatomy has yet to tell us. J Anat. 2005;207:271. https://doi.org/10.1111/J.1469-7580.2005.00446.X.
49. O'Mara SM, Commins S, Anderson M, Gigg J. The subiculum: a review of form, physiology and function. Prog Neurobiol. 2001;64:129–55. https://doi.org/10.1016/S0301-0082(00)00054-X.
50. McNaughton N. The role of the subiculum within the behavioural inhibition system. Behav Brain Res. 2006;174:232–50. https://doi.org/10.1016/j.bbr.2006.05.037.
51. Zhu Y, Gao H, Tong L, Li ZL, Wang L, Zhang C, et al. Emotion regulation of hippocampus using real-time fMRI neurofeedback in healthy human. Front Hum Neurosci. 2019;13:444712. https://doi.org/10.3389/FNHUM.2019.00242/BIBTEX.
52. Yarwood M. The relationship between emotions and common brain structures hippocampus. In: Anxiety, depression and emotion. Oxford University Press; 2000. p. 52.
53. Richter-Levin G. The amygdala, the hippocampus, and emotional modulation of memory. Neuroscientist. 2004;10:31–9. https://doi.org/10.1177/1073858403259955.
54. Kamali A, Milosavljevic S, Gandhi A, Lano KR, Shobeiri P, Sherbaf FG, et al. The cortico-limbo-thalamo-cortical circuits: an update to the original Papez circuit of the human limbic system. Brain Topogr. 2023;36:371–89. https://doi.org/10.1007/s10548-023-00955-y.
55. Romero-Barragán MT, Gruart A, Delgado-García JM. Transsynaptic long-term potentiation in the hippocampus of behaving mice. Front Synaptic Neurosci. 2022;13:811806. https://doi.org/10.3389/FNSYN.2021.811806/BIBTEX.
56. Dringenberg HC. The history of long-term potentiation as a memory mechanism: controversies, confirmation, and some lessons to remember. Hippocampus. 2020;30:987–1012. https://doi.org/10.1002/HIPO.23213.
57. Grosser S, Buck N, Braunewell KH, Gilling KE, Wozny C, Fidzinski P, et al. Loss of long-term potentiation at hippocampal output synapses in experimental temporal lobe epilepsy. Front Mol Neurosci. 2020;13:554949. https://doi.org/10.3389/FNMOL.2020.00143/BIBTEX.
58. Stacho M, Manahan-Vaughan D. The intriguing contribution of hippocampal long-term depression to spatial learning and long-term memory. Front Behav Neurosci. 2022;16:806356. https://doi.org/10.3389/FNBEH.2022.806356.
59. Contreras MP, Fechner J, Born J, Inostroza M. Accelerating maturation of spatial memory systems by experience: evidence from sleep oscillation signatures of memory processing. J Neurosci. 2023;43:3509–19. https://doi.org/10.1523/JNEUROSCI.1967-22.2023.
60. Wood R, Chan D. The hippocampus, spatial memory and Alzheimer's disease. Adv Clin Neurosci Rehabil. 2015; https://doi.org/10.47795/PAEA2610.
61. King JA, Trinkler I, Hartley T, Vargha-Khadem F, Burgess N. The hippocampal role in spatial memory and the familiarity-recollection distinction: a case study. Neuropsychology. 2004;18:405–17. https://doi.org/10.1037/0894-4105.18.3.405.
62. Zhu Y, Gao H, Tong L, Li ZL, Wang L, Zhang C, et al. Emotion regulation of hippocampus using real-time fMRI neurofeedback in healthy human. Front Hum Neurosci. 2019;13:242. https://doi.org/10.3389/FNHUM.2019.00242.
63. Marek J, Maj E, Przybyla OK, Skrzynski W, Pasicz K, Fabiszewska E, et al. The impact of studying on the hippocampal volume in medical students and its correlation with the results of the Final Medical Examination: a single-centre, prospective observational cohort study. Pol J Radiol. 2023;88:e22. https://doi.org/10.5114/PJR.2023.124433.
64. Mok RM, Love BC. A multilevel account of hippocampal function in spatial and concept learning: bridging models of behavior and neural assemblies. Sci Adv. 2023;9:6903. https://doi.org/10.1126/SCIADV.ADE6903.
65. Ghasemi M, Navidhamidi M, Rezaei F, Azizikia A, Mehranfard N. Anxiety and hippocampal neuronal activity: relationship and potential mechanisms. Cogn Affect Behav Neurosci. 2021;22:431–49. https://doi.org/10.3758/S13415-021-00973-Y.

66. Dale E, Pehrson AL, Jeyarajah T, Li Y, Leiser SC, Smagin G, et al. Effects of serotonin in the hippocampus: how SSRIs and multimodal antidepressants might regulate pyramidal cell function. CNS Spectr. 2016;21:143–61. https://doi.org/10.1017/S1092852915000425.
67. Shi HJ, Wang S, Wang XP, Zhang RX, Zhu LJ. Hippocampus: molecular, cellular, and circuit features in anxiety. Neurosci Bull. 2023;39:1009–26. https://doi.org/10.1007/S12264-023-01020-1.
68. Altman J. Autoradiographic investigation of cell proliferation in the brains of rats and cats. Anat Rec. 1963;145:573–91. https://doi.org/10.1002/ar.1091450409.
69. Altman J, Das GD. Autoradiographic and histological evidence of postnatal hippocampal neurogenesis in rats. J Comp Neurol. 1965;124:319–35. https://doi.org/10.1002/cne.901240303.
70. Nottebohm F. A brain for all seasons: cyclical anatomical changes in song control nuclei of the canary brain. Science (1979). 1981;214:1368–70. https://doi.org/10.1126/science.7313697.
71. Eriksson PS, Perfilieva E, Björk-Eriksson T, Alborn A-M, Nordborg C, Peterson DA, et al. Neurogenesis in the adult human hippocampus. Nat Med. 1998;4:1313–7.
72. Gonçalves JT, Schafer ST, Gage FH. Adult neurogenesis in the hippocampus: from stem cells to behavior. Cell. 2016;167:897–914. https://doi.org/10.1016/j.cell.2016.10.021.
73. Kempermann G. Environmental enrichment, new neurons and the neurobiology of individuality. Nat Rev Neurosci. 2019;20:235–45. https://doi.org/10.1038/s41583-019-0120-x.
74. Gebara E, Bonaguidi MA, Beckervordersandforth R, Sultan S, Udry F, Gijs PJ, et al. Heterogeneity of radial glia-like cells in the adult hippocampus. Stem Cells. 2016;34:997–1010. https://doi.org/10.1002/STEM.2266.
75. Sierra A, Martín-Suárez S, Valcárcel-Martín R, Pascual-Brazo J, Aelvoet SA, Abiega O, et al. Neuronal hyperactivity accelerates depletion of neural stem cells and impairs hippocampal neurogenesis. Cell Stem Cell. 2015;16:488. https://doi.org/10.1016/J.STEM.2015.04.003.
76. Shin J, Berg DA, Zhu Y, Shin JY, Song J, Bonaguidi MA, et al. Single-cell RNA-Seq with waterfall reveals molecular cascades underlying adult neurogenesis. Cell Stem Cell. 2015;17:360–72. https://doi.org/10.1016/J.STEM.2015.07.013.
77. Calzolari F, Michel J, Baumgart EV, Theis F, Götz M, Ninkovic J. Fast clonal expansion and limited neural stem cell self-renewal in the adult subependymal zone. Nat Neurosci. 2015;18:490–2. https://doi.org/10.1038/nn.3963.
78. Li G, Fang L, Fernández G, Pleasure SJ. The ventral hippocampus is the embryonic origin for adult neural stem cells in the dentate gyrus. Neuron. 2013;78:658–72. https://doi.org/10.1016/j.neuron.2013.03.019.
79. Anacker C, Hen R. Adult hippocampal neurogenesis and cognitive flexibility-linking memory and mood. Nat Rev Neurosci. 2017;18:335–46. https://doi.org/10.1038/nrn.2017.45.
80. Toda T, Parylak SL, Linker SB, Gage FH. The role of adult hippocampal neurogenesis in brain health and disease. Mol Psychiatry. 2019;24:67–87. https://doi.org/10.1038/s41380-018-0036-2.
81. Lucassen PJ, Fitzsimons CP, Salta E, Maletic-Savatic M. Adult neurogenesis, human after all (again): classic, optimized, and future approaches. Behav Brain Res. 2020;381:112458. https://doi.org/10.1016/J.BBR.2019.112458.
82. Boldrini M, Underwood MD, Hen R, Rosoklija GB, Dwork AJ, John Mann J, et al. Antidepressants increase neural progenitor cells in the human hippocampus. Neuropsychopharmacology. 2009;34:2376–89. https://doi.org/10.1038/npp.2009.75.
83. Moreno-Jiménez EP, Terreros-Roncal J, Flor-García M, Rábano A, Llorens-Martín M. Evidences for adult hippocampal neurogenesis in humans. J Neurosci. 2021;41:2541–53. https://doi.org/10.1523/JNEUROSCI.0675-20.2020.
84. Cho S, Joh TH, Baik HH, Dibinis C, Volpe BT. Melatonin administration protects CA1 hippocampal neurons after transient forebrain ischemia in rats. Brain Res. 1997;755:335–8. https://doi.org/10.1016/S0006-8993(97)00188-1.
85. Carneiro RCG, Reiter RJ. Melatonin protects against lipid peroxidation induced by δ-aminolevulinic acid in rat cerebellum, cortex and hippocampus. Neuroscience. 1997;82:293–9. https://doi.org/10.1016/S0306-4522(97)00262-5.
86. Albendea CD, Gómez Trullén EM, Fuentes-Broto L, Miana-Mena FJ, Millán-Plano S, Reyes Gonzales MC, et al. Melatonin reduces lipid and protein oxidative damage in synaptosomes due to aluminium. J Trace Elem Med Biol. 2007;21:261–8. https://doi.org/10.1016/j.jtemb.2007.04.002.
87. Hsu C, Chi B, Casida JE. Melatonin reduces phosphine-induced lipid and DNA oxidation in vitro and in vivo in rat brain. J Pineal Res. 2002;32:53–8. https://doi.org/10.1034/j.1600-079x.2002.10809.x.
88. Pang R, Avdic-Belltheus A, Meehan C, Martinello K, Mutshiya T, Yang Q, et al. Melatonin and/or erythropoietin combined with hypothermia in a piglet model of perinatal asphyxia. Brain Commun. 2021;3:211. https://doi.org/10.1093/braincomms/fcaa211.
89. Yawno T, Castillo-Melendez M, Jenkin G, Wallace EM, Walker DW, Miller SL. Mechanisms of melatonin-induced protection in the brain of late gestation fetal sheep in response to hypoxia. Dev Neurosci. 2012;34:543–51. https://doi.org/10.1159/000346323.
90. Li P, Ji X, Shan M, Wang Y, Dai X, Yin M, et al. Melatonin regulates microglial polarization to M2 cell via RhoA/ROCK signaling pathway in epi-

lepsy. Immun Inflamm Dis. 2023;11:900. https://doi.org/10.1002/iid3.900.

91. Carloni S, Facchinetti F, Pelizzi N, Buonocore G, Balduini W. Melatonin acts in synergy with hypothermia to reduce oxygen-glucose deprivation-induced cell death in rat hippocampus organotypic slice cultures. Neonatology. 2018;114:364–71. https://doi.org/10.1159/000491859.
92. Leeboonngam T, Pramong R, Sae-ung K, Govitrapong P, Phansuwan-Pujito P. Neuroprotective effects of melatonin on amphetamine-induced dopaminergic fiber degeneration in the hippocampus of postnatal rats. J Pineal Res. 2018;64:12456. https://doi.org/10.1111/jpi.12456.
93. Pinato L, da Silveira Cruz-Machado S, Franco DG, Campos LMG, Cecon E, Fernandes PACM, et al. Selective protection of the cerebellum against intracerebroventricular LPS is mediated by local melatonin synthesis. Brain Struct Funct. 2015;220:827–40. https://doi.org/10.1007/s00429-013-0686-4.
94. Hoppe JB, Frozza RL, Horn AP, Comiran RA, Bernardi A, Campos MM, et al. Amyloid-β neurotoxicity in organotypic culture is attenuated by melatonin: involvement of GSK-3β, *tau* and neuroinflammation. J Pineal Res. 2010;48:230–8. https://doi.org/10.1111/j.1600-079X.2010.00747.x.
95. Feng Z, Cheng Y, Zhang J. Long-term effects of melatonin or 17 *β*-estradiol on improving spatial memory performance in cognitively impaired, ovariectomized adult rats. J Pineal Res. 2004;37:198–206. https://doi.org/10.1111/j.1600-079X.2004.00158.x.
96. Yang L, Wang J, Deng Y, Gong C, Li Q, Chen Q, et al. Melatonin improves neurological outcomes and preserves hippocampal mitochondrial function in a rat model of cardiac arrest. PLoS One. 2018;13:e0207098. https://doi.org/10.1371/journal.pone.0207098.
97. Parisotto EB, Vidal V, García-Cerro S, Lantigua S, Wilhelm Filho D, Sanchez-Barceló EJ, et al. Chronic melatonin administration reduced oxidative damage and cellular senescence in the hippocampus of a mouse model of Down syndrome. Neurochem Res. 2016;41:2904–13. https://doi.org/10.1007/s11064-016-2008-8.
98. Olcese JM, Cao C, Mori T, Mamcarz MB, Maxwell A, Runfeldt MJ, et al. Protection against cognitive deficits and markers of neurodegeneration by long-term oral administration of melatonin in a transgenic model of Alzheimer disease. J Pineal Res. 2009;47:82–96. https://doi.org/10.1111/j.1600-079X.2009.00692.x.
99. Stewart LS, Leung LS. Hippocampal melatonin receptors modulate seizure threshold. Epilepsia. 2005;46:473–80. https://doi.org/10.1111/j.0013-9580.2005.30204.x.
100. Mei Y, Duan C, Li X, Zhao Y, Cao F, Shang S, et al. Reduction of endogenous melatonin accelerates cognitive decline in mice in a simulated occupational formaldehyde exposure environment. Int J Environ Res Public Health. 2016;13:258. https://doi.org/10.3390/ijerph13030258.
101. Lee S, Chun W, Kong P, Han JA, Cho BP, Kwon O, et al. Sustained activation of Akt by melatonin contributes to the protection against kainic acid-induced neuronal death in hippocampus. J Pineal Res. 2006;40:79–85. https://doi.org/10.1111/j.1600-079X.2005.00283.x.
102. Behan WMH, McDonald M, Darlington LG, Stone TW. Oxidative stress as a mechanism for quinolinic acid-induced hippocampal damage: protection by melatonin and deprenyl. Br J Pharmacol. 1999;128:1754–60. https://doi.org/10.1038/sj.bjp.0702940.
103. Baydas G, Özer M, Yasar A, Tuzcu M, Koz ST. Melatonin improves learning and memory performances impaired by hyperhomocysteinemia in rats. Brain Res. 2005;1046:187–94. https://doi.org/10.1016/j.brainres.2005.04.011.
104. Sotomoyano R, Burgos H, Flores F, Valladares L, Sierralta W, Fernandez V, et al. Melatonin administration impairs visuo-spatial performance and inhibits neocortical long-term potentiation in rats. Pharmacol Biochem Behav. 2006;85:408–14. https://doi.org/10.1016/j.pbb.2006.09.009.
105. Talaei SA, Sheibani V, Salami M. Light deprivation improves melatonin related suppression of hippocampal plasticity. Hippocampus. 2010;20:447–55. https://doi.org/10.1002/hipo.20650.
106. Benleulmi-Chaachoua A, Chen L, Sokolina K, Wong V, Jurisica I, Emerit MB, et al. Protein interactome mining defines melatonin MT_1 receptors as integral component of presynaptic protein complexes of neurons. J Pineal Res. 2016;60:95–108. https://doi.org/10.1111/jpi.12294.
107. Jilg A, Bechstein P, Saade A, Dick M, Li TX, Tosini G, et al. Melatonin modulates daytime-dependent synaptic plasticity and learning efficiency. J Pineal Res. 2019;66:12553. https://doi.org/10.1111/jpi.12553.
108. Sung J-Y, Bae J-H, Lee J-H, Kim Y-N, Kim D-K. The melatonin signaling pathway in a long-term memory in vitro study. Molecules. 2018;23:737. https://doi.org/10.3390/molecules23040737.
109. Liu X-J, Yuan L, Yang D, Han W-N, Li Q-S, Yang W, et al. Melatonin protects against amyloid-β-induced impairments of hippocampal LTP and spatial learning in rats. Synapse. 2013;67:626–36. https://doi.org/10.1002/syn.21677.
110. Suwannakot K, Sritawan N, Naewla S, Aranarochana A, Sirichoat A, Pannangrong W, et al. Melatonin attenuates methotrexate-induced reduction of antioxidant activity related to decreases of neurogenesis in adult rat hippocampus and prefrontal cortex. Oxidative Med Cell Longev. 2022;2022:1–13. https://doi.org/10.1155/2022/1596362.
111. Chen D, Zhang T, Lee TH. Cellular mechanisms of melatonin: insight from neurodegenerative diseases. Biomolecules. 2020;10:1–26. https://doi.org/10.3390/biom10081158.

112. Hosseinzadeh A, Changizi-Ashtiyani S, Koosha F, Amiri S, Karimi-Behnagh A, Mehrzadi S. Melatonin: therapeutic potential for stroke and other neurodegenerative diseases. Melatonin Res. 2023;6:192–34. https://doi.org/10.32794/mr112500144.
113. Wei M, Wu T, Chen N. Bridging neurotrophic factors and bioactive peptides to Alzheimer's disease. Ageing Res Rev. 2024;94:102177. https://doi.org/10.1016/j.arr.2023.102177.
114. Shukla M, Govitrapong P, Boontem P, Reiter RJ, Satayavivad J. Mechanisms of melatonin in alleviating Alzheimer's disease. Curr Neuropharmacol. 2017;15:1010–31. https://doi.org/10.2174/1570159x15666170313123454.
115. Cardinali DP. Melatonin: clinical perspectives in neurodegeneration. Front Endocrinol (Lausanne). 2019;10:480. https://doi.org/10.3389/fendo.2019.00480.
116. Srinivasan V, Pandi-Perumal SR, Maestroni GJM, Esquifino AI, Hardeland R, Cardinali DP. Role of melatonin in neurodegenerative diseases. Neurotox Res. 2005;7:293–318. https://doi.org/10.1007/BF03033887.
117. Wang X, Wang R, Li J. Influence of sleep disruption on protein accumulation in neurodegenerative diseases. Ageing Neurodegen Dis. 2022; https://doi.org/10.20517/and.2021.10.
118. Knight E, Geetha T, Broderick TL, Babu JR. The role of dietary antioxidants and their potential mechanisms in Alzheimer's disease treatment. Metabolites. 2023;13:438. https://doi.org/10.3390/metabo13030438.
119. Polimeni G, Esposito E, Bevelacqua V, Guarneri C, Cuzzocrea S. Role of melatonin supplementation in neurodegenerative disorders. Front Biosci. 2014;19:429–46.
120. Tseng P-T, Zeng B-Y, Chen Y-W, Yang C-P, Su K-P, Chen T-Y, et al. The dose and duration-dependent association between melatonin treatment and overall cognition in Alzheimer's dementia: a network meta-analysis of randomized placebo-controlled trials. Curr Neuropharmacol. 2022;20:1816. https://doi.org/10.2174/1570159X20666220420122322.
121. Laudon M, Wade AG, Farmer M, Harari G, Fund N, Nir T, et al. Add-on prolonged-release melatonin for cognitive function and sleep in mild to moderate Alzheimer's disease: a 6-month, randomized, placebo-controlled, multicenter trial. Clin Interv Aging. 2014;9:947. https://doi.org/10.2147/CIA.S65625.
122. Wang Y, Zheng W, Ng CH, Ungvari GS, Wei W, Xiang Y. Meta-analysis of randomized, double-blind, placebo-controlled trials of melatonin in Alzheimer's disease. Int J Geriatr Psychiatry. 2017;32:50–7. https://doi.org/10.1002/gps.4571.
123. Hossain MF, Uddin MS, Uddin GMS, Sumsuzzman DM, Islam MS, Barreto GE, et al. Melatonin in Alzheimer's disease: a latent endogenous regulator of neurogenesis to mitigate Alzheimer's neuropathology. Mol Neurobiol. 2019;56:8255–76. https://doi.org/10.1007/s12035-019-01660-3.
124. Liu Y, Niu L, Liu X, Cheng C, Le W. Recent progress in non-motor features of Parkinson's disease with a focus on circadian rhythm dysregulation. Neurosci Bull. 2021;37:1010–24. https://doi.org/10.1007/S12264-021-00711-X/FIGURES/2.
125. Radad K, Moldzio R, Krewenka C, Kranner B, Rausch W-D. Pathophysiology of non-motor signs in Parkinson's disease: some recent updating with brief presentation. Open Explor. 2023;3:24–46. https://doi.org/10.37349/ENT.2023.00036.
126. Ma H, Yan J, Sun W, Jiang M, Zhang Y. Melatonin treatment for sleep disorders in Parkinson's disease: a meta-analysis and systematic review. Front Aging Neurosci. 2022;14:784314. https://doi.org/10.3389/fnagi.2022.784314.
127. Haapaniemi TH, Pursiainen V, Korpelainen JT, Huikuri HV, Sotaniemi KA, Myllylä VV. Ambulatory ECG and analysis of heart rate variability in Parkinson's disease. J Neurol Neurosurg Psychiatry. 2001;70(3):305–10.
128. Milanowski J, Kozerawski K, Falęcka W, Dudek D, Lisewska B, Lisewski P, et al. Changes in the secretion of melatonin and selected adipokines during the progression of Parkinson's disease—preliminary studies. Metabolites. 2023;13:668. https://doi.org/10.3390/metabo13050668.
129. Li T, Cheng C, Jia C, Leng Y, Qian J, Yu H, et al. Peripheral clock system abnormalities in patients with Parkinson's disease. Front Aging Neurosci. 2021;13:736026. https://doi.org/10.3389/fnagi.2021.736026.
130. Delgado-Lara DL, González-Enríquez GV, Torres-Mendoza BM, González-Usigli H, Cárdenas-Bedoya J, Macías-Islas MA, et al. Effect of melatonin administration on the PER1 and BMAL1 clock genes in patients with Parkinson's disease. Biomed Pharmacother. 2020;129:110485. https://doi.org/10.1016/j.biopha.2020.110485.
131. Hadi F, Agah E, Tavanbakhsh S, Mirsepassi Z, Mousavi SV, Talachi N, et al. Safety and efficacy of melatonin, clonazepam, and trazodone in patients with Parkinson's disease and sleep disorders: a randomized, double-blind trial. Neurol Sci. 2022;43:6141. https://doi.org/10.1007/s10072-022-06188-4.
132. Daneshvar Kakhaki R, Ostadmohammadi V, Kouchaki E, Aghadavod E, Bahmani F, Tamtaji OR, et al. Melatonin supplementation and the effects on clinical and metabolic status in Parkinson's disease: a randomized, double-blind, placebo-controlled trial. Clin Neurol Neurosurg. 2020;195:105878. https://doi.org/10.1016/j.clineuro.2020.105878.
133. Ahn JH, Kim M, Park S, Jang W, Park J, Oh E, et al. Prolonged-release melatonin in Parkinson's disease patients with a poor sleep quality: a randomized trial. Parkinsonism Relat Disord. 2020;75:50–4. https://doi.org/10.1016/j.parkreldis.2020.03.029.

134. Jiménez-Delgado A, Ortiz GG, Delgado-Lara DL, González-Usigli HA, González-Ortiz LJ, Cid-Hernández M, et al. Effect of melatonin administration on mitochondrial activity and oxidative stress markers in patients with Parkinson's disease. Oxidative Med Cell Longev. 2021; https://doi.org/10.1155/2021/5577541.
135. Givler D, Givler A, Luther PM, Wenger DM, Ahmadzadeh S, Shekoohi S, et al. Chronic administration of melatonin: physiological and clinical considerations. Neurol Int. 2023;15:518–33. https://doi.org/10.3390/neurolint15010031.
136. Wirtz PH, Spillmann M, Bärtschi C, Ehlert U, Von Känel R. Oral melatonin reduces blood coagulation activity: a placebo-controlled study in healthy young men. J Pineal Res. 2008;44(2):127–33. https://doi.org/10.1111/J.1600-079X.2007.00499.X.
137. Tabrizi SJ, Flower MD, Ross CA, Wild EJ. Huntington disease: new insights into molecular pathogenesis and therapeutic opportunities. Nat Rev Neurol. 2020;16:529–46. https://doi.org/10.1038/s41582-020-0389-4.
138. Bates GP, Dorsey R, Gusella JF, Hayden MR, Kay C, Leavitt BR, et al. Huntington disease. Nat Rev Dis Primers. 2015;1:15005. https://doi.org/10.1038/nrdp.2015.5.
139. Verma AK, Singh S, Rizvi SI. Therapeutic potential of melatonin and its derivatives in aging and neurodegenerative diseases. Biogerontology. 2023;24:183–206. https://doi.org/10.1007/s10522-022-10006-x.
140. Jauhari A, Baranov SV, Suofu Y, Kim J, Singh T, Yablonska S, et al. Melatonin inhibits cytosolic mitochondrial DNA–induced neuroinflammatory signaling in accelerated aging and neurodegeneration. J Clin Invest. 2020;130:3124–36. https://doi.org/10.1172/JCI135026.
141. Kalliolia E, Silajdžić E, Nambron R, Hill NR, Doshi A, Frost C, et al. Plasma melatonin is reduced in Huntington's disease. Mov Disord. 2014;29:1511–5. https://doi.org/10.1002/mds.26003.
142. Kim J, Li W, Wang J, Baranov SV, Heath BE, Jia J, et al. Biosynthesis of neuroprotective melatonin is dysregulated in Huntington's disease. J Pineal Res. 2023;75:12909. https://doi.org/10.1111/jpi.12909.
143. Wang X, Sirianni A, Pei Z, Cormier K, Smith K, Jiang J, et al. The melatonin MT1 receptor axis modulates mutant Huntingtin-mediated toxicity. J Neurosci. 2011;31:14496–507. https://doi.org/10.1523/JNEUROSCI.3059-11.2011.
144. Xue F, Shi C, Chen Q, Hang W, Xia L, Wu Y, et al. Melatonin mediates protective effects against kainic acid-induced neuronal death through safeguarding ER stress and mitochondrial disturbance. Front Mol Neurosci. 2017;10:49. https://doi.org/10.3389/fnmol.2017.00049.
145. Valdés-Tovar M, Estrada-Reyes R, Solís-Chagoyán H, Argueta J, Dorantes-Barrón AM, Quero-Chávez D, et al. Circadian modulation of neuroplasticity by melatonin: a target in the treatment of depression. Br J Pharmacol. 2018;175:3200–8. https://doi.org/10.1111/bph.14197.
146. Chang H-M, Lin H-C, Cheng H-L, Liao C-K, Tseng T-J, Renn T-Y, et al. Melatonin successfully rescues the hippocampal molecular machinery and enhances anti-oxidative activity following early-life sleep deprivation injury. Antioxidants. 2021;10:774. https://doi.org/10.3390/antiox10050774.
147. Walker WH, Walton JC, DeVries AC, Nelson RJ. Circadian rhythm disruption and mental health. Transl Psychiatry. 2020;10:28. https://doi.org/10.1038/s41398-020-0694-0.
148. McCarthy MJ, Gottlieb JF, Gonzalez R, McClung CA, Alloy LB, Cain S, et al. Neurobiological and behavioral mechanisms of circadian rhythm disruption in bipolar disorder: a critical multi-disciplinary literature review and agenda for future research from the ISBD task force on chronobiology. Bipolar Disord. 2022;24:232–63. https://doi.org/10.1111/bdi.13165.
149. Gao Z, Luo K, Hu Y, Niu Y, Zhu X, Li S, et al. Melatonin alleviates chronic stress-induced hippocampal microglia pyroptosis and subsequent depression-like behaviors by inhibiting Cathepsin B/NLRP3 signaling pathway in rats. Transl Psychiatry. 2024;14:166. https://doi.org/10.1038/s41398-024-02887-y.
150. Yang Y, Jiang W, Feng Y, Liu J, Chen H, Wang D, et al. Melatonin alleviates hippocampal GR inhibition and depression-like behavior induced by constant light exposure in mice. Ecotoxicol Environ Saf. 2021;228:112979. https://doi.org/10.1016/j.ecoenv.2021.112979.
151. Liu J, Clough SJ, Dubocovich ML. Role of the MT1 and MT2 melatonin receptors in mediating depressive- and anxiety-like behaviors in C3H/HeN mice. Genes Brain Behav. 2017;16:546–53. https://doi.org/10.1111/gbb.12369.
152. Thomson DM, Mitchell EJ, Openshaw RL, Pratt JA, Morris BJ. Mice lacking melatonin MT2 receptors exhibit attentional deficits, anxiety and enhanced social interaction. J Psychopharmacol. 2021;35:1265–76. https://doi.org/10.1177/02698811211032439.
153. Weil ZM, Hotchkiss AK, Gatien ML, Pieke-Dahl S, Nelson RJ. Melatonin receptor (MT1) knockout mice display depression-like behaviors and deficits in sensorimotor gating. Brain Res Bull. 2006;68:425–9. https://doi.org/10.1016/j.brainresbull.2005.09.016.
154. Klosen P, Lapmanee S, Schuster C, Guardiola B, Hicks D, Pevet P, et al. MT1 and MT2 melatonin receptors are expressed in nonoverlapping neuronal populations. J Pineal Res. 2019;67:12575. https://doi.org/10.1111/jpi.12575.
155. Adamah-Biassi E, Zhang Y, Jung H, Vissapragada S, Miller R, Dubocovich M. Distribution of MT_1 melatonin receptor promoter-driven RFP expression in the brains of BAC C3H/HeN transgenic mice. J Histochem Cytochem. 2014;62:70–84. https://doi.org/10.1369/0022155413507453.

156. Zhang C, Clough SJ, Adamah-Biassi EB, Sveinsson MH, Hutchinson AJ, Miura I, et al. Impact of endogenous melatonin on rhythmic behaviors, reproduction, and survival revealed in melatonin-proficient C57BL/6J congenic mice. J Pineal Res. 2021;71:12748. https://doi.org/10.1111/jpi.12748.
157. Pfeffer M, von Gall C, Wicht H, Korf H-W. The role of the melatoninergic system in circadian and seasonal rhythms—insights from different mouse strains. Front Physiol. 2022;13:883637. https://doi.org/10.3389/fphys.2022.883637.
158. Metzger J, Wicht H, Korf H-W, Pfeffer M. Seasonal variations of locomotor activity rhythms in melatonin-proficient and -deficient mice under seminatural outdoor conditions. J Biol Rhythm. 2020;35:58–71. https://doi.org/10.1177/0748730419881922.
159. Lima-Cabello E, Díaz-Casado ME, Guerrero JA, Otalora BB, Escames G, López LC, et al. A review of the melatonin functions in zebrafish physiology. J Pineal Res. 2014;57:1–9. https://doi.org/10.1111/jpi.12149.
160. Aranda-Martínez P, Fernández-Martínez J, Ramírez-Casas Y, Guerra-Librero A, Rodríguez-Santana C, Escames G, et al. The zebrafish, an outstanding model for biomedical research in the field of melatonin and human diseases. Int J Mol Sci. 2022;23:7438. https://doi.org/10.3390/ijms23137438.
161. Ang MJ, Kang S, Moon C. Melatonin alters neuronal architecture and increases cysteine-rich protein 1 signaling in the male mouse hippocampus. J Neurosci Res. 2020;98:2333–48. https://doi.org/10.1002/jnr.24708.
162. Zhang S, Wang P, Ren L, Hu C, Bi J. Protective effect of melatonin on soluble Aβ1–42-induced memory impairment, astrogliosis, and synaptic dysfunction via the Musashi1/Notch1/Hes1 signaling pathway in the rat hippocampus. Alzheimers Res Ther. 2016;8:40. https://doi.org/10.1186/s13195-016-0206-x.
163. Cheng X-P, Sun H, Ye Z-Y, Zhou J-N. Melatonin modulates the GABAergic response in cultured rat hippocampal neurons. J Pharmacol Sci. 2012;119:177–85. https://doi.org/10.1254/jphs.11183FP.
164. Elkamhawy A, Woo J, Gouda NA, Kim J, Nada H, Roh EJ, et al. Melatonin analogues potently inhibit MAO-B and protect PC12 cells against oxidative stress. Antioxidants. 2021;10:1604. https://doi.org/10.3390/antiox10101604.
165. Lee W, Chen L, Lin J, Cheng T, Kuo C, Wu C, et al. Melatonin promotes neuroblastoma cell differentiation by activating hyaluronan synthase 3-induced mitophagy. Cancer Med. 2019;8:4821–35. https://doi.org/10.1002/cam4.2389.
166. Luchetti F, Nasoni MG, Burattini S, Mohammadi A, Pagliarini M, Canonico B, et al. Melatonin attenuates ischemic-like cell injury by promoting autophagosome maturation via the Sirt1/FoxO1/Rab7 axis in hippocampal HT22 cells and in organotypic cultures. Cells. 2022;11:3701. https://doi.org/10.3390/cells11223701.
167. Singrang N, Nopparat C, Panmanee J, Govitrapong P. Melatonin inhibits hypoxia-induced Alzheimer's disease pathogenesis by regulating the amyloidogenic pathway in human neuroblastoma cells. Int J Mol Sci. 2024;25:5225. https://doi.org/10.3390/ijms25105225.
168. Merlo S, Caruso GI, Korde DS, Khodorovska A, Humpel C, Sortino MA. Melatonin activates anti-inflammatory features in microglia in a multicellular context: evidence from organotypic brain slices and HMC3 cells. Biomolecules. 2023;13:373. https://doi.org/10.3390/biom13020373.
169. Caruso GI, Korde DS, Humpel C. Melatonin supports the survival of cholinergic neurons in organotypic brain slices of the basal nucleus of Meynert. Pharmacology. 2023;108:204–12. https://doi.org/10.1159/000527887.
170. Clinton J, McWilliams-Koeppen P. Initiation, expansion, and cryopreservation of human primary tissue-derived normal and diseased organoids in embedded three-dimensional culture. Curr Protoc Cell Biol. 2019;82:66. https://doi.org/10.1002/cpcb.66.
171. Khavinson VK, Linkova NS, Chalisova NI, Dudkov AV, Koncevaya EA. Effect of short peptides on expression of signaling molecules in organotypic pineal cell culture. Bull Exp Biol Med. 2011;152:138–41. https://doi.org/10.1007/s10517-011-1473-y.
172. Kamfar WW, Khraiwesh HM, Ibrahim MO, Qadhi AH, Azhar WF, Ghafouri KJ, et al. Comprehensive review of melatonin as a promising nutritional and nutraceutical supplement. Heliyon. 2024;10:e24266. https://doi.org/10.1016/J.HELIYON.2024.E24266.
173. Cachán-Vega C, Vega-Naredo I, Potes Y, Bermejo-Millo JC, Rubio-González A, García-González C, et al. Chronic treatment with melatonin improves hippocampal neurogenesis in the aged brain and under neurodegeneration. Molecules. 2022;27:5543. https://doi.org/10.3390/MOLECULES27175543.
174. Song C, Suo Z, Wang Z, Cao J, Dong Y, Chen Y. Melatonin modulates neuroinflammatory response and microglial activation in mice exposed to dim blue light at night. Front Pharmacol. 2024;15:1416350. https://doi.org/10.3389/FPHAR.2024.1416350/BIBTEX.

Index

M. Turgut et al. (eds.), *The Human Hippocampus*, https://doi.org/10.1007/978-3-031-88907-3

W

X

Z

MIX
Papier aus verantwortungsvollen Quellen
Paper from responsible sources
FSC® C105338

If you have any concerns about our products,
you can contact us on
ProductSafety@springernature.com

In case Publisher is established outside the EU,
the EU authorized representative is:
Springer Nature Customer Service Center GmbH
Europaplatz 3, 69115 Heidelberg, Germany

Printed by Libri Plureos GmbH
in Hamburg, Germany